Trefftz and Collocation Methods

Trefftz and Collocation Methods

Z.-C. Li
National Sun Yat-sen University, Taiwan and
National Center for Theoretical Science, Taiwan

T.-T. Lu
National Sun Yat-sen University, Taiwan and
National Center for Theoretical Science, Taiwan

H.-Y. Hu
Tunghai University, Taiwan

A. H.-D. Cheng
University of Mississippi, USA

WIT*PRESS* Southampton, Boston

Z.-C. Li
National Sun Yat-sen University, Taiwan and
National Center for Theoretical Science, Taiwan
T.-T. Lu
National Sun Yat-sen University, Taiwan and
National Center for Theoretical Science, Taiwan
H.-Y. Hu
Tunghai University, Taiwan
A. H.-D. Cheng
University of Mississippi, USA

Published by

WIT Press
Ashurst Lodge, Ashurst, Southampton, SO40 7AA, UK
Tel: 44 (0) 238 029 3223; Fax: 44 (0) 238 029 2853
E-Mail: witpress@witpress.com
http://www.witpress.com

For USA, Canada and Mexico

WIT Press
25 Bridge Street, Billerica, MA 01821, USA
Tel: 978 667 5841; Fax: 978 667 7582
E-Mail: infousa@witpress.com
http://www.witpress.com

British Library Cataloguing-in-Publication Data

A Catalogue record for this book is availablefrom the British Library

ISBN: 978-1-84564-153-5

Library of Congress Catalog Card Number: 2007922336

Printed in Great Britain by Cambridge Printing.

To our friends,

Wen-Jang Huang and Mong-Na Lo Huang.

Contents

Erich Trefftz
21/02/1888 – 21/01/1937

The bust is on display in the Willers building of the Technical University of Dresden
(Photo courtesy of Professor Andrezej P. Zieliñski)

Preface

This book covers a class of numerical methods that are generally referred to as *Collocation Methods*. Different from the finite element and the finite difference methods, the discretization and approximation of the collocation method is based on a set of unstructured points in space. This *meshless* feature is attractive because it eliminates the bookkeeping requirements of the element-based methods, particularly, if the basis functions used satisfy the governing equation; the collocation is conducted only on the boundary. The boundary collocation methods are also known as Trefftz methods. The main advantages of these methods include the flexible representation of the irregular and deforming geometry, ease of data input and preprocessing, high accuracy of the numerical solution, and the efficient computation.

This book contains an introduction, an appendix, and eleven chapters in which several types of collocation methods are discussed. These include the radial basis function method, the Trefftz method, and the coupled collocation and finite element method. Governing equations investigated include Laplace, Poisson, Helmholtz, and bi-harmonic equations. Regular boundary value problems, boundary value problems with singularity, and eigenvalue problems are also examined. Rigorous mathematical proofs are contained in these chapters, and numerical experiments are also provided to support the algorithms and to verify the theory. A tutorial on the applications of these methods is provided in the introduction and a historic review of boundary methods in the appendix.

This book is an extension of the leading author's earlier books on combined methods based on the theoretical analysis of finite element method (FEM). However, this book has several distinct features, which are addressed as follows:

1. In this book, the boundary collocation method, which is a form of Trefftz method (TM) [1], is presented, and referred to as the collocation Trefftz method (CTM). The boundary approximation method (BAM) as discussed in [2], which involves numerical integration, is also classified as a CTM. New analysis of exponential convergence and excellent numerical results are demonstrated. The CTM is shown to be the most accurate numerical method not only for the global solutions, but also for the leading coefficient of the singularity expansion, which is important for problems like fracture mechanics.

2. This book also covers the original TM, the hybrid TM, the direct TM and the indirect TM. There was a special journal issue published in 1995 celebrating 70 years of Trefftz method [3], [4]. Besides, the first and the second International Workshops of Trefftz methods held in Cracow, Poland, 1996, organized by A.P. Zielinski, and in Sintra, Portugal, 1999, organized by J.A.T. Treitas and J.P.M. Almeida, and the invited talks are published in Computer Assisted Mechanics and Engineering Sciences (CAMES) in Vol. 4 (1977) and Vol. 8 (2001). Although a number of papers on TMs were collected, only a few involved analysis. The analysis of TMs lags behind that of the FEM and the boundary element method (BEM). For the TM, there is a significant gap between the excellent computation and the theoretical analysis to support the results. This book presents a systematic analysis for the CTM, the hybrid TM, the indirect TM, and the direct TM to bridge the gap.
3. It also demonstrates the advantages of the CTM over other TMs. The CTM is the simplest algorithm because the collocation equations can be assembled in a straightforward way. For solving Motz's problem, the CTM provides the most accurate solutions not only in the global H^1 sense, but also in its leading singular term. More importantly, the condition number of the stiffness matrix from the CTM is significantly smaller than that of the other TMs. It should be mentioned that the application of CTM is limited to those PDEs whose particular solutions or local particular solutions can be found explicitly. Particular solutions are used in this book in a wide sense, to satisfy the homogeneous or the non-homogeneous elliptic equation with partial or no boundary conditions.
4. More topics are explored in this book, such as the biharmonic equation, the Helmholtz equation, and eigenvalue problems by means of particular solutions of elliptic equations. The combinations of the collocation Trefftz method with high order FEM are also discussed, as compared to the linear and bilinear FEMs reported earlier [5], [6].
5. Particular solutions are essential to the Trefftz methods. We provide the series expansion solutions for the Laplace equation on a polygon, particularly those involving mild singularity.
6. The collocation method (CM) on the entire domain, in contrast to boundary collocation, is studied. The CM can be interpreted as the least squares method with numerical integration. The analysis can be conducted by means of the FEM approach, and optimal weights for different collocation equations resulting from the PDE and different boundary conditions can be found theoretically.
7. The radial basis functions (RBF) are a new approximation tool for smooth functions. In this book the RBF has been developed to solve the elliptic equation with singularities. Moreover, the convergence of Kansa's method is proved with error estimates in H^1 norm.

8. To enhance the education value, a historical review of the boundary methods is provided as an appendix.

This book is organized as follows. The introduction reviews the fundamentals of the collocation and the Trefftz methods from several viewpoints. The remainder of the book is divided into three parts, Part I: The Collocation Trefftz Method; Part II: The Collocation Methods; and Part III: Advanced Topics. Part I is mainly concerned with the algorithms, the error estimates, and stability analysis of both the Trefftz method and the collocation Trefftz method. Several popular examples of PDEs with singularities, including Poisson's equation (Motz's and its variants), and the biharmonic equations with crack singularities are examined. Part II gives a unified framework of combinations of collocation methods with other numerical methods. Part III introduces advanced topics for the collocation Trefftz method.

To appeal to both applied mathematicians and engineers, we have carefully selected only the important analyses and the significant numerical experiments. The mathematics retained is necessary to provide a deeper insight into the numerical algorithms proposed. For easy reading of the book, each chapter can be treated as an independent unit; hence, readers can directly refer to the chapters that interest them. We hope that through this book we can bring the engineering and applied mathematics community a step closer to recognizing the power of the collocation and Trefftz methods.

The Authors, 2008

References

[1] Trefftz, E., Konvergenz und fehlerabschatzung beim Ritz'schen verfahren, *Math. Ann.*, **100**, pp. 503–521, 1928.

[2] Li, Z.C., Mathon, R. & Sermer, P., Boundary methods for solving elliptic problem with singularities and interfaces, *SIAM J. Numer. Anal.*, **24**, pp. 487–498, 1987.

[3] Kamiya, N. & Kita, E., Trefftz method 70 years, *Adv. Eng. Softw.*, **24** (1-3), pp. 1, 1995.

[4] Kita, E. & Kamiya, N., Trefftz method, An overview, *Adv. Eng. Softw.*, **24** (1-3), pp. 3-12, 1995.

[5] Li, Z.C., *Numerical Methods for Elliptic Equations with Singularities, Boundary Methods and Nonconforming Combinations*, World Scientific, Singapore, 1990.

[6] Li, Z.C., *Combined Methods for Elliptic Equations with Singularities, Interfaces and Infinities*, Kluwer Academic Publishers, Boston, London, 1998.

Acknowledgements

We are obliged to Professors C.S. Huang, H.T. Huang, I. Herrera, G.G. Georgiou, C. Xenohontos, J. Huang, T. Lü, S. Wang, L. Ling, and R.C.D. Chen for their collaboration or valuable suggestions on the research works related to this book, and to the graduate students Y.L. Chen, H.S. Tsai, C.H. Huang, S.Y. Tsai, W.L. Jhou, S.L. Huang, S.J. Jian and J.R. Wang for reading the manuscript and giving their comments and suggestions. We also are indebted to S.Y. Wang, W.L. Jhou and S.L. Huang for typing the manuscript.

We are grateful to Professor C.A. Brebbia and the editorial staff at the WIT Press for their encouragement and assistance in publishing this book. We also wish to thank Professors C.C. Chang, J.C. Yao, N.C. Wong and C.K. Law at National Sun Yat-sen University, Taiwan, for their support during the writing process.

This book and the related works were supported in part by the National Science Council of Taiwan, and National Center for Theoretical Sciences, Mathematics Division, Taiwan.

Tutorial introduction

In this introduction, we survey the algorithms of the collocation method (CM), the Trefftz method (TM), and the collocation Trefftz method (CTM), and summarize the coupling techniques for the exterior and interior boundary conditions, which include the very original TM, the hybrid TM, the indirect TM, and the direct TM. Also the boundary element methods (BEMs) and other kinds of boundary methods are briefly described. In the last section, discussions and comparisons of the CTM with other numerical methods are made.

I.1 Algorithms of CM, TM, and CTM

We choose the solution of a boundary value problem to be a continuously differentiable function, and then enforce it to exactly satisfy the partial differential equation (PDE) and the boundary conditions at a set of points. This leads to CM, or in engineering literature, it is called the residual method in Percell and Wheeler [361, 362]. By an equivalence between the Galerkin–finite element method (FEM) and the CM, e.g., see Swartz and Wendroff [429], the linear algebraic equations can be easily constructed for the CM, which leads to an approximate solution of the boundary value problem.

I.1.1 Algorithms of CM

Consider the second-order elliptic boundary problem

$$\begin{cases} \pounds u = f, & \text{in } S, \\ Bu = g, & \text{on } \partial S, \end{cases} \tag{I.1.1}$$

where

$$£ = -\frac{\partial}{\partial x} p \frac{\partial}{\partial x} - \frac{\partial}{\partial y} p \frac{\partial}{\partial y} + c,$$

and the operators, $B = 1, p\frac{\partial}{\partial \nu}$, and $p\frac{\partial}{\partial \nu} + \alpha$, represent the Dirichlet, the Neumann, and the Robin boundary conditions, respectively, where ν is the unit outward normal of ∂S. In this book, the constant c in $£$ may be chosen as $c = 0$, $c > 0$, and $c < 0$. Choose the smooth basis functions $\{\Psi_i\}$, which are complete, linearly independent, and at least twice differentiable. The orthogonal polynomials and the Fourier functions, as well as piecewise cubic splines are often chosen as the admissible functions. Then, we can express the solution by the series form

$$u = \sum_{i=1}^{\infty} \overline{a}_i \Psi_i, \tag{I.1.2}$$

where $\overline{a}_i$ are the true expansion coefficients. If only finite terms are chosen in eqn. (I.1.2), we obtain the approximate solution

$$v = \sum_{i=1}^{N} a_i \Psi_i, \tag{I.1.3}$$

where the coefficients a_i are an approximation of $\overline{a}_i$. Hence, the eqn. (I.1.1) leads to

$$\begin{cases} \sum_{i=1}^{N} a_i £\Psi_i = f, \\ \sum_{i=1}^{N} a_i B\Psi_i = g. \end{cases}$$

Since there are N unknowns, we may choose $M(\geq N)$ collocation points to enforce eqn. (I.1.3) such that

$$\sum_{i=1}^{N} a_i £\Psi_i(Q_j) = f(Q_j), \quad j = 1, 2, \ldots, M_1, \quad Q_j \in S,$$

$$\sum_{i=1}^{N} a_i B\Psi_i(Q_j) = g(Q_j), \quad j = M_1 + 1, \ldots, M, \quad Q_j \in \partial S. \tag{I.1.4}$$

Since the differential equation and the boundary conditions play different roles in the boundary value problem, different weights should be imposed on eqn. (I.1.4),

$$\sum_{i=1}^{N} a_i £\Psi_i(Q_j) = f(Q_j), \quad j = 1, 2, \ldots, M_1, \quad Q_j \in S,$$

$$\sum_{i=1}^{N} w_B a_i B\Psi_i(Q_j) = w_B g(Q_j), \quad j = M_1 + 1, \ldots, M, \quad Q_j \in \partial S, \tag{I.1.5}$$

where the weight functions $w_{\mathrm{B}} > 0$ may be different for different boundary conditions, i.e., $w_{\mathrm{B}} = w_D, w_N, w_R$, respectively. Hence, we obtain the discrete equations

$$\mathbf{Ax} = \mathbf{b}, \tag{I.1.6}$$

where $\mathbf{x} = (a_1, \ldots, a_N)^T$ is the unknown vector, $\mathbf{b}$ a known vector, and the stiffness matrix $\mathbf{A} = \mathbf{A}_{M\times N} = (a_{i,j})$. The matrix entries are given by

$$a_{j,i} = \pounds\Psi_i(Q_j), \quad j \le M_1; \qquad a_{j,i} = w_{\mathrm{B}} B\Psi_i(Q_j), \quad j > M_1.$$

When $M > N$, the eqn. (I.1.6) is an overdetermined system, which can be solved by the least squares method (LSM), to provide the approximate coefficients a_i.

When $f = 0$ and the basis functions Ψ_i are particular solutions of $\pounds\Psi_i = 0$, the boundary equations are reduced to (i.e., $M_1 = 0$ in eqn. (I.1.5))

$$\sum_{i=1}^{N} w_{\mathrm{B}} a_i B\Psi_i(Q_j) = w_{\mathrm{B}} g(Q_j), \quad Q_j \in \partial S, \; j = 1, 2, \ldots, M.$$

This is called the CTM in this book, which is also referred as the indirect TM in Kita and Kamiya [249], or the boundary solution procedure in Zienkiewicz, Kelley, and Bettess [488], and the TM in Zieliński and Zienkiewicz [486], Zieliński and Herrera [485], and Lefeber [270].

Particular solutions are used in this book in a wide sense, to satisfy the homogeneous or the non-homogeneous elliptic equations with or without part of boundary conditions. In some literature, e.g., Ref. [85], the particular solutions are referred only to the solutions satisfying the non-homogeneous equations.

We may represent eqn. (I.1.1) in the residual form

$$\iint_S R_{\mathrm{E}} v \, ds = \iint_S (\pounds u - f) v \, ds = 0,$$
$$\int_{\partial S} w_{\mathrm{B}} R_{\mathrm{B}} v \, d\ell = \int_{\partial S} w_{\mathrm{B}} (Bu - g) v \, d\ell = 0, \tag{I.1.7}$$

where R_{E} and R_{B} are the residuals of the PDE and their boundary conditions. Let the test function v be the Dirac's delta function,

$$v = \delta(P - Q_i) = \begin{cases} \infty, & P = Q_i, \\ 0, & P \ne Q_i. \end{cases}$$

Since

$$\iint_S \delta(P - Q_i) f \, ds = f(Q_i),$$

the eqn. (I.1.7) leads to the CM, i.e., eqn. (I.1.5).

In the early time, the CM was used for solving ordinary differential equation (ODE), see Ascher, Mattheij, and Russell [7], Russell and Shampine [397], Lucas and Reddien [319], Russell [395, 396], Dunn and Wheeler [132], Diaz [127], and

Wheeler [465]. Later the CM is applied for solving PDE as well, see Bialecki and Cai [39], Prenter and Russell [371], Sun [428], and Quarteroni and Zampieri [375]. Besides, the residual method was studied by Percell and Wheeler [361] and Oliveira [352]. Recently, there have been a number of reports on the CM, see Bialecki [38], Bialecki and Fernandes [42], Bialecki and Dryja [40], Bialecki, Fairweather, and Kargeorghis [41], Brunner, Pedas, and Vainikko [65], Cao, Herdman, and Xu [74], Chen, Micchelli, and Xu [88], Laubin and Baiwir [266], Layton [267], Li, Fairweather, and Bialecki [272], Ma and Sun [320], Parter [357, 358], and Russell and Sun [398]. Moreover, textbooks involving the CM have appeared, such as Ascher, Mattheij, and Russell [7] for ODE, Quarteroni and Valli [374] for PDE, and Canuto et al. [72] for spectral methods.

The method using the orthogonal polynomials is also called the spectral method in Bernardi and Maday [35]. The spectral method as the Galerkin–FEM involving integration approximation may lead to the CM if some special nodes are chosen, such as the Gauss–Lobatto nodes (i.e., the roots of the polynomials). (Note that since the Galerkin–FEM leads to the CM, which is different from eqn. (I.1.4), the error analysis is also different from this book.) References of spectral methods include Kreiss and Oliger [254], Gottlieb and Orszag [170], Boyd [50], and Canuto and Quarteroni [73].

I.1.2 Viewpoint of boundary approximation methods

Consider the non-homogeneous elliptic equation with the Robin boundary condition,

$$\pounds u = f, \quad \text{in } S,$$

$$p\frac{\partial u}{\partial \nu} + \alpha u = g, \quad \text{on } \partial S,$$

where $\alpha \geq 0$ and ν is the outward normal of ∂S. Suppose that a particular solution Ψ_0 can be found to satisfy

$$\pounds \Psi_0 = f.$$

By the transformation $w = u - \Psi_0$, we obtain the homogeneous elliptic equation with the revised Robin condition

$$\pounds w = 0, \quad \text{in } S,$$

$$p\frac{\partial w}{\partial \nu} + \alpha w = \overline{g}, \quad \text{on } \partial S,$$

where $\overline{g} = g - (p\frac{\partial \Psi_0}{\partial \nu} + \alpha \Psi_0)$. Assuming that particular solutions can be found below, without the loss of generality, we may only consider the homogeneous equation,

$$\pounds u = 0, \quad \text{in } S,$$

$$p\frac{\partial u}{\partial \nu} + \alpha u = g, \quad \text{on } \partial S. \tag{I.1.8}$$

Suppose that a complete set of particular solutions Ψ_i of eqn. (I.1.8) can be found explicitly,

$$\pounds\Psi_i = 0, \quad i = 1, 2, \dots$$

Therefore, the solution of eqn. (I.1.8) can be approximated by

$$u_{\rm L} = \sum_{i=1}^{L} a_i \Psi_i, \tag{I.1.9}$$

where the coefficients a_i are to be determined. Let $V_{\rm L}$ denote the finite-dimensional collection of $u_{\rm L}$ in eqn. (I.1.9). The weak solution of eqn. (I.1.8) can be expressed by

$$a(u, v) = f(v),$$

where

$$a(u, v) = \iint_S (p \nabla u \cdot \nabla v + cuv)\, ds,$$
$$f(v) = \oint_{\partial S} gv\, d\ell.$$

The notation $\nabla = \vec{i}\frac{\partial}{\partial x} + \vec{j}\frac{\partial}{\partial y}$, and $\vec{i}$ and $\vec{j}$ are the unit vectors along the x and the y directions, respectively.

Since the admissible functions $u_{\rm L}$ in eqn. (I.1.9) already satisfy the equation $\pounds u = 0$, for v satisfying $\pounds v = 0$, we obtain from the Green theorem

$$\begin{aligned} a(u, v) &= \iint_S (p \nabla u \cdot \nabla v + cuv)\, ds + \int_{\partial S} \alpha uv\, d\ell \\ &= \oint_{\partial S} \left(p \frac{\partial u}{\partial \nu} + \alpha u \right) v\, d\ell = \oint_{\partial S} gv\, d\ell. \end{aligned}$$

This gives

$$\oint_{\partial S} \left(p \frac{\partial u_{\rm L}}{\partial \nu} + \alpha u_{\rm L} \right) v\, d\ell = \oint_{\partial S} gv\, d\ell, \quad \forall v \in V_{\rm L}. \tag{I.1.10}$$

Since only the boundary condition is involved in eqn. (I.1.10), it is called the boundary method or the boundary approximation method (BAM) in Li [274, 276, 280].

Below, we provide a variational approach. Since the admissible functions have satisfied the equation $\pounds u = 0$ already, the coefficients a_i in eqn. (I.1.9) are chosen to satisfy the boundary conditions as best as possible. The solution of eqn. (I.1.10) can also be obtained as: To find $u_{\rm L} \in V_{\rm L}$ such that

$$I_{\rm R}(u_{\rm L}) = \min_{\forall v \in V_{\rm L}} I_{\rm R}(v),$$

where

$$I_{\mathrm{R}}(v) = \oint_{\partial S} \left(p\frac{\partial v}{\partial \nu} + \alpha v - g \right)^2 d\ell.$$

When the Dirichlet condition also occurs, we shall utilize the boundary penalty technique. Consider the elliptic boundary value problem

$$\begin{cases} \pounds u = 0, & \text{in } S, \\ u = g_1, & \text{on } \Gamma_D, \\ p\frac{\partial u}{\partial \nu} + \alpha u = g_3, & \text{on } \Gamma_M \cup \Gamma_N, \end{cases} \tag{I.1.11}$$

where $\alpha > 0$ on Γ_M, $\alpha = 0$ on Γ_N, and $\Gamma_D \neq \emptyset$ for uniqueness of the solution. Define the energy

$$I_{\mathrm{B}}(v) = \int_{\Gamma_D} (v - g_1)^2 \, d\ell + w^2 \int_{\Gamma_M \cup \Gamma_N} \left(p\frac{\partial v}{\partial \nu} + \alpha v - g_3 \right)^2 d\ell, \tag{I.1.12}$$

where w is a positive weight to balance two different boundary conditions. Since the solution derivatives are usually larger than the solutions themselves, we choose $w < 1$. Of course, a better choice of w should depend upon error analysis (see Chapter 1). The BAM for eqn. (I.1.12) is designed as: To seek $u_{\mathrm{L}} \in V_{\mathrm{L}}$ such that

$$I_{\mathrm{B}}(u_{\mathrm{L}}) = \min_{\forall v \in V_{\mathrm{L}}} I_{\mathrm{B}}(v). \tag{I.1.13}$$

Equation (I.1.13) leads to

$$\mathbf{Ax} = \mathbf{b}, \tag{I.1.14}$$

where $\mathbf{A}$ is symmetric and positive definite, but not sparse, and $\mathbf{x}$ the unknown vector consisting of components a_i. The eqn. (I.1.14) is called the algebraic normal equation.

A remarkable advantage of BAM is high accuracy, due to its exponential convergence rates. The BAM is much simpler and more efficient than the original Ritz–Galerkin method (RGM), because the approximate process is reduced to that on ∂S only. The relation between RGM and BAM is analogous to that between FEM and BEM. However, the BAM is limited to elliptic equations with constant coefficients where particular solutions can be found. Also, the solution can be unstable if the discretized system is large. A remedy is that the solution domain may be divided into several or many subdomains with or without overlaps, and different particular solutions or different numerical methods can be used in different subdomains, see Herrera [191, 193].

In practical computation, the BAM eqn. (I.1.13) always involves the integration approximation of eqn. (I.1.12). Choose the simplest middle-point rule of integration, e.g.,

$$\int_a^b f(x)\, dx \approx \widetilde{\int_a^b} f(x)\, dx = (b-a) f\left(\frac{a+b}{2}\right).$$

We partition ∂S into subintervals with the following partition nodes:

$$\Gamma_D : s_0 < s_1 < \cdots < s_{M_2},$$
$$\Gamma_M \cup \Gamma_N : s_{M_2} < s_{M_2+1} < \cdots < s_M.$$

Hence, the integrals in eqn. (I.1.12) become

$$\begin{aligned} I_{\rm B}(v) \approx \widetilde{I}_{\rm B}(v) &= \widetilde{\int_{\Gamma_D}} (v-g_1)^2\, d\ell + w^2 \widetilde{\int_{\Gamma_M\cup\Gamma_N}} \left(p\frac{\partial v}{\partial \nu} + \alpha v - g_3\right)^2 d\ell \\ &= \sum_{i=1}^{M_2} (f_{i-\frac{1}{2}})^2 \delta s_i + w^2 \sum_{i=M_2+1}^{M} (f^*_{i-\frac{1}{2}})^2 \delta s_i, \end{aligned} \tag{I.1.15}$$

where $\delta s_i = s_i - s_{i-1}$, $f = v - g_1$, $f^* = p\frac{\partial v}{\partial \nu} + \alpha v - g_3$, and $f_{i-\frac{1}{2}} = f(s_{i-\frac{1}{2}})$ with $s_{i-\frac{1}{2}} = \frac{s_i + s_{i-1}}{2}$. The BAM involving integration approximation is designed as: To seek $\widetilde{u}_{\rm L} \in V_{\rm L}$ such that

$$\widetilde{I}_{\rm B}(\widetilde{u}_{\rm L}) = \min_{\forall v \in V_{\rm L}} \widetilde{I}_{\rm B}(v).$$

On the other hand, for eqn. (I.1.11) we may enforce functions, i.e., eqn. (I.1.3) directly to satisfy the collocation equations at the midpoints of $[s_{i-1}, s_i]$

$$\begin{cases} f_{i-\frac{1}{2}} = 0, & i = 1, 2, \ldots, M_2, \\ f^*_{i-\frac{1}{2}} = 0, & i = M_2+1, M_2+2, \ldots, M. \end{cases} \tag{I.1.16}$$

Suppose that $M > N$, the number of equations is larger than that of unknown coefficients a_i. In the sense of the weighted LSM, we should add some weights to the equations in eqn. (I.1.16) to balance the effects of different boundary conditions and different subsections. We multiply eqn. (I.1.16) by $\sqrt{\delta s_i}$ and $w\sqrt{\delta s_i}$ on Γ_D and $\Gamma_M \cup \Gamma_N$, respectively, to lead to

$$\begin{cases} \sqrt{\delta s_i} f_{i-\frac{1}{2}} = 0, & i = 1, 2, \ldots, M_2, \\ w\sqrt{\delta s_i} f^*_{i-\frac{1}{2}} = 0, & i = M_2+1, M_2+2, \ldots, M. \end{cases} \tag{I.1.17}$$

This is just the discrete CTM of eqn. (I.1.11), which is just the indirect TM in Kamiya and Kita [235] and Kita and Kamiya [249]. Summing the squares of all left sides in eqn. (I.1.17) yields exactly the energy approximation $\widetilde{I}_{\rm B}(v)$ in eqn. (I.1.15). For this reason, we call the BAM involving integration approximation as the CTM in this book.

In fact, we rewrite eqn. (I.1.17) as the overdetermined system

$$\mathbf{Fx} - \mathbf{d} = 0, \tag{I.1.18}$$

where $\mathbf{F} \in R^{M\times N}$, $M > N$, $\mathbf{x} \in R^N$, and $\mathbf{d} \in R^M$. Then, from the above arguments we obtain for $v \in V_L$

$$\widetilde{I}_B(v) = (\mathbf{Fx} - \mathbf{d})^T(\mathbf{Fx} - \mathbf{d})$$
$$= \mathbf{x}^T\mathbf{F}^T\mathbf{Fx} - (\mathbf{d}^T\mathbf{Fx} + \mathbf{x}^T\mathbf{F}^T\mathbf{d}) + \mathbf{d}^T\mathbf{d}.$$

Minimization of $\widetilde{I}_B(v)$ yields the normal equation

$$0 = \frac{\partial}{\partial \mathbf{x}}\widetilde{I}_B(v) = \mathbf{F}^T\mathbf{Fx} - \mathbf{F}^T\mathbf{d},$$

i.e.,

$$\mathbf{Ax} = \mathbf{F}^T\mathbf{Fx} = \mathbf{F}^T\mathbf{d}. \tag{I.1.19}$$

Denote $\mathbf{b} = \mathbf{F}^T\mathbf{d}$, then the normal eqn. (I.1.19) is just eqn. (I.1.14).

There are two methods to seek the coefficients a_i: (a) to solve eqn. (I.1.18), and (b) to solve eqn. (I.1.19). Although the solution method (b) is simpler, the condition number is quadratically larger, given by

$$\text{Cond}(\mathbf{A}) = \frac{\lambda_{\max}(\mathbf{A})}{\lambda_{\min}(\mathbf{A})} = \frac{\sigma^2_{\max}(\mathbf{F})}{\sigma^2_{\min}(\mathbf{F})} = \{\text{Cond}(\mathbf{F})\}^2,$$

where $\sigma_{\max}$ and $\sigma_{\min}$ are the maximal and minimal singular values of $\mathbf{F}$, respectively. For method (a), we may employ the LSM using the QR method or the singular value decomposition method in Golub and van Loan [168], with the small condition number $\text{Cond}(\mathbf{F}) = \frac{\sigma_{\max}(\mathbf{F})}{\sigma_{\min}(\mathbf{F})}$. Hence, method (a) solving the overdetermined system, i.e., eqn. (I.1.18) *is strongly recommended.*

The references on BAM are included in Delves [122], Delves and Hall [124], Delves and Freeman [123], Li [274, 276, 280], Li, Mathon, and Sermer [306], Li and Mathon [304, 305], and Zieliński and Zienkiewicz [486].

In summary, the BAM is classified as TM in this book, and the BAM involving integration approximation is equivalent to the CTM. They are also called the indirect TM in Kita and Kamiya [249]. The theoretical analysis for the TM (i.e., BAM) was first given in the Doctoral dissertation Ref. [274] of the first author in 1986, and the analysis for the CTM is provided in Chapter 2 in this book, also see Lu, Hu, and Li [315].

I.1.3 Viewpoint of least squares methods

Consider the three boundary conditions,

$$\pounds u = f \quad \text{in } S,$$
$$u = g_1 \quad \text{on } \Gamma_D,$$
$$p\frac{\partial u}{\partial \nu} = g_2 \quad \text{on } \Gamma_N,$$
$$p\frac{\partial u}{\partial \nu} + \alpha u = g_3 \quad \text{on } \Gamma_M. \tag{I.1.20}$$

For solving eqn. (I.1.20), we may choose the objective functional as

$$I(v) = \iint_S (\pounds v - f)^2\,ds + w_1^2 \int_{\Gamma_D} (v - g_1)^2\,d\ell + w_2^2 \int_{\Gamma_N} \left(p\frac{\partial v}{\partial \nu} - g_2\right)^2 d\ell$$
$$+ w_3^2 \int_{\Gamma_M} \left(p\frac{\partial v}{\partial \nu} + \alpha v - g_3\right)^2 d\ell, \qquad \text{(I.1.21)}$$

where w_i are positive weight constants. The LSM is described as: To seek the solution $u \in V_{\mathrm{H}}$ such that

$$I(u) = \min_{\forall v \in V_{\mathrm{H}}} I(v),$$

where $V_{\mathrm{H}}(\subset H^2(S))$ is the finite-dimensional collection of admissible functions without satisfying any boundary conditions, and $H^2(S)$ is the Sobolev space. The LSM can also be viewed as the squares residual method.

The advantages of LSM are that the associated matrix $\mathbf{A}$ is always symmetric and positive definite (or semi-definite), and that rather complicated constraints of the solutions can be easily incorporated with the numerical method. However, the disadvantage of LSM is that the condition number will increase significantly, compared with FEM. Moreover, both essential and natural boundary conditions must be expressed in the energy explicitly.

One way to overcome the large condition number is to divide the domain into subregions. Given the continuity of solutions and flux along the interior boundary $\Gamma_0 = \partial S^+ \cap \partial S^-$ of two subdomains S^+ and S^-, the problem can be solved as follows. Let

$$u = \begin{cases} u^+ & \text{in } S^+, \\ u^- & \text{in } S^-. \end{cases}$$

Then, the objective functional is modified as

$$\widehat{I}(v) = \iint_{S^+} (\pounds v - f)^2\,ds + \iint_{S^-} (\pounds v - f)^2\,ds + D(v),$$

where

$$D(v) = w_4^2 \int_{\Gamma_0} (v^+ - v^-)^2\,d\ell + w_5^2 \int_{\Gamma_0} \left(p\frac{\partial v^+}{\partial \nu} - p\frac{\partial v^-}{\partial \nu}\right)^2 d\ell$$
$$+ w_1^2 \int_{\Gamma_D} (v - g_1)^2\,d\ell + w_2^2 \int_{\Gamma_N} \left(p\frac{\partial v}{\partial \nu} - g_2\right)^2 d\ell$$
$$+ w_3^2 \int_{\Gamma_M} \left(p\frac{\partial v}{\partial \nu} + \alpha v - g_3\right)^2 d\ell,$$

and $w_i > 0$ are the weight constants. Hence, the LSM, the CM, and the CTM can handle the interior boundary condition as well.

LSM may deal with complicated problems easily, such as Navier–Stokes and even hyperbolic equations, see Aziz, Kellogg, and Stephen [12], Bramble and Schatz [54, 55, 56], Bramble and Nitsche [53], Zhou and Feng [480], Aziz, Kellogg, and Stephen [12], Bohmer and Locker [49], Huffel and Vandewalle [218], Bjorck [46], and Bochev and Gunzburger [47]. Note that these algorithms require high smoothness of admissible functions. Besides, reduction from PDEs to first-order systems of equations makes LSM reactive, see Carey and Shen [76]. More reports of the LSMs are given by Kim, Lee, and Shin [247], Liu [312], Jiang [224], and Aziz and Liu [13].

In application, since the integrals in eqn. (I.1.21) cannot be evaluated exactly, the numerical integration should be used. Then the integration nodes of quadrature rules should be chosen as the collocation nodes, and the LSM leads to exactly the CM. This linkage of the CM to the LSM is a key for error analysis in this book, which is different from most of the existing literature of the CM cited in Section I.1.1. In the traditional CM, the special rules with special nodes such as the Gauss–Lobatto nodes are chosen so that the domain is limited to rectangles, and the Galerkin–FEM involving the integration approximation leads to the CM. The error bounds are obtained by the Strang lemma in Strang and Fix [426], where the true errors include both the interpolation errors and the integration errors. In this book, the integration errors play a role only to satisfy the uniformly V_h-elliptic inequality, so that the simplest central rule may retain high convergence rates, such as exponential convergence rates of the CTM, even for polygonal domains or other bounded domains.

I.1.4 Complete systems of solutions

The algorithm for the CTM requires the explicit knowledge of the particular solutions. For the standard PDEs with the constant coefficients on the regular domains, such as the rectangular and sectorial domains, the particular solutions can be found by the techniques of separation of variables. A number of particular solutions are provided in the PDE textbooks, and the explicit harmonic solutions of Laplace's equation are provided in Chapter 11. In some cases, a special analysis as in Li [275] is also necessary in order to derive the particular solutions needed. In this subsection, we collect the particular solutions of a few typical equations in 2D, which are often used in practice. The problems on unbounded domains and in 3D can be similarly solved by following the approaches in this book.

First, consider the disk

$$S = \{(r, \theta) \,|\, 0 \le r < R, \quad 0 \le \theta \le 2\pi\}, \tag{I.1.22}$$

and the unbounded domain outside a disk

$$S_\infty = \{(r, \theta) \,|\, R < r < \infty, \quad 0 \le \theta \le 2\pi\}. \tag{I.1.23}$$

We list the following particular solutions for a few PDEs, see also Tikhonov and Samarskii [438], Herrera [191], and Zieliński [484],

I. For the Laplace equation, $\Delta u = 0$, the particular solutions for eqns. (I.1.22) and (I.1.23) are

$$1,\ r^n \cos n\theta, \qquad r^n \sin n\theta, \qquad n = 1, 2, \ldots, \quad (r, \theta) \in S,$$
$$\ln \frac{1}{r},\ r^{-n} \cos n\theta, \qquad r^{-n} \sin n\theta, \qquad n = 1, 2, \ldots, \quad (r, \theta) \in S_\infty,$$

respectively.

II. For the equation, $-\Delta u + k^2 u = 0$ with real $k > 0$, the particular solutions for eqns. (I.1.22) and (I.1.23) are

$$I_0(kr),\ I_n(kr) \cos n\theta, \qquad I_n(kr) \sin n\theta, \qquad n = 1, 2, \ldots, \quad (r, \theta) \in S,$$
$$K_0(kr),\ K_n(kr) \cos n\theta, \qquad K_n(kr) \sin n\theta, \qquad n = 1, 2, \ldots, \quad (r, \theta) \in S_\infty,$$

respectively, where $I_\mu(r)$ and $K_\mu(r)$ are the Bessel and Hankel functions for a purely imaginary argument, respectively, defined by

$$I_\mu(r) = \sum_{k=0}^{\infty} \frac{1}{\Gamma(k+1)\Gamma(k+\mu+1)} \left(\frac{r}{2}\right)^{2k+\mu},$$
$$K_\mu(r) = \frac{1}{2} \int_{-\infty}^{\infty} \exp(-r \cosh \eta - \mu\eta)\, d\eta.$$

When $k = 1$, $-\Delta u + u = 0$ is the Debye–Huckel equation.

III. For the Helmholtz equation, $\Delta u + k^2 u = 0$ with real $k > 0$. Suppose that k^2 is not an eigenvalue of Laplace's operator, then the particular solutions for eqns. (I.1.22) and (I.1.23) are

$$J_0(kr),\ J_n(kr) \cos n\theta, \qquad J_n(kr) \sin n\theta, \qquad n = 1, 2, \ldots, \quad (r, \theta) \in S,$$
$$H_0^{(1)}(kr),\ H_n^{(1)}(kr) \cos n\theta, \qquad H_n^{(1)}(kr) \sin n\theta, \qquad n = 1, 2, \ldots, \quad (r, \theta) \in S_\infty,$$

respectively, where $J_\mu(r)$ and $H_\mu^{(1)}(r)$ are the Bessel and Hankel functions of the first kind, respectively, defined by

$$J_\mu(r) = \sum_{k=1}^{\infty} \frac{(-1)^k}{\Gamma(k+1)\Gamma(k+\mu+1)} \left(\frac{r}{2}\right)^{2k+\mu},$$
$$H_\mu^{(1)}(r) = \frac{i}{\sin \mu\pi} \{\exp(-\mu\pi i) J_\mu(r) - J_{-\mu}(r)\}.$$

For the unbounded domain problem on S_∞, the Sommerfeld radiation condition is satisfied.

IV. For the biharmonic equation, $\Delta^2 u = 0$, the particular solutions for eqn. (I.1.22) are

$$\begin{aligned} &1,\ r^n \cos n\theta, && r^n \sin n\theta, && n = 1,2,\ldots, \\ &r^2,\ r^{n+2} \cos n\theta, && r^{n+2} \sin n\theta, && n = 1,2,\ldots,\ (r,\theta) \in S. \end{aligned}$$

In application, we should consider the particular solutions on a sector with different boundary conditions. In Chapter 11, detailed results are provided for the Laplace equation. Here, we only give the particular solutions on

$$S^* = \{(r,\theta)\,|\,0 \le r < R, \quad 0 \le \theta \le \Theta\}, \ \ \Theta \le 2\pi, \tag{I.1.24}$$

and the unbounded domain outside a sector

$$S^*_\infty = \{(r,\theta)\,|\,R < r < \infty, \quad 0 \le \theta \le \Theta\}, \ \ \Theta \le 2\pi. \tag{I.1.25}$$

Let $\Theta = \pi$, then the S^* is a semi-disk. Suppose that the mixed type of the Dirichlet and Neumann conditions are provided as

$$\frac{\partial u}{\partial \nu} = 0 \quad \text{at } \theta = 0; \qquad u = 0 \quad \text{at } \theta = \pi, \tag{I.1.26}$$

which result in Motz's problem and the interior crack problems. We also list their particular solutions without proof.

V. For the Laplace equation, $\Delta u = 0$, the particular solutions for eqns. (I.1.24) and (I.1.25) are

$$r^{n+\frac{1}{2}} \cos\left(n + \frac{1}{2}\right)\theta, \qquad n = 0,1,\ldots,\ (r,\theta) \in S^*, \tag{I.1.27}$$

$$r^{-(n+\frac{1}{2})} \cos\left(n + \frac{1}{2}\right)\theta, \quad n = 0,1,\ldots,\ (r,\theta) \in S^*_\infty,$$

respectively.

VI. For the equation, $-\Delta u + k^2 u = 0$ with real $k > 0$, the particular solutions for eqns. (I.1.24) and (I.1.25) are

$$I_{n+\frac{1}{2}}(kr) \cos\left(n + \frac{1}{2}\right)\theta, \quad n = 0,1,\ldots,\ (r,\theta) \in S^*,$$

$$K_{n+\frac{1}{2}}(kr) \cos\left(n + \frac{1}{2}\right)\theta, \quad n = 0,1,\ldots,\ (r,\theta) \in S^*_\infty,$$

respectively.

VII. For the Helmholtz equation, $\Delta u + k^2 u = 0$ with real $k > 0$, the particular solutions for eqns. (I.1.24) and (I.1.25) are

$$J_{n+\frac{1}{2}}(kr)\cos\left(n+\frac{1}{2}\right)\theta, \quad n = 0,1,\ldots, \quad (r,\theta)\in S^*,$$

$$H^{(1)}_{n+\frac{1}{2}}(kr)\cos\left(n+\frac{1}{2}\right)\theta, \quad n = 0,1,\ldots, \quad (r,\theta)\in S^*_\infty,$$

respectively.

VIII. For the biharmonic equation, $\Delta^2 u = 0$, the symmetric and clamped boundary conditions are given on the boundary of S^* as

$$\frac{\partial u}{\partial \nu} = \frac{\partial^3 u}{\partial \nu^3} = 0 \quad \text{at } \theta = 0; \qquad u = \frac{\partial u}{\partial \nu} = 0 \quad \text{at } \theta = \pi.$$

Then, the particular solutions are given by

$$r^{n+1}\left[\cos(n-1)\theta - \cos(n+1)\theta\right], \tag{I.1.28}$$

$$r^{n+\frac{1}{2}}\left[\cos\left(n-\frac{3}{2}\right)\theta - \frac{n-\frac{3}{2}}{n+\frac{1}{2}}\cos\left(n+\frac{1}{2}\right)\theta\right], \quad n = 1,2,\ldots, \quad (r,\theta)\in S^*.$$

The TM using eqn. (I.1.28) is reported in Chapter 4.

There are some systematic methods for deriving the complete systems of particular solutions. Of course, the separation of variables is most popular. Maybe, the most extensive one stems from the function theoretic approach, which was pioneered by Bergman [31] and Vekua [448], and then further developed by Colton, Gilbert, Kracht-Kreyzig, Lanckau, and others. A good survey of these methods is given in Begehr and Gilbert [28]. The other approach of T-complete systems proposed by Herrera and his colleagues (see [189, 190]) has been applied for a variety of problems. For instance, the T-complete systems for Stokes problems and biharmonic equations were given in Gourgeon and Herrera [171] and Herrera and Gourgeon [197], respectively, and a remarkably simple T-complete system developed for plane waves of the Helmholtz equation (or called the reduced wave equation) was also reported in Sanchez-Sesma, Herrera, and Aviles [400]. Other developments of T-complete systems include Jirousek and Wróblewski [231] and Jirousek and Zielinski [232].

I.2 Coupling techniques

For matching different particular solutions in the TM, or matching the TM with other methods, an effective coupling strategy is essential, in order to yield optimal convergence rates and good stability of the numerical solutions. To minimize the errors, we use an additional integral equation along the common boundary and apply a penalty plus hybrid techniques. Such techniques have been applied in the standard FEM in Arnold [5], Baker [20], Barrett and Elliott [25], Nitsche [346], Fairweather [142], and Gatica, Harbrecht, and Schneider [155].

I.2.1 Six combinations

Below, we list six efficient coupling techniques for combinations of the TM and FEM. Details are given in Refs. [280, 286, 288]. Consider a general elliptic equation

$$-\frac{\partial}{\partial x}\left(p\frac{\partial u}{\partial x}\right)-\frac{\partial}{\partial y}\left(p\frac{\partial u}{\partial y}\right)+cu=f, \quad (x,y)\in S, \tag{I.2.1}$$

with the boundary conditions

$$u = g_1 \quad \text{on } \Gamma_D, \tag{I.2.2}$$

$$\frac{\partial u}{\partial \nu}+qu = g_2 \quad \text{on } \Gamma_M, \tag{I.2.3}$$

where the domain S is a bounded polygon with the exterior boundary $\Gamma=\Gamma_D\cup\Gamma_M$ and $\text{Meas}(\Gamma_D)>0$. The functions p, c, f, q, g_1, and g_2 are sufficiently smooth, and

$$c=c(x,y)\geq 0, \quad q=q(x,y)\geq 0, \quad p=p(x,y)\geq p_0>0,$$

where p_0 is a constant.

The solution of the problem, i.e., eqns. (I.2.1) to (I.2.3) can be equivalently expressed by minimizing a quadratic functional $I(v)$:

$$I(u)=\min_{v\in H_*^1(S)} I(v), \tag{I.2.4}$$

where the quadratic functional is

$$I(v)=\frac{1}{2}\iint_S[p(\nabla v\cdot\nabla v)+cv^2]\,ds+\frac{1}{2}\int_{\Gamma_M}qv^2\,d\ell-\iint_S fv\,ds-\int_{\Gamma_M}g_2v\,d\ell,$$

and $H_*^1(S)$ is a set of the space defined as

$$H_*^1(S)=\{v \mid v, v_x, v_y\in L^2(S), \quad \text{and } v\mid_{\Gamma_D}=g_1\}.$$

In some cases, e.g., $P_c=0$, $\alpha=1$, and $\beta=0$ for the simplified hybrid combined method, the minimum eqn. (I.2.4) does not hold. However, the first variation of $I_h^{(1,0)}(v)$ provides the desired equation of the solution.

Let S be divided by Γ_0, the piecewise straight line, into S_1 and S_2. The TM is used in S_2, and the $k(\geq 1)$-order Lagrange FEM is used in S_1, where the true solution is supposed to be smooth enough such that $u \in H^{k+1}(S_1)$. Therefore, the admissible functions can be written as follows:

$$v = \begin{cases} v^- = V_1^k & \text{in } S_1, \\ v^+ = \sum_{i=1}^{N} a_i \Psi_i & \text{in } S_2, \end{cases} \tag{I.2.5}$$

where V_1^k are piecewise k-order Lagrange interpolation polynomials on the quasi-uniform triangulation of S_1 with the maximal boundary length h, $\{\Psi_i\}$ are analytic, complete, and linearly independent basis functions on S_2, and a_i are unknown coefficients to be sought. Note that the admissible functions v^+ in eqn. (I.2.5) may not satisfy the elliptic eqn. (I.2.1) exactly.

Define a space

$$H = \{v \mid v \in L^2(S), v \in H^1(S_1), v \in H^1(S_2), \quad \text{and } v \mid_{\Gamma_D \cap \partial S_1} = g_1\}.$$

Let $V_h^* \subset H$ be a finite-dimensional collection of the functions, i.e., eqn. (I.2.5). For simplicity, we assume that the functions $v \in V_h^*$ will strictly satisfy the Dirichlet boundary condition, i.e., eqn. (I.2.2) on $\Gamma_D \cap \partial S_1$ where the FEM is used (otherwise, see Strang and Fix [426]). In this subsection, the combination using the penalty plus hybrid techniques is designed as: To seek an approximate solution $u_h \in V_h^*$ such that

$$I_h^{(\alpha,\beta)}(u_h) = \min_{v \in V_h^*} I_h^{(\alpha,\beta)}(v), \tag{I.2.6}$$

where

$$\begin{aligned} I_h^{(\alpha,\beta)}(v) = &\frac{1}{2}\iint_{S_1} (p(\nabla v \cdot \nabla v) + cv^2)\, ds + \frac{1}{2}\iint_{S_2} (p(\nabla v \cdot \nabla v) + cv^2)\, ds \\ &+ \frac{1}{2}\int_{\Gamma_M} qv^2\, d\ell + \frac{P_c}{2h^{*\sigma}} \int_{\Gamma_D \cap S_2} (v^+ - g_1)^2\, d\ell \\ &- \int_{\Gamma_D \cap S_2} p\frac{\partial v^+}{\partial n}(v^+ - g_1)\, d\ell + \frac{P_c}{2h^{*\sigma}} \int_{\Gamma_0} (v^+ - v^-)^2\, d\ell \\ &- \int_{\Gamma_0} p\left(\alpha\frac{\partial v^+}{\partial n} + \beta\frac{\partial v^-}{\partial n}\right)(v^+ - v^-)\, d\ell - \iint_S fv\, ds - \int_{\Gamma_M} g_2 v\, d\ell, \end{aligned} \tag{I.2.7}$$

where α and β are two real parameters, $P_c > 0$, $\sigma > 0$, and h^* is the maximal boundary length of finite elements on Γ_0. The right-hand side of eqn. (I.2.7) shows the general approach of using additional integral for interface coupling, which contains the following six possible combinations, based on different α and β values:

1. Combination I: $P_c > 0$, and $\alpha = 0$, $\beta = 1$.
2. Combination II: $P_c > 0$, and $\alpha = 1$, $\beta = 0$.
3. Symmetric Combination: $P_c > 0$, and $\alpha = \beta = \frac{1}{2}$.

4. The Simplified Hybrid Combined Method: $P_c = 0$, $\alpha = 1$ and $\beta = 0$.
5. The Penalty Combination in Li [279]: $P_c > 0$, and $\alpha = \beta = 0$.
6. The Non-conforming Combination in Li [273]: as $P_c \to \infty$, or $\sigma \to \infty$.

For the simplified hybrid combined method, the eqn. (I.2.6) should be replaced by seeking the critical point, $\frac{\partial I_h^{(1,0)}(v)}{\partial v} = 0$.

The important analysis in Ref. [280] and the references cited therein to prove that the optimal convergence rates of

$$\|\varepsilon\|_h = \{\|\varepsilon\|_{1,S_1}^2 + \|\varepsilon\|_{1,S_2}^2\}^{\frac{1}{2}} = O(h^k)$$

can be achieved for coupling the k-order FEM with the TM. Also a stability analysis indicates that the condition number of the stiffness matrix is

$$\text{Cond} = O(h^{-2} + P_c h^{-(2+\sigma)}), \quad \sigma \in (0, 1].$$

I.2.2 The original TM

Consider the Laplace equation with the Dirichlet boundary condition $u = g$ on Γ, and the TM is used on the whole domain with no interior boundary Γ_0. The simplified hybrid method is given from eqn. (I.2.6) as: To seek an approximate solution $u_{\text{L}} \in V_{\text{L}}$ such that

$$I_{\text{Simp}}(u_{\text{L}}) = \min_{v \in V_{\text{L}}} I_{\text{Simp}}(v), \tag{I.2.8}$$

where

$$I_{\text{Simp}}(v) = \frac{1}{2} \iint_S (\nabla v \cdot \nabla v)\, ds - \int_\Gamma \frac{\partial v}{\partial n}(v - g)\, d\ell.$$

Suppose that the admissible functions are also harmonic,

$$v_{\text{L}} = \sum_{i=1}^{L} a_i \Psi_i, \qquad \nabla^2 \Psi_i = 0, \quad \text{in } S,$$

where a_i are real coefficients, and V_{L} is the finite-dimensional collection of v_{L}. The eqn. (I.2.8) is just the original TM in Trefftz [441], see Section A.6, and more discussions are given in Section 3.4. The error analysis of such a method is reported in Li and Liang [298], Li and Bui [285, 287], Li [282], and Li and Huang [292]. The indirect TM is reported in Kita, Ikeda, and Kamiya [248] and Chang et al. [83]. The TM with FEM and BEM is discussed in Qin [373].

I.2.3 The direct TM

In order to accommodate

$$u^+ = u^-, \quad \text{on } \Gamma_0, \tag{I.2.9}$$

in the non-conforming combination, the direct constraints of the admissible functions as the collocation equations,

$$v^+(Q_i) = v^-(Q_i), \quad \forall Q_i \in \Gamma_0 \tag{I.2.10}$$

can be employed, without using the additional integral on Γ_0. In eqn. (I.2.10), Q_i denote the element nodes on Γ_0. Define by $\overline{V}_h \subset H$ a finite-dimensional collection of the admissible functions, i.e., eqn. (I.2.5) that satisfy the constraint conditions, i.e., eqn. (I.2.10) and the Dirichlet boundary condition, i.e., eqn. (I.2.2). We obtain the non-conforming combination from eqn. (I.2.7) as $P_c = \alpha = \beta = 0$

$$\frac{\partial}{\partial v} I_h^{(0,0)}(u_h) = 0, \quad v \in \overline{V}_h.$$

Now, we turn to the continuity requirement, i.e., eqn. (I.2.9), by introducing an additional integral $\int_{\Gamma_0} \lambda(v^+ - v^-)\, d\ell$, where λ is a continuous function of Lagrange multipliers. We describe the Lagrange multiplier coupling as: To seek $(u_h, \lambda) \in \overline{V}_h \times R$ such that

$$B(u_h, v; \lambda, \mu) = f(v), \quad (v, \mu) \in \overline{V}_h \times R, \tag{I.2.11}$$

where

$$\begin{aligned} B(u, v; \lambda, \mu) = & \iint_{S_1} (p \nabla u \cdot \nabla v + cuv)\, ds \\ & + \iint_{S_2} (p \nabla u \cdot \nabla v + cuv)\, ds + D(u, v; \lambda, \mu). \end{aligned} \tag{I.2.12}$$

The notation $(v, \mu) \in \overline{V}_h \times R$ denotes that $v \in \overline{V}_h$ and $\mu \in R$, and the boundary integrals are

$$\begin{aligned} D(u, v; \lambda, \mu) = & \int_{\Gamma_N} quv\, d\ell - \int_{\Gamma_D \cap S_2} \lambda v\, d\ell - \int_{\Gamma_D \cap S_2} \mu u\, d\ell \\ & - \int_{\Gamma_0} \lambda(v^+ - v^-)\, d\ell - \int_{\Gamma_0} \mu(u^+ - u^-)\, d\ell, \end{aligned}$$

and the Lagrange multiplier λ has the true solution, $\lambda = p\frac{\partial u}{\partial n}\big|_{\Gamma_0}$. In eqn. (I.2.12), the λ is treated as an extra variable.

Consider the Laplace equation with the Dirichlet condition $u = g$ on Γ, and the TM is used in the entire domain without interior boundary. We have from eqn. (I.2.11)

$$\iint_S \nabla u \cdot \nabla v\, ds - \int_\Gamma \mu(u - g)\, d\ell - \int_\Gamma \lambda v\, d\ell = 0, \tag{I.2.13}$$

where $p=1, \lambda=\frac{\partial u}{\partial n}$ is treated as unknown, and $\lambda, \mu \in H^{-\frac{1}{2}}(\Gamma)$. $H^{-\frac{1}{2}}(\Gamma)$ is equipped with the negative norm in the Sobolev space. In fact, we may define a function

$$F(v) = \frac{1}{2}\iint_S \nabla v \cdot \nabla v \, ds - \int_\Gamma \mu(v-g)\, d\ell.$$

The variational equation, $\frac{\partial F(v)}{\partial v}=0$, leads to eqn. (I.2.13). This is called the Lagrange multiplier method in Ref. [280], which is just the direct TM in Jin and Cheung [226] and Kamiya and Kita [235]. The direct TM is explored in Li [280, 281], and in Section 3.6.

The Lagrange multiplier method was first introduced by Babuska [14] to treat the constraint Dirichlet boundary condition as a natural boundary condition, and to relax the limitation on the admissible functions used. Since then the techniques of Lagrange multipliers have drawn much attention. Such techniques have been adopted to mixed and hybrid methods, see Brezzi and Fortin [62] and Raviart and Thomas [377].

The boundary condition using Lagrange multipliers is also extended to that involving flux in Bramble [52]. More analysis and applications are given in Fix [145], Pitkaranta [365, 366], Lee [269], Barbosa and Hughes [24], and particularly for domain decomposition methods in Liang and Liang [309], Liang and He [308], and Mandel and Tezaur [326].

I.2.4 Trefftz–Herrera approaches for coupling problems

By the Green formulas, the TM can be extended to general elliptic equations, the framework of which was given by the algebraic approaches in Herrera [191] in 1984. Since the algebraic notations and operations are simple and easily understood (see Section A.5 in Appendix), the TM (or called the Trefftz–Herrera approaches in this book) has been applied to many engineering problems, in particular, the coupling problems where there exist the jumps of both the solutions and their derivatives along the interior boundary Γ_0 (see Section A.7). A great progress has been made by Herrera and his colleagues, and numerous papers have been published. Here, we only mention a few important works, Herrera [190, 191, 192, 193], Herrera and Diaz [195, 196], and Herrera and Solano [198], and a complete list of references for the Trefftz–Herrera approaches can be found in Refs. [190, 193].

Consider the self-adjoint second-order elliptic equations,

$$-\nabla \cdot (p^\pm \nabla u^\pm) + c^\pm u^\pm = f^\pm \quad \text{in } S^\pm, \tag{I.2.14}$$

$$u = g \quad \text{on } \Gamma, \tag{I.2.15}$$

where $\Gamma=\partial S$, $S=S^+\cup S^-\cup\Gamma_0$, and Γ_0 is the interior boundary with the jumps

$$[u]_{\Gamma_0} = \delta, \quad [pu_\nu]_{\Gamma_0} = \delta_\nu. \tag{I.2.16}$$

In eqn. (I.2.16), the notations are

$$[u]_{\Gamma_0} = u^+ \mid_{\Gamma_0} - u^- \mid_{\Gamma_0}, \quad [pu_\nu]_{\Gamma_0} = p^+ u_\nu^+ \mid_{\Gamma_0} - p^- u_\nu^- \mid_{\Gamma_0},$$

and δ and δ_ν are the known jump functions. In eqns. (I.2.14)–(I.2.16), the coefficient functions satisfy $p^\pm > p_0 > 0$ and $c^\pm \geq 0$. We assume that the functions $p^\pm, c^\pm$, $f^\pm, \delta$, and δ_ν are smooth enough so that the solutions within S^+ and S^- are also smooth (such a smoothness of the solutions does not include the case of crossing Γ_0). When $\delta \neq 0$ or $\delta_\nu \neq 0$, the eqns. (I.2.14)–(I.2.16) are also called the coupling problems in some engineering literature.

Denote the subsets

$$H_*^{\#}(S) = \{v = v^\pm \in H^1(S^\pm), v \mid_\Gamma = g, \quad \text{and } [v] \mid_{\Gamma_0} = \delta\},$$
$$H_0^{\#}(S) = \{v = v^\pm \in H^1(S^\pm), v \mid_\Gamma = 0, \quad \text{and } [v] \mid_{\Gamma_0} = 0\}.$$

The solution u of eqns. (I.2.14)–(I.2.16) can be expressed in a weak form: To seek $u \in H_*^{\#}(S)$ such that

$$A(u, v) = f(v), \quad \forall v \in H_0^{\#}(S), \tag{I.2.17}$$

where

$$A(u, v) = \iint_{S^+} (p^+ \nabla u^+ \cdot \nabla v^+ + c^+ u^+ v^+) + \iint_{S^-} (p^- \nabla u^- \cdot \nabla v^- + c^- u^- v^-),$$
$$f(v) = \iint_{S^+} f^+ v^+ + \iint_{S^-} f^- v^- + \frac{1}{2} \int_{\Gamma_0} \delta_\nu (v^+ + v^-).$$

First let us choose the FEM. Denote the finite-dimensional collections $V_h^*(\subset H_*^{\#}(S))$ and $V_h^0(\subset H_0^{\#}(S))$ for the piecewise k-order polynomials. The FEM can be expressed by: To seek $u_h \in V_h^*$ such that

$$A(u_h, v) = f(v), \quad \forall v \in V_h^0. \tag{I.2.18}$$

Since to formulate the admissible functions v satisfying $[v] \mid_{\Gamma_0} = \delta$ and $v \mid_\Gamma = g$ is rather complicated, we resort to the coupling techniques. Take the penalty coupling for example. The admissible functions do not necessarily satisfy $[v] \mid_{\Gamma_0} = \delta$ and $v \mid_\Gamma = g$. Denote $H_*(S) = \{v^\pm \in H^1(S^\pm)\}$, and the finite-dimensional collection $V_h(\subset H_*(S))$ of piecewise k-order polynomials. The penalty method can be expressed by: To seek $u_h^P \in V_h$ such that

$$A_h(u_h^P, v) = f(v), \quad \forall v \in V_h, \tag{I.2.19}$$

where

$$A_h(u, v) = A(u, v) + \frac{P_c}{h^{2\sigma}} \widehat{\int_{\Gamma_0}} (u^+ - u^- - \delta)(v^+ - v^-) + \frac{P_c}{h^{2\sigma}} \widehat{\int_\Gamma} (u - g) v, \tag{I.2.20}$$

$A(u,v)$ and $f(v)$ are given in eqn. (I.2.17), and $\widehat{\int}_{\Gamma_0}$ is the approximation of $\int_{\Gamma_0}$ by some rules. In eqn. (I.2.20), the parameters $\sigma(>0)$ may be suitably chosen to be independent of h, where h is the the maximal boundary of triangular elements $\triangle_{ij}$ or rectangular elements $\square_{ij}$. Suppose that the solutions $u^\pm$ within $S^\pm$ are smooth enough, and that the $\triangle_{ij}$ or the $\square_{ij}$ are quasi-uniform, the optimal convergence $O(h^k)$ and the superconvergence $O(h^{k+p})$, $p=1,2$ in H^1 norm can be achieved. When there exists a singularity of solutions on $\partial S^\pm$, the combinations of the TM and the FEMs can be employed, to reach the same optimal convergence and the same superconvergence, see Chapter 8.

Suppose that $f^\pm \equiv 0$, we may use the CTM in Section I.1.2. Let $\Phi_i^\pm$ satisfy the elliptic equations exactly

$$\nabla(p^\pm \nabla \Phi_i^\pm) + c^\pm u^\pm = 0 \quad \text{in } S^\pm.$$

Then, the solution can be approximated by the linear combination of $\Phi_i^\pm$,

$$u_n = u_n^\pm = \sum_{i=0}^{n^\pm} a_i^\pm \Phi_i^\pm \quad \text{in } S^\pm, \tag{I.2.21}$$

where $n^\pm$ are positive integers, and the coefficients $a_i^\pm$ can be sought by fitting the exterior and interior boundary conditions as best as possible. Denote the energy

$$I(v) = \widehat{\int_{\Gamma_0}} (v^+ - v^- - \delta)^2 + w^2 \widehat{\int_{\Gamma_0}} (p^+ v_\nu^+ - p^- v_\nu^- - \delta_\nu)^2 + \widehat{\int_{\Gamma}} (v-g)^2,$$

where $w > 0$ is a suitable weight. The solution $\hat{u}_n$ can be obtained by

$$I(\hat{u}_n) = \min_{v \in V_n} I(v), \tag{I.2.22}$$

where V_n is the finite-dimensional collection of eqn. (I.2.21). Suppose that the simplest central rule is applied to get $\widehat{\int}_{\Gamma_0}$. Then the eqn. (I.2.22) can be reduced to the least squares problem of the following collocation equations,

$$\begin{aligned} &[u_n] \,|\, _{Q_i} = \delta \,|\, _{Q_i}, \qquad w[p(u_n)_\nu] \,|\, _{Q_i} = w\delta_\nu \,|\, _{Q_i}, \quad Q_i \in \Gamma_0, \\ &u_n \,|\, _{P_i} = g \,|\, _{P_i}, \quad P_i \in \Gamma, \end{aligned}$$

where Q_i and P_i are the middle nodes of small intervals of Γ_0 and Γ, respectively. Obviously, this CTM is simpler and more efficient than the FEM in eqn. (I.2.18). Its error and stability analysis is demonstrated in this book, and the exponential convergence rates can be achieved, see Chapters 1 and 2. When the piecewise k-order polynomials are replaced by eqn. (I.2.21), the eqn. (I.2.19) may lead to the penalty TM. Other kinds of TMs can be further developed.

The Trefftz–Herrera domain decomposition is described in Herrera and Diaz [196] for the general second-order elliptic equation

$$-\nabla\cdot(p^{\pm}\nabla u^{\pm})+\nabla\cdot(\mathbf{b}^{\pm}u^{\pm})+c^{\pm}u^{\pm}=f^{\pm},\quad \text{in } S^{\pm}, \tag{I.2.23}$$

with interior jump conditions, where $\mathbf{b}^{\pm}$ is a vector. The non-self-adjoint term $\nabla\cdot(\mathbf{b}^{\pm}u^{\pm})$ results from the convection flow. When the norm $\|\mathbf{b}^{\pm}\|$ is large, the solution of eqn. (I.2.23) may represent the convection-dominant flow with a boundary layer singularity. Moreover, for the general $2m$-order elliptic equations, the Green formulas are given in Oden and Reddy [348], p. 165, in which the interior jump conditions can be obtained, and the algorithms of CTM may be easily formulated. Based on the regularity of solutions in Chapter 5 in Ref. [348], the error bounds of the solutions by the CTM for the general elliptic equations involving the solution jumps on Γ_0 may be further developed; details will be reported later.

I.3 Boundary element methods

In this section, we provide a concise introduction of the BEM for smooth solutions. The systematic and comprehensive introduction of BEM can be found in Brebbia [58], Brebbia and Dominguez [61], and Atkinson [10].

I.3.1 Green theorem

Let ∂S be piecewise smooth, and functions $u,\ v\in C^2(S)$. Then, the Green formula is given by

$$\iint_S (vTu-uTv)\,ds=\oint_{\partial S}\left(vp\frac{\partial u}{\partial \nu}-up\frac{\partial v}{\partial \nu}\right)d\ell,$$

where

$$Tu=\left(\frac{\partial}{\partial x}p\frac{\partial u}{\partial x}+\frac{\partial}{\partial y}p\frac{\partial u}{\partial y}\right).$$

For simplicity, let us consider Poisson's equation.

$$-\Delta u=f\quad \text{in } S,\qquad u=g_1\quad \text{on } \Gamma_D,\qquad \frac{\partial u}{\partial \nu}=g_2\quad \text{on } \Gamma_N, \tag{I.3.1}$$

where $\Delta=\dfrac{\partial^2}{\partial x^2}+\dfrac{\partial^2}{\partial y^2}$. Choose the fundamental solution of Laplace's equation

$$w=\frac{1}{2\pi}\ln\frac{1}{r},\quad r=\sqrt{(x-x_0)^2+(y-y_0)^2},\quad M_0=(x_0,\,y_0)\in S.$$

We provide two important lemmas without proofs.

Lemma I.3.1

Let $\frac{\partial u}{\partial \nu} \in C(\partial S)$*, then the solution*

$$u(M_0) = \frac{1}{2\pi}\left\{\int_{\partial S}\left[\left(\ln\frac{1}{r}\right)\frac{\partial u}{\partial \nu} - u\frac{\partial}{\partial \nu}\left(\ln\frac{1}{r}\right)\right] d\ell + \iint_S \left(\ln\frac{1}{r}\right) f\, ds\right\}, \quad \forall M_0 \in S. \tag{I.3.2}$$

It can be seen from eqn. (I.3.2) that, if both u and $\frac{\partial u}{\partial \nu}$ on ∂S are known, the interior solutions may be obtained immediately. However, if only one of u and $\frac{\partial u}{\partial \nu}$ at each point of ∂S is known from the boundary conditions in eqn. (I.3.1), then the other boundary values can be obtained as below.

Lemma I.3.2

Let $\frac{\partial u}{\partial \nu} \in C(\partial S)$ *and* ∂S *be a smooth boundary, then*

$$u(M_0) = \frac{1}{\pi}\left\{\int_{\partial S}\left[\ln\left(\frac{1}{r}\right)\frac{\partial u}{\partial \nu} - u\frac{\partial}{\partial \nu}\left(\ln\frac{1}{r}\right)\right] d\ell + \iint_S \left(\ln\frac{1}{r}\right) f\, ds\right\}, \quad \forall M_0 \in \partial S. \tag{I.3.3}$$

From the boundary conditions in eqn. (I.3.1), we have

$$\int_{\partial S} u\frac{\partial}{\partial \nu}\ln\left(\frac{1}{r}\right) d\ell = \int_{\Gamma_D} g_1\frac{\partial}{\partial \nu}\ln\left(\frac{1}{r}\right) d\ell + \int_{\Gamma_N} u\frac{\partial}{\partial \nu}\ln\left(\frac{1}{r}\right) d\ell,$$

$$\int_{\partial S} \ln\left(\frac{1}{r}\right)\frac{\partial u}{\partial \nu} d\ell = \int_{\Gamma_D} \ln\left(\frac{1}{r}\right)\frac{\partial u}{\partial \nu} d\ell + \int_{\Gamma_N} g_2\ln\left(\frac{1}{r}\right) d\ell.$$

Hence, we obtain from Lemma I.3.2

$$u(M_0) = \frac{1}{\pi}\left\{\int_{\Gamma_D}\frac{\partial u}{\partial \nu}\ln\left(\frac{1}{r}\right) d\ell - \int_{\Gamma_N} u\frac{\partial}{\partial \nu}\ln\left(\frac{1}{r}\right) d\ell - \int_{\Gamma_D} g_1\frac{\partial}{\partial \nu}\ln\left(\frac{1}{r}\right) d\ell + \int_{\Gamma_N} g_2\ln\left(\frac{1}{r}\right) d\ell + \iint_S \ln\left(\frac{1}{r}\right) f\, ds\right\}, \quad \forall M_0 \in \partial S. \tag{I.3.4}$$

I.3.2 Discrete approximation

Below, let us describe the simplest BEM discretization form of eqn. (I.3.3) by following the FEM approaches. Let ∂S be divided into quasi-uniform subintervals (s_i, s_{i+1}) with the arc length h_{i+1}, where $s_{i+1} = s_i + h_{i+1}$ and $s_0 = s_N$. Also choose piecewise constant and linear functions as the admissible functions of $u_\nu = \frac{\partial u}{\partial \nu}$ and u, respectively,

$$\frac{\partial u}{\partial \nu} = \sum_{i=1}^{N} (u_\nu)_i \Psi_i(s), \quad u = \sum_{i=1}^{N} u_i \Phi_i(s),$$

where $\Psi_i(s)$ and $\Phi_i(s)$ are the basis functions given by

$$\Psi_i(s) = \begin{cases} 1, & s_i - \frac{h_i}{2} < s < s_i + \frac{h_{i+1}}{2}, \\ 0, & \text{otherwise,} \end{cases}$$

and

$$\Phi_i(s) = \begin{cases} \frac{1}{h_i}(s - s_{i-1}), & s_{i-1} \le s \le s_i, \\ 1 - \frac{1}{h_{i+1}}(s - s_i), & s_i \le s \le s_{i+1}, \\ 0, & \text{otherwise.} \end{cases}$$

Substitute these two series into eqn. (I.3.4) and use the integration approximation in eqn. (I.3.4); then there come up N equations and N unknowns including either u_i or $(\frac{\partial u}{\partial \nu})_i$, leading to a linear algebraic system,

$$\mathbf{A}_\mathrm{B}\mathbf{x}_\mathrm{B} = \mathbf{b}, \tag{I.3.5}$$

where $\mathbf{A}_\mathrm{B}$ is an $N \times N$ matrix, which is neither symmetric nor sparse. Once the solution of eqn. (I.3.5) is obtained, we may seek the solution in S from Lemma I.3.1

$$u(M_0) = \frac{1}{2\pi}\left[\sum_{i=1}^{N} (u_\nu)_i \int_{\partial S} \Psi_i(t) \ln\left(\frac{1}{r}\right) d\ell - \sum_{i=1}^{N} u_i \int_{\partial S} \Phi_i(t) \frac{\partial}{\partial \nu} \ln\left(\frac{1}{r}\right) d\ell \right.$$
$$\left. + \iint_S \ln\left(\frac{1}{r}\right) f \, ds \right].$$

Note that in the BEM, both u_i and $(u_\nu)_i$ are treated as unknowns. We may employ other FEM approaches to obtain the BEM solutions. Both BEM and TM are of one-dimensional solver, one dimension lower than that in FEM, so the unknown number in BEM is substantially smaller than that in FEM. This is a remarkable advantage. However, when $f \neq 0$, the approximation integration of $\iint_S \ln(\frac{1}{r}) f \, ds$ may consume much more CPU time than the solution procedure of eqn. (I.3.5). Hence, BEM is strongly recommended for $f \equiv 0$. This can be done by a transformation $v = u - \overline{u}$, where $\overline{u}$ is a special solution of $\mathcal{L}\overline{u} = f$. Note that the BEM is even more beneficial for 3D elliptic problems.

I.3.3 Natural BEM

Consider the pure Neumann–Laplace problem

$$\Delta u = 0 \quad \text{in } S, \qquad \frac{\partial u}{\partial \nu} = g \quad \text{on } \partial S.$$

The consistent condition

$$\int_{\partial S} \frac{\partial u}{\partial \nu}\, d\ell = \oint_{\partial S} g\, d\ell = 0$$

guarantees existence of the solutions. There are multiple solutions, and each differs by a constant. From eqn. (I.3.3), we have

$$u(Y) = \frac{1}{\pi}\left\{\int_{\partial S} \frac{\partial u(X)}{\partial \nu_X} \ln \frac{1}{|X-Y|}\, d\ell(X) - \int_{\partial S} u(X) \frac{\partial}{\partial \nu_X} \ln \frac{1}{|X-Y|}\, d\ell(X)\right\},$$

which lead to

$$\begin{aligned}\frac{\partial u(Y)}{\partial \nu_Y} = \frac{1}{\pi}\Bigg\{&\int_{\partial S} \frac{\partial u(X)}{\partial \nu_X} \frac{\partial}{\partial \nu_Y} \ln \frac{1}{|X-Y|}\, d\ell(X) \\ &- \int_{\partial S} u(X) \frac{\partial}{\partial \nu_X} \frac{\partial}{\partial \nu_Y} \ln \frac{1}{|X-Y|}\, d\ell(X)\Bigg\}.\end{aligned}$$

By applying the Neumann boundary condition $\frac{\partial u}{\partial \nu_X} = g$ on Γ, we obtain

$$g = \frac{1}{\pi}\left\{\int_{\partial S} g \frac{\partial}{\partial \nu_Y} \ln \frac{1}{|X-Y|}\, d\ell(Y) - \int_{\partial S} u(X) \frac{\partial}{\partial \nu_Y} \frac{\partial}{\partial \nu_X} \ln \frac{1}{|X-Y|}\, d\ell(Y)\right\}.$$

Hence, the solutions on ∂S can be sought by the natural BEM: Find $u \in H^{\frac{1}{2}}(\Gamma)/P_0$ such that

$$D(u, v) = \int_{\partial S} g v\, d\ell,$$

where P_0 is the space of an arbitrary constant, and

$$\begin{aligned}D(u, v) = \frac{1}{\pi}\Bigg\{&\int_{\partial S}\int_{\partial S} \left(g \frac{\partial}{\partial \nu_Y} \ln \frac{1}{|X-Y|}\right) v\, d\ell(Y)\, d\ell(X) \\ &- \int_{\partial S}\int_{\partial S} \left(\frac{\partial}{\partial \nu_Y} \frac{\partial}{\partial \nu_X} \ln \frac{1}{|X-Y|}\right) uv\, d\ell(Y)\, d\ell(X)\Bigg\}.\end{aligned}$$

The discretization schemes of the natural BEM can be found in Yu [476].

The key difference between the BEM and the CTM is that the fundamental functions and the particular solutions are chosen in the related boundary equations for the BEM and the CTM, respectively. References on BEM are given in Brebbia [58], Wendland [464], Chen and Zhou [91], and Brebbia and Dominguez [61]. More study can be found in Cruse [114], Bayliss, Gunzburger, and Turkel [27], Han and Wu [184], Hsiao [202, 203], and Feng and Yu [144].

I.4 Other kinds of boundary methods

Other kinds of boundary methods are found in the integral solutions of the PDE. Take the Poisson integral on a unit disk for example. A harmonic function satisfies

$$u(r,\theta) = \int_0^{2\pi} K(r,\theta,\xi)u(1,\xi)\,d\xi, \quad r \leq 1, \tag{I.4.1}$$

where the Poisson kernel is

$$K(r,\theta,\xi) = \frac{1-r^2}{2\pi}\frac{1}{1-2r\cos(\theta-\xi)+r^2}.$$

When the Dirichlet condition for the Laplace equation is given, the solution within the unit disk can be obtained from eqn. (I.4.1). Next, consider a sector $S^* = \{(r,\theta)\,|\,0 \leq \theta \leq \Theta, 0 \leq r \leq 1\}$, with the Dirichlet, the Neumann, or their mixed boundary conditions on two edges $\theta = 0$ and $\theta = \Theta$ where $\Theta \leq \pi$. The general Poisson integrals are derived by Volkov [451, 452] as

$$u(r,\theta) = \int_0^{\Theta} K^*(r,\theta,\xi)u(1,\xi)\,d\xi, \quad (r,\theta) \in S^*, \tag{I.4.2}$$

with the explicit kernels $K^*(r,\theta,\xi)$. Then the block method is developed to seek the Laplace's solutions on polygons, by dividing the solution domain into several sectors with overlaps, and by employing the Schwarz alternating method (SAM).

In fact, the solution on the unit disk can be expressed by the particular solutions,

$$u(r,\theta) = \frac{\overline{a}_0}{2} + \sum_{i=0}^{\infty}(\overline{a}_i r^i \cos i\theta + \overline{b}_i r^i \sin i\theta), \tag{I.4.3}$$

where the true coefficients are

$$\overline{a}_i = \frac{1}{\pi}\int_{-\pi}^{\pi} u(1,\theta)\cos i\theta\,d\theta, \quad \overline{b}_i = \frac{1}{\pi}\int_{-\pi}^{\pi} u(1,\theta)\sin i\theta\,d\theta. \tag{I.4.4}$$

Substituting eqn. (I.4.4) into eqn. (I.4.3) leads to Poisson's integral eqn. (I.4.1).

Next, consider the particular solutions, i.e., eqn. (I.1.27) on the unit semi-disk, to give the solution

$$u(r,\theta) = \sum_{i=0}^{\infty} \overline{a}_i r^{i+\frac{1}{2}} \cos\left(i+\frac{1}{2}\right)\theta, \quad 0 \leq r \leq 1, \tag{I.4.5}$$

where the true coefficients are

$$\bar{a}_i = \frac{2}{\pi}\int_0^{\pi} u(1,\theta)\cos\left(i+\frac{1}{2}\right)\theta\, d\theta. \tag{I.4.6}$$

Substituting eqn. (I.4.6) into eqn. (I.4.5) gives the following kernel for the harmonic functions satisfying the boundary conditions, i.e., eqn. (I.1.26)

$$K^*(r,\theta,\xi)$$

$$= \frac{\sqrt{r}(1-r)}{\pi}\left\{\frac{\cos\left(\frac{\theta+\xi}{2}\right)}{1-2r\cos(\theta+\xi)+r^2} + \frac{\cos\left(\frac{\theta-\xi}{2}\right)}{1-2r\cos(\theta-\xi)+r^2}\right\}, \quad r \le 1.$$

The above analysis displays that the series solution consisting of particular solutions is linked to Poisson's kernel. The errors of the TM are mainly from the truncation of the finite series, but the errors of the block method as well as the BEM are from the approximation of the solution by low-order polynomials.

The block method is developed for singularity problems, such as Motz's problem in Volkov and Kornoukhov [453], and Dosiyev [130]. Note that from Li et al. [301, 302], for solving Poisson's equation, the CTM is much simplier than the Poisson integral eqn. (I.4.2).

It is well known that the potential of a double layer of density μ over a closed surface Γ defined by

$$u(P) = -\frac{1}{2\pi}\int_\Gamma \mu(Q)\frac{\partial}{\partial \nu_Q}\ln\|P-Q\|\, dS_Q, \quad P \notin \Gamma \tag{I.4.7}$$

is harmonic, where $\|\cdot\|$ is the Euclidean norm. For Laplace's equation with the Dirichlet condition,

$$\Delta u = 0 \quad \text{in } S, \qquad u = g \quad \text{on } \Gamma, \tag{I.4.8}$$

where S is a bounded domain and Γ is its exterior boundary. The density function $\mu(P)$ on Γ can be found by

$$-\frac{1}{2}\mu(P) - \frac{1}{2\pi}\int_\Gamma \mu(Q)\frac{\partial}{\partial \nu_Q}\ln\|P-Q\|\, dS_Q = g(P), \quad P \in \Gamma. \tag{I.4.9}$$

For the exterior problem of the Laplace equation,

$$\Delta u = 0 \quad \text{in } S_\infty, \qquad u = g \quad \text{on } \Gamma, \tag{I.4.10}$$

where S_∞ is an unbounded domain, and Γ is its interior boundary. The density function $\mu(P)$ on Γ can be found by

$$\frac{1}{2}\mu(P) - \frac{1}{2\pi}\int_\Gamma \mu(Q)\frac{\partial}{\partial \nu_Q}\ln\|P-Q\|\, dS_Q = g(P), \quad P \in \Gamma. \tag{I.4.11}$$

There exist only the sign differences between eqns. (I.4.9) and (I.4.11). The existence of the solutions for eqns. (I.4.9) and (I.4.11) was proven in Courant and Hilbert [109], p. 301. Both eqns. (I.4.9) and (I.4.11) are Fredholm equations of the second kind. Note that the eqns. (I.4.7) and (I.4.9) (or (I.4.11)) are analogous to Lemmas I.3.1 and I.3.2, respectively. Once function $\mu(P)$ is obtained from eqns. (I.4.9) or (I.4.11), the solution in S or S_∞ is given by eqn. (I.4.7).

The potential of a single layer of density σ over a closed surface Γ defined by

$$u(P) = -\frac{1}{2\pi}\int_\Gamma \sigma(Q)\ln\|P-Q\|\,dS_Q, \quad P\in\Gamma, \tag{I.4.12}$$

is also harmonic. Based on eqn. (I.4.12), similar boundary integral equations can be found in textbooks for the Laplace equation with the Neumann boundary condition. However, we may directly apply eqn. (I.4.12) for the Laplace equation with the Dirichlet boundary condition. Since $u(P)$ in eqn. (I.4.12) is continuous, the solution of both eqns. (I.4.8) and (I.4.10) is then expressed by

$$g(P) = -\frac{1}{2\pi}\int_\Gamma \sigma(Q)\ln\|P-Q\|\,dS_Q, \quad P\in\Gamma. \tag{I.4.13}$$

Note that the eqn. (I.4.13) is the Fredholm equation of the first kind, and there exist some difficulties. The unique solution of eqn. (I.4.13) exists if the logarithmic capacity (i.e., the transfinite diameter) $C_\Gamma \neq 1$. When Γ is a circle, C_Γ is its diameter, see Yan [475]. When $C_\Gamma \neq 1$, for solving eqn. (I.4.13), several numerical methods are proposed, such as the Galerkin method in Stephan and Wendland [422] and Sloan and Spence [415], the CM in Costabel, Ervin, and Stephan [107], Elschner and Graham [136], and Yan [475], and the quadrature method in Sidi and Israrli [413], Saranen [401], and Saranen and Sloan [402]. Other numerical reports are given in Carstensen and Praetorius [77] and Huang and Shaw [217]. Recently, a new quadrature method, called the mechanical quadrature method (MQM), is developed by Huang [215] and Huang and Lü [216]. The MQM has the $O(h^3)$ order of convergence rates, and the excellent stability with Cond $= O(h^{-1})$, where h is the maximal meshspacing of the quadrature nodes, and Cond is the traditional condition number. Furthermore, the high order $O(h^{3+p})(0<p\le 3)$ can also be achieved for both smooth and singular solutions by the MQM using the extrapolation and the splitting extrapolation techniques, see Refs. [215, 216].

In eqn. (I.4.13), the source points Q are located on Γ, to cause a logarithmic singularity. If we choose Q outside of S, the eqn. (I.4.13) can easily be solved by the Galerkin, the collocation, and the quadrature methods. We may interpret this method as the TM using fundamental solutions, which is also called the method of fundamental solutions (MFS). Such a method was first proposed by Kupradze [258]. Choose the admissible functions

$$u_N = \sum_{i=1}^{N} c_i \ln|\overline{PQ}_i|, \quad P\in S\cup\Gamma, \tag{I.4.14}$$

where c_i are coefficients, and the source points Q_i are located uniformly on an outside circle (see Bogomolny [48]),

$$Q_i = \{(x_i, y_i) \,|\, x_i = R\cos(i\Delta\theta), \quad y_i = R\sin(i\Delta\theta)\}, \tag{I.4.15}$$

where $\Delta\theta = \frac{2\pi}{N}$ and $R > \max_S r$. For the Dirichlet boundary condition in eqn. (I.4.8), the collocation equations are obtained as

$$w_j \sum_{i=1}^{N} c_i \ln |\overline{P_j Q_i}| = w_j g(P_j), \quad P_j \in \Gamma, \tag{I.4.16}$$

where $w_j(>0)$ are suitable weights. If P_i are chosen as the quadrature nodes, and if w_j are the quadrature weights, the eqn. (I.4.16) is just the quadrature method of eqn. (I.4.13). On the other hand, the eqn. (I.4.16) can be regarded as the CTM using fundamental solutions, instead of particular solutions.

For the MFS, the polynomial convergence rates are proved for smooth solutions in Bogomolny [48], and the exponential convergence rates are provided in Katsurda and Okamoto [240]. However, the ill-conditioning of MFS is severe, to have the exponential growth rates of Cond, which are proven for Dirichlet problems on disk domains in Christiansen [100]. In fact, the exponential rates of Cond can be proved for other kinds of boundary conditions, and on non-disk domains. Since the coefficients c_i in eqn. (I.4.14) obtained from the MFS are large and highly oscillating, the subtraction cancellation of instability occurs in the final harmonic solutions, i.e., eqn. (I.4.14). To reduce the large coefficients c_i, the truncated singular value decomposition method (TSVD) and the Tikhonov regularization can be employed. Details of stability by the MFS will appear elsewhere.

I.5 Comparisons

To close this introduction, let us make brief comparisons of different numerical methods. First, the FEM gains the maximum popularity owing to its high flexibility, in particular for arbitrary geometric shapes of S, variable coefficients, and different elliptic equations. Since the triangular partition and piecewise low-order polynomials (e.g., Lagrange or Hermite elements) are very flexible, the FEM may fit into very wide scale of elliptic problems. Therefore, FEM has been well developed in both practical numerical algorithms and rigorous theoretical analysis. However, the FEM has a few deficiencies. First, the mesh generation cannot be fully automatized, despite the help of auto mesh generator. Second, the evaluation of the stiffness matrix is CPU time-consuming. Third, its accuracy is limited by the low-order polynomial approximation and interpolation.

For certain types of simple elliptic equations, other numerical methods, e.g., BEM, TM, etc., may be more efficient. The BEM and the TM are confined to certain linear and constant coefficient elliptic equations, where the fundamental functions and the particular solutions can be found in textbooks or by analytical work. The high efficiency of TM is not surprising, because the particular solutions are best to approximate the true solutions. Moreover, the high accuracy with the exponential convergence rates can be achieved by both TM and CTM. However, such particular solutions may not be found for rather arbitrary S.

The BEM and the TM are simpler and more efficient than FEM, finite difference method (FDM), and finite volume method (FVM). The TM is also simpler than the BEM, because the expansion solutions from the TM are more explicit in use. In particular, their leading coefficient displays explicitly the singularity of the solutions and gives the fracture intensity factor. The advantages of the TM are also given in Jirousek and Wróblewski [230, 231].

Let us compare the CTM with the direct TM as referred in Ref. [226]. In the CM and the CTM, there is no Lagrangian multiplier λ needed and the error analysis is rather easy, see this book. On the other hand, in the direct TM, the stiffness matrix is symmetric but indefinite and the difficult Ladyzhenskaya–Babuska–Brezzi (LBB) condition must be verified for the error analysis. Moreover, the CTM using collocation equation is also simpler than the original TM, i.e., the simplified hybrid method. Overall, the CTM is the simplest and most accurate and the most stable method, and detailed discussions and numerical results are reported in Chapters 2 and 3.

However, *there is no perfect method*, and *merits and drawbacks are twins*. The high efficiency of the TM relies on the following:

1. The explicit particular solutions must exist, and have high convergence rates.
2. The solution domain should be divided into several subdomains with or without overlaps, in order to reduce instability and to retain the high accuracy. The Trefftz–Herrera method and the Trefftz–Herrera domain decomposition are developed by Herrera [191, 193]. The shape of the domain and its partition may greatly affect the efficiency of the TM. The stability analysis on the domain shapes is explored in Li and Mathon [305], and its computational aspects in Kita, Kamiya, and Ikeda [251].
3. The suitable coupling techniques are also imperative to enforcement of the exterior and interior boundary conditions, see Qin [373]. In the TM, we may choose the central or the Gaussian rules to formulate directly the collocation equations, see Chapter 2.
4. For rather complicated PDE, with no explicit particular solutions on the entire domain, we may still apply the TM locally, by combining it with FEM, FDM, and FVM. The examples are given in Chapter 8.

The TM has been developing for many engineering problems since the important work by Zienkiewicz, Kelley, and Bettess [488] and Jirousek and Leon [229] in 1977. Before 1995, TM was investigated by Shaw, Huang, and Zhao [408], Cheung, Jin, and Zienkiewicz [97], Ruge [394], and Zieliński and Herrera [485]. After 1995, it was studied by Dong et al. [129], Sladek, Sladek, and van Keer [414], Domingues, Portela, and de Castro [128], Portela and Charafi [369], Reutskiy [382], Abou-Dina [1], de Freitas and Leitao [121], Herrera [194], Jin and Cheung [226],

Kita, Kamiya, and Iio [250], Leitao [271], Zieliński [484], Jirousek and Wróblewski [230, 231], Freitas and Cismasiu [152], and Herrera and Solano [198]. Moreover, an equivalence of the TM and the boundary method using the fundamental solutions is explored in Chen et al. [90], and the TM is extended to time-dependent PDE in Cho et al. [99]. Only a few reports of the TM involve analysis, see Christiansen and Hansen [101], Melenk and Babuska [333], Ohnimus, Rüter, and Stein [350], Comodi and Mathon [105], and Herrera and Diaz [195]. Hence, the analysis of the TM is behind that of the BEM. There exists a gap between its excellent numerical results (see Piltner [363] and Zheng and Yao [479]) and the theoretical analysis. In 1995, there was a special issue on the TM in Advances in Engineering Software, edited by Kamiya and Kita [235] and Kita and Kamiya [249], which has drawn significant attention. Later on May 30–June 1, 1996, the first International Workshop on the Trefftz method was held in Cracow, Poland, and Zieliński [483] and Zienkiewicz [487] reported the history and development of TM. The second International Workshop on the Trefftz method was held in Sintra, Portugal, in September 1999. The invited talks of workshops were published in Computer Assisted Mechanics and Engineering Sciences (CAMES), vol. 4 (1997) and vol. 8 (2001). In 2002, the third International Workshop on the Trefftz method was held at University of Exeter, U.K., and the Proceeding is given in CAMES, vol. 10 (2003). Inspired by the special issue and the workshops of TM, we have carried out new studies on the BAM (i.e., the TM), and summarized the recent new results of the CM and the CTM in this book. We hope that the existing gap of the TM between theory and computation is narrowed by the present effort. Hence, this book is also devoted to celebration of the 80 years of developments of TM. A bust image of Erich Trefftz (1888–1937) is given on page xiii in memory of him.

The boundaries which divide Life from Death
are at best shadowy and vague.
Who shall say where the one ends,
and where the other begins?
——— *The Premature Burial (1844)* ———

Edgar Allan Poe
(1809–1849)

Part I
Collocation Trefftz method

The Trefftz method (TM) was first proposed in 1926 [441]. The method can be classified as a boundary-type solution procedure. The main idea of the TM is to use particular solutions as the admissible functions, which satisfy the partial differential equation (PDE) exactly. The numerical effort is required only to approximate the boundary conditions.

There exists an extensive list of literatures on the TM in engineering journals. In 1995, there was a special issue on the TM, edited by Kamiya and Kita [235], concerning the fundamentals, applications, and analysis schemes (cf. [226, 249, 482]). Other applications include Jin, Cheung, and Zienkiewicz [227], Herrera [191], Jirousek and Guex [228], Zieliński [481], Zieliński and Zienkiewicz [486], Zienkiewicz, Kelly, and Bettess [488], Piltner and Taylor [364], Kolodziej [252], Herrera and Diaz [195], and Leitão [271]. The TM has also been applied to the interface problems and the unbounded domain problems in Refs. [304, 305]. Besides, there are the International Workshops of Trefftz method; the first one was held in Cracow, Poland, 1996, organized by A. P. Zieliński.

Here, let us summarize the terminologies used in the literature and in this book. The boundary approximation method (BAM) was studied in Li [274, 276, 280], Li and Mathon [304, 305], and Li, Mathon, and Sermer [306]. Its numerical implementation leads to the collocation Trefftz method (CTM) called in this book. In Kita and Kamiya [249] the TM is classified into a direct and an indirect method, and the direct method is also referred to the method of fundamental solutions (MFS) [96, 226], which has a close relation with the discretized form of the boundary element method (BEM). Because the theory of BEM has been well developed, this book focuses on the analysis of the indirect TM, i.e., the CTM.

The theory of the indirect TM without integration approximation was first established in the Doctoral dissertation of the first author in 1986 as the BAM, see Li [274] also Li [276], whose materials were published in Li, Mathon, and Sermer [306] and Li and Mathon [304, 305]. Only recently it was realized that the BAM and the indirect TM not only share the same theoretical basis, but also are identical in algorithm. Since the TM has been used in the engineering community for a long time, in this book we choose the term TM for communication with more engineering researchers.

The singularity problems have drawn much attention in the last several decades, and reported in numerous papers. Most of them deal with the second-order PDEs including the point singularity [280, 341, 433] and singular boundary layers [338, 392]. There exist a few books and papers for the fourth-order PDEs, such as the biharmonic equations with crack singularities, see Grisvard [178], Lefeber [270], Schiff, Fishelov, and Whiteman [404], Whiteman [467], Russo [399], and Karageorghis [239]. Textbooks and papers on biharmonic equations by the finite element method (FEM), the finite difference method (FDM), and the BEM include Chien [98], Carey and Oden [75], Birkhoff and Lynch [45], Arad, Yakhot, and Ben-Dor [4], and Brebbia and Dominguez [61]. In Part I, we pursue the more accurate solutions by using series expansion around the angular singularity of very high convergence rates. At the same time, we employ highly accurate numerical methods, such as CTMs, for obtaining their solutions. There are three key issues in the approaches: (1) Find the particular solutions of PDEs, even locally in some special regions, see Herrera [191], Zieliński [482], Li [280], etc. (2) Split the domain S into subdomains and find piecewise particular solutions. This technique provides better stability and accuracy; some interesting examples can be found in Li [280]. (3) The coupling techniques on the exterior and interior boundary conditions are important in the CTM. In Li [280], a number of efficient coupling techniques are introduced for matching the particular solutions for the TM and for matching of other methods, such as the FEM, FDM, finite volume method (FVM), etc.

Part I consists of four chapters:

Chapter 1: Basic Algorithms and Theory.
Chapter 2: Motz's Problem and Its Variants.
Chapter 3: Coupling Techniques.
Chapter 4: Biharmonic Equations with Singularities.

A brief description of Part I is given as follows.

Chapter 1 gives the theoretical foundation of the TM. The error analysis is made on the entire domain by the Sobolev norm, and the exponential convergence rates can be achieved. However, the traditional condition number of the stiffness matrix is also exponentially increasing when the number of particular solutions increases. When the solution domain is divided into several subdomains, and when suitable piecewise particular solutions are chosen, the condition number will decline significantly. Because of the high accuracy of the method, only a few terms of particular solutions are needed for practical application, so the instability of the TM is not severe. Hence, the TM can be used with a good balance between accuracy and instability.

Chapter 2 uses the CTM to seek the approximate solutions of benchmark singularity problems. The Gaussian rules with high order (in contrast to the central rule) are used to approximate the integrals in variational approaches. Such algorithms will provide the exact solution, which is much more accurate than any published literatures [280, 306].

Chapter 3 is a continuation of the study of Li [280], to explore the generalized boundary approximation methods (GBAMs) for PDE with singularities. GBAMs use the local particular solutions of PDEs, but adopt other coupling strategies to deal with the boundary conditions, which are different from the classic BAM in Li [280]. Three new GBAMs are discussed, and a new analysis is explored. Since suitable coupling techniques for interior and exterior boundary conditions are essential to a wide range of success of the method, this chapter enriches the BAM with variant formulations which arc beneficial to wide applications. The CTM is most recommended due to its high accuracy and less computational cost, although all TMs are efficient. The effective condition number is also discussed. New computation formulas for the effective condition numbers are derived to provide a reasonable bound of the relative errors of the solutions by the TM. The new formulas are casy to apply in practical application. Numerical experiments are conducted for the TMs with the effective condition number formulas.

Chapter 4 extends the CTM to biharmonic equations with singularity. First, we derive the Green formulas for biharmonic equations on bounded domains with non-smooth boundary, and the corner terms are studied. Next, three models are provided, in which a brief analysis of error bounds is performed. Two of those models are shown to be superior, and can be used as benchmarks of biharmonic problems.

Due to the historical development, different names have been assigned to the same (or basically the same) method. Before starting Part I, the equivalence of terminologies used in Parts I and III is summarized below for easy reference.

Boundary Method

- Trefftz Method (TM)
  - Traditional Trefftz Method: Similar to Ritz method, it minimizes an energy functional, but uses homogeneous particular solutions as trial functions.
  - Boundary Approximation Method (BAM): When particular solutions satisfying the governed equation are chosen, the approximate solutions are obtained by satisfying the interior and exterior boundary conditions as best as possible in the least squares sense, as described in Li [276, 280].
  - Collocation Trefftz Method (CTM or collocation TM), also called
    - Indirect Trefftz Method (Indirect TM),
    - BAM with quadrature rule.
  - Generalized Trefftz Method (GTM): Use penalty and other methods to satisfy the boundary conditions. There are four different TMs discussed in this book:
    1. Penalty TM,
    2. Hybrid TM,
    3. Penalty plus Hybrid TM,
    4. Lagrangian Multiplier TM, also called as
       - Direct Trefftz Method (Direct TM).
  - Least squares method (LSM).
- Boundary Element Method (BEM), based on
  - Direct Method: Green's formula.
    - Similar to direct TM, but uses fundamental solution as particular solutions.
- Boundary Methods
  - Indirect Method: Fredholm integral equations.
  - Method of Fundamental Solutions (MFS), called Kupradze's method.
    - Similar to CTM (Indirect TM), but uses fundamental solutions as particular solutions.

1 Basic algorithms and theory

For solving homogeneous elliptic equations, the Trefftz method (TM) uses particular solutions to approximate the boundary conditions as accurately as possible, usually in a least squares sense. In the interior of a given region, the approximate solution satisfies the differential equation exactly. The TM can cope easily with complicated boundaries and boundary conditions, as well as with singularities and infinite domains.

Approximation by particular solutions using harmonic polynomials was first applied by Kantorovich and Krylov [238] to solve Laplace's equation. Fox, Henrici, and Moler [148] used particular solutions to find eigenvalues of the Laplace operator. Bergman's (Bergman [31], Bergman and John [32, 33]) and Vekua's [448] integral representations of solutions yielded particular solution expansions for a large class of elliptic equations. The error analysis for the Bergman–Vekua method was worked out by Eisenstat [134].

From a computational point of view, the TM is easy to use, and benefiting from the reduced complexity of boundary approximation. In addition, it is often possible to control the errors of the approximate solutions by the computable errors on the boundary, even for elliptic problems that do not possess the maximum principle (see Mathon and Sermer [330]).

However, difficulties may arise in the situations where a large number of particular solutions are needed to achieve a satisfactory approximation, which can lead to ill-conditioning of the associated least squares matrices. For the problems with material interfaces, singularities, and unbounded domains, it may not be possible to find a single, uniform expansion as particular solution, which is valid in the entire domain. It is therefore necessary to divide the domain into several subdomains and to use different expansions in each of them. A solution is then obtained by approximating both the exterior boundary conditions and the interior continuity conditions across the various interfaces. We note that for a good subdivision, only a few terms are needed in each subdomain to achieve a highly accurate approximation, and to mitigate the numerical instability. Moreover, the numerical stability is greatly improved and the least squares matrices have many zero entries.

We will study the TM for solving homogeneous self-adjoint elliptic equations that combine several expansions of particular solutions from different parts of the region. In this chapter, we give an error analysis and stability analysis of the TM, which provide the basic theory of the TM in this book. The contents in this chapter are adopted from Li [276, 278, 280], Li and Lu [299], Li, Mathon, and Sermer [306], and Li and Mathon [304, 305].

1.1 Notations and preliminaries

Let S be a bounded polygon in the plane R^2 with a piecewise smooth boundary ∂S. We will consider the problem

$$\Delta u = \frac{\partial^2 u}{\partial x^2} + \frac{\partial^2 u}{\partial y^2} = 0 \quad \text{in } S, \tag{1.1.1}$$

$$u = f \quad \text{on } \Gamma_D, \qquad u_\nu = g \quad \text{on } \Gamma_N, \tag{1.1.2}$$

where f and g are sufficiently smooth functions, Γ_D and Γ_N are non-overlapped compact subsets of ∂S, $\Gamma_D \neq \phi$, $\Gamma_D \cup \Gamma_N = \partial S$, and u_ν is the normal derivative.

Let Γ_0 be an interface that divides S into two subdomains, S^+ and S^-, and let u^+ and u^- be defined in S^+ and S^-, respectively. If u^+ and u^- satisfy eqns. (1.1.1) and (1.1.2) on the corresponding subdomains and boundaries, and

$$u^+ = u^-, \qquad u^+_\nu = u^-_\nu \quad \text{on } \Gamma_0, \tag{1.1.3}$$

then $u^+ = u$ in S^+ and $u^- = u$ in S^-.

For an integer $k \geq 0$, $H^k(S)$ denote the Sobolev spaces of order k of real-valued functions with the Sobolev norms $\|\cdot\|_{k,S}$ and seminorms $|\cdot|_{k,S}$:

$$\|v\|_{k,S} = \left\{ \sum_{|\alpha|\leq k} \iint_S |D^\alpha v|^2 \, ds \right\}^{\frac{1}{2}}, \qquad |v|_{k,S} = \left\{ \sum_{|\alpha|=k} \iint_S |D^\alpha v|^2 \, ds \right\}^{\frac{1}{2}}.$$

If S is divided by an interface Γ_0 into S^+ and S^-, we will use a new space H:

$$H = H(S) = \{v \in L_2(S) \,|\, v \in H^1(S^+), v \in H^1(S^-), \text{ and } \Delta v = 0 \text{ in } S^+ \text{ and } S^-\},$$

where $L_2(S)$ is the usual space of square integrable functions in S with the inner product $(u, v) = \iint_S uv \, ds$ and the norm $\|u\|_0 = (u, u)^{\frac{1}{2}}$. The space H is equipped with the norm and seminorm:

$$\|v\|_1 = \{\|v\|^2_{1,S^+} + \|v\|^2_{1,S^-}\}^{\frac{1}{2}}, \qquad |v|_1 = \{|v|^2_{1,S^+} + |v|^2_{1,S^-}\}^{\frac{1}{2}}.$$

If γ is a curve, then $|\cdot|_{k,\gamma}$ denotes the norms in the Sobolev spaces $H^k(\gamma)$ of functions defined on γ.

For $v \in H$, $f \in L_2(\Gamma_D)$ and $g \in L_2(\Gamma_N)$ given by eqn. (1.1.2), we define a functional

$$I(v) = \int_{\Gamma_D} (v - f)^2 \, d\ell + w^2 \int_{\Gamma_N} (v_\nu - g)^2 \, d\ell + \int_{\Gamma_0} (v^+ - v^-)^2 \, d\ell + w^2 \int_{\Gamma_0} (v_\nu^+ - v_\nu^-)^2 \, d\ell,$$

where w is some positive weight. On $H \times H$, we will consider a bilinear form $[\cdot, \cdot]$ given by

$$[u, v] = \int_{\Gamma_D} uv \, d\ell + w^2 \int_{\Gamma_N} u_\nu v_\nu \, d\ell + \int_{\Gamma_0} (u^+ - u^-)(v^+ - v^-) \, d\ell + w^2 \int_{\Gamma_0} (u_\nu^+ - u_\nu^-)(v_\nu^+ - v_\nu^-) \, d\ell.$$

Clearly, $[\cdot, \cdot]$ is an inner product which induces the norm on H

$$|v|_{\mathrm{B}}^2 = [v, v]. \tag{1.1.4}$$

We need the following lemma.

Lemma 1.1.1

If $v \in H$, then there exists a positive constant C independent of v such that

$$\|v\|_1 \leq C\{|v|_1 + |v|_{0,\Gamma_D} + |v^+ - v^-|_{0,\Gamma_0}\}.$$

Proof.

For a curve $\gamma \in \partial S$ and $v \in H^1(S)$, we have (Sobolev [417], p. 68)

$$\|v\|_{1,S} \leq C\{|v|_{1,S} + |v|_{0,\gamma}\}. \tag{1.1.5}$$

Applying this inequality to S^+ and S^-, we obtain

$$\begin{aligned} \|v\|_1 &\leq C\{|v|_1 + |v|_{0,\Gamma_D \cap S^-} + |v^+|_{0,\Gamma_0}\} \\ &\leq C\{|v|_1 + |v|_{0,\Gamma_D} + |v^-|_{0,\Gamma_0} + |v^+ - v^-|_{0,\Gamma_0}\}, \end{aligned} \tag{1.1.6}$$

where we assume, without loss of generality, that $\mathrm{Meas}(\Gamma_D \cap S^-) \neq 0$. Using the Sobolev imbedding theorem (Sobolev [417], p. 84) and eqn. (1.1.5), we obtain

$$\begin{aligned} |v^-|_{0,\Gamma_0} &\leq C\|v\|_{1,S^-} \leq C\{|v|_{1,S^-} + |v|_{0,\Gamma_D \cap S^-}\} \\ &\leq C\{|v|_1 + |v|_{0,\Gamma_D}\}. \end{aligned} \tag{1.1.7}$$

The desired result follows from eqns. (1.1.6) and (1.1.7). ■

1.2 Approximation problems

Let $\{\psi_i^+\}$ and $\{\psi_i^-\}$ $(i=1,2,\ldots)$ be complete sets of particular solutions of eqn. (1.1.1) in S^+ and S^-, respectively. Consider finite-dimensional subspaces $S_{m,n} \subseteq H$ defined by

$$S_{m,n} = \left\{ v \,\middle|\, v = v^+ = \sum_{i=1}^{m} c_i \psi_i^+ \text{ in } S^+, \quad \text{and } v = v^- = \sum_{i=1}^{n} d_i \psi_i^- \text{ in } S^- \right\},$$

where c_i and d_i are real coefficients.

A function $u_{m,n} \in S_{m,n}$ will be called a TM approximation to the solution u of eqns. (1.1.1)–(1.1.3) if it minimizes I over $S_{m,n}$, i.e.,

$$I(u_{m,n}) \leq I(v), \quad \forall v \in S_{m,n}. \tag{1.2.1}$$

The space $S_{m,n}$ is assumed to satisfy the following two properties:

Inverse Property: For any $v \in S_{m,n}$, we have

$$|v_\nu|_{0,\Gamma_D} \leq k_{m,n}\|v\|_1, \qquad |v_\nu^+|_{0,\Gamma_0} \leq k_{m,n}\|v\|_1, \tag{1.2.2}$$

where $k_{m,n}$ is unbounded as $m,n \to \infty$.

Approximatability Property: For any $u \in H$, there exists a function $v \in S_{m,n}$ such that

$$\|u^+ - v^+\|_{l,\partial S^+} \leq \alpha^+_{m,k,l}|u^+|_{k,\partial S^+}, \quad 0 \leq l \leq k, \tag{1.2.3}$$

and

$$\|u^- - v^-\|_{l,\partial S^-} \leq \alpha^-_{n,t,l}|u^-|_{t,\partial S^-}, \quad 0 \leq l \leq t, \tag{1.2.4}$$

where $\alpha^+_{m,k,l}, \alpha^-_{n,t,l} \to 0$ as $m,n \to \infty$. The bounds of the constants, $\alpha^+_{m,k,l}$ and $\alpha^-_{n,t,l}$, can be obtained from Cheney [92] and Eisenstat [134]. In eqns. (1.2.3) and (1.2.4), the Sobolev norms on the boundary ∂S are defined by

$$\|v\|_{l,\partial S} = \left\{ \sum_{|\alpha| \leq l} \int_{\partial S} \left(\frac{\partial^\alpha v}{\partial s^\alpha} \right)^2 d\ell \right\}^{\frac{1}{2}}, \qquad |v|_{l,\partial S} = \left\{ \sum_{|\alpha| = l} \int_{\partial S} \left(\frac{\partial^\alpha v}{\partial s^\alpha} \right)^2 d\ell \right\}^{\frac{1}{2}}.$$

We require the following estimates.

Lemma 1.2.1

If $v \in S_{m,n}$ and the inverse property, i.e., eqn. (1.2.2) holds, then

$$\|v\|_1 \leq C(k_{m,n} + w^{-1})|v|_{\mathrm{B}}, \tag{1.2.5}$$

where C is a positive constant independent of m and n.

Proof.

By using Green's theorem and that $\Delta v = 0$ in S^+ and S^-, we have

$$0 = \iint_{S^+} v\,\Delta v\,ds + \iint_{S^-} v\,\Delta v\,ds = -|v|_1^2 + \int_{\partial S^+} v v_\nu\,d\ell + \int_{\partial S^-} v v_\nu\,d\ell.$$

Hence, from Hölder's inequality it follows that

$$\begin{aligned}|v|_1^2 &= \int_{\Gamma_D} v v_\nu\,d\ell + \int_{\Gamma_N} v v_\nu\,d\ell + \int_{\Gamma_0} [(v^+ - v^-)v_\nu^+ + v^-(v_\nu^+ - v_\nu^-)]\,d\ell \\ &\le |v|_{0,\Gamma_D}|v_\nu|_{0,\Gamma_D} + |v|_{0,\Gamma_N}|v_\nu|_{0,\Gamma_N} + |v^+ - v^-|_{0,\Gamma_0}|v_\nu^+|_{0,\Gamma_0} \\ &\quad + |v^-|_{0,\Gamma_0}|v_\nu^+ - v_\nu^-|_{0,\Gamma_0}. \end{aligned} \tag{1.2.6}$$

By applying the Sobolev imbedding theorem and the inverse property, we obtain $|v|_1^2 \le Cq\|v\|_1$, where

$$q = k_{m,n}(|v|_{0,\Gamma_D} + |v^+ - v^-|_{0,\Gamma_0}) + |v_\nu|_{0,\Gamma_N} + |v_\nu^+ - v_\nu^-|_{0,\Gamma_0}.$$

From the imbedding theorem, Lemma 1.1.1, and that $k_{m,n} \ge 1$ we have

$$\begin{aligned}\|v\|_1^2 &\le C\{|v|_1^2 + |v|_{0,\Gamma_D}^2 + |v^+ - v^-|_{0,\Gamma_0}(|v^+|_{0,\Gamma_0} + |v^-|_{0,\Gamma_0})\} \\ &\le C\{|v|_1^2 + (|v|_{0,\Gamma_D} + |v^+ - v^-|_{0,\Gamma_0})\|v\|_1\} \\ &\le Cq\|v\|_1.\end{aligned}$$

Dividing both sides by $\|v\|_1$ and noting that $q \le (k_{m,n} + w^{-1})|v|_{\mathrm{B}}$, we obtain the desired result. ■

Concerning the existence, uniqueness, and stability of $u_{m,n}$ defined by eqn. (1.2.1), we have the following lemma.

Lemma 1.2.2

Suppose that u is the solution of eqns. (1.1.1) and (1.1.2). Then for any $w > 0$, there exists a unique function $u_{m,n} \in S_{m,n}$ which satisfies

$$[u_{m,n}, v] = \int_{\Gamma_D} f v\,d\ell + w^2 \int_{\Gamma_N} g v_\nu\,d\ell, \quad \forall v \in S_{m,n}, \tag{1.2.7}$$

$$[u - u_{m,n}, v] = 0, \quad \forall v \in S_{m,n}, \tag{1.2.8}$$

and

$$\|u_{m,n}\|_1 \le C(k_{m,n} + w^{-1})\{|f|_{0,\Gamma_D} + w|g|_{0,\Gamma_N}\}. \tag{1.2.9}$$

Also, $u_{m,n}$ minimizes $I(v)$ over $S_{m,n}$ if and only if it satisfies eqn. (1.2.7).

Proof.

Since $u=f$ on Γ_D, $u_\nu=g$ on Γ_N, and u and u_ν are continuous across Γ_0, the right-hand side expression in eqn. (1.2.7) equals $[u, v]$, and so eqn. (1.2.8) easily follows from eqn. (1.2.7). In eqn. (1.2.7), let $v=u_{m,n}$ and apply the Schwarz inequality to obtain

$$|u_{m,n}|_{\mathrm{B}}^2 = [u_{m,n}, u_{m,n}] = [u, u_{m,n}] \le |u|_{\mathrm{B}} |u_{m,n}|_{\mathrm{B}}.$$

Consequently,

$$|u_{m,n}|_{\mathrm{B}} \le |u|_{\mathrm{B}} = \{|f|_{0,\Gamma_D}^2 + w^2 |g|_{0,\Gamma_N}^2\}^{\frac{1}{2}},$$

and therefore there exists a unique solution of eqn. (1.2.7) by the Riesz representation theory, since $S_{m,n}$ is finite-dimensional. By applying the last inequality to eqn. (1.2.5), we obtain eqn. (1.2.9) which expresses stability of the TM solutions with respect to perturbations of the boundary data f and g.

To prove the last part of the lemma, assume that $\eta \in S_{m,n}$. Then, we have

$$I(u_{m,n}+\eta) = I(u_{m,n}) + 2[u_{m,n}-u, \eta] + |\eta|_{\mathrm{B}}^2.$$

If $u_{m,n}$ satisfies eqn. (1.2.8), then

$$I(u_{m,n}+\eta) = I(u_{m,n}) + |\eta|_{\mathrm{B}}^2 \ge I(u_{m,n}),$$

and hence it minimizes $I(v)$ over $S_{m,n}$. On the other hand, if $u_{m,n}$ minimizes I, then for any $v \in S_{m,n}$ and scalar α:

$$\frac{\partial I\,(u_{m,n}+\alpha v)}{\partial \alpha} = 2[u_{m,n}-u, v] + 2\alpha |v|_{\mathrm{B}}^2.$$

Letting $\alpha=0$ shows that $u_{m,n}$ satisfies eqn. (1.2.8). ■

1.3 Error estimates

In this section, error estimates are derived for $u-u_{m,n}$ in the space H. Since for any $v \in H$, $\|v\|_0 \le \|v\|_1$, this will automatically imply convergence of the approximation in the L_2-norm.

Theorem 1.3.1

Let $u \in H^1(S)$ be the solution of eqns. (1.1.1) and (1.1.2), and let $u_{m,n} \in S_{m,n}$ be the TM approximation satisfying eqn. (1.2.7). If the inverse property holds, then for any $w>0$, there exists a positive constant C independent of m, n, and u such that

$$\|u-u_{m,n}\|_1 \le \inf_{\forall v \in S_{m,n}} \{\|u-v\|_1 + C(k_{m,n}+w^{-1})|u-v|_{\mathrm{B}}\}, \tag{1.3.1}$$

where $k_{m,n}$ is defined in eqn. (1.2.2).

Proof.

For $v \in S_{m,n}$, let $\eta = v - u_{m,n}$. Then, since $\eta \in S_{m,n}$, Lemma 1.2.1 implies that

$$\begin{aligned}\|u - u_{m,n}\|_1 &\leq \|u - v\|_1 + \|\eta\|_1 \\ &\leq \|u - v\|_1 + C(k_{m,n} + w^{-1})|\eta|_{\mathrm{B}}. \qquad (1.3.2)\end{aligned}$$

Applying the orthogonality property, i.e., eqn. (1.2.8), we obtain

$$|\eta|_{\mathrm{B}}^2 = [\eta, \eta] = [v - u, \eta] \leq |u - v|_{\mathrm{B}} |\eta|_{\mathrm{B}}.$$

Hence, $|\eta|_{\mathrm{B}} \leq |u - v|_{\mathrm{B}}$ and eqn. (1.3.1) follows directly from eqn. (1.3.2). ■

Next we estimate bounds of the error in H, in terms of errors on the boundary and the interface.

Theorem 1.3.2

Let u and $u_{m,n}$ be the same as in Theorem 1.3.1 and let $v \in S_{m,n}$ be an approximation to u. If the inverse property holds and $w = 1/k_{m,n}$, then there exists a positive constant C independent of m, n, u, and v such that

$$\begin{aligned}\|u - u_{m,n}\|_1 \leq C\{&(|r|_{0,\Gamma_D} |r_\nu|_{0,\Gamma_D})^{\frac{1}{2}} + (|r|_{0,\Gamma_N} |r_\nu|_{0,\Gamma_N})^{\frac{1}{2}} \\ &+ k_{m,n}(|r|_{0,\Gamma_D} + |r^+|_{0,\Gamma_0} + |r^-|_{0,\Gamma_0}) \\ &+ |r_\nu|_{0,\Gamma_N} + |r_\nu^+|_{0,\Gamma_0} + |r_\nu^-|_{0,\Gamma_0}\}, \qquad (1.3.3)\end{aligned}$$

where $r = u - v$.

Proof.

From eqn. (1.3.1), we obtain the following bounds:

$$\|u - u_{m,n}\|_1 \leq \|r\|_1 + Ck_{m,n}|r|_{\mathrm{B}}. \qquad (1.3.4)$$

To estimate $\|r\|_1$, we have from Lemma 1.1.1 and the triangle inequality

$$\|r\|_1 \leq C\{|r|_1 + |r|_{0,\Gamma_D} + |r^+|_{0,\Gamma_0} + |r^-|_{0,\Gamma_0}\}.$$

Since $\Delta r = 0$ on S^+ and S^-, the inequality eqn. (1.2.6) is also valid for r and so

$$\begin{aligned}|r|_1^2 \leq\ & |r|_{0,\Gamma_D} |r_\nu|_{0,\Gamma_D} + |r|_{0,\Gamma_N} |r_\nu|_{0,\Gamma_N} + 2(|r^+|_{0,\Gamma_0} + |r^-|_{0,\Gamma_0}) \\ &\times (|r_\nu^+|_{0,\Gamma_0} + |r_\nu^-|_{0,\Gamma_0}).\end{aligned}$$

Using the Minkowski inequality and the inequality $\sqrt{ab} \le \frac{a+b}{2}$, we obtain

$$\begin{aligned} \|r\|_1 \le C\{(|r|_{0,\Gamma_D}|r_\nu|_{0,\Gamma_D})^{\frac{1}{2}} + (|r|_{0,\Gamma_N}|r_\nu|_{0,\Gamma_N})^{\frac{1}{2}} + |r|_{0,\Gamma_D} \\ + |r^+|_{0,\Gamma_0} + |r^-|_{0,\Gamma_0} + |r_\nu^+|_{0,\Gamma_0} + |r_\nu^-|_{0,\Gamma_0}\}, \end{aligned} \tag{1.3.5}$$

$$\begin{aligned} |r|_{\mathrm{B}} \le |r|_{0,\Gamma_D} + |r^+|_{0,\Gamma_0} + |r^-|_{0,\Gamma_0} \\ + w\{|r_\nu|_{0,\Gamma_N} + |r_\nu^+|_{0,\Gamma_0} + |r_\nu^-|_{0,\Gamma_0}\}. \end{aligned} \tag{1.3.6}$$

Substitution of eqns. (1.3.5) and (1.3.6) into eqn. (1.3.4) and the fact that $k_{m,n} \ge 1$ lead to the desired result, i.e., eqn. (1.3.3). ■

Assuming that eqns. (1.2.2)–(1.2.4) are valid in $S_{m,n}$, we obtain the following theorem.

Theorem 1.3.3

Let u and $u_{m,n}$ be the same as in Theorem 1.3.1. If both the inverse and approximatability properties hold and $w = k_{m,n}^{-1}$, then there exists a positive constant C independent of m, n, and u such that

$$\begin{aligned} \|u - u_{m,n}\|_1 \le C\{(k_{m,n}\alpha_{m,k,0}^+ + \alpha_{m,k,1}^+)|u^+|_{k,\partial S^+} \\ + (k_{m,n}\alpha_{n,t,0}^- + \alpha_{n,t,1}^-)|u^-|_{t,\partial S^-}\}, \end{aligned} \tag{1.3.7}$$

where k, $t > 1$.

Proof.

Note that both square roots on the right-hand side of eqn. (1.3.3) are bounded from above by

$$\begin{aligned} \{(|r^+|_{0,\partial S^+} + |r^-|_{0,\partial S^-})(|r_\nu^+|_{0,\partial S^+} + |r_\nu^-|_{0,\partial S^-})\}^{\frac{1}{2}} \\ \le C\{k_{m,n}(|r^+|_{0,\partial S^+} + |r^-|_{0,\partial S^-}) + |r_\nu^+|_{0,\partial S^+} + |r_\nu^-|_{0,\partial S^-}\}, \end{aligned}$$

where again we use the inequality $\sqrt{ab} \le \frac{a+b}{2}$. Consequently, from eqn. (1.3.3) it follows that

$$\|u - u_{m,n}\|_1 \le C\{k_{m,n}(|r^+|_{0,\partial S^+} + |r^-|_{0,\partial S^-}) + |r_\nu^+|_{0,\partial S^+} + |r_\nu^-|_{0,\partial S^-}\}. \tag{1.3.8}$$

Finally, the eqn. (1.3.7) is obtained from eqns. (1.2.3), (1.2.4), and (1.3.8). ■

Note that the approximation $u_{m,n}$ may have a different number of terms in S^+ from that in S^-, depending on the smoothness of u on ∂S^+ and ∂S^-. Hence, k and t in eqns. (1.2.3) and (1.2.4) may have different values.

Let us now investigate two special cases in which the error estimates, i.e., eqns. (1.3.3) and (1.3.7) apply.

First, assume that the admissible functions $v \in S_{m,n}$ satisfy also the continuity conditions, i.e., eqn. (1.1.3) on Γ_0. Then, we can modify eqn. (1.3.3) as follows.

Corollary 1.3.1

Suppose that in addition to the conditions of Theorem 1.3.2, the functions in $S_{m,n}$ satisfy eqn. (1.1.3). Then, there exists a positive constant C such that

$$\|u - u_{m,n}\|_1 \leq C\{(|r|_{0,\Gamma_D}|r_\nu|_{0,\Gamma_D})^{\frac{1}{2}} + (|r|_{0,\Gamma_N}|r_\nu|_{0,\Gamma_N})^{\frac{1}{2}} + k_{m,n}|r|_{0,\Gamma_D} + |r_\nu|_{0,\Gamma_N}\},$$

where $r = u - v$.

Second, assume that the admissible functions $v \in S_{m,n}$ also satisfy the boundary conditions, i.e., eqn. (1.1.2) on ∂S. Then, we can write eqn. (1.3.7) as follows.

Corollary 1.3.2

Suppose that u is a solution of eqns. (1.1.1) and (1.1.2). Let $u_{m,n}$ be the TM approximation satisfying eqn. (1.2.7) in $S_{m,n}$ and eqn. (1.1.2) on ∂S. If the inverse property is satisfied, $w = k_{m,n}^{-1}$, and for any $v \in S_{m,n}$ and $0 \leq l \leq k$,

$$\|u - v^+\|_{l,\Gamma_0} \leq \beta^+_{m,k,l}|u|_{k,\Gamma_0}, \tag{1.3.9}$$

and

$$\|u - v^-\|_{l,\Gamma_0} \leq \beta^-_{n,k,l}|u|_{k,\Gamma_0}. \tag{1.3.10}$$

Then, there exists a positive constant C such that for $k \geq 1$,

$$\|u - u_{m,n}\|_1 \leq C\{k_{m,n}(\beta^+_{m,k,0} + \beta^-_{n,k,0}) + \beta^+_{m,k,1} + \beta^-_{n,k,1}\}|u|_{k,\Gamma_0}.$$

Note that the bounds, i.e., eqns. (1.3.9) and (1.3.10) can be conveniently applied to problems with singularities on the boundary $\partial S = \Gamma_D \cup \Gamma_N$, because the differentiability of u is required only on the interior boundary Γ_0.

We will close this section with an estimate of the factor $k_{m,n}$ in eqn. (1.2.2) as a function of m and n. Consider the special case when Γ_0 is a circular arc.

A harmonic function can be written as

$$v = \sum_{i=1}^{n} \rho^{\mu_i}(a_i \cos \mu_i\theta + b_i \sin \mu_i\theta), \quad (\rho, \theta) \in S_0, \tag{1.3.11}$$

where S_0 is a simply connected region, a_i and b_i are coefficients, and the powers μ_i (not necessarily integers) are arranged in ascending order.

Let Γ_0 be a circular arc ($\rho = R_0, 0 \le \theta \le \Theta \le 2\pi$), and denote by S_0 the corresponding sector ($0 \le \rho \le R_0$, $0 \le \theta \le \Theta$). Suppose that the trigonometric functions $\cos \mu_i \theta$ and $\sin \mu_i \theta$ form an orthogonal system on $[0, \Theta]$. For a function v of the form, i.e., eqn. (1.3.11), the orthogonality and Sobolev's imbedding theorem imply that

$$R_0^2 |v_\nu|_{0,\Gamma_0}^2 = \Theta \sum_{i=1}^{n} \mu_i^2 R_0^{2\mu_i+1} \frac{(a_i^2 + b_i^2)}{2} \le \mu_n^2 \Theta \sum_{i=1}^{n} R_0^{2\mu_i+1} \frac{(a_i^2 + b_i^2)}{2}$$

$$= \mu_n^2 |v|_{0,\Gamma_0}^2 \le C \mu_n^2 \|v\|_{1,S_0}^2.$$

Consequently, we have

$$|v_\nu|_{0,\Gamma_0} \le C \mu_n \|v\|_{1,S_0}.$$

Hence, if S^+ is a sector with circular boundary Γ_0 and the admissible functions v^+ are of the form, i.e., eqn. (1.3.11), then

$$k_{m,n} = C \mu_n. \tag{1.3.12}$$

We will use eqn. (1.3.12) for the weight selections in the numerical experiments in computation.

1.4 Debye–Huckel equation

Consider an elliptic boundary value problem on a domain divided into several subdomains by artificial or material interfaces. If the admissible functions consist of particular solutions of the underlying elliptic equation on the subdomains, a boundary approximation solution can be obtained by satisfying the exterior boundary conditions and the continuity conditions on the interior interface as much as possible in a least squares sense. Since the boundary approximation is performed only on the interior and exterior boundaries, we call this method the *Trefftz method* (TM).

The advantages of the TM are again summarized as follows:

1. It is easy to solve the problems with corners and interface singularities as well as with unbounded domains, with which the standard FEMs and FDMs have difficulties to cope with.
2. The solution procedure is simple, because only the interior and exterior boundary conditions are taken into account in the solution process.
3. A very accurate solution can be obtained by using relatively few expansion terms of particular solutions (approximation in one lower dimension), thus greatly saving on CPU time and storage space.
4. It is possible to estimate errors of the approximate solutions, even though the exact solution of the physical problem is unknown. In this section, a useful

relation for the error behavior will be established:

$$\|\varepsilon\|_1 = O(M^\alpha |\varepsilon|_B), \tag{1.4.1}$$

where $\alpha = 1$ or $\frac{1}{2}$, and M is the total number of unknown coefficients in the piecewise expansions used. The formula, i.e., eqn. (1.4.1) is significant in practical calculation because we can evaluate error $\|\varepsilon\|_1$ in the domain in terms of the errors on the boundary, $|\varepsilon|_B$, which is naturally obtained from the TM. Once the errors of solutions are known, we can easily control the calculation procedure.

However, the following two difficulties arise in using the TM.

1. Piecewise particular solutions of elliptic equations have to be known. For the most important elliptic equations in application, we may find useful particular solutions in textbooks of PDEs, e.g., Tikhonov and Samarskii [438]. Nevertheless quite often, it is necessary to find asymptotic expansions near singular points and infinity, in order to obtain accurate expansion solutions for all numerical methods.
2. Stability of numerical solutions is also important. In fact, stability will rely substantially on both the choice of piecewise particular solutions and the partition of the solution domain. Our intention here is to use stability analysis to guide us in choosing partitions so that better solutions can be obtained via the TM.

It is worth pointing out that the TM may fall into the class of general weighted least squares methods (LSMs) for elliptic systems of Aziz, Kellogg, and Stephen [12], which are applied both within small elements (viewed here as subdomains) and on their boundaries (viewed as interfaces). Since only particular solutions are chosen to be admissible functions, the number of unknown coefficients decreases drastically. A good TM approximation can be obtained even if several subdomains are used. It would also be interesting to develop an adaptive LSMs in which the number of trial functions (as in this part) and the number of subdomains (as in Ref. [12]) are both changing, as done in the *h–p* version of FEM.

In this and the next sections, we will present an analysis of errors and in particular stability for the Debye–Huckel equation

$$-\Delta u + u = 0. \tag{1.4.2}$$

It should be noted that the infinite elements of Han [183] and Thatcher [435, 436] cannot be applied to eqn. (1.4.2).

Kellogg's singular functions for interface problems (Kellogg [243, 244, 245]) are not complete when the intersection angles of interfaces are $\Theta = \pi/n$, $n = 2, 3, \ldots$. The additional analytic functions found in Li [275] together with Kellogg's function form a complete set for interface problems. Of course, a complete system of particular solutions is essential not only to theoretical research, but also to numerical methods, such as the TM and the combined methods discussed in this book.

Moreover, we will establish an error norm relation, i.e., eqn. (1.4.1), analyze the stability of the TM approximations, and investigate different shapes of subdomains divided by circular arcs and straight lines.

Let the bounded domain S be divided by Γ_0 into two non-overlapped subdomains S^+ and S^-, i.e., $S = S^+ \cup S^-$ and $\Gamma_0 = S^+ \cap S^-$. In the rest of this chapter, we will consider the piecewise equations

$$-\Delta u + u = 0 \quad \text{in } S^+ \text{ and } S^-, \tag{1.4.3}$$

with the interior and exterior boundary conditions

$$u^+ = u^-, \qquad \frac{\partial u^+}{\partial \nu} = \frac{\partial u^-}{\partial \nu} \quad \text{on } \Gamma_0, \tag{1.4.4}$$

$$u = f \quad \text{on } \Gamma_D, \qquad u_\nu = g \quad \text{on } \Gamma_N. \tag{1.4.5}$$

For eqns. (1.4.3)–(1.4.5), the TM is exactly the same as eqn. (1.2.7).

For the space $S_{m,n}$, we assume that the inverse property, i.e., eqn. (1.2.2), and the approximatability property, i.e., eqns. (1.2.3) and (1.2.4) still hold. Approximatability properties may be found in Cheney [92] for some spaces, or in Eisenstat [134] for the Bergman–Vekua space. For the equation $-\Delta u + u = 0$, the approximatability properties of $S_{m,n}$ can be obtained only when the given subdomains are embedded into sectors, which are within the solution domains. In other cases, further study of the approximating spaces needs to be done (refer to the density study in Aziz, Dorr, and Kellogg [11] and Browder [63]).

To provide more precise estimates, we assume that $S_{m,n}$ also satisfies the following inverse property.

Second Inverse Property: For any $v \in S_{m,n}$,

$$|v|_{0,\Gamma_N} \le q_{m,n} \|v\|_1, \qquad |v|_{0,\Gamma_0} \le q_{m,n} \|v\|_1, \tag{1.4.6}$$

where $q_{m,n}$ is constant.

Obviously, the constant $q_{m,n}$ is bounded because of the Sobolev imbedding theorem (Sobolev [417]); but in some cases, there may exist better estimates:

$$q_{m,n} = o(1) \quad \text{as } m, n \to \infty,$$

for the errors $v = u - u_{m,n}$.

Under these assumptions, we will provide error estimates and establish the relation, i.e., eqn. (1.4.1). First, we give two lemmas.

Lemma 1.4.1

Let $v \in S_{m,n}$ and suppose that the inverse property, i.e., eqn. (1.2.2) and the second inverse property hold. Then when $w > 0$, there exist the norm bounds

$$\|v\|_1 \le (k_{m,n} + q_{m,n}/w)|v|_B.$$

Proof.

By using Green's theorem and the fact

$$-\Delta v + v = 0 \quad \text{in } S^+ \text{ and } S^-,$$

we obtain

$$\begin{aligned}
0 &= \iint_{S^+} (\Delta v^+ - v^+)v^+ \, ds + \iint_{S^-} (\Delta v^- - v^-)v^- \, ds \\
&= -\iint_{S^+} [(v_x^+)^2 + (v_y^+)^2 + (v^+)^2] \, ds - \iint_{S^-} [(v_x^-)^2 + (v_y^-)^2 + (v^-)^2] \, ds \\
&\quad + \int_{\partial S^+} (v_\nu^+)v^+ \, d\ell + \int_{\partial S^-} (v_\nu^-)v^- \, d\ell.
\end{aligned}$$

Then, we have for $v = v^\pm$ in $S^\pm$

$$\begin{aligned}
\|v\|_1^2 &= \left| \int_{\partial S^+} (v_\nu^+)v^+ \, d\ell + \int_{\partial S^-} (v_\nu^-)v^- \, d\ell \right| \\
&\le \left| \int_{\Gamma_D} (v_\nu)v \, d\ell \right| + \left| \int_{\Gamma_N} (v_\nu)v \, d\ell \right| \\
&\quad + \left| \int_{\Gamma_0} [(v^+ - v^-)(v_\nu^+) + v^-(v_\nu^+ - v_\nu^-)] \, d\ell \right| \\
&\le |v|_{0,\Gamma_D} |v_\nu|_{0,\Gamma_D} + |v|_{0,\Gamma_N} |v_\nu|_{0,\Gamma_N} \\
&\quad + |v^+ - v^-|_{0,\Gamma_0} |v_\nu^+|_{0,\Gamma_0} + |v^-|_{0,\Gamma_0} |v_\nu^+ - v_\nu^-|_{0,\Gamma_0}.
\end{aligned}$$

Therefore, from the inverse property, i.e., eqn. (1.2.2) and the second inverse property, i.e., eqn. (1.4.6), we can obtain

$$\begin{aligned}
\|v\|_1^2 \le \{ & k_{m,n} |v|_{0,\Gamma_D} + q_{m,n} |v_\nu|_{0,\Gamma_N} + k_{m,n} |v^+ - v^-|_{0,\Gamma_0} \\
& + q_{m,n} |v_\nu^+ - v_\nu^-|_{0,\Gamma_0} \} \|v\|_1.
\end{aligned}$$

The desired result is obtained by dividing both sides by $\|v\|_1$ and by noting the definition $|v|_{\mathrm{B}}$ in eqn. (1.1.4). ■

We can prove the following lemma by using the same arguments as in Lemma 1.2.2.

Lemma 1.4.2

Let u be the solution of eqns. (1.4.3)–(1.4.5). Then for any $w > 0$, there exists a unique function, $u_{m,n} \in S_{m,n}$, such that

$$[u_{m,n}, v] = \int_{\Gamma_D} f v \, d\ell + w^2 \int_{\Gamma_N} g v_\nu \, d\ell, \quad \forall v \in S_{m,n}, \tag{1.4.7}$$

$$[u - u_{m,n}, v] = 0, \quad \forall v \in S_{m,n}, \tag{1.4.8}$$

and

$$\|u_{m,n}\|_1 \leq (k_{m,n} + q_{m,n}/w)\{|f|_{0,\Gamma_D} + w|g|_{0,\Gamma_N}\}.$$

Also, $u_{m,n}$ minimizes $I(v)$ over $S_{m,n}$ if and only if eqn. (1.4.7) holds.

Now, we have an error estimate theorem.

Theorem 1.4.1

Let $u \in H^1(S)$ be the solution of eqns. (1.4.3)–(1.4.5), and let $u_{m,n} \in S_{m,n}$ be the TM approximation satisfying eqn. (1.4.7). If the inverse property, i.e., eqn. (1.2.2) and the second inverse property, i.e., eqn. (1.4.6) hold, then for any $w > 0$ there exists a constant C independent of m, n, and u such that

$$\|u - u_{m,n}\|_1 \leq \inf_{v \in S_{m,n}} \{\|u - v\|_1 + (k_{m,n} + q_{m,n}/w)|u - v|_{\mathrm{B}}\}, \tag{1.4.9}$$

where $k_{m,n}$ is defined in eqn. (1.2.2), and $q_{m,n}$ in eqn. (1.4.6).

Proof.

Let $v \in S_{m,n}$ and $\eta = v - u_{m,n}$. We have from Lemma 1.4.1

$$\begin{aligned} \|u - u_{m,n}\|_1 &\leq \|u - v\|_1 + \|\eta\|_1 \\ &\leq \|u - v\|_1 + (k_{m,n} + q_{m,n}/w)|\eta|_{\mathrm{B}}. \end{aligned} \tag{1.4.10}$$

Applying the orthogonality property, i.e., eqn. (1.4.8), we obtain

$$|\eta|_{\mathrm{B}}^2 = [\eta, \eta] = [v - u, \eta] \leq |u - v|_{\mathrm{B}} |\eta|_{\mathrm{B}}.$$

Hence, $|\eta|_{\mathrm{B}} \leq |u - v|_{\mathrm{B}}$, and the desired inequality eqn. (1.4.9) follows from eqn. (1.4.10). ■

Obviously, Theorem 1.3.1 is a special case of Theorem 1.4.1 because the constant $q_{m,n} \leq C$.

Similar error bounds as Theorems 1.3.2 and 1.3.3, and Corollaries 1.3.1 and 1.3.2 can be easily derived. However, here we only investigate the relation between $\|u - u_{m,n}\|_1$ and $|u - u_{m,n}|_{\mathrm{B}}$.

Let u be the solution of eqns. (1.4.3)–(1.4.5), then the error norms satisfy

$$
\begin{aligned}
I(u_{m,n}) &= |\varepsilon|_{\mathrm{B}}^2 = |u - u_{m,n}|_{\mathrm{B}}^2 \\
&= \int_{\Gamma_D} (u_{m,n} - f)^2 \, d\ell + w^2 \int_{\Gamma_N} \left(\frac{\partial u_{m,n}}{\partial \nu} - g \right)^2 d\ell + \int_{\Gamma_0} (u_{m,n}^+ - u_{m,n}^-)^2 d\ell \\
&\quad + w^2 \int_{\Gamma_0} \left(\frac{\partial u_{m,n}^+}{\partial \nu} - \frac{\partial u_{m,n}^-}{\partial \nu} \right)^2 d\ell.
\end{aligned} \tag{1.4.11}
$$

We note that the explicit, true solution u disappears in $|\varepsilon|_{\mathrm{B}}$ (see eqn. (1.4.11)) and the values of $|\varepsilon|_{\mathrm{B}}$ are easily computed in the least squares procedure employed in the TM. We are then interested in evaluating $\|\varepsilon\|_1$ in terms of $|\varepsilon|_{\mathrm{B}}$. Such a relation of these two norms is given in the following theorem:

Theorem 1.4.2

Let $u \in H^1(S)$ be the solution of eqns. (1.4.3)–(1.4.5) and $u_{m,n} \in S_{m,n}$ be the TM approximation eqn. (1.4.7). If the inverse property, i.e., eqn. (1.2.2) and the second inverse property, i.e., eqn. (1.4.6) hold for the difference $u - u_{m,n}$, then for any $w > 0$ there exists a constant C independent of m, n, and u such that

$$
\|u - u_{m,n}\|_1 \leq (k_{m,n} + q_{m,n}/w)|u - u_{m,n}|_{\mathrm{B}}. \tag{1.4.12}
$$

Proof.

Letting $v = u_{m,n} \in S_{m,n}$, we have from eqn. (1.4.4)

$$
\begin{aligned}
\|u - v\|_1^2 &\leq \left| \int_{\partial S^+} (u_\nu^+ - v_\nu^+)(u^+ - v^+) \, d\ell + \int_{\partial S^-} (u_\nu^- - v_\nu^-)(u^- - v^-) \, d\ell \right| \\
&\leq \left| \int_{\Gamma_D} (u_\nu - v_\nu)(f - v) \, d\ell \right| + \left| \int_{\Gamma_N} (g - v_\nu)(u - v) \, d\ell \right| \\
&\quad + \left| \int_{\Gamma_0} [(v^+ - v^-)(u_\nu^+ - v_\nu^+) + (u^- - v^-)(v_\nu^+ - v_\nu^-)] \, d\ell \right| \\
&\leq |u_\nu - v_\nu|_{0,\Gamma_D} |f - v|_{0,\Gamma_D} + |g - v_\nu|_{0,\Gamma_N} |u - v|_{0,\Gamma_N} \\
&\quad + |v^+ - v^-|_{0,\Gamma_0} |u_\nu^+ - v_\nu^+|_{0,\Gamma_0} + |u^- - v^-|_{0,\Gamma_0} |v_\nu^+ - v_\nu^-|_{0,\Gamma_0} \\
&\leq \{k_{m,n}|f - v|_{0,\Gamma_D} + q_{m,n}|g - v_\nu|_{0,\Gamma_N} + k_{m,n}|v^+ - v^-|_{0,\Gamma_0} \\
&\quad + q_{m,n}|v_\nu^+ - v_\nu^-|_{0,\Gamma_0}\} \|u - v\|_1.
\end{aligned}
$$

Dividing both sides by $\|u - v\|_1$ and using eqn. (1.4.11), we obtain

$$
\|u - v\|_1 \leq (k_{m,n} + q_{m,n}/w)|u - v|_{\mathrm{B}}.
$$

The inequality eqn. (1.4.12) follows from the substitution $v = u_{m,n}$. ■

Based on Theorem 1.4.2, we can easily obtain the following two corollaries as the desired result, i.e., eqn. (1.4.1). By noting $q_{m,n} \le C$, Theorem 1.4.2 leads to the following corollary.

Corollary 1.4.1

Let the weight $w = 1/M$, where $M = Max(m,n)$ and $m = O(n)$, and all the conditions in Theorem 1.4.2 hold. Also, suppose that the constant $k_{m,n}$ in the inverse property satisfies

$$k_{m,n} \le CM,$$

where C is a bounded constant independent of m, n, and u. Then

$$\|u - u_{m,n}\|_1 = O(M|u - u_{m,n}|_B).$$

Corollary 1.4.2

Let the weight $w = \frac{1}{M}$, where $M = Max(m, n)$ and $m = O(n)$, and all the conditions in Theorem 1.4.2 hold. Also, suppose that for $u - u_{m,n}$ the constants $k_{m,n}$ and $q_{m,n}$ in the inverse property and the second inverse property satisfy

$$k_{m,n} \le C\sqrt{M}, \qquad q_{m,n} \le C/\sqrt{M}.$$

Then,

$$\|u - u_{m,n}\|_1 = O(M^{\frac{1}{2}}|u - u_{m,n}|_B).$$

1.5 Stability analysis

In this section, we present stability analysis for TM based on domain decomposition and discuss the choice of geometric shapes for the subdomains used. In order to discuss stability of the solution $u_{m,n}$, we need to estimate the values of condition numbers using the LSM (see Golub and van Loan [168]). The square root in the Cond is for solving the collocation equations $\mathbf{Fx} = \mathbf{b}$, where $\mathbf{B} = \mathbf{F}^T\mathbf{F}$, see Chapter 2.

$$\text{Cond} = \left(\frac{\lambda_{\max}(\mathbf{B})}{\lambda_{\min}(\mathbf{B})}\right)^{\frac{1}{2}}.$$

Here, the associated coefficient matrix $\mathbf{B}$ is defined by $\mathbf{x}^T\mathbf{Bx} = |v|^2_B$, $v \in S_{m,n}$, where the vector $\mathbf{x}$ is composed of the unknown coefficients.

Let $\underline{S}^{\pm}$ and $\overline{S}^{\pm}$ be the bounded domains such that

$$\underline{S}^{+} \subseteq S^{+} \subseteq \overline{S}^{+}, \qquad \underline{S}^{-} \subseteq S^{-} \subseteq \overline{S}^{-}. \tag{1.5.1}$$

We define two matrices as follows:

$$\mathbf{F}_{S^+} = (f_{i,j}^+), \qquad \mathbf{F}_{S^-} = (f_{i,j}^-),$$

where the matrix elements $f_{i,j}^{\pm}$ are

$$f_{i,j}^+ = (\psi_i^+, \psi_j^+)_{S^+} + \left(\frac{\partial \psi_i^+}{\partial x}, \frac{\partial \psi_j^+}{\partial x}\right)_{S^+} + \left(\frac{\partial \psi_i^+}{\partial y}, \frac{\partial \psi_j^+}{\partial y}\right)_{S^+}$$

and

$$f_{i,j}^- = (\psi_i^-, \psi_j^-)_{S^-} + \left(\frac{\partial \psi_i^-}{\partial x}, \frac{\partial \psi_j^-}{\partial x}\right)_{S^-} + \left(\frac{\partial \psi_i^-}{\partial y}, \frac{\partial \psi_j^-}{\partial y}\right)_{S^-}.$$

Then, we have

$$(\mathbf{x}^+)^T \mathbf{F}_{S^+} \mathbf{x}^+ = \|v^+\|_{1,S^+}^2, \qquad (\mathbf{x}^-)^T \mathbf{F}_{S^-} \mathbf{x}^- = \|v^-\|_{1,S^-}^2,$$

where the notations are

$$v^+ = \sum_{i=1}^{m} c_i \psi_i^+, \qquad v^- = \sum_{i=1}^{n} d_i \psi_i^-,$$

$$\mathbf{x}^+ = (c_1,\ c_2, \ldots, c_m)^T, \qquad \mathbf{x}^- = (d_1,\ d_2, \ldots, d_n)^T.$$

We now prove a main theorem.

Theorem 1.5.1

Suppose that for any $v \in S_{m,n}$, the following bounds are satisfied:

$$|v_\nu|_{0,\Gamma_N} \ \text{and}\ \left|v_\nu^{\pm}\right|_{0,\Gamma_0} \le k_{m,n}\|v\|_1, \tag{1.5.2}$$

$$|v|_{0,\Gamma_D} \ \text{and}\ \left|v^{\pm}\right|_{0,\Gamma_0} \le q_{m,n}\|v\|_1, \tag{1.5.3}$$

with the positive constants, $k_{m,n}$ and $q_{m,n}$. Then for any $w > 0$, there exists a bounded constant C independent of m, n, and u such that

$$\text{Cond} \le w[k_{m,n} + q_{m,n}/w]^2 \times \left\{\frac{\max[\lambda_{\max}(\mathbf{F}_{\overline{S}^+}), \lambda_{\max}(\mathbf{F}_{\overline{S}^-})]}{\min[\lambda_{\min}(\mathbf{F}_{\underline{S}^+}), \lambda_{\min}(\mathbf{F}_{\underline{S}^-})]}\right\}^{\frac{1}{2}}.$$

Proof.

We have from Lemma 1.4.1

$$|v|_{\mathrm{B}}^2 \ge \frac{\|v\|_1^2}{(k_{m,n} + q_{m,n}/w)^2} \ge \frac{\|v^+\|_{1,\underline{S}^+}^2 + \|v^-\|_{1,\underline{S}^-}^2}{(k_{m,n} + q_{m,n}/w)^2}.$$

Then,

$$\lambda_{\min}(\mathbf{B}) = \min_{\mathbf{x}\neq 0} \frac{|v|_{\mathrm{B}}^2}{\mathbf{x}^T\mathbf{x}} \geq \frac{1}{(k_{m,n}+q_{m,n}/w)^2} \min_{\mathbf{x}\neq 0} \frac{\|v^+\|^2_{1,\underline{S}^+} + \|v^-\|^2_{1,\underline{S}^-}}{\mathbf{x}^T\mathbf{x}}.$$

Let the matrix $\mathbf{T}$ be denoted by

$$\mathbf{T} = \begin{bmatrix} \mathbf{F}_{\underline{S}^+} & 0 \\ 0 & \mathbf{F}_{\underline{S}^-} \end{bmatrix},$$

then we obtain a relation for the smallest eigenvalues of the matrices $\mathbf{T}$, $\mathbf{F}_{\underline{S}^+}$, and $\mathbf{F}_{\underline{S}^-}$,

$$\lambda_{\min}(\mathbf{T}) = \min_{\mathbf{x}\neq 0} \frac{\|v^+\|^2_{1,\underline{S}^+} + \|v^-\|^2_{1,\underline{S}^-}}{\mathbf{x}^T\mathbf{x}} = \min[\lambda_{\min}(\mathbf{F}_{\underline{S}^+}), \lambda_{\min}(\mathbf{F}_{\underline{S}^-})].$$

Obviously,

$$\lambda_{\min}(\mathbf{B}) \geq \frac{1}{(k_{m,n}+q_{m,n}/w)^2} \min[\lambda_{\min}(\mathbf{F}_{\underline{S}^+}), \lambda_{\min}(\mathbf{F}_{\underline{S}^-})]. \tag{1.5.4}$$

Similarly, from the assumptions (1.5.2), (1.5.3), and the Sobolev imbedding theorem we can see that

$$|v|_{\mathrm{B}}^2 \leq (q_{m,n}+wk_{m,n})^2\|v\|_1^2 \leq (q_{m,n}+wk_{m,n})^2[\|v^+\|^2_{1,\overline{S}^+} + \|v^-\|^2_{1,\overline{S}^-}].$$

Therefore,

$$\lambda_{\max}(\mathbf{B}) \leq (q_{m,n}+wk_{m,n})^2 \max[\lambda_{\max}(\mathbf{F}_{\overline{S}^+}), \lambda_{\max}(\mathbf{F}_{\overline{S}^-})]. \tag{1.5.5}$$

It follows by computing eqns. (1.5.4) and (1.5.5) that

$$\frac{\lambda_{\max}(\mathbf{B})}{\lambda_{\min}(\mathbf{B})} \leq w^2[k_{m,n}+q_{m,n}/w]^4 \times \frac{\max[\lambda_{\max}(\mathbf{F}_{\overline{S}^+}), \lambda_{\max}(\mathbf{F}_{\overline{S}^-})]}{\min[\lambda_{\min}(\mathbf{F}_{\underline{S}^+}), \lambda_{\min}(\mathbf{F}_{\underline{S}^-})]}.$$

Now eqn. (1.5.4) is obtained by $\mathrm{Cond} = \sqrt{\frac{\lambda_{\max}(\mathbf{B})}{\lambda_{\min}(\mathbf{B})}}$ for solving the overdetermined system, see Chapter 2. ■

Below, we will apply Theorem 1.5.1 to the problem using the division displayed in fig. 1.1 and the admissible functions:

$$v^+ = \sum_{i=1}^{m} c_i \frac{I_{\mu_i^+}(r)}{I_{\mu_i^+}(R_0^+)} \sin \mu_i^+\theta, \quad (r,\theta) \in S^+ \tag{1.5.6}$$

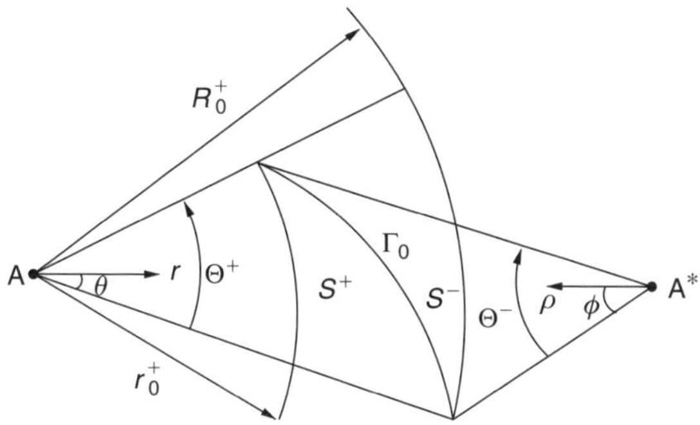

Figure 1.1: A division with $S = S^+ \cup S^-$.

and

$$v^- = \sum_{i=1}^{n} d_i \frac{I_{\mu_i^-}(\rho)}{I_{\mu_i^-}(R_0^-)} \sin \mu_i^- \phi, \quad (\rho, \phi) \in S^-, \tag{1.5.7}$$

where c_i and d_i are unknown coefficients, $\mu_i^\pm = i\pi/\Theta^\pm$, $\Theta^\pm$ are the intersection angles; (r, θ) and (ρ, ϕ) are the polar coordinates with the origins at A and A^*, respectively, and the radii $R_0^\pm$ are defined by the following formulae, i.e., eqn. (1.5.9). In eqns. (1.5.6) and (1.5.7), the Bessel functions $I_\mu(r)$ for a purely imaginary argument can be expressed by (Abramowitz and Stegun [2])

$$I_\mu(r) = \frac{(\frac{r}{2})^\mu}{\Gamma(\frac{1}{2})\Gamma(\mu + \frac{1}{2})} \int_{-1}^{1} e^{\pm rt}(1 - t^2)^{\mu - \frac{1}{2}}\, dt. \tag{1.5.8}$$

Let the sectors $\overline{S}^\pm$ and $\underline{S}^\pm$ satisfy eqn. (1.5.1) such that

$$\begin{aligned} \overline{S}^- &= \{0 < \rho < R_0^- \quad \text{and} \quad 0 < \phi < \Theta^-\}, \\ \underline{S}^- &= \{0 < \rho < r_0^- \quad \text{and} \quad 0 < \phi < \Theta^-\}, \\ \overline{S}^+ &= \{0 < r < R_0^+ \quad \text{and} \quad 0 < \theta < \Theta^+\}, \\ \underline{S}^+ &= \{0 < r < r_0^+ \quad \text{and} \quad 0 < \theta < \Theta^+\}. \end{aligned} \tag{1.5.9}$$

Then, we have the following corollary.

Corollary 1.5.1

Let $v^\pm$ be admissible functions given by eqns. (1.5.6) and (1.5.7) on the division in fig. 1.1. If $m = O(n)$ and the conditions in Theorem 1.5.1 are satisfied, then there *exists a bounded constant C independent of m, n, and u such that*

$$\text{Cond} \le Cw[k_{m,n} + q_{m,n}/w]^2 \max\left\{\left[\frac{R_0^+}{r_0^+}\right]^{\mu_m^+}, \left[\frac{R_0^-}{r_0^-}\right]^{\mu_n^-}\right\}, \tag{1.5.10}$$

where the radii, $R_0^{\pm}$ and $r_0^{\pm}$, are defined in eqn. (1.5.9).

Proof.

Using the orthogonality of $\sin \mu_i^+ \theta$, we can see that

$$\|v^+\|^2_{1,\overline{S}^+} = \frac{\Theta^+}{2}\sum_{i=1}^{m} c_i^2 \int_0^{R_0^+} rG_i^+(r)\,dr, \quad \|v^+\|^2_{1,\underline{S}^+} = \frac{\Theta^+}{2}\sum_{i=1}^{m} c_i^2 \int_0^{r_0^+} rG_i^+(r)\,dr,$$

where

$$G_i^+(r) = \frac{\left[[I'_{\mu_i^+}(r)]^2 + \frac{(\mu_i^+)^2}{r^2} I^2_{\mu_i^+}(r) + I^2_{\mu_i^+}(r)\right]}{I^2_{\mu_i^+}(R_0^+)}.$$

On the other hand, there exist the bounds in terms of the definition $I_\mu(r)$ in eqn. (1.5.8),

$$\beta_\mu r^\mu e^{-\max r} \le I_\mu(r) \le \beta_\mu r^\mu e^{\max r},$$

where the constants are

$$\beta_\mu = \frac{(\frac{1}{2})^\mu}{\Gamma(\frac{1}{2})\Gamma(\mu + \frac{1}{2})} \int_{-1}^{1} (1 - t^2)^{\mu - \frac{1}{2}}\, dt,$$

with $\beta_{\mu+1} \le \beta_\mu$. Moreover, by noting the formula (see Gradshteyn and Ryzhik [173]),

$$I'_\mu(r) = \frac{\mu}{r} I_\mu(r) + I_{\mu+1}(r),$$

we can obtain the following bounds,

$$\int_0^{R_0^+} rG_i^+(r)\,dr \le C\mu_i^+ e^{4R_0^+}, \qquad \int_0^{r_0^+} rG_i^+(r)\,dr \ge \delta_0 \mu_i^+ e^{-4R_0^+} \left[\frac{r_0^+}{R_0^+}\right]^{2\mu_i^+},$$

with constants $0 < \delta_0 < C < \infty$. This yields

$$\lambda_{\max}(\mathbf{F}_{\overline{S}^+}) \le C\mu_m^+, \qquad \lambda_{\min}(\mathbf{F}_{\underline{S}^+}) \ge \delta_0 \min\left[1, \mu_m^+ \left(\frac{r_0^+}{R_0^+}\right)^{2\mu_m^+}\right].$$

Similarly, we obtain

$$\lambda_{\max}(\mathbf{F}_{\overline{S}^-}) \leq C\mu_n^-, \qquad \lambda_{\min}(\mathbf{F}_{\underline{S}^-}) \geq \delta_0 \min\left[1, \mu_n^- \left(\frac{r_0^-}{R_0^-}\right)^{2\mu_n^-}\right].$$

At least, one of the following two inequalities:

$$r_0^+ < R_0^+ \quad \text{and} \quad r_0^- < R_0^-$$

must hold. Therefore,

$$\begin{aligned}
&\min[\lambda_{\min}(\mathbf{F}_{\underline{S}^+}), \lambda_{\min}(\mathbf{F}_{\underline{S}^-})] \\
&= \min \delta_0 \left(1, \mu_m^+ \left[\frac{r_0^+}{R_0^+}\right]^{2\mu_m^+}, \mu_n^- \left[\frac{r_0^-}{R_0^-}\right]^{2\mu_n^-}\right) \\
&= \min \delta_0 \left(\mu_m^+ \left[\frac{r_0^+}{R_0^+}\right]^{2\mu_m^+}, \mu_n^- \left[\frac{r_0^-}{R_0^-}\right]^{2\mu_n^-}\right),
\end{aligned}$$

for some large numbers μ_m^+ and μ_n^-. Consequently, Theorem 1.5.1 yields

$$\begin{aligned}
\text{Cond} &\leq Cw(k_{\min} + q_{m,n}/w)^2 \\
&\times \left[\frac{\max[\mu_m^+, \mu_n^-]}{\min[\mu_m^+(r_0^+/R_0^+)^{2\mu_m^+}, \mu_n^-(r_0^-/R_0^-)^{2\mu_n^-}]}\right]^{\frac{1}{2}}.
\end{aligned}$$

The desired inequality eqn. (1.5.10) is obtained from the fact

$$\mu_m^+ = O(\mu_n^-),$$

which results from $\mu_i^{\pm} = i\pi/\Theta^{\pm}$ and the assumption $m = O(n)$. ■

As a result of Corollary 1.5.1, the following formula holds for $w = 1/M$,

$$\text{Cond} = O\left\{M \times \max\left[\left(\frac{R_0^+}{r_0^+}\right)^{\mu_m^+}, \left(\frac{R_0^-}{r_0^-}\right)^{\mu_n^-}\right]\right\} \tag{1.5.11}$$

provided that the bounds $k_{m,n} \leq CM$ are satisfied. To end this section, let us consider the geometric shapes of solution domains while using the TM. For the solution domain as a rhomboid (fig. 1.2), the values of Cond in the case of B with a straight line Γ_0 are smaller than those in the case of A with a circular arc Γ_0, based on eqn. (1.5.10) or (1.5.11). Also the divisions in fig. 1.3 will yield a good stability of numerical solutions by TMs.

(A) A circular arc Γ_0 (B) A straight line Γ_0

Figure 1.2: Division of a rhomboid.

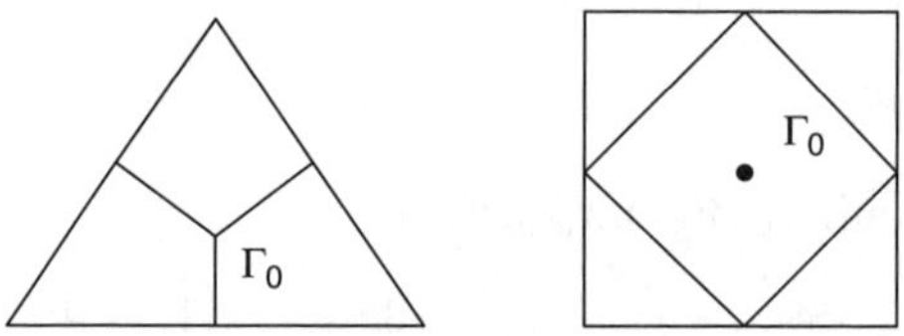

Figure 1.3: Good divisions for the TM.

2 Motz's problem and its variants

In this chapter, we develop the TM in Chapter 1 involving integration approximation, to lead to the collocation Trefftz method (called the collocation TM, or the CTM). We choose the high-order Gaussian rules and the central rule, link the collocation method and the least squares method (LSM), and demonstrate exponential convergence rates of the obtained solutions. For Motz's problem and its variants, the CTM is used to seek their approximate solutions $u_N = \sum_{i=0}^{N} D_i r^{i+\frac{1}{2}} \cos(i+\frac{1}{2})\theta$, where D_i are the expansion coefficients. Compared with the solutions in the previous literature, the present Motz's solutions are more accurate and the leading coefficient D_0 using the Gaussian quadrature rule with six nodes can achieve 17 significant (decimal) digits. This chapter proves that when the rules of quadrature involved have the relative errors less than three quarters, the solution from the CTM may converge exponentially. Such an analysis provides the theoretical support that the CTM is the most accurate method for Motz's problem and its variants.

2.1 Introduction

Motz's problem was first discussed by Motz [341] in 1947 for the relaxation method. Since then, many researchers have selected Motz's problem as a prototype of singularity problems for verifying efficiency of numerical methods (see Ref. [280]).

Motz's problem solves the Laplace equation on the rectangle $S = \{(x, y) \mid -1 < x < 1, 0 < y < 1\}$

$$\Delta u = \frac{\partial^2 u}{\partial x^2} + \frac{\partial^2 u}{\partial y^2} = 0 \quad \text{in } S, \tag{2.1.1}$$

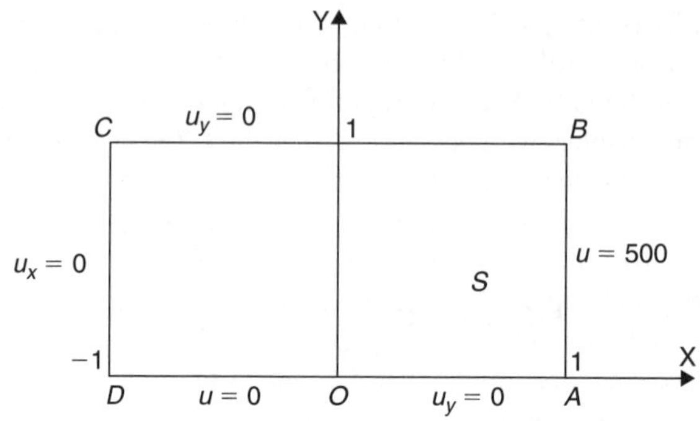

Figure 2.1: Motz's problem.

with the mixed Neumann–Dirichlet boundary conditions, see fig. 2.1,

$$u\,|_{-1<x<0\cap y=0} = 0, \qquad u\,|_{x=1} = 500, \tag{2.1.2}$$

$$\left.\frac{\partial u}{\partial y}\right|_{y=1} = \left.\frac{\partial u}{\partial y}\right|_{0<x<1\cap y=0} = \left.\frac{\partial u}{\partial x}\right|_{x=-1} = 0. \tag{2.1.3}$$

Note that there exists a singularity at the origin (0, 0) due to the intersection of the Neumann–Dirichlet boundary conditions. In fact, the singular solutions of eqns. (2.1.1)–(2.1.3) are found as

$$u(r,\theta) = \sum_{i=0}^{\infty} d_i r^{i+\frac{1}{2}} \cos\left(i+\frac{1}{2}\right)\theta, \tag{2.1.4}$$

where d_i are the true expansion coefficients, and (r,θ) are the polar coordinates with the origin at $(0,0)$. Since its convergence radius, $r=2$, is analyzed in Rosser and Papamichael [393], the series expansions, i.e., eqn. (2.1.4) are well suited to the entire solution domain S. Hence, the admissible functions with finite terms,

$$u_N(r,\theta) = \sum_{i=0}^{N} D_i r^{i+\frac{1}{2}} \cos\left(i+\frac{1}{2}\right)\theta, \tag{2.1.5}$$

where D_i are the unknown coefficients, are most efficient as numerical Motz's solutions. The exponential convergence rates $O(e^{-cN})$ can be obtained for eqn. (2.1.5) with some positive constant c. When functions, i.e., eqn. (2.1.5) are chosen, the eqn. (2.1.1), $u\,|_{-1<x<0\cap y=0} = 0$ and $\frac{\partial u}{\partial y}\,|_{0<x<1\cap y=0} = 0$ are satisfied. Then the coefficients D_i are sought by satisfying the rest of boundary conditions in eqns. (2.1.2) and (2.1.3). This is called the boundary approximation method (BAM) in Refs. [280, 306] or the CTM in this chapter. Under the computation in double precision and $N=34$, the maximal absolute error at $x=1$ (e.g., on $\overline{AB}$) of Motz's solution in Ref. [306] reaches up to 5.47×10^{-9}. Also the leading coefficient D_0 in Ref. [306] has 12 significant digits by the central rule. The solutions in Ref. [306] have been recognized to be the very accurate solutions for Motz's problem, see

Refs. [160, 161, 318]. In this chapter, to pursue the better leading coefficient D_0, we choose the Gaussian rules of high orders. Surprisingly, the obtained D_0 may have 17 significant digits by Fortran programs in double precision. Based on the new results in this chapter, we may address that the CTM (i.e., the BAM) is the highly accurate method for Motz's problem, not only in the global solutions but also in the leading coefficient D_0. As for Motz's problem, the conformal transformation method of Rosser and Papamichael [393] developing from Whiteman and Papamichael [468] and Papamichael and Whiteman [355] can also yield the most accurate leading coefficient D_0.

The same approaches are applied to its variants; one of them is called Model A, which is another frequently used benchmark model for testing new numerical methods [146, 160, 161, 353, 426]. Its highly accurate solutions are also provided with the leading D_0 having 17 significant digits. The advantage of Model A over Motz's problem is that half of the expansion coefficients are zero.

This chapter is organized as follows. In Section 2.2, basic algorithms of the CTM are provided for Motz's problem, and the highly accurate solutions are obtained in double precision. In Section 2.3, a new analysis is made for the quadrature involved. In Section 2.4, Model A is discussed, and its highly accurate solutions and the leading coefficient D_0 with 17 significant digits can also be achieved. In the last section, concluding remarks are addressed.

2.2 Basic algorithms of CTM

Since the expansions, i.e., eqn. (2.1.5) satisfy the Laplace equation and boundary conditions at $y=0$, the coefficients D_i should be chosen to satisfy the rest of the boundary conditions,

$$u\,|_{x=1} = u\,|_{\overline{AB}} = 500, \tag{2.2.1}$$

$$\left.\frac{\partial u}{\partial y}\right|_{y=1} = \left.\frac{\partial u}{\partial \nu}\right|_{\overline{BC}} = 0, \qquad \left.\frac{\partial u}{\partial \nu}\right|_{\overline{CD}} = -\left.\frac{\partial u}{\partial x}\right|_{x=-1} = 0, \tag{2.2.2}$$

as best as possible, where $u_\nu = \frac{\partial u}{\partial \nu}$ is the outward normal derivative to ∂S, and $\overline{AB}$, $\overline{BC}$, and $\overline{CD}$ are shown in fig. 2.1. Hence, the LSM may be designed as follows. Denote

$$[u, v] = \int_{\overline{AB}} uv\, d\ell + w^2 \int_{\overline{BC}\cup\overline{CD}} u_\nu v_\nu\, d\ell,$$

where w is a positive weight constant, and a good choice of the weight,

$$w = \frac{1}{N+1}, \tag{2.2.3}$$

can be found in Chapter 1. Denote by V_N the collection of finite-dimensional functions, i.e., eqn. (2.1.5). Then, we may seek $u_N \in V_N$ such that

$$[u_N, v] = 500 \int_{\overline{AB}} v\, d\ell, \quad \forall v \in V_N. \tag{2.2.4}$$

Denote the energy

$$I(v) = \int_{\overline{AB}} (v - 500)^2 \, d\ell + w^2 \int_{\overline{BC} \cup \overline{CD}} v_\nu^2 \, d\ell. \tag{2.2.5}$$

The solution of eqn. (2.2.4) can also be expressed by: To seek $u_N \in V_N$ such that

$$I(u_N) = \min_{v \in V_N} I(v). \tag{2.2.6}$$

Both eqns. (2.2.4) and (2.2.6) lead to the same linear algebraic system

$$\mathbf{Ax} = \mathbf{b}, \tag{2.2.7}$$

where $\mathbf{x} \in R^{N+1}$ is the unknown vector consisting of coefficients $D_i, i = 0, \ldots, N$, $\mathbf{b} \in R^{N+1}$ is the known vector resulting from the non-homogeneous Dirichlet condition, i.e., eqn. (2.2.1), and the matrix, $\mathbf{A} \in R^{(N+1)\times(N+1)}$, is symmetric and positive definite, but not sparse. By the Gaussian elimination without pivoting [168], the coefficients D_i (i.e., $\mathbf{x}$) can be obtained. Once the coefficients D_i are known, the errors on $\overline{AB} \cup \overline{BC} \cup \overline{CD}$

$$\begin{aligned} \|u - u_N\|_{\mathrm{B}} &= \left[\int_{\overline{AB}} (500 - u_N)^2 \, d\ell + w^2 \int_{\overline{BC} \cup \overline{CD}} (u_N)_\nu^2 \, d\ell \right]^{\frac{1}{2}} \\ &= \sqrt{I(u_N)} \end{aligned}$$

are computable, where the notation

$$\|v\|_{\mathrm{B}} = |v|_{\mathrm{B}} = \sqrt{[v, v]}. \tag{2.2.8}$$

Suppose that certain rules of integration are adopted to the integrals in eqn. (2.2.5). Let $\overline{AB}$ be divided into small segments $\overline{Z_i Z_{i+1}}$, i.e., $\overline{AB} = \bigcup_i \overline{Z_i Z_{i+1}}$. Then the integral is approximated by some rules,

$$\int_{\overline{AB}} v^2 \, d\ell \approx \widetilde{\int_{\overline{AB}}} v^2 \, d\ell = \sum_i \widetilde{\int_{\overline{Z_i Z_{i+1}}}} v^2 \, d\ell. \tag{2.2.9}$$

For example, the central and the trapezoidal rules are given by

$$\widetilde{\int_{\overline{Z_i Z_{i+1}}}} v^2 \, d\ell = v_{i+\frac{1}{2}}^2 h_i, \text{ and} \tag{2.2.10}$$

$$\widetilde{\int_{\overline{Z_i Z_{i+1}}}} v^2 \, d\ell = \frac{1}{2}(v_i^2 + v_{i+1}^2) h_i,$$

respectively, where $h_i = \overline{Z_i Z_{i+1}}$, $v_i = v(Z_i)$, $v_{i+\frac{1}{2}} = v(Z_{i+\frac{1}{2}})$ and $Z_{i+\frac{1}{2}} = \frac{Z_i + Z_{i+1}}{2}$. Other kinds of the Newton–Cotes and the Gaussian rules can also be employed and will be discussed later. Hence, for the numerical quadrature, we may seek $\widetilde{u}_N \in V_N$ such that

$$\widetilde{I}(\widetilde{u}_N) = \min_{v \in V_N} \widetilde{I}(v), \tag{2.2.11}$$

where

$$\widetilde{I}(v) = \widetilde{\int_{\overline{AB}}} (v - 500)^2 \, d\ell + w^2 \widetilde{\int_{\overline{BC} \cup \overline{CD}}} v_\nu^2 \, d\ell. \tag{2.2.12}$$

The minimization of $\widetilde{I}(v)$ also leads to a linear system like eqn. (2.2.7). This is a direct implementation of the LSM involving numerical integration, called the normal method (NM).

Now, we turn to the CTM, which can be regarded as a certain kind of the LSM involving specific quadratures. For simplicity in exposition, let us first consider the central rule, i.e., eqn. (2.2.10). Divide the boundaries $\overline{AB}$, $\overline{BC}$, and $\overline{CD}$ into uniform sub-intervals (see fig. 2.1). Then

$$h = \frac{\overline{AB}}{M} = \frac{\overline{CD}}{M} = \frac{\overline{CB}}{2M}.$$

Equations (2.2.1) and (2.2.2) can be transformed to the boundary collocation equations,

$$u_N(P_i) = 500, \qquad i = 1, 2, \ldots, M, \tag{2.2.13}$$

$$\frac{\partial u_N}{\partial x}(P_i^*) = -\frac{\partial u_N}{\partial \nu}(P_i^*) = 0, \quad i = 1, 2, \ldots, M, \tag{2.2.14}$$

$$\frac{\partial u_N}{\partial y}(Q_i^\pm) = \frac{\partial u_N}{\partial \nu}(Q_i^\pm) = 0, \quad i = 1, 2, \ldots, M. \tag{2.2.15}$$

Let $x_i^\pm = \pm(i - \frac{1}{2})h$ and $y_i = (i - \frac{1}{2})h$. The nodes $P_i = (1, y_i) \in \overline{AB}$, $P_i^* = (-1, y_i) \in \overline{CD}$, and $Q_i^\pm = (x_i^\pm, 1) \in \overline{BC}$, and their polar coordinates are computed by

$$P_i = (r_i, \theta_i), \qquad r_i = \sqrt{1 + y_i^2}, \qquad \theta_i = \cos^{-1}\left(\frac{1}{\sqrt{1 + y_i^2}}\right),$$

$$P_i^* = (r_i, \theta_i^*), \qquad \theta_i^* = \pi - \theta_i,$$

where $0 < \theta_i < \frac{\pi}{4}$. Besides,

$$Q_i^\pm = (\bar{r}_i, \theta_i^\pm), \qquad \bar{r}_i = \sqrt{1 + x_i^2}, \qquad \theta_i^+ = \sin^{-1}\left(\frac{1}{\sqrt{1 + x_i^2}}\right),$$

where $0 < \theta_i^+ < \frac{\pi}{4}$ and $\theta_i^- = \pi - \theta_i^+$.

In eqns. (2.2.13)–(2.2.15), there are $m = 4M$ equations, but $N+1$ unknown coefficients. Usually, we select $m > N+1$. We invoke the standard LSM in Ref. [168] to solve the overdetermined system of eqns. (2.2.13)–(2.2.15). Denote eqns. (2.2.13)–(2.2.15) by

$$\mathbf{F}_i \mathbf{x} = \mathbf{b}_i, \quad i = 1, 2, 3, \tag{2.2.16}$$

respectively, where $\mathbf{F}_i$ and $\mathbf{b}_i$ are the known matrices and vectors, respectively. Since the equations in eqn. (2.2.16) result from different boundary conditions, different weights should be assigned. When the weights $\sqrt{h}$ and $w\sqrt{h}$ are applied to the first and the other two equations in eqn. (2.2.16), respectively, the global target function becomes

$$T(\mathbf{x}) = h\|\mathbf{F}_1\mathbf{x} - \mathbf{b}_1\|^2 + w^2 h \sum_{i=2}^{3} \|\mathbf{F}_i\mathbf{x} - \mathbf{b}_i\|^2, \tag{2.2.17}$$

where $\|\cdot\|$ is the Euclidean norm, and w is a suitable weight constant given in eqn. (2.2.3). We can easily verify the following lemma by direct manipulation.

Lemma 2.2.1

Let the central rule, i.e., eqn. (2.2.10) be used in eqn. (2.2.12), and eqn. (2.2.16) be the collocation eqns. (2.2.13)–(2.2.15). Then we have

$$\widetilde{I}(\widetilde{u}_N) = T(\mathbf{x}),$$

where $\widetilde{I}(\widetilde{u}_N)$ and $T(\mathbf{x})$ are defined in eqns. (2.2.12) and (2.2.17), respectively.

Note that the admissible functions and their derivatives are given by

$$u_N = u_N(r,\theta) = \sum_{l=0}^{N} D_l r^{l+\frac{1}{2}} \cos\left(l + \frac{1}{2}\right)\theta,$$

$$\frac{\partial u_N}{\partial x} = \sum_{l=0}^{N} D_l \left(l + \frac{1}{2}\right) r^{l-\frac{1}{2}} \cos\left(l - \frac{1}{2}\right)\theta,$$

$$\frac{\partial u_N}{\partial y} = \sum_{l=0}^{N} D_l \left(l + \frac{1}{2}\right) r^{l-\frac{1}{2}} \sin\left(\frac{1}{2} - l\right)\theta.$$

Then, Lemma 2.2.1 enables us to obtain the solutions D_l by solving the following overdetermined system of the equations

$$\sqrt{h}\sum_{l=0}^{N} D_l r_i^{l+\frac{1}{2}} \cos\left(l + \frac{1}{2}\right)\theta_i = 500\sqrt{h}, \quad 1 \le i \le M, \tag{2.2.18}$$

$$w\sqrt{h}\sum_{l=0}^{N} D_l\left(l+\frac{1}{2}\right)(r_i)^{l-\frac{1}{2}}\cos\left(l-\frac{1}{2}\right)\theta_i^* = 0, \quad 1 \le i \le M, \tag{2.2.19}$$

$$w\sqrt{h}\sum_{l=0}^{N} D_l\left(l+\frac{1}{2}\right)(\overline{r}_i)^{l-\frac{1}{2}}\sin\left(\frac{1}{2}-l\right)\theta_i^{\pm} = 0, \quad 1 \le i \le M, \tag{2.2.20}$$

where $m = 4M > N+1$. Denote the overdetermined system of eqns. (2.2.18)–(2.2.20) by

$$\mathbf{F}\mathbf{x} = \mathbf{b}^*, \tag{2.2.21}$$

where the associated matrix $\mathbf{F} \in R^{m\times(N+1)}$, $\mathbf{b}^* \in R^m$, and $\mathbf{x} \in R^{N+1}$. In fact, the entries of $\mathbf{F} = (F_{i,l})$ are given by

$$F_{i,l} = \begin{cases} \sqrt{h}r_i^{l+\frac{1}{2}}\cos(l+\frac{1}{2})\theta_i, & 1 \le i \le M, \quad 0 \le l \le N, \\ w\sqrt{h}(l+\frac{1}{2})r_{i-M}^{l-\frac{1}{2}}\cos(l-\frac{1}{2})\theta_{i-M}^*, & M < i \le 2M, \quad 0 \le l \le N, \\ w\sqrt{h}(l+\frac{1}{2})\overline{r}_{i-2M}^{l-\frac{1}{2}}\sin(\frac{1}{2}-l)\theta_{i-2M}^+, & 2M < i \le 3M, \quad 0 \le l \le N, \\ w\sqrt{h}(l+\frac{1}{2})\overline{r}_{i-3M}^{l-\frac{1}{2}}\sin(\frac{1}{2}-l)\theta_{i-3M}^-, & 3M < i \le 4M, \quad 0 \le l \le N. \end{cases} \tag{2.2.22}$$

In general, we can rewrite the overdetermined system of eqns. (2.2.18)–(2.2.20) as

$$\alpha_i(u_N(P_i) - 500) = 0, \quad P_i \in \overline{AB}, \tag{2.2.23}$$

$$w\beta_i\frac{\partial u_N}{\partial x}(P_i^*) = 0, \quad P_i^* \in \overline{CD}, \tag{2.2.24}$$

$$w\gamma_i\frac{\partial u_N}{\partial y}(Q_i) = 0, \quad Q_i \in \overline{BC}, \tag{2.2.25}$$

where P_i, P_i^*, and Q_i are the nodes of integration rules, and α_i, β_i, and γ_i are positive weights. Equations (2.2.23)–(2.2.25) may be obtained from other quadratures. Take the Gaussian rules for example. Denote $h_k = \overline{Z_kZ}_{k+1}$. By using an affine transformation, the interval $[Z_k, Z_{k+1}]$ can be converted to $[-1, 1]$. Hence by this transformation, $x \in \overline{Z_kZ}_{k+1}$ is mapped to $t \in [-1, 1]$, $f(x)$ to $f\left(\frac{Z_{k+1}-Z_k}{2}t + \frac{Z_{k+1}+Z_k}{2}\right)$, and the integral on $\overline{Z_kZ}_{k+1}$ is changed to

$$\int_{\overline{Z_kZ}_{k+1}} f(x)\,dx = \frac{h_k}{2}\int_{-1}^{1} f\left(\frac{Z_{k+1}-Z_k}{2}t + \frac{Z_{k+1}+Z_k}{2}\right)dt.$$

The Gaussian rules with r nodes are given by

$$\int_{-1}^{1} f(t)\,dt \approx \widetilde{\int_{-1}^{1}} f(t)\,dt = \sum_{i=1}^{r}\omega_i f(t_i), \tag{2.2.26}$$

where the locations of nodes $t_i \in [-1, 1]$ and positive weights ω_i are provided in textbooks (e.g., Ref. [9]). For eqn. (2.2.23), a point P_i located at the j-th node of $\overline{Z_k Z_{k+1}} \in \overline{AB}$ has the weights $\alpha_i = \sqrt{\omega_j h_k/2}$. The weights β_i and γ_i can be obtained similarly. When $r = 1$, the Gaussian rule is just the central rule with $t_1 = 0$ and $\omega_1 = 2$. For the Gaussian rules, we have the following proposition, similar to Lemma 2.2.1.

Proposition 2.2.1

Let the Gaussian rules, i.e., eqn. (2.2.26) be used in eqn. (2.2.12), and $m \geq N + 1$. Then the coefficients D_ℓ from the corresponding collocation eqns. (2.2.23)–(2.2.25) are just the solutions from eqn. (2.2.11).

Let us consider the computer complexity of this method. In eqn. (2.2.22) we may employ the recursive formulas to save CPU time:

$$\cos\left(l + \frac{1}{2}\right)\theta = 2\cos\theta\cos\left(l - \frac{1}{2}\right)\theta - \cos\left(l - \frac{3}{2}\right)\theta,$$

$$r_i^{l+\frac{1}{2}} = r_i \cdot r_i^{l-\frac{1}{2}}.$$

To solve the least squares solution of eqn. (2.2.21) with full rank $\mathbf{F}$, we may use the QR method by the Householder orthogonalization with the flops ([168], p. 248),

$$T_L = 2mn^2 - \frac{2n^3}{3}, \quad n = N + 1.$$

On the other hand, the normal equations from eqn. (2.2.21) are

$$\mathbf{Ax} = \mathbf{F}^T\mathbf{Fx} = \mathbf{F}^T\mathbf{b}^* = \mathbf{b}, \tag{2.2.27}$$

where $\mathbf{A}$ is symmetric and positive definite. Then the flops for $\mathbf{F}^T\mathbf{F}$ and the Gaussian elimination of symmetric matrices are $m(n^2 + n)$ and $\frac{1}{3}n^3$, respectively. So, the main flops needed are

$$T_N = mn^2 + \frac{1}{3}n^3.$$

In our case, $n = N + 1$ and $m = 4M$. Evidently, when $m \gg n$, we have $T_L \leq 2T_N$, and when $m \geq n$,

$$T_L - T_N = (m - n)n^2 \geq 0.$$

Then, we conclude the following:

Corollary 2.2.1

The flops needed to solve a least squares problem, i.e., eqn. (2.2.21) by the Householder QR method are larger than, but at most double of, those by the NM, i.e., eqn. (2.2.27).

Besides the QR method, the singular value decomposition (SVD) can also be used to solve the overdetermined system, i.e., eqn. (2.2.21). A comparison between the QR method and the SVD is given in Ref. [126]. In general, the latter uses more flops than the former. So, the SVD is not recommended here.

Since the condition number of matrix **A** is nearly square of that of matrix **F** (see Ref. [168]), using the normal equations incurs a serious loss of solution accuracy for Motz's problem. Some numerical experiments of the normal equation were reported in Lefeber [270], where only four and five significant digits of Motz's solutions were obtained from the computation in double precision. Hence to obtain the numerical solutions of Motz's problem, we always choose the QR method to solve eqn. (2.2.21). Such a numerical approach is called the BAM in Refs. [280, 306] and the CTM in this chapter. (Strictly speaking, the description of BAM in Refs. [280, 306] does not involve numerical quadrature; the computed results in Refs. [280, 306] are, indeed, obtained by the algorithms described in this chapter using the central rule. The BAM involving numerical integration leads to the CTM. We join the BAM into the Trefftz family recently for easy communication with others.) Note that stability analysis of the CTM has been explored in Refs. [280, 305].

To close this section, we provide numerical experiments for Motz's problem. First, for the central rule, errors of the solutions and condition numbers are listed in tables 2.1 and 2.2, where $\varepsilon = u - u_N$, M denotes the number of collocation nodes along $\overline{AB}$, and the total number of all collocation nodes used is $4M$. In these tables, $\Delta D_i = d_i - D_i$, $\|\varepsilon\|_{\infty,\overline{AB}} = \max_{\overline{AB}} |\varepsilon|$, and the condition number is defined by

$$\text{Cond} = \left\{ \frac{\lambda_{\max}(\mathbf{F}^T\mathbf{F})}{\lambda_{\min}(\mathbf{F}^T\mathbf{F})} \right\}^{\frac{1}{2}} = \left\{ \frac{\lambda_{\max}(\mathbf{A})}{\lambda_{\min}(\mathbf{A})} \right\}^{\frac{1}{2}}, \tag{2.2.28}$$

where $\lambda_{\max}(\mathbf{A})$ and $\lambda_{\min}(\mathbf{A})$ are the maximal and the minimal eigenvalues of **A**, respectively. It can be seen from table 2.2 that M should be chosen as $M \geq \frac{N}{2}$ for $N = 34$. Tables 2.1–2.5 are all computed by means of Fortran programs in double precision.

Moreover, for the Gaussian rule with six nodes and those with 1, 2, 4, 6, 8, and 10 nodes, the results are listed in tables 2.3 and 2.4, respectively, and the best leading coefficients in table 2.5 by the Gaussian rule with six nodes as $N = 34$ and

Table 2.1: The error norms and condition numbers from the CTM for Motz's problem by the central rule.

N	M	$\|\varepsilon\|_B$	$\|\varepsilon\|_{\infty,\overline{AB}}$	Cond	$\left\|\frac{\Delta D_0}{D_0}\right\|$	$\left\|\frac{\Delta D_1}{D_1}\right\|$	$\left\|\frac{\Delta D_2}{D_2}\right\|$	$\left\|\frac{\Delta D_3}{D_3}\right\|$
10	8	0.250(−1)	0.149(−1)	94.3	0.189(−5)	0.491(−5)	0.601(−5)	0.928(−3)
18	12	0.133(−3)	0.811(−4)	0.193(4)	0.158(−7)	0.113(−6)	0.290(−6)	0.502(−6)
26	16	0.973(−6)	0.734(−6)	0.366(5)	0.216(−9)	0.155(−8)	0.380(−8)	0.202(−8)
34	24	0.839(−8)	0.459(−8)	0.666(6)	0.169(−11)	0.121(−10)	0.296(−10)	0.152(−10)

Table 2.2: The error norms and condition numbers from the CTM for Motz's problem by the central rule as $N = 34$.

M	$\|\varepsilon\|_B$	$\|\varepsilon\|_{\infty,\overline{AB}}$	Cond	$\left\|\frac{\Delta D_0}{D_0}\right\|$	$\left\|\frac{\Delta D_1}{D_1}\right\|$	$\left\|\frac{\Delta D_2}{D_2}\right\|$	$\left\|\frac{\Delta D_3}{D_3}\right\|$
9	0.135(−8)	0.496(−6)	0.267(8)	0.377(−9)	0.266(−8)	0.641(−8)	0.342(−8)
12	0.587(−8)	0.713(−7)	0.992(6)	0.337(−10)	0.239(−9)	0.578(−9)	0.305(−9)
16	0.772(−8)	0.189(−7)	0.679(6)	0.729(−11)	0.520(−10)	0.127(−9)	0.655(−10)
24	0.839(−8)	0.459(−8)	0.669(6)	0.169(−11)	0.121(−11)	0.296(−10)	0.152(−10)
32	0.849(−8)	0.462(−8)	0.669(6)	0.769(−11)	0.550(−11)	0.134(−10)	0.695(−11)

Table 2.3: The error norms and condition numbers from the CTM for Motz's problem as $N = 34$ by the Gaussian rule with six nodes.

M	$\|\varepsilon\|_B$	$\|\varepsilon\|_{\infty,\overline{AB}}$	Cond	$\left\|\frac{\Delta D_0}{D_0}\right\|$	$\left\|\frac{\Delta D_1}{D_1}\right\|$	$\left\|\frac{\Delta D_2}{D_2}\right\|$	$\left\|\frac{\Delta D_3}{D_3}\right\|$
12	0.359(−8)	0.721(−8)	0.675(6)	0.531(−13)	0.646(−12)	0.405(−11)	0.868(−11)
18	0.494(−8)	0.629(−8)	0.679(6)	0.468(−14)	0.211(−14)	0.620(−13)	0.352(−14)
24	0.491(−8)	0.530(−8)	0.679(6)	0.567(−15)	0.324(−15)	0.103(−14)	0.337(−13)
30	0.493(−8)	0.520(−8)	0.676(6)	0*	0.162(−15)	0.124(−14)	0.317(−13)
36	0.494(−8)	0.520(−8)	0.679(6)	0.850(−15)	0.324(−15)	0.103(−14)	0.308(−13)

* Denotes errors less than computer rounding errors in double precision.

Table 2.4: The error norms and condition numbers from the CTM for Motz's problem by different Gaussian rules with r nodes as $N = 34$.

r	M	$\|\varepsilon\|_B$	$\|\varepsilon\|_{\infty,\overline{AB}}$	Cond	$\left\|\frac{\Delta D_0}{D_0}\right\|$	$\left\|\frac{\Delta D_1}{D_1}\right\|$	$\left\|\frac{\Delta D_2}{D_2}\right\|$	$\left\|\frac{\Delta D_3}{D_3}\right\|$
1	24	0.839(−8)	0.459(−8)	0.669(6)	0.169(−11)	0.121(−11)	0.296(−10)	0.152(−10)
2	24	0.854(−8)	0.369(−8)	0.672(6)	0.708(−13)	0.512(−12)	0.133(−11)	0.106(−11)
4	24	0.610(−8)	0.540(−8)	0.679(6)	0.425(−15)	0.535(−14)	0.641(−13)	0.755(−13)
6	30	0.493(−8)	0.520(−8)	0.676(6)	0*	0.162(−15)	0.124(−14)	0.317(−13)
8	24	0.428(−8)	0.519(−8)	0.679(6)	0.142(−15)	0.648(−15)	0.618(−15)	0.315(−13)
10	20	0.639(−8)	0.521(−8)	0.679(6)	0.142(−15)	0*	0.412(−15)	0.308(−13)

* Denotes errors less than computer rounding errors in double precision.

$M = 30$. Note that the central rule is the simplest Gaussian rule with $r = 1$. When using the Gaussian rule, there seems no significant effect to reduce the errors $\|\varepsilon\|_B$ and $\|\varepsilon\|_{\infty,\overline{AB}}$ (e.g., from $\|\varepsilon\|_B = 0.839(-8)$ down to $0.428(-8)$, see table 2.4). From table 2.4, however, the Gaussian rules of high order do improve evidently the accuracy of leading coefficients. For $N = 34$, $M = 30$, and the Gaussian rule of

Table 2.5: The leading coefficients D_i from the CTM for Motz's problem by the Gaussian rule with six nodes as $N = 34$ and $M = 30$.

i	All digits	Sig. digits	No. of Sig. digits
0	401.162453745234416	401.16245374523442	17
1	87.6559201950879299	87.6559201950879	15
2	17.2379150794467897	17.2379150794468	15
3	−8.0712152596987790	−8.07121525970	12
4	1.44027271702238968	1.44027271702	12
5	0.331054885920006037	0.33105488592	12
6	0.275437344507860671	0.27543734451	11
7	−0.869329945041107943(−1)	−0.869329945(−1)	9
8	0.336048784027428854(−1)	0.336048784(−1)	9
9	0.153843744594011413(−1)	0.153843745(−1)	9
10	0.730230164737157971(−2)	0.7302302(−2)	7
11	−0.318411361654662899(−2)	−0.3184114(−2)	7
12	0.122064586154974736(−2)	0.1220646(−2)	7
13	0.530965295822850803(−3)	0.530965(−3)	6
14	0.271512022889081647(−3)	0.271512(−3)	6
15	−0.120045043773287966(−3)	−0.12005(−3)	5
16	0.505389241414919585(−4)	0.5054(−4)	4
17	0.231662561135488172(−4)	0.2317(−4)	4
18	0.115348467265589439(−4)	0.11535(−4)	5
19	−0.529323807785491411(−5)	−0.529(−5)	3
20	0.228975882995988624(−5)	0.229(−5)	3
21	0.106239406374917051(−5)	0.106(−5)	3
22	0.530725263258556923(−6)	0.531(−6)	3
23	−0.245074785537844696(−6)	−0.25(−6)	2
24	0.108644983229739802(−6)	0.11(−6)	2
25	0.510347415146524412(−7)	0.5(−7)	1
26	0.254050384217598898(−7)	0.3(−7)	1
27	−0.110464929421918792(−7)	−0.1(−7)	1
28	0.493426255784041972(−8)	/	0
29	0.232829745036186828(−8)	/	0
30	0.115208023942516515(−8)	/	0
31	−0.345561696019388690(−9)	/	0
32	0.153086899837533823(−9)	/	0
33	0.722770554189099639(−10)	/	0
34	0.352933005315648864(−10)	/	0

six nodes, the highly accurate solutions are listed in table 2.5 with the best leading coefficient

$$D_0 = 401.162453745234416. \tag{2.2.29}$$

On comparing with more accurate values in Ref. [299] as well as Ref. [84] using Mathematica, the relative error of D_0 in eqn. (2.2.29) is less than the rounding error of double precision. (This seems impossible! In fact, there exist some guard digits in computer for arithmetic of floating point numbers by noting that there are 18 digits in computer outputs of double precision, and some cancellation of rounding errors in statistics may also happen in the computation. In fact, the effective condition number is much smaller. For Motz's problem, the effective condition number is about 30, see Section 3.7. Hence, occasionally excellent results may happen at random.) Note that D_0 in eqn. (2.2.29) has 17 significant digits, while the D_0 in Refs. [280, 306] has only 12 significant digits. This is an improvement over Refs. [280, 306]. Besides, we also list D_i with significant digits (Sig. digits) in table 2.5, which are obtained from D_i with all digits by rounding. The errors of significant digits occur only at the last digit at most with a half unit, compared with the more accurate coefficients in Ref. [299]. Although $D_{28}-D_{34}$ are incorrect, they are indispensable to reach the global optimal solutions. Hence, the solutions from this chapter are optimal in the global errors, and the highly accurate leading coefficients are its natural consequences.

2.3 Error bounds and integration approximation

Define the norm

$$\|v\|_1 = \|v\|_{1,S} = \left[\iint_S (v^2 + v_x^2 + v_y^2)\, ds\right]^{\frac{1}{2}}.$$

We obtain a lemma based on the theories developed in Chapter 1.

Lemma 2.3.1

Let $u \in H^1(S)$ be the solution of Motz's problem. If the following inverse property holds:

$$\|v_\nu\|_{0,\overline{AB}} \le K_N \|v\|_1, \quad v \in V_N, \tag{2.3.1}$$

where $\|v_\nu\|_{0,\overline{AB}} = \{\int_{\overline{AB}} v_\nu^2 d\ell\}^{\frac{1}{2}}$. Then for any $w > 0$, there exists a constant C independent of N and u such that

$$\|u - u_N\|_1 \le C\left(K_N + \frac{1}{w}\right)\|u - u_N\|_B.$$

The constant C in this chapter is used as a generic, positive constant; their values may be different in different contexts.

Below, new analysis is devoted to the CTM involving numerical approximation of integration. Denote

$$[u, v]_{\widetilde{B}} = \widetilde{\int}_{\overline{AB}} uv\, d\ell + w^2 \widetilde{\int}_{\overline{BC} \cup \overline{CD}} u_\nu v_\nu \, d\ell,$$

$$\|v\|_{\widetilde{B}} = |v|_{\widetilde{B}} = \sqrt{[v, v]_{\widetilde{B}}} = \left\{ \widetilde{\int}_{\overline{AB}} v^2 \, d\ell + w^2 \widetilde{\int}_{\overline{BC} \cup \overline{CD}} v_\nu^2 \, d\ell \right\}^{\frac{1}{2}}. \tag{2.3.2}$$

The solutions $\widetilde{u}_N$ of eqn. (2.2.11) will satisfy

$$\|u - \widetilde{u}_N\|_{\widetilde{B}} = \min_{v \in V_N} \|u - v\|_{\widetilde{B}} = \min_{v \in V_N} \sqrt{\widetilde{I}(v)}.$$

For the integration rules involved, we denote

$$\|v\|^2_{\widetilde{B}} = \|\hat{v}\|^2_{\mathrm{B}},$$

where $\widehat{v^2}$ are piecewise interpolation polynomials of v^2 with order k along $\Gamma = \partial S$. Note that the true solution u satisfies eqns. (2.2.13)–(2.2.15) exactly. Then for eqn. (2.3.2), we have

$$[u - \widetilde{u}_N, v]_{\widetilde{B}} = 0. \tag{2.3.3}$$

We can prove the following lemma, similarly based on Chapter 1.

Equation (2.3.3) holds for the collocation method; this is different from the finite element method (FEM), the finite volume method (FVM), etc., where the true solution does not satisfy the algorithms involving integration approximations of numerical solutions. Hence, the same conclusions involving integration approximation made for the collocation method in this book may not be applied to other numerical methods such as FEM.

Lemma 2.3.2

The solutions $\widetilde{u}_N$ obtained by the CTMs with integral approximation satisfy

$$[u - \widetilde{u}_N, v]_{\widetilde{B}} = 0, \quad \forall v \in V_N, \tag{2.3.4}$$

and

$$\|v - \widetilde{u}_N\|_{\widetilde{B}} \leq \|u - v\|_{\widetilde{B}}, \quad \forall v \in V_N. \tag{2.3.5}$$

Next, let us examine the errors from integration rules. Suppose that the rules are chosen to have the following relative errors for v and $u - v$, where $v \in V_N$,

$$\frac{\left|\left(\int_{\overline{AB}} - \widetilde{\int_{\overline{AB}}}\right) v^2 \, d\ell\right|}{\int_{\overline{AB}} v^2 \, d\ell} \le b < \frac{3}{4}, \tag{2.3.6}$$

$$\frac{\left|\left(\int_{\overline{BC}} - \widetilde{\int_{\overline{BC}}}\right) v_\nu^2 \, d\ell\right|}{\int_{\overline{BC}} v_\nu^2 \, d\ell} \le b < \frac{3}{4}, \tag{2.3.7}$$

$$\frac{\left|\left(\int_{\overline{CD}} - \widetilde{\int_{\overline{CD}}}\right) v_\nu^2 \, d\ell\right|}{\int_{\overline{CD}} v_\nu^2 \, d\ell} \le b < \frac{3}{4}, \tag{2.3.8}$$

where b is a constant. Then, we have the following proposition.

Proposition 2.3.1

For those rules of quadrature satisfying eqns. (2.3.6)–(2.3.8), the following bound holds:

$$\left|\frac{\|v\|_{\mathrm{B}} - \|v\|_{\widetilde{B}}}{\|v\|_{\mathrm{B}}}\right| \le a < \frac{1}{2}, \quad v \in V_N, \tag{2.3.9}$$

where $a = 1 - \sqrt{1-b}$ is a constant.

Proof.

We have from the assumptions,

$$\frac{\left|\|v\|_{\mathrm{B}}^2 - \|v\|_{\widetilde{B}}^2\right|}{\|v\|_{\mathrm{B}}^2} \le \frac{\left|\left(\int_{\overline{AB}} - \widetilde{\int_{\overline{AB}}}\right) v^2 \, d\ell + \left(\int_{\overline{BC} \cup \overline{CD}} - \widetilde{\int_{\overline{BC} \cup \overline{CD}}}\right) v_\nu^2 \, d\ell\right|}{\int_{\overline{AB}} v^2 \, d\ell + \int_{\overline{BC} \cup \overline{CD}} v_\nu^2 \, d\ell} \le b. \tag{2.3.10}$$

We obtain

$$1 - b \le \frac{\|v\|_{\widetilde{B}}^2}{\|v\|_{\mathrm{B}}^2} \le 1 + b.$$

The above equation gives

$$\sqrt{1-b} \le \frac{\|v\|_{\widetilde{B}}}{\|v\|_{\mathrm{B}}} \le \sqrt{1+b}. \tag{2.3.11}$$

Next, we have from eqns. (2.3.10) and (2.3.11),

$$\frac{\left|\|v\|_{\mathrm{B}}-\|v\|_{\widetilde{B}}\right|}{\|v\|_{\mathrm{B}}} \leq \frac{b}{\|v\|_{\mathrm{B}}+\|v\|_{\widetilde{B}}}\|v\|_{\mathrm{B}}$$

$$\leq \frac{b}{1+\frac{\|v\|_{\widetilde{B}}}{\|v\|_{\mathrm{B}}}} \leq \frac{b}{1+\sqrt{1-b}} = 1-\sqrt{1-b} = a < \frac{1}{2}. \quad \blacksquare$$

Take the central rule in eqns. (2.2.9) and (2.2.10) for example. We have from Ref. [9]

$$\left(\int_{\overline{AB}} - \widetilde{\int_{\overline{AB}}}\right) f\, d\ell = \frac{h^2}{24} f''(\xi), \tag{2.3.12}$$

where $f = v^2$ or $f = (u-v)^2$, and $\xi \in \overline{AB}$. Since $f'' = 2\{(v')^2 + vv''\}$ for $f = v^2$, the requirements of quadrature errors in Proposition 2.3.1 imply that

$$\frac{1}{4}\int_{\overline{AB}} v^2\, d\ell \leq \widetilde{\int_{\overline{AB}}} v^2\, d\ell \leq \frac{7}{4}\int_{\overline{AB}} v^2\, d\ell,$$

or equivalently

$$\frac{h^2}{12}\left|((v')^2 + vv'')(\xi)\right| \leq \frac{3}{4}\int_{\overline{AB}} v^2\, d\ell.$$

Next, we give a new lemma.

Lemma 2.3.3

Suppose that the rules of integrations in eqn. (2.2.12) are chosen to satisfy the bound, i.e., eqn. (2.3.9). Then, the norms $\|\cdot\|_{\mathrm{B}}$ and $\|\cdot\|_{\widetilde{B}}$ defined in eqns. (2.2.8) and (2.3.2) are equivalent to each other:

$$C_1\|v\|_{\mathrm{B}} \leq \|v\|_{\widetilde{B}} \leq C_2\|v\|_{\mathrm{B}}, \quad v \in V_N, \tag{2.3.13}$$

where C_1 and C_2 are two positive constants independent of v and N.

Proof.

We have from eqn. (2.3.9)

$$\|v\|_{\mathrm{B}} - \|v\|_{\widetilde{B}} \leq a\|v\|_{\mathrm{B}},$$

and then

$$\|v\|_{\mathrm{B}} \leq \frac{1}{1-a}\|v\|_{\widetilde{B}}. \tag{2.3.14}$$

Also from eqn. (2.3.9)

$$\|v\|_{\widetilde{B}} - \|v\|_{\mathrm{B}} \le a\|v\|_{\mathrm{B}},$$

and then

$$\|v\|_{\widetilde{B}} \le (1+a)\|v\|_{\mathrm{B}}. \tag{2.3.15}$$

Hence, the desired result, i.e., eqn. (2.3.13) follows from eqns. (2.3.14) and (2.3.15). ■

Accordingly, we have a new important theorem.

Theorem 2.3.1

Let the condition, i.e., eqn. (2.3.1) hold, and the rules of integrations involved in eqn. (2.2.12) satisfy eqn. (2.3.9) for v and $u-v$, $\forall v \in V_N$. Then,

$$\|u - \widetilde{u}_N\|_1 \le \inf_{v\in V_N} \left\{ \|u-v\|_1 + C\left(K_N + \frac{1}{w}\right)\|u-v\|_{\mathrm{B}} \right\}, \tag{2.3.16}$$

where C is a bounded constant independent of u, v, and N. Moreover,

$$\|u - \widetilde{u}_N\|_1 \le \|R_N\|_1 + C\left(K_N + \frac{1}{w}\right)\|R_N\|_{\mathrm{B}}, \tag{2.3.17}$$

where

$$R_N = \sum_{i=N+1}^{\infty} d_i r^{i+\frac{1}{2}} \cos\left(i+\frac{1}{2}\right)\theta, \tag{2.3.18}$$

and d_i are the true expansion coefficients.

Proof.

From Lemma 1.2.1 in Chapter 1, we have

$$\|v\|_1 \le C\left(K_N + \frac{1}{w}\right)\|v\|_{\mathrm{B}}, \quad \forall v \in V_N. \tag{2.3.19}$$

Let $\eta = v - \widetilde{u}_N$, then $\eta \in V_N$ if $v \in V_N$. In view of eqn. (2.3.19) and the norm equivalence, i.e., eqn. (2.3.13),

$$\begin{aligned}\|u - \widetilde{u}_N\|_1 \le \|u-v\|_1 + \|\eta\|_1 &\le \|u-v\|_1 + C\left(K_N + \frac{1}{w}\right)\|\eta\|_{\mathrm{B}} \\ &\le \|u-v\|_1 + \frac{C}{C_1}\left(K_N + \frac{1}{w}\right)\|\eta\|_{\widetilde{B}}. \end{aligned} \tag{2.3.20}$$

From the orthogonal property eqn. (2.3.4), we obtain

$$\|\eta\|_{\widetilde{B}}^2 = [\eta, \eta]_{\widetilde{B}} = [v - u, \eta]_{\widetilde{B}} \le \|u - v\|_{\widetilde{B}} \|\eta\|_{\widetilde{B}}.$$

The above bound and the norm equivalence for $u - v$ leads to

$$\|\eta\|_{\widetilde{B}} \le \|u - v\|_{\widetilde{B}} \le C\|u - v\|_{\mathrm{B}}. \tag{2.3.21}$$

Combining eqns. (2.3.20) and (2.3.21) gives the first desired result, i.e., eqn. (2.3.16).

Next, the solution, i.e., eqn. (2.1.4) with the true coefficients d_i can be split into

$$u = \overline{u}_N + R_N,$$

where

$$\overline{u}_N = \sum_{i=0}^{N} d_i r^{i+\frac{1}{2}} \cos\left(i + \frac{1}{2}\right)\theta,$$

and the remainder R_N is given by eqn. (2.3.18). Then, letting $v = \overline{u}_N$ in eqn. (2.3.16) we obtain

$$\begin{aligned} \|u - \widetilde{u}_N\|_1 &\le \|u - \overline{u}_N\|_1 + C\left(K_N + \frac{1}{w}\right)\|u - \overline{u}_N\|_{\mathrm{B}} \\ &\le \|R_N\|_1 + C\left(K_N + \frac{1}{w}\right)\|R_N\|_{\mathrm{B}}. \end{aligned}$$

This is the second bound, i.e., eqn. (2.3.17) as desired. ■

Even for a rough quadrature like the simplest central rule, the relative errors of its approximate integrals have no difficulty to be less than three quarters. So, the conditions, i.e., eqn. (2.3.6)–(2.3.9) can be satisfied easily. Hence, the solutions $\widetilde{u}_N$ may still have the exponential convergence rates. This is a significant difference of global errors from the traditional role of integration in the finite element analysis. Besides, from Theorem 2.3.1, there is not much difference between lower-order and higher-order quadratures. However, for the accuracy of the leading coefficient D_0, the high-order rules, such as the Gaussian quadratures with six and eight nodes, may raise its accuracy, based on tables 2.3 and 2.4. Note that the new analysis of quadratures in this section provides an excellent theoretical foundation for the high accuracy of the CTM.

2.4 Variants of Motz's problem

As a variant of Motz's problem, Model A is discussed in the rest of this chapter. Its highly accurate solution can be sought similarly by the CTM. Not only its highly

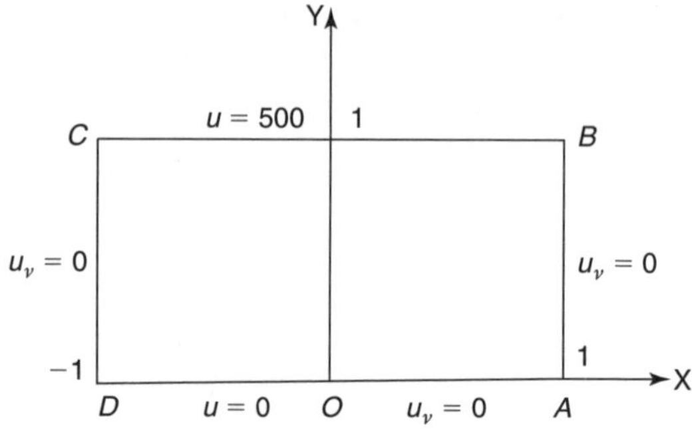

Figure 2.2: Model A of Motz's variant problems.

accurate solutions are obtained in this chapter, but also the highly accurate leading coefficient with 17 significant digits can be achieved by the Gaussian rule. Half of its expansion coefficients are zero, which is supported by *a posteriori* analysis. Hence, as a singularity model, Model A given in this section seems to be superior to Motz's problem in Sections 2.2 and 2.3.

When the boundary conditions on $\overline{AB}$ and on $\overline{BC}$ in fig. 2.1 are exchanged as

$$u\,|_{\overline{BC}} = 500, \qquad u\,|_{\overline{OD}} = 0, \qquad u_\nu\,|_{\overline{OA}} = 0, \qquad u_\nu\,|_{\overline{AB}\cup\overline{CD}} = 0, \tag{2.4.1}$$

this Laplace boundary value problem gives Model A, see fig. 2.2. Its original model in Refs. [146, 161, 353] was defined on the domain $\hat{S} = \{(x,y)\,|\,-\frac{1}{2} \le x \le \frac{1}{2}, 0 \le y \le \frac{1}{2}\}$ in fig. 2.3. These two models, in S and $\hat{S}$, are of the same nature. In fact, their solutions can be scaled from one to the other, which will be explained later. Since functions, i.e., eqn. (2.1.4) are also the solutions of Model A, we choose

$$u_N(r,\theta) = \sum_{i=0}^{N} \hat{D}_i r^{(i+\frac{1}{2})} \cos\left(i + \frac{1}{2}\right)\theta, \tag{2.4.2}$$

where the notation $\hat{D}_i$ with a hat on the head is used to distinguish itself from the D_i of Motz's problem in Section 2.2.

When the conditions on $\overline{OA} \cup \overline{OD}$ of Motz's problem are retained, and when some of the rest boundary conditions are changed, we obtain many variants of Motz's problem. Since the angular singularity as $O(r^{1/2})$ remains, the basic approaches of the CTM for Motz's problem in Sections 2.2 and 2.3 are valid. However, based on the analysis of Chapter 11, there may exist the mild singularities $O(\rho^k \ln \rho)$ at the corners A, B, C, and D; the CTM using piecewise singular functions is recommended, see Chapter 11 and Ref. [301]. Besides, a number of variants of Motz's problem are investigated in Tang [433]. The traditional problem as in fig. 2.3 was first studied in Fix [145] for the torsion of the cracked beam with square cross-section, and was called the cracked-beam problem in Georgiou, Boudouvis, and

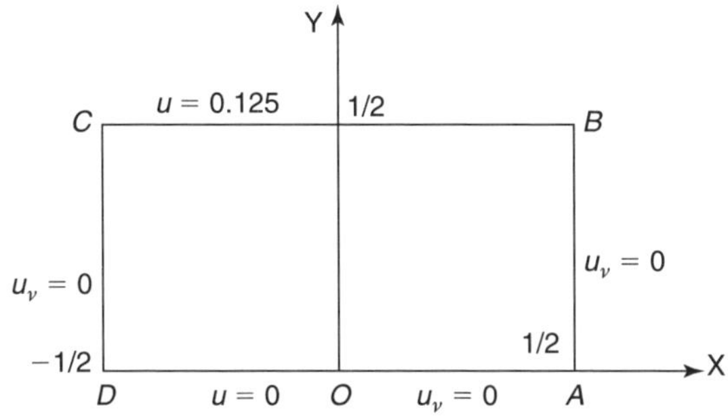

Figure 2.3: The traditional problem in $\hat{S}$.

Poullikkas [160], and then in Lu, Hu, and Li [315] and Lu and Li [316]. In this book, we classify such a problem with eqn. (2.4.1) into the variants (e.g., the family) of Motz's problem, and call it Model A of Motz's variant problems (or simply Model A), because it is most significant, and because it is popular (see Refs. [145, 161, 315, 316, 353]). Model B of Motz's variant problem is studied in Section 11.5 of Chapter 11, which involves both angular and mild singularities. Other interesting variants of Motz's problem appear elsewhere.

We also use V_N as the finite collection of functions, i.e., eqn. (2.4.2). Since u_N satisfies the Laplace equation in S and the boundary conditions on $\overline{OD} \cup \overline{OA}$ already, the coefficients $\hat{D}_i$ should be chosen to satisfy the rest of the boundary conditions as best as possible. Define the error norm on $\overline{AB} \cup \overline{BC} \cup \overline{CD}$:

$$\|u - v\|_{\mathrm{B}} = \left\{ \int_{\overline{BC}} (v - 500)^2 + w^2 \int_{\overline{AB} \cup \overline{CD}} v_\nu^2 \right\}^{\frac{1}{2}}, \quad w = \frac{1}{N+1}.$$

The solution u_N can be obtained by

$$\|u - u_N\|_{\widetilde{B}} = \inf_{v \in V_N} \|u - v\|_{\widetilde{B}},$$

where

$$\|v\|_{\widetilde{B}} = \left\{ \widetilde{\int_{\overline{BC}}} v^2 + w^2 \widetilde{\int_{\overline{AB} \cup \overline{CD}}} v_\nu^2 \right\}^{\frac{1}{2}}. \tag{2.4.3}$$

We first employ the central rule with a uniform distributed points P_i on $\overline{AB} \cup \overline{BC} \cup \overline{CD}$. We may require $\sqrt{h}\, v = \sqrt{h}\, 500$ at $P_i \in \overline{BC}$ and $\sqrt{h}\, w u_\nu = 0$ at $P_i \in \overline{AB} \cup \overline{CD}$. Let the number $4M$ of all collocation nodes P_i be larger than $N+1$, then we obtain an overdetermined system of linear algebraic equations $\mathbf{Fx} = \mathbf{b}$, where $\mathbf{F}$ is a matrix of $4M \times (N+1)$, and $\mathbf{x}$ is the unknown vector consisting of $\hat{D}_i$. We employ the CTM in Section 2.2 to solve it. The errors, condition numbers, and the leading coefficients are given in Ref. [315]. It is interesting from

Refs. [315, 433] to note that $\hat{D}_{4\ell+2} \approx \hat{D}_{4\ell+3} \approx 0$. Hence, we may simply seek a solution of the following simplified forms:

$$u_N^* = \sum_{\ell=0}^{L} \sum_{k=0}^{1} \hat{D}_{4\ell+k} r^{4\ell+k+\frac{1}{2}} \cos\left(4\ell + k + \frac{1}{2}\right)\theta, \tag{2.4.4}$$

where $N = 4L + 1$. Denote by V_N^* the finite collection of functions in eqn. (2.4.4). Hence, another CTM can be formulated as in Section 2.2: to seek the solution $u_N^* \in V_N^*$ such that

$$\|u - u_N^*\|_{\widetilde{B}} = \inf_{v \in V_N^*} \|u - v\|_{\widetilde{B}},$$

where $\|v\|_{\widetilde{B}}$ is defined in eqn. (2.4.3). From Ref. [315], we have observed the asymptotes:

$$\|u - u_N\|_{\mathrm{B}} = O(0.553^N), \qquad \|u - u_N\|_{\infty,\overline{BC}} = O(0.564^N), \tag{2.4.5}$$
$$\mathrm{Cond} = O(1.42^N),$$
$$\|u - u_N^*\|_{\mathrm{B}} = O(0.558^N), \qquad \|u - u_N^*\|_{\infty,\overline{BC}} = O(0.558^N), \tag{2.4.6}$$
$$\mathrm{Cond} = O(1.39^N).$$

Note that the convergence rates and the condition numbers in eqn. (2.4.6) are close to those in eqn. (2.4.5), but only half coefficients of u_N in eqn. (2.4.2) are needed. Hence, for the computational purpose, the solutions, i.e., eqn. (2.4.4) may be better. From this point of view, Model A using eqn. (2.4.4) may serve as a better testing model of singularity problems than Motz's problem.

The numerical experiments of Model A are reported in Lu, Hu, and Li [315], where the Gaussian rule is used in the CTM, to give the leading coefficient $\hat{D}_0$ with 17 significant digits, which is more accurate than $\hat{D}_0$ with 15 significant digits given in Ref. [316] using the central rule. Besides, the integrated singular basis method (ISBM) and the integral method are used in Georgiou, Boudouvis, and Poullikkas [160] for solving the Model A, to give the leading coefficient with only 12 significant digits. Numerical comparisons are made in Ref. [315], to display the advantages of CTM for Model A over other methods (e.g., ISBM in Ref. [160]).

To confirm the admissible functions as eqn. (2.4.4), we only prove the following proposition.

Proposition 2.4.1

Let the errors $\varepsilon_N = u - u_N^*, N = 4L + 1$ *and*

$$\|(\varepsilon_N)_\nu\|_{0,\overline{BC}} \le K_N \|\varepsilon_N\|_{1,S},$$

where the constant $K_N (\ge 1)$ *may be unbounded as* $N \to \infty$. *Suppose*

$$\left(K_N + \frac{1}{w}\right) \|\varepsilon_N\|_{\mathrm{B}} \to 0, \quad \text{as } N \to \infty. \tag{2.4.7}$$

Then the solution of Model A can be expressed by

$$u = \sum_{\ell=0}^{\infty} \sum_{k=0}^{1} \hat{D}_{4\ell+k} r^{4\ell+k+\frac{1}{2}} \cos\left(4\ell + k + \frac{1}{2}\right)\theta.$$

Proof.

From the bounds similar to Lemma 2.3.1, we have

$$\|\varepsilon_N\|_{1,S} = \|u - u_N^*\|_{1,S} \leq C\left(K_N + \frac{1}{w}\right)\|\varepsilon_N\|_{\mathrm{B}}, \tag{2.4.8}$$

where C is a bounded constant independent of N. From eqns. (2.4.7) and (2.4.8), $\{\varepsilon_N\}$ is a bounded sequence. Based on the Kandrasov or Rellich theorem [103], any bounded sequence in the space $H^1(S)$ contains a subsequence that converges in $H^0(S)$. Then there must exist a subsequence $\{\varepsilon_N^+\}$ in $H^0(S)$ such that $\lim_{N\to\infty} \varepsilon_N^+ = \overline{\varepsilon}$. Since $\{\varepsilon_N^+\}$ are bounded in $H^1(S)$, the convergent limit $\overline{\varepsilon} \in H^1(S)$. This implies that

$$\lim_{N\to\infty} u_N^+ = \lim_{N\to\infty} (u - \varepsilon_N) = u - \overline{\varepsilon} = \overline{u} \in H^1(S).$$

Moreover, since $K_N \geq 1$ and $w = \frac{1}{N+1}$, we conclude that $\|\overline{u} - 500\|_{0,\overline{BC}} = 0$ and $\|\overline{u}_\nu\|_{0,\overline{AB}\cup\overline{CD}} = 0$. Hence, $\overline{u}$ must be the unique solution of Model A. Obviously, the entire sequence u_N^* also converges to $\overline{u}(=u)$ based on $\|u - u_N^*\|_{\mathrm{B}} \to 0$ as $N \to \infty$ from eqns. (2.4.7) and (2.4.8). ∎

When $w = \frac{1}{N+1}$, the empirical exponential convergent rates in eqn. (2.4.6) guarantee eqn. (2.4.7). The analysis of Proposition 2.4.1 is made based on the *a posteriori* numerical results, so we call it *a posteriori* analysis. Proposition 2.4.1 implies that $\hat{D}_{4\ell+2} = \hat{D}_{4\ell+3} = 0, \forall \ell \geq 0$. We also note that condition, i.e., eqn. (2.4.7) is stronger than that $\|\varepsilon_N\|_{\mathrm{B}} \to 0$ as $N \to \infty$.

To close this section, let us explore the coefficients for the scaled Model A. Consider Model A on a scaled domain, $\hat{S} = \{(\xi, \eta) \mid -a < \xi < a, 0 < \eta < a\}$, where the parameter satisfies $0 < a \leq 1$. The scaled Model A is described by the Laplace equation $\Delta w = 0$ on $\hat{S}$ satisfying the following boundary conditions:

$$w(\xi, a) = b, \quad -a < \xi < a, \tag{2.4.9}$$

$$w(\xi, 0) = 0, \quad -a < \xi < 0, \qquad \frac{\partial w}{\partial \nu}(\xi, 0) = 0, \quad 0 < \xi < a, \tag{2.4.10}$$

$$\frac{\partial w}{\partial \nu}(\pm a, \eta) = 0, \quad 0 < \eta < a, \tag{2.4.11}$$

where b is a constant and ν the outward normal to $\partial\hat{S}$. Here, another Cartesian coordinate system (ξ, η) is chosen. For fig. 2.2, $a = 1$ and $b = 500$, and for fig. 2.3

from the traditional model [146, 160, 161, 353, 426], $a=\frac{1}{2}$ and $b=0.125$. The Laplace solution satisfying eqns. (2.4.9)–(2.4.11) can also be expressed by

$$w(\xi,\eta)=\sum_{i=0}^{\infty}\alpha_i\rho^{i+\frac{1}{2}}\cos\left(i+\frac{1}{2}\right)\theta,$$

where α_i are the coefficients, (ρ,θ) are the polar coordinates at the origin O, and $\rho=\sqrt{\xi^2+\eta^2}$. There exist the relations for the coefficients of $\hat{D}_i$ in Ref. [315] and α_i:

$$\alpha_i=\frac{b}{500}a^{-(i+\frac{1}{2})}\hat{D}_i. \tag{2.4.12}$$

Now, let us prove eqn. (2.4.12). Under the affine transformation T: $(x,y)\to(\xi,\eta)$, where $\xi=ax$ and $\eta=ay$, domain S is converted to $\hat{S}$, and the boundary conditions, i.e., eqn. (2.4.1) are transformed to

$$\begin{aligned}
&u(\xi,a)=b, \quad -a<\xi<a,\\
&u(\xi,0)=0, \quad -a<\xi<0, \qquad \frac{\partial u}{\partial\nu}(\xi,0)=0, \quad 0<\xi<a,\\
&\frac{\partial u}{\partial\nu}(\pm a,\eta)=0, \quad 0<\eta<a.
\end{aligned} \tag{2.4.13}$$

On comparing eqn. (2.4.13) with $u\,|_{\overline{BC}}=500$ in eqn. (2.4.1), we find the relations between w and u,

$$w=\frac{b}{500}u.$$

This gives

$$\sum_{i=0}^{\infty}\alpha_i\rho^{i+\frac{1}{2}}\cos\left(i+\frac{1}{2}\right)\theta=\frac{b}{500}\sum_{i=0}^{\infty}\hat{D}_i r^{i+\frac{1}{2}}\cos\left(i+\frac{1}{2}\right)\theta. \tag{2.4.14}$$

Since $r=\sqrt{x^2+y^2}$, we have $r=\frac{\rho}{a}$. The eqn. (2.4.14) is reduced to

$$\sum_{i=0}^{\infty}\left\{\alpha_i-\frac{b}{500}a^{-(i+\frac{1}{2})}\hat{D}_i\right\}\rho^{i+\frac{1}{2}}\cos\left(i+\frac{1}{2}\right)\theta=0.$$

Since functions $\{\rho^{i+\frac{1}{2}}\cos(i+\frac{1}{2})\theta\}$ are linearly independent, we obtain

$$\alpha_i-\frac{b}{500}a^{-(i+\frac{1}{2})}\hat{D}_i=0,$$

which is the desired eqn. (2.4.12).

2.5 Concluding remarks

To close this chapter, let us make a few remarks.

1. Computational algorithms of the CTM are provided in Section 2.2. The overdetermined system, i.e., eqn. (2.2.21) is recommended in computation since its algorithm is simple, which is, indeed, just the collocation method at the boundary nodes, based on Proposition 2.2.1. The remarkable advantage of eqn. (2.2.21) is that the condition numbers of the stiffness matrix can be dramatically reduced, compared to eqn. (2.2.27) of the normal equation.
2. Different quadratures, such as the central and Gaussian rules, are investigated for the TM. Theorem 2.3.1 reveals that different integration rules do not make much differences in the global errors over the entire domain S. However, the rules used may affect significantly the accuracy of the leading coefficient, based on numerical experiments in this chapter.
3. The quadrature is used to link the collocation method and the LSM. However from our error analysis, the accuracy of a quadrature may be very rough, in the sense that its relative errors are less than three quarters! This feature is significantly different from the traditional integral approximation in error analysis, e.g., the FEM analysis, where the integration errors should be chosen to balance the optimal errors of the solutions. Based on the analysis in Section 2.3, the solutions of Motz's and Model A solved by the CTM have the exponential convergence rates. Note that Theorem 2.3.1 and Proposition 2.3.1 are new, which provide an excellent theoretical foundation for high accuracy of the CTM (e.g., the BAM). This is also a justification for the CTM to become the most accurate method for Motz's and Model A. The collocation methods both in S and on ∂S, on the other hand, are explored in Chapter 5.
4. The numerical results in Section 2.2 are better than those in Refs. [280, 306]. The Gaussian rule with six nodes are used to raise the accuracy of the leading coefficient to

$$D_0 = 401.162453745234416 \tag{2.5.1}$$

in the CTM. Compared with the more accurate value of D_0 in Refs. [280, 299], this D_0 has exactly 17 significant digits, the error of which happens to coincide with the rounding errors of double precision. Note that coefficient D_0 in Ref. [306] has only 12 significant digits. This new discovery will change the evaluation of the BAM (i.e., the collocation TM) given in Ref. [280]. Based on the numerical results in Ref. [306] using the central rule, it is pointed out in Ref. [280], p. 133, that "*BAM may produce the best global solutions*," but "*the conformal transformation method is the highly accurate method for leading coefficients.*" Now, we may address that for Motz's problem, the CTM (i.e., the BAM) by the Gaussian rule of high order is a highly accurate method, not only for the global solutions but also for the leading coefficient D_0.
5. Although the condition number of the CTM is large, the effective condition number is, indeed, small. For Motz's problem, the effective condition number

about 30 for the solution in table 2.5 in the next chapter explains well the highly accurate solution and the D_0 with 17 significant digits. More study on stability is given later for Motz's problem by the CTM.

6. Motz's problem and Model A of Motz's variant problems are linked and compared by considering the same domain S. The traditional Model A in Refs. [146, 160, 161, 353, 426] is formulated as a special case of the scaled Model A in this chapter, whose solutions can be obtained straightforwardly by eqn. (2.4.12). Besides, numerical experiments by the approaches using eqn. (2.4.12) and by direct computation of the problem defined in fig. 2.3 have been reported in Ref. [315]. The former is superior due to its smaller condition numbers. Motz's problem and its variants satisfy the Laplace equation on S with two different boundary conditions along the edges. Hence, different boundary conditions on ∂S may have different impacts on the singular behavior of the Laplace solutions in S. These computational results will be reported in Ref. [301].
7. There was a special issue on the TM, i.e., Advances in Engineering Software, vol. 24, 1995. The first and second International Workshops on the Trefftz method were held in Cracow, Poland, May 30–June 1, 1996 and in Sintra, Portugal, September 1999, and the invited talks are published in CAMES, vol. 4 (1997) and vol. 8 (2001), respectively. Overviews of the method can be found in Zienkiewicz [487], Kita and Kamiya [249], Zieliński [482, 483], and Jin and Cheung [226]. In Refs. [249, 226], the TM are classified into the indirect and direct methods. The CTM in this chapter is the indirect TM. The direct TM is analogous to the method of fundamental solutions (MFS), in which the fundamental functions are replaced by the singular function in the trial space. We use the same terminology, the Trefftz collocation method, as in Leitao [271]. We report in this chapter the new computational results and the new analysis of the indirect TM. These results have narrowed the gap existing before between the excellent computation and the theoretical support of this method.

3 Coupling techniques

We begin with solving Laplace's equation with singularities by the boundary approximation method (BAM), i.e., the collocation Trefftz method (CTM). Different coupling techniques to match the boundary conditions are explored. This chapter also examines the generalized Trefftz methods (GTMs) for partial differential equations (PDEs) with singularities. GTMs use the local particular solutions of PDEs, but adopt the coupling strategies to deal with the boundary conditions, which are different from the classic BAM in Chapter 1 and the CTM in Chapter 2. Three new kinds of GTMs are discussed: (1) the hybrid TM, (2) the penalty plus hybrid TM, and (3) the multiplier TM. An *a priori* error analysis of GTMs is provided, to choose the optimal parameters used, and to derive the optimal exponential order of convergence rates. To study the stability, the condition number is replaced by the effective condition number to provide a better upper bound of relative errors for the TM solutions resulting from rounding errors. New computational formulas are derived for the effective condition number and its simplified forms. Numerical experiments are carried out for Motz's problem, to verify the error analysis. Comparisons are made to show that the CTM is the best among all TMs in accuracy, stability, and simplicity of the algorithm, although all TMs are efficient. On the other hand, the boundary element method (BEM) uses the fundamental solutions satisfying the PDEs. Both the GTMs and the BEM deal with only the boundary conditions. However, the GTMs are easier for handling the solutions with singularity and other complicated boundary conditions. Since proper coupling techniques for interior and exterior boundary conditions are imperative, the study of this chapter may enrich not only the Trefftz method (TM) with variant formulations, but also a wide range of other numerical methods.

3.1 Introduction

We consider Motz's problem that solves the Laplace equation on the rectangle $S=\{(x,y) \mid -1<x<1, 0<y<1\}$

$$\Delta u = \frac{\partial^2 u}{\partial x^2} + \frac{\partial^2 u}{\partial y^2} = 0 \quad \text{in } S, \tag{3.1.1}$$

with the mixed Dirichlet–Neumann boundary conditions, see fig. 2.1,

$$u\,|_{\overline{OD}} = 0, \qquad u\,|_{\overline{AB}} = 500, \tag{3.1.2}$$

$$\left.\frac{\partial u}{\partial y}\right|_{\overline{OA}} = \left.\frac{\partial u}{\partial y}\right|_{\overline{BC}} = \left.\frac{\partial u}{\partial x}\right|_{\overline{CD}} = 0. \tag{3.1.3}$$

In fact, the singular solutions of eqns. (3.1.1)–(3.1.3) are found as

$$u(r,\theta) = \sum_{i=0}^{\infty} d_i r^{i+\frac{1}{2}} \cos\left(i+\frac{1}{2}\right)\theta, \tag{3.1.4}$$

where d_i are the true expansion coefficients, and (r,θ) are the polar coordinates with the origin at $(0,0)$ (see fig. 2.1). Hence, the admissible functions of finite terms,

$$u_N(r,\theta) = \sum_{i=0}^{N} D_i r^{i+\frac{1}{2}} \cos\left(i+\frac{1}{2}\right)\theta \tag{3.1.5}$$

with the unknown coefficients D_i, are most efficient in matching Motz's solutions numerically, to yield the exponential convergence rates $O(e^{-cN})$, where c is a positive constant. When functions, i.e., eqn. (3.1.5) are chosen, the eqn. (3.1.1) and $u\,|_{\overline{OD}}=0$ and $\frac{\partial u}{\partial y}|_{\overline{OA}}=0$ are satisfied automatically. Then the coefficients D_i are sought by the collocation equations of the rest of boundary conditions in eqns. (3.1.2) and (3.1.3), which is called the TM in this book.

The highly accurate solution, obtained in double precision with the 35 leading coefficients, $D_0 - D_{34}$, was provided in [306]. It should be noted that an error at the power of D_{31} was discovered by Lucas and Oh [318]. This highly accurate solution can be regarded as a true solution for testing other numerical methods. In Chapter 2, the more accurate leading coefficients are achieved by the Gaussian rule. In Georgiou, Olson, and Smyrlis [161], the same singular functions as in Ref. [306] are chosen, but the boundary conditions are matched by Lagrange multiplier method. Here questions arise: Are there other matching techniques to deal with the boundary conditions, when the same particular solutions have been chosen? If yes, what are the error bounds of the approximate solutions by the new techniques? And how about their computational performance? To answer these questions is one of the objectives of this chapter.

In Li [280], a solution domain S is split into two subdomains S_1 and S_2; the particular solutions are used in just one subdomain S_2, and other numerical methods are used in S_1, such as the finite element method (FEM), the finite difference method (FDM), finite volume method (FVM), etc. The comprehensive coupling strategies are investigated therein. In this chapter, we employ these coupling techniques for matching different particular solutions in S_1 and S_2 along their interfaces $\Gamma_0 = \partial S_1 \cap \partial S_2$, which have not been discussed earlier. Four new TMs are explored in this chapter: (1) the penalty TM, (2) the hybrid TM, (3) the penalty plus hybrid TM, and (4) the Lagrange multiplier TM in Ref. [161]. Brief analysis is conducted to derive error bounds and to find optimal parameters used. Note that the conclusions drawn in this chapter are rather different from those in Ref. [280]. For instance, it is concluded in the present work that the penalty combination is quite efficient in matching the TM using the particular solutions and the FEM using piecewise polynomials of low order; the penalty TM, on the other hand, should not be chosen due to inferior accuracy of the solutions obtained. Moreover, since a worse instability occurs in the hybrid TM and the penalty plus hybrid TM, the leading coefficients $D_0 - D_3$ are less accurate than those by the CTM discussed in Chapter 2. In this chapter, if the straightforward CTM by the central rule or the Gaussian rule is used, the CTM is called. If the particular solutions of PDEs are chosen, and if the approximate solutions are obtained by satisfying only the boundary conditions, the GTMs are called.

In numerical computations for the PDE solutions, accuracy and stability of numerical solutions are two critical issues, which are related to each other.

The FEM provides good numerical stability, but low accuracy. Whenever feasible, we should first try one of the high-accuracy methods presented in this book. If the present methods do not apply, then methods, such as the FEM, or a coupling of the FEM with TMs should be used.

This chapter is organized as follows. In the next section, the GTMs are described. In Sections 3.3–3.5, the penalty TM, hybrid TM, and penalty plus hybrid TM are explored for Motz's problem, with a concise error analysis. In Section 3.6, a brief analysis is made for the Lagrange multiplier TM. In Section 3.7, formulas are derived for the effective condition numbers. In the last section, numerical experiments are carried out, and comparisons and conclusions are made. The materials of this chapter are adapted from Li et al. [289, 302].

3.2 Description of generalized TMs

Consider Poisson's equation

$$\Delta u = \frac{\partial^2 u}{\partial x^2} + \frac{\partial^2 u}{\partial y^2} = f \quad \text{in } S, \tag{3.2.1}$$

with the mixed type of boundary conditions

$$u = g_1 \quad \text{on } \Gamma_D, \tag{3.2.2}$$

$$\frac{\partial u}{\partial n} + qu = g_2 \quad \text{on } \Gamma_N, \tag{3.2.3}$$

where the domain S is bounded, simply connected polygon with the exterior boundary Γ, $\Gamma = \Gamma_D \cup \Gamma_N$, $\text{Meas}(\Gamma_D) > 0$, the functions q, g_1, and g_2 are sufficiently smooth, and $q = q(x,y) \geq 0$.

The solution of the problem, i.e., eqns. (3.2.1)–(3.2.3) can be equivalently expressed by minimizing an energy $I(v)$:

$$I(u) = \min_{v \in H_*^1(S)} I(v),$$

where the energy is given by

$$I(v) = \frac{1}{2} \iint_S |\nabla v|^2 \, ds + \frac{1}{2} \int_{\Gamma_N} q v^2 \, d\ell - \int_{\Gamma_N} g_2 v \, d\ell,$$

and $H_*^1(S)$ is the subset of the Sobolev space defined by

$$H_*^1(S) = \{v \mid v, v_x, v_y \in L^2(S), \quad \text{and } v \mid_{\Gamma_D} = g_1\}.$$

More precisely, the Sobolev space is only for the linear space, e.g., defined by

$$H_0^1(S) = \{v \mid v, v_x, v_y \in L^2(S), \quad \text{and } v \mid_{\Gamma} = 0\},$$

since the linear combinations must be in the same space. On the other hand, $H_*^1(S)$ is the space to $H_0^1(S)$ by a translation.

Let S be divided by Γ_0 into S_1 and S_2, see fig. 3.1. The Ritz–Galerkin method is used in both S_1 and S_2, with the admissible functions as

$$v = \begin{cases} v^- = \Phi_0 + \sum_{i=1}^{M} a_i \Phi_i & \text{in } S_1, \\ v^+ = \Psi_0 + \sum_{i=1}^{N} b_i \Psi_i & \text{in } S_2, \end{cases} \tag{3.2.4}$$

where

$$\Delta\Phi_0 = f, \quad \text{in } S_1, \qquad \Delta\Psi_0 = f, \quad \text{in } S_2,$$
$$\Delta\Phi_i = 0, \quad \text{in } S_1, \quad \text{for } i \geq 1, \qquad \Delta\Psi_i = 0, \quad \text{in } S_2, \quad \text{for } i \geq 1.$$

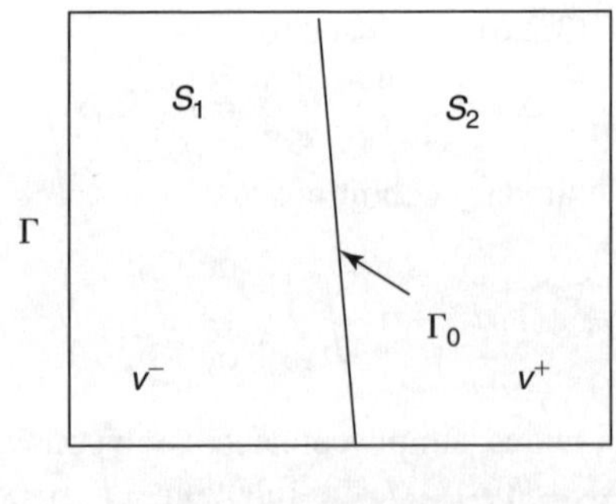

Figure 3.1: Partition of S.

The functions $\{\Phi_i\}$ and $\{\Psi_i\}$ are analytic, complete, and linearly independent basis functions in S_1 and S_2, respectively, and a_i and b_i are unknown coefficients to be determined.

Define a space

$$H = \{v \mid v \in L^2(S), v \in H^1(S_1), v \in H^1(S_2)\}.$$

Let $V_N^* \subset H$ be a finite-dimensional collection of eqn. (3.2.4). The TMs using the generalized coupling techniques are designed to seek an approximate solution $u_N \in V_N^*$ such that

$$I_h(u_N) = \min_{v \in V_N^*} I_h(v), \tag{3.2.5}$$

where the energy

$$\begin{aligned} I_h(v) = \frac{1}{2}\iint_{S_1} |\nabla v|^2\, ds + \frac{1}{2}\iint_{S_2} |\nabla v|^2\, ds + \frac{1}{2}\int_{\Gamma_N} qv^2\, d\ell \\ + w^2 \int_{\Gamma_D} (v - g_1)^2\, d\ell - \alpha^* \int_{\Gamma_D} \frac{\partial v}{\partial n}(v - g_1)\, d\ell \\ + w^2 \int_{\Gamma_0} (v^+ - v^-)^2\, d\ell - \int_{\Gamma_0} \left(\alpha \frac{\partial v^+}{\partial n} + \beta \frac{\partial v^-}{\partial n}\right)(v^+ - v^-)\, d\ell \\ - \int_{\Gamma_N} g_2 v\, d\ell - \iint_S f v, \end{aligned} \tag{3.2.6}$$

where $w(\geq 0)$ is a weight, n is the outward normal of ∂S_2 and ∂S, and α, α^*, and β are parameters. Since the solutions v^+ and v^- satisfy the Laplace equation already, we have from the Green formula

$$\iint_{S_1} |\nabla v^-|^2\, ds + \iint_{S_2} |\nabla v^+|^2\, ds = -\int_{\partial S_1} v^- \frac{\partial v^-}{\partial n}\, d\ell + \int_{\partial S_2} v^+ \frac{\partial v^+}{\partial n}\, d\ell.$$

Hence, the energy eqn. (3.2.6) is reduced to

$$\begin{aligned} I_h(v) = -\frac{1}{2}\int_{\partial S_1} v^- \frac{\partial v^-}{\partial n}\, d\ell + \frac{1}{2}\int_{\partial S_2} v^+ \frac{\partial v^+}{\partial n}\, d\ell + \frac{1}{2}\int_{\Gamma_N} qv^2\, d\ell \\ + w^2 \int_{\Gamma_D} (v - g_1)^2\, d\ell - \alpha^* \int_{\Gamma_D} \frac{\partial v}{\partial n}(v - g_1)\, d\ell + w^2 \int_{\Gamma_0} (v^+ - v^-)^2\, d\ell \\ - \int_{\Gamma_0} \left(\alpha \frac{\partial v^+}{\partial n} + \beta \frac{\partial v^-}{\partial n}\right)(v^+ - v^-)\, d\ell - \int_{\Gamma_N} g_2 v\, d\ell - \iint_S f v. \end{aligned} \tag{3.2.7}$$

Note that in eqn. (3.2.7), the integrals are involved only in the interior and exterior boundaries. In fact, the combinations as in eqn. (3.2.7) are general approaches using additional integrals, which contain the following three variants of TMs:

1. The Penalty TM: $w > 0$, $\alpha = \beta = 0$, and $\alpha^* = 0$.
2. The Simplified Hybrid TM: $w = 0$, $\alpha = 1, \beta = 0$, and $\alpha^* = 1$.
3. The Penalty plus Hybrid TM: $w > 0$, $\alpha = \beta = \frac{1}{2}$, and $\alpha^* = 1$.

In addition, the Lagrange multipliers method is given by Refs. [161, 280] as

$$B(\lambda, \mu; u, v) = \int_{\Gamma_N} g_2 v \, d\ell + \int_{\Gamma_D} \mu g_1 \, d\ell,$$

where

$$\begin{aligned} B(\lambda, \mu; u, v) &= \iint_{S_1} \nabla u \cdot \nabla v \, ds + \iint_{S_2} \nabla u \cdot \nabla v \, ds + D(\lambda, \mu; u, v) \\ &= -\int_{\partial S_1} \left(v^- \frac{\partial u}{\partial n} \right) d\ell + \int_{\partial S_2} \left(v^+ \frac{\partial u}{\partial n} \right) d\ell + D(\lambda, \mu; u, v), \end{aligned}$$

and the boundary integrals are given by

$$\begin{aligned} D(\lambda, \mu; u, v) = &\int_{\Gamma_N} quv \, d\ell - \int_{\Gamma_D} \lambda v \, d\ell - \int_{\Gamma_D} \mu u \, d\ell \\ &- \int_{\Gamma_0} \lambda (v^+ - v^-) \, d\ell - \int_{\Gamma_0} \mu (u^+ - u^-) \, d\ell. \end{aligned}$$

Note that the λ is treated as an extra variable.

3.3 Penalty TMs

We will study the case of $S = S_2$ and $S_1 = \Gamma_0 = \emptyset$, i.e., no division of S, and provide a brief analysis of the GTMs for the Motz's problem in the rest of this chapter. The analysis of the case the $\Gamma_0 \neq \emptyset$ for eqn. (3.2.4) can be similarly developed, see Li and Huang [293].

The admissible functions, i.e., eqn. (3.1.4) satisfy Laplace eqn. (3.1.1) and two boundary conditions $u|_{\overline{OD}} = 0$ and $\frac{\partial u}{\partial y}|_{\overline{OA}} = 0$ already, see fig. 2.1. We should choose the approximate coefficients D_i in eqn. (3.1.5) to satisfy the rest of the boundary conditions of eqns. (3.1.2)–(3.1.3) as best as possible. Therefore, the optimal solution u_N can be sought by the penalty TMs, which is given from eqn. (3.2.7) by noting $S = S_2$, $S_1 = \Gamma_0 = \emptyset$, $\alpha^* = 0$, $f = q = g_2 = 0$, and $g_1 = 500$. We seek $\widetilde{u}_N^P$ from

$$I_h(\widetilde{u}_N^P) = \min_{v \in V_N} I_h(v),$$

where

$$I_h(v) = \frac{1}{2} \int_{\overline{AB} \cup \overline{BC} \cup \overline{CD}} v \frac{\partial v}{\partial n} \, d\ell + w^2 \int_{\overline{AB}} (v - 500)^2 \, d\ell,$$

in which V_N is a finite-dimensional collection of the admissible functions in eqn. (3.1.5) and $w > 0$ is a weight parameter.

The solutions of the minimal energy is given by

$$T_1^*(\widetilde{u}_N^P) = \min_{v \in V_N} T_1^*(v), \tag{3.3.1}$$

where

$$T_1^*(v) = \int_{\Gamma_N^*} v \frac{\partial v}{\partial n} \, d\ell + w^2 \int_{\overline{AB}} (v - 500)^2 \, d\ell,$$

and $\Gamma_N^* = \overline{AB} \cup \overline{BC} \cup \overline{CD}$.

In this section, we use the equivalence notations $a \asymp b$ or $a \asymp O(b)$ of a and $b(>0)$, to indicate that there exist two constants $C_1(>0)$ and $C_2(>0)$ such that

$$C_1 b \le |a| \le C_2 b, \quad b > 0.$$

From the theory of the Sobolev space, we cite the following lemma from Sobolev [417].

Lemma 3.3.1

For Meas$(\Gamma_D) \neq 0$*, there exists an equivalence relation:*

$$|v|_{1,S}^2 + \|v\|_{0,\Gamma_D}^2 \asymp \|v\|_{1,S}^2.$$

Define a new norm

$$\|v\|_H = \left\{ \int_{\Gamma_N^*} v \frac{\partial v}{\partial n} \, d\ell + w^2 \int_{\overline{AB}} v^2 \, d\ell \right\}^{\frac{1}{2}}, \quad w \ge 1. \tag{3.3.2}$$

Since $\int_{\Gamma_N^*} v \frac{\partial v}{\partial n} \, d\ell = |v|_{1,S}^2, \forall v \in V_N$, we have the following lemma.

Lemma 3.3.2

Let Meas$(\Gamma_D) \neq 0$ *and* $w = 1$*,*

$$\|v\|_H \asymp \|v\|_{1,S}, \quad \forall v \in V_N.$$

Lemma 3.3.3

Let Meas$(\Gamma_D) \neq 0$ *and* $w \geq 1$, *there exist the bounds for* $\forall v \in V_N$,

$$C_1 \|v\|_{1,S} \leq \|v\|_H \leq C_2 w \|v\|_{1,S}, \tag{3.3.3}$$

where C_1 *and* C_2 *are two constants independent of* v.

Proof.

Since for $w \geq 1$, we have from Lemma 3.3.1,

$$\|v\|_H^2 \geq |v|_{1,S}^2 + \|v\|_{0,\Gamma_D}^2 \geq C_0 \|v\|_{1,S}^2,$$

and from Lemma 3.3.2

$$\|v\|_H^2 = w^2 \left\{ \frac{1}{w^2} |v|_{1,S}^2 + \|v\|_{0,\Gamma_D}^2 \right\} \leq w^2 \{ |v|_{1,S}^2 + \|v\|_{0,\Gamma_D}^2 \} \leq C w^2 \|v\|_{1,S}^2.$$

The desired results, i.e., eqn. (3.3.3) follow. ■

Next let us give an *a priori* error analysis.

Theorem 3.3.1

The solution $\widetilde{u}_N^P$ *from eqn. (3.3.1), the penalty TM, has the error bound,*

$$\|u - \widetilde{u}_N^P\|_H \leq C \left(\inf_{v \in V_N} \|u - v\|_H + \frac{1}{w} \left\| \frac{\partial u}{\partial n} \right\|_{0,\overline{AB}} \right).$$

Proof.

We may rewrite the penalty TM in eqn. (3.3.1) as

$$B_1(\widetilde{u}_N^P, v) = f_1(v), \quad \forall v \in V_N, \tag{3.3.4}$$

where

$$B_1(u, v) = \int_{\Gamma_N^*} \frac{\partial u}{\partial n} v \, d\ell + w^2 \int_{\overline{AB}} uv \, d\ell,$$

$$f_1(v) = 500 w^2 \int_{\overline{AB}} v \, d\ell.$$

Since the true solution u of Motz's problem satisfies

$$B_1(u,v) = \int_{\overline{AB}} \frac{\partial u}{\partial n} v \, d\ell + f_1(v),$$

we obtain from eqn. (3.3.4)

$$B_1(u - \widetilde{u}_N^P, v) = \int_{\overline{AB}} \frac{\partial u}{\partial n} v \, d\ell. \tag{3.3.5}$$

Let $\delta = \widetilde{u}_N^P - v$, $v \in V_N$ and then $\delta \in V_N$. From Lemma 3.3.3, eqns. (3.3.2) and (3.3.5), and the bound $|B(u,v)| \le C\|u\|_H \|v\|_H$, we have

$$\begin{aligned} \|\delta\|_H^2 = B_1(\widetilde{u}_N^P - v, \delta) &= B_1(u - v, \delta) - \int_{\overline{AB}} \frac{\partial u}{\partial n} \delta \, d\ell \\ &\le C\|u - v\|_H \|\delta\|_H + \left\| \frac{\partial u}{\partial n} \right\|_{0,\overline{AB}} \|\delta\|_{0,\overline{AB}} \\ &\le C\left(\|u - v\|_H + \frac{1}{w} \left\| \frac{\partial u}{\partial n} \right\|_{0,\overline{AB}} \right) \|\delta\|_H. \end{aligned} \tag{3.3.6}$$

Dividing two sides of eqn. (3.3.6) by $\|\delta\|_H$ gives

$$\|\delta\|_H \le C\left(\|u - v\|_H + \frac{1}{w} \left\| \frac{\partial u}{\partial n} \right\|_{0,\overline{AB}} \right).$$

Then, we have

$$\begin{aligned} \|u - \widetilde{u}_N^P\|_H &\le \|u - v\|_H + \|v - \widetilde{u}_N^P\|_H \\ &\le C\left(\|u - v\|_H + \frac{1}{w} \left\| \frac{\partial u}{\partial n} \right\|_{0,\overline{AB}} \right). \end{aligned}$$

■

Choose

$$v = u_N = \sum_{i=0}^{N} d_i r^{i+\frac{1}{2}} \cos\left(i + \frac{1}{2}\right)\theta,$$

where d_i are the true coefficients. Let $u = u_N + r_N$, where the remainder

$$r_N = \sum_{i=N+1}^{\infty} d_i r^{i+\frac{1}{2}} \cos\left(i + \frac{1}{2}\right)\theta.$$

We have the following lemma.

Lemma 3.3.4

There exists the bound for the remainder,

$$\|r_N\|_H \leq \|r_N\|_{0,\Gamma_N^*}^{\frac{1}{2}} \left\| \frac{\partial r_N}{\partial n} \right\|_{0,\Gamma_N^*}^{\frac{1}{2}} + w\|r_N\|_{0,\overline{AB}}. \tag{3.3.7}$$

Proof.

We have

$$\|r_N\|_H^2 \leq \|r_N\|_{0,\Gamma_N^*} \left\| \frac{\partial r_N}{\partial n} \right\|_{0,\Gamma_N^*} + w^2\|r_N\|_{0,\overline{AB}}^2.$$

The desired results, i.e., eqn. (3.3.7) are obtained from $\sqrt{a^2+b^2} \leq |a|+|b|$. ■

Corollary 3.3.1

Suppose that for $0 < a < 1$,

$$\|r_N\|_{0,\overline{AB}} \leq Ca^N, \qquad \|r_N\|_{0,\Gamma_N^*} \leq Ca^N,$$

$$\left\| \frac{\partial r_N}{\partial n} \right\|_{0,\Gamma_N^*} \leq CNa^N.$$

Then, when the optimal parameter $w = a^{-\frac{N}{2}}$, *the error of the numerical solution* $\widetilde{u}_N^P$ *by the penalty TM has the bound:*

$$\|u - \widetilde{u}_N^P\|_H \leq C\sqrt{N}a^{\frac{N}{2}}. \tag{3.3.8}$$

Proof.

Let $v = u_N$. Then from $u = u_N + r_N$, Theorem 3.3.1, and Lemma 3.3.4, we have

$$\|u - \widetilde{u}_N^P\|_H \leq C\left(\inf_{v\in V_N} \|u-v\|_H + \frac{1}{w}\left\| \frac{\partial u}{\partial n} \right\|_{0,\overline{AB}} \right) \leq C\left(\|r_N\|_H + \frac{1}{w}\left\| \frac{\partial u}{\partial n} \right\|_{0,\overline{AB}} \right)$$

$$\leq C\left\{ \|r_N\|_{0,\Gamma_N^*}^{\frac{1}{2}} \left\| \frac{\partial r_N}{\partial n} \right\|_{0,\Gamma_N^*}^{\frac{1}{2}} + w\|r_N\|_{0,\overline{AB}} + \frac{1}{w}\left\| \frac{\partial u}{\partial n} \right\|_{0,\overline{AB}} \right\}$$

$$\leq C\left(\sqrt{N}a^N + wa^N + \frac{1}{w} \right).$$

The optimal parameter $w = O(a^{-\frac{N}{2}})$ and the desired result, i.e., eqn. (3.3.8) is obtained. ■

It is worth noting that the huge parameter $w = O(a^{-\frac{N}{2}})$ damages stability and lowers the convergence rates down to $O(\sqrt{N}a^{\frac{N}{2}})$, although they are still exponential. Below, we will pursue better TMs, to lead to optimal convergence rates and better stability.

3.4 Simplified hybrid TMs

In this section, we consider the hybrid TM for Motz's problem in fig. 2.1. Denote

$$H_0^1(S) = \left\{ v \,\middle|\, v, v_x, v_y \in L^2(S), \quad v|_{\overline{OD}} = 0, \ \left.\frac{\partial v}{\partial n}\right|_{\overline{OA}} = 0 \right\}. \tag{3.4.1}$$

We seek $u \in H_0^1(S)$ such that

$$\iint_S \nabla u \cdot \nabla v \, ds - \int_{\overline{AB}} \frac{\partial u}{\partial n} v \, d\ell + \int_{\overline{AB}} \frac{\partial v}{\partial n}(u - 500) \, d\ell = 0, \quad \forall v \in H_0^1(S). \tag{3.4.2}$$

From the Green Theorem, eqn. (3.4.2) is reduced to

$$\int_{\overline{BC} \cup \overline{CD}} \frac{\partial u}{\partial n} v \, d\ell + \int_{\overline{AB}} \frac{\partial v}{\partial n}(u - 500) \, d\ell = 0. \tag{3.4.3}$$

Let V_N be the space of functions v in eqn. (3.1.5). Then, the hybrid TM is designed to seek $\widetilde{u}_N^H \in V_N$ such that

$$\int_{\overline{BC} \cup \overline{CD}} \frac{\partial \widetilde{u}_N^H}{\partial n} v \, d\ell + \int_{\overline{AB}} \frac{\partial v}{\partial n}(\widetilde{u}_N^H - 500) \, d\ell = 0, \quad \forall v \in V_N. \tag{3.4.4}$$

A relation between eqns. (3.4.4) and (3.2.7) is explored in Li [280]. We may also rewrite the above equation as: To seek $\widetilde{u}_N^H \in V_N$ such that

$$B_2(\widetilde{u}_N^H, v) = f_2(v), \quad \forall v \in V_N, \tag{3.4.5}$$

where

$$B_2(u, v) = \int_{\overline{BC} \cup \overline{CD}} \frac{\partial u}{\partial n} v d\ell + \int_{\overline{AB}} u \frac{\partial v}{\partial n} \, d\ell,$$

and

$$f_2(v) = 500 \int_{\overline{AB}} \frac{\partial v}{\partial n} \, d\ell.$$

We have the following theorem.

Theorem 3.4.1

For the solutions by the hybrid TM, there exists the error bound,

$$|u - \widetilde{u}_N^H|_{1,S} \le 2 \inf_{v \in V_N} |u - v|_{1,S}. \tag{3.4.6}$$

Proof.

First we note from eqn. (3.4.2)

$$B_2(v, v) = \int_{\Gamma_N^*} \frac{\partial v}{\partial n} v \, d\ell = \iint_S |\nabla v|^2 \, ds = |v|_{1,S}^2.$$

Also it follows from eqns. (3.4.3) and (3.4.4)

$$B_2(u - \widetilde{u}_N^H, v) = 0, \quad \forall v \in V_N, \tag{3.4.7}$$

where u is the true solution of Motz's problem. Let $\delta = \widetilde{u}_N^H - v$ for $v \in V_N$, then $\delta \in V_N$. We have from eqn. (3.4.7)

$$\begin{aligned} |\delta|_{1,S}^2 &= B_2(\widetilde{u}_N^H - v, \delta) = B_2(u - v, \delta) \\ &\le \{B_2(u - v, u - v) B_2(\delta, \delta)\}^{\frac{1}{2}} = |u - v|_{1,S} |\delta|_{1,S}. \end{aligned}$$

This leads to $|\delta|_{1,S} \le |u - v|_{1,S}$. Therefore, for $v \in V_N$, we obtain

$$|u - \widetilde{u}_N^H|_{1,S} \le |u - v|_{1,S} + |\delta|_{1,S} \le 2|u - v|_{1,S}.$$

The desired result, i.e., eqn. (3.4.6) is obtained. ■

By following the analysis in Section 3.3, we have the following corollary.

Corollary 3.4.1

Let the conditions in Corollary 3.3.1 hold. Then there exists the bound

$$\|u - \widetilde{u}_N^H\|_{1,S} \le C\sqrt{N} a^N. \tag{3.4.8}$$

Compared to the error bounds $O(\sqrt{N} a^{\frac{N}{2}})$ by the penalty TM in Section 3.3, the convergence rate in eqn. (3.4.8) by the hybrid TM is higher.

Let us consider the integration approximation in eqn. (3.4.4). The hybrid TM involving integral approximation is designed to seek $\hat{u}_N^H \in V_N$ such that

$$\widetilde{\int}_{\overline{BC}\cup\overline{CD}} \frac{\partial \hat{u}_N^H}{\partial n} v \, d\ell + \widetilde{\int}_{\overline{AB}} \frac{\partial v}{\partial n} (\hat{u}_N^H - 500) \, d\ell = 0, \quad \forall v \in V_N, \tag{3.4.9}$$

where $\widetilde{\int}$ is an approximation of $\int$ by some rules. We have the following theorem.

Theorem 3.4.2

For the solutions by the hybrid TM, there exists the error bound,

$$|u - \hat{u}_N^H|_{1,S} \le C \inf_{v \in V_N} |u - v|_{1,S} + C \sup_{v \in V_N} \frac{1}{\|v\|_{1,S}} \times \left\{ \left| \left(\int_{\overline{BC}\cup\overline{CD}} - \widetilde{\int}_{\overline{BC}\cup\overline{CD}} \right) \frac{\partial u}{\partial n} v \, d\ell \right| + \left| \left(\int_{\overline{AB}} - \widetilde{\int}_{\overline{AB}} \right) \frac{\partial v}{\partial n} u \, d\ell \right| \right\}. \tag{3.4.10}$$

The proof of Theorem 3.4.2 can be completed by following the analysis in Ref. [293]. Note that the additional errors from the integration approximation in eqn. (3.4.10) are analogous to those in FEM, but not to the CTM in Section 2.3, where the integration errors are fairly small to guarantee the uniformly V_N elliptic inequality. Since the integration rules are formulated based on k-order polynomials, such as the Gaussian and Newton–Cotes rules in Ref. [9], only the polynomial convergence rates can be obtained from eqn. (3.4.10).

3.5 Penalty plus hybrid TMs

The numerical solutions $\widetilde{u}_N^{HP}$ may be sought by minimizing the following energy:

$$T_3^*(\widetilde{u}_N^{HP}) = \min_{v \in V_N} T_3^*(v), \tag{3.5.1}$$

where

$$T_3^*(v) = \frac{1}{2} \int_{\Gamma_N^*} \frac{\partial v}{\partial n} v \, d\ell - \alpha \int_{\overline{AB}} \frac{\partial v}{\partial n} (v - 500) \, d\ell + w^2 \int_{\overline{AB}} (v - 500)^2 \, d\ell, \tag{3.5.2}$$

in which $0 \le \alpha \le 1$ and $w \ge 0$. The eqns. (3.5.1) and (3.5.2) are obtained from eqns. (3.2.5) and (3.2.7) by choosing $S = S_2, S_1 = \Gamma_0 = \emptyset, f = q = g_2 = 0$, and $g_1 = 500$.

Below, let us consider how to choose two parameters α and w. First, w^2 should be chosen to balance the first and the second terms on the right-hand side in eqn. (3.5.2). For the simple case of $S = S_R$ as in Lemma 3.5.2 below, it is better to choose $w^2 = O(N+1)$, i.e., $w^2 = P_c(N+1)$, where P_c is a constant independent of N. Also we shall find the optimal choice $\alpha = 1$, by means of *a priori* error estimates given in Theorem 3.5.1.

For simplicity, choose a semi-disk $S_R = \{(r,\theta) \mid 0 \le r \le R, 0 \le \theta \le \pi\}$, and a semi-circle $l_R = \{(r,\theta) \mid r = R, 0 \le \theta \le \pi\}$. We have the following lemma from the orthogonality of $\cos(i+\frac{1}{2})\theta$ on l_R.

Lemma 3.5.1

There exist the equalities for $v = \sum_{i=0}^{N} D_i r^{i+\frac{1}{2}} \cos(i+\frac{1}{2})\theta$ *with arbitrary* D_i,

$$\int_{l_R} v^2\, d\ell = \frac{\pi}{2} \sum_{i=0}^{N} D_i^2 R^{2i+2},$$

$$\int_{l_R} v \frac{\partial v}{\partial n}\, d\ell = \frac{\pi}{2} \sum_{i=0}^{N} D_i^2 \left(i+\frac{1}{2}\right) R^{2i+1},$$

$$\int_{l_R} \left(\frac{\partial v}{\partial n}\right)^2 d\ell = \frac{\pi}{2} \sum_{i=0}^{N} D_i^2 \left(i+\frac{1}{2}\right)^2 R^{2i},$$

$$\iint_{S_R} v^2\, ds = \frac{\pi}{2} \sum_{i=0}^{N} \frac{D_i^2}{2i+3} R^{2i+3}.$$

Now, we prove a lemma.

Lemma 3.5.2

For $S = S_R$ *and there exists the bound for* $v \in V_N$

$$\int_{l_R} v \frac{\partial v}{\partial n}\, d\ell \le \frac{N+1}{R} \int_{l_R} v^2\, d\ell.$$

Proof.

From Lemma 3.5.1, we have

$$\int_{l_R} v \frac{\partial v}{\partial n}\, d\ell = \frac{\pi}{2} \sum_{i=0}^{N} D_i^2 \left(i+\frac{1}{2}\right) R^{2i+1}$$

$$\le \frac{(N+1)}{R} \frac{\pi}{2} \sum_{i=0}^{N} D_i^2 R^{2i+2} = \frac{(N+1)}{R} \int_{l_R} v^2\, d\ell. \qquad \blacksquare$$

Based on Lemmas 3.5.1 and 3.5.2 we may choose the parameter $w^2 = P_c(N+1)$, where P_c is a positive constant but independent of N as shown in Lemma 3.5.3 below. The eqn. (3.5.1) may be rewritten as: To seek $\tilde{u}_N^{PH} \in V_N$ such that

$$B_3(\tilde{u}_N^{PH}, v) = f_3(v), \quad \forall v \in V_N, \tag{3.5.3}$$

where

$$B_3(u,v) = \int_{\Gamma_N^*} \frac{\partial u}{\partial n} v\, d\ell - \alpha \int_{\overline{AB}} \left(\frac{\partial u}{\partial n} v + \frac{\partial v}{\partial n} u \right) d\ell + P_c(N+1) \int_{\overline{AB}} uv\, d\ell$$

$$= \int_{\overline{BC} \cup \overline{CD}} \frac{\partial u}{\partial n} v\, d\ell + (1-\alpha) \int_{\overline{AB}} \frac{\partial u}{\partial n} v\, d\ell - \alpha \int_{\overline{AB}} \frac{\partial v}{\partial n} u\, d\ell$$

$$+ P_c(N+1) \int_{\overline{AB}} uv\, d\ell,$$

and

$$f_3(v) = 500 P_c(N+1) \int_{\overline{AB}} v\, d\ell - 500\alpha \int_{\overline{AB}} \frac{\partial v}{\partial n} d\ell.$$

Define a norm

$$\|v\|_h = \{|v|_{1,S}^2 + P_c(N+1)\|v\|_{0,\overline{AB}}^2\}^{\frac{1}{2}}. \tag{3.5.4}$$

We have the following lemma.

Lemma 3.5.3

Suppose that there exists a constant C independent of N such that

$$\left\| \frac{\partial v}{\partial n} \right\|_{0,\overline{AB}} \leq C(N+1)\|v\|_{0,\overline{AB}}, \quad \forall v \in V_N. \tag{3.5.5}$$

Then, when $0 < \alpha \leq 1$ and P_c is chosen to be large enough but independent of N, there exists the uniformly V_N-elliptic inequality,

$$B_3(v,v) \geq \frac{1}{2}\|v\|_h^2, \quad \forall v \in V_N, \tag{3.5.6}$$

and

$$B_3(u,v) \leq C\|u\|_h \|v\|_h,$$

where C is a constant independent of N.

Proof.

We have

$$B_3(v,v) = \iint_S |\nabla v|^2\, ds - 2\alpha \int_{\overline{AB}} \frac{\partial v}{\partial n} v\, d\ell + P_c(N+1) \int_{\overline{AB}} v^2\, d\ell.$$

From eqn. (3.5.5),

$$\int_{\overline{AB}} \frac{\partial v}{\partial n} v \, d\ell \le \left\| \frac{\partial v}{\partial n} \right\|_{0,\overline{AB}} \|v\|_{0,\overline{AB}} \le C(N+1)\|v\|^2_{0,\overline{AB}},$$

where C is a constant independent of N. Then, we obtain

$$B_3(v,v) \ge \iint_S |\nabla v|^2 \, ds + (P_c - 2C\alpha)(N+1)\|v\|^2_{0,\overline{AB}}.$$

When P_c is chosen large enough to satisfy $P_c - 2C\alpha \ge \frac{P_c}{2}$, i.e., $P_c \ge 4C\alpha$, we have

$$B_3(v,v) \ge \iint_S |\nabla v|^2 \, ds + \frac{P_c}{2}(N+1)\|v\|^2_{0,\overline{AB}} \ge \frac{1}{2}\|v\|^2_h.$$

This is eqn. (3.5.6).

Next, we have similarly,

$$\begin{aligned} B_3(u,v) &= \iint_S \nabla u \cdot \nabla v \, ds - \alpha \int_{\overline{AB}} \left(\frac{\partial u}{\partial n} v + \frac{\partial v}{\partial n} u \right) d\ell + P_c(N+1) \int_{\overline{AB}} uv \, d\ell \\ &\le \left| \iint_S \nabla u \cdot \nabla v \, ds \right| + \alpha \left(\left| \int_{\overline{AB}} \frac{\partial u}{\partial n} v \, d\ell \right| + \left| \int_{\overline{AB}} \frac{\partial v}{\partial n} u \, d\ell \right| \right) \\ &\quad + P_c(N+1) \left| \int_{\overline{AB}} uv \, d\ell \right|. \end{aligned}$$

Moreover,

$$\left| \int_{\overline{AB}} \frac{\partial u}{\partial n} v \, d\ell \right| \le \left\| \frac{\partial u}{\partial n} \right\|_{-\frac{1}{2},\overline{AB}} \|v\|_{\frac{1}{2},\overline{AB}}, \tag{3.5.7}$$

where the norms

$$\|v\|_{\frac{1}{2},\Gamma} = \left\{ \|v\|^2_{0,\Gamma} + \int_\Gamma \int_\Gamma \frac{(v(P) - v(Q))^2}{\|P - Q\|^2} \, d\ell(P) \, d\ell(Q) \right\}^{\frac{1}{2}},$$

$$\|u\|_{-\frac{1}{2},\Gamma} = \sup_{v \ne 0} \frac{|\int_\Gamma uv \, d\ell|}{\|v\|_{\frac{1}{2},\Gamma}}.$$

Since the Laplacian solutions have the following bounds,

$$\left\| \frac{\partial u}{\partial n} \right\|_{-\frac{1}{2},\overline{AB}} \le \left\| \frac{\partial u}{\partial n} \right\|_{-\frac{1}{2},\partial S} \le C\|u\|_{1,S} \le C_1|u|_{1,S} \le C_1\|u\|_h, \tag{3.5.8}$$

and

$$\|v\|_{\frac{1}{2},\overline{AB}} \le C\|v\|_{1,S} \le C_1\|v\|_h, \tag{3.5.9}$$

where C_1 and C are also constants independent of N. Combining eqns. (3.5.7)–(3.5.9) gives

$$\left|\int_{\overline{AB}} \frac{\partial u}{\partial n} v\, d\ell\right| \le C\|u\|_h\|v\|_h.$$

Similarly,

$$\left|\int_{\overline{AB}} \frac{\partial v}{\partial n} u\, d\ell\right| \le C\|u\|_h\|v\|_h.$$

Therefore, we obtain from the Schwarz inequality

$$\begin{aligned} B_3(u,v) &\le \left|\iint_S \nabla u \cdot \nabla v\, ds\right| + C\alpha\|u\|_h\|v\|_h + P_c(N+1)\left|\int_{\overline{AB}} uv\, d\ell\right| \\ &\le (1+C\alpha)\|u\|_h\|v\|_h \le C\|u\|_h\|v\|_h. \end{aligned}$$

■

Theorem 3.5.1

Let the condition, i.e., eqn. (3.5.5) hold. Then there exists the error bound for $\widetilde{u}_N^{PH}$

$$\|u - \widetilde{u}_N^{PH}\|_h \le C\left\{\inf_{v\in V_N}\|u-v\|_h + \frac{|1-\alpha|}{\sqrt{P_c(N+1)}}\left\|\frac{\partial u}{\partial n}\right\|_{0,\overline{AB}}\right\}. \tag{3.5.10}$$

Proof.

For the true solution of Motz's problems satisfying eqns. (3.1.1) and (3.1.3), we have from eqn. (3.5.3)

$$B_3(u,v) = (1-\alpha)\int_{\overline{AB}} \frac{\partial u}{\partial n} v\, d\ell + f_3(v),$$

and then from eqn. (3.5.4)

$$\begin{aligned} B_3(u-\widetilde{u}_N^{PH}, v) &= (1-\alpha)\int_{\overline{AB}} \frac{\partial u}{\partial n} v\, d\ell \\ &\le |1-\alpha|\left\|\frac{\partial u}{\partial n}\right\|_{0,\overline{AB}}\|v\|_{0,\overline{AB}} \le C\frac{|1-\alpha|}{\sqrt{P_c(N+1)}}\left\|\frac{\partial u}{\partial n}\right\|_{0,\overline{AB}}\|v\|_h. \end{aligned}$$

For $\delta = v - \widetilde{u}_N^{PH}$ with $v \in V_N$ and $\delta \in V_N$, we have from Lemma 3.5.3

$$\begin{aligned}
\frac{1}{2}\|\delta\|_h^2 &\le B_3(v - \widetilde{u}_N^{PH}, \delta) \\
&= B_3(v - u, \delta) + (1-\alpha)\left(\int_{\overline{AB}} \frac{\partial u}{\partial n}\delta\, d\ell\right) \\
&\le C\left\{\|u - v\|_h \|\delta\|_h + \frac{|1-\alpha|}{\sqrt{P_c(N+1)}}\left\|\frac{\partial u}{\partial n}\right\|_{0,\overline{AB}} \|\delta\|_h\right\}.
\end{aligned}$$

Therefore, we have

$$\|\delta\|_h \le C\left\{\|u - v\|_h + \frac{|1-\alpha|}{\sqrt{P_c(N+1)}}\left\|\frac{\partial u}{\partial n}\right\|_{0,\overline{AB}}\right\},$$

and

$$\|u - \widetilde{u}_N^{PH}\|_h \le \|u - v\|_h + \|\delta\|_h \le C\left\{\|u - v\|_h + \frac{|1-\alpha|}{\sqrt{P_c(N+1)}}\left\|\frac{\partial u}{\partial n}\right\|_{0,\overline{AB}}\right\}.$$

The desired result, i.e., eqn. (3.5.10) is obtained. ■

Based on Theorem 3.5.1, we should choose $\alpha = 1$, to raise the accuracy of the solutions. In this case, the penalty plus hybrid TM, i.e., eqn. (3.5.3) is reduced to

$$B_3(\widetilde{u}_N^{PH}, v) = f_3(v), \tag{3.5.11}$$

where

$$B_3(u, v) = \int_{\overline{BC}\cup\overline{CD}} \frac{\partial u}{\partial n} v\, d\ell - \int_{\overline{AB}} \frac{\partial v}{\partial n} u\, d\ell + P_c(N+1)\int_{\overline{AB}} uv\, d\ell,$$

and

$$f_3(v) = 500 P_c(N+1)\int_{\overline{AB}} v\, d\ell - 500\int_{\overline{AB}} \frac{\partial v}{\partial n}\, d\ell,$$

where P_c is large enough but independent of N. For Motz's problem, $P_c = 1$ is a good choice by trial computation.

Theorem 3.5.2

Let eqn. (3.5.5) and the conditions of Corollary 3.3.1 hold, $\alpha = 1$ and $w^2 = P_c(N+1)$, where P_c is chosen large enough but independent of N. Then there exists the error bound for $\widetilde{u}_N^{PH}$

$$\|u - \widetilde{u}_N^{PH}\|_h \le C \inf_{v \in V_N} \|u - v\|_h = C_1\sqrt{N}a^N,$$

where C_1 is also a constant independent of N.

Remark 3.5.1

From the analysis in Sections 3.3–3.5, the penalty TM is less efficient. Such a conclusion is different from Ref. [280], where the penalty combination is quite efficient. Note that when $\alpha < 1$ is chosen in eqn. (3.5.3), the same reduced convergence rate as that of the penalty TM in Section 3.3 is obtained, based on Theorem 3.5.1. Both the hybrid TM and the penalty plus hybrid TM with $\alpha = 1$ are efficient for singularity problems.

3.6 Lagrange multiplier TM

For Motz's problem, we may regard the Dirichlet condition $u\big|_{\overline{AB}} = 500$ as a constraint, to minimize the energy, $\frac{1}{2}\iint_S |\nabla v|^2\, ds$, and then define a functional

$$I(v) = \frac{1}{2}\iint_S |\nabla v|^2\, ds - \int_{\overline{AB}} \lambda(v - 500)\, d\ell, \tag{3.6.1}$$

with the Lagrange multiplier λ. The critical point of eqn. (3.6.1) is given by: To seek $(u, \lambda) \in H_0^1(S) \times H^{-\frac{1}{2}}(\overline{AB})$ such that

$$\iint_S \nabla u \cdot \nabla v\, ds - \int_{\overline{AB}} \lambda v\, d\ell - \int_{\overline{AB}} \mu(u - 500)\, d\ell = 0, \tag{3.6.2}$$

$$\forall (v, \mu) \in H_0^1(S) \times H^{-\frac{1}{2}}(\overline{AB}),$$

where $H_0^1(S)$ is defined in eqn. (3.4.1).

Let λ_L be the L-order polynomials on $\overline{AB}$, which can be expressed as the Chebyshev polynomials,

$$\lambda_L = \sum_{i=0}^{L} A_i T_i(1 - 2y), \quad 0 \le y \le 1, \tag{3.6.3}$$

where A_i are the coefficients to be sought, and the Chebyshev polynomials are defined by

$$T_i(x) = \cos(i \arccos(x)), \quad -1 \le x \le 1. \tag{3.6.4}$$

Denote by $V_N \times T_L$ the collection of finite dimensions of eqns. (3.1.5) and (3.6.3). The discrete Lagrange multiplier method (i.e., the direct TM) is given by: To seek $(\widetilde{u}_N, \lambda_L) \in V_N \times T_L$ such that

$$A(\widetilde{u}_N, v) + b(\widetilde{u}_N, v; \lambda_L, \mu) = 0, \quad \forall (v, \mu) \in V_N \times T_L,$$

$$A(u, v) = \iint_S \nabla u \cdot \nabla v\, ds,$$

$$b(u, v; \lambda, \mu) = -\int_{\overline{AB}} \lambda v\, d\ell - \int_{\overline{AB}} \mu(u - 500)\, d\ell.$$

Below, we give a brief justification for Laplace's equation only. In contrast, the analysis of the multiplier methods in Babuska [14] and Li [280] is made for the equation, $-\Delta u + u = 0$.

We obtain a theorem from Ref. [280].

Theorem 3.6.1

Suppose the following three assumptions hold.
(A1) *For $A(u, v)$, there exist two positive constants C_0 and C independent of N such that*

$$C_0 \|v\|_{1,S}^2 \le A(v, v), \qquad \forall v \in V_N, \tag{3.6.5}$$

$$|A(u, v)| \le C\|u\|_{1,S}\|v\|_{1,S}, \quad \forall v \in V_N.$$

(A2) *For $\int_{\overline{AB}} \mu v \, d\ell$, there exists the Ladyzhenskaya–Babuska–Brezzi (LBB) condition: $\forall \mu_L \in T_L$, $\exists v_N \in V_N$, $v_N \neq 0$ such that*

$$\left| \int_{\overline{AB}} \mu_L v_N \, d\ell \right| \ge \beta \|v_N\|_{1,S} \|\mu_L\|_{-\frac{1}{2},\overline{AB}}, \tag{3.6.6}$$

where $\beta > 0$ is a constant independent of N and L.
(A3) *Also the following bound holds*

$$\left| \int_{\overline{AB}} \lambda v \, d\ell \right| \le C\|\lambda\|_{-\frac{1}{2},\overline{AB}} \|v\|_{1,S}, \quad \forall v \in V_N.$$

Then, there exist the error bounds,

$$\|u - \tilde{u}_N^L\|_{1,S} \le C\{ \inf_{v \in V_N} \|u - v\|_{1,S} + \inf_{\eta \in T_L} \|\lambda - \eta\|_{-\frac{1}{2},\overline{AB}}\},$$

$$\|\lambda - \lambda_L\|_{-\frac{1}{2},\overline{AB}} \le C\{ \inf_{v \in V_N} \|u - v\|_{1,S} + \inf_{\eta \in T_L} \|\lambda - \eta\|_{-\frac{1}{2},\overline{AB}}\},$$

where C is a constant independent of N and L.

Here, we only check the condition, i.e., eqn. (3.6.5) and the LBB condition. We have

$$A(v, v) = |v|_{1,S}^2. \tag{3.6.7}$$

Since $v \,|_{\overline{OD}} = 0$, we obtain from the Poincare–Friedrichs inequality in Ciarlet [103],

$$|v|_{1,S} \ge C_0 \|v\|_{1,S}, \quad \forall v \in H_0^1(S). \tag{3.6.8}$$

Combining eqns. (3.6.7) and (3.6.8) gives eqn. (3.6.5).

Next, we verify the LBB condition. First we consider an auxiliary problem of Motz's problem, see fig. 2.1,

$$\begin{aligned} &\Delta w = 0, \quad \text{in } S, \\ &\left.\frac{\partial w}{\partial n}\right|_{\overline{AB}} = \mu_L, \\ &w\big|_{\overline{OD}} = 0, \qquad \left.\frac{\partial w}{\partial n}\right|_{\overline{OA}\cup\overline{BC}\cup\overline{CD}} = 0, \end{aligned} \tag{3.6.9}$$

where $\mu_L \in T_L$ and $w \in H_0^1(S)$. From eqns. (3.6.9) and (3.6.8), we have

$$\int_{\overline{AB}} \mu_L w \, d\ell = \int_{\overline{AB}} w \frac{\partial w}{\partial n} \, d\ell = \iint_S |\nabla w|^2 \, ds = |w|_{1,S}^2 \geq C_0 \|w\|_{1,S}^2. \tag{3.6.10}$$

Also, since $\Delta w = 0$ in S, there exists the bound,

$$\|\mu_L\|_{-\frac{1}{2},\overline{AB}} = \left\| \frac{\partial w}{\partial n} \right\|_{-\frac{1}{2},\overline{AB}} \leq C \|w\|_{1,S}. \tag{3.6.11}$$

Therefore, we obtain from eqns. (3.6.10) and (3.6.11)

$$\int_{\overline{AB}} \mu_L w \, d\ell \geq C_0 \|w\|_{1,S}^2 \geq \beta \|w\|_{1,S} \|\mu_L\|_{-\frac{1}{2},\overline{AB}}, \tag{3.6.12}$$

where β is a constant independent of w and L.

Let $w = w_N + r_N$, where $w_N \in V_N$, and r_N is the remainder. Then, we have from eqn. (3.6.12)

$$\begin{aligned} \int_{\overline{AB}} \mu_L w_N \, d\ell &= \int_{\overline{AB}} \mu_L w \, d\ell - \int_{\overline{AB}} \mu_L r_N \, d\ell \\ &\geq \beta \|w\|_{1,S} \|\mu_L\|_{-\frac{1}{2},\overline{AB}} - C_1 \|r_N\|_{1,S} \|\mu_L\|_{-\frac{1}{2},\overline{AB}} \\ &\geq \beta \|w_N\|_{1,S} \|\mu_L\|_{-\frac{1}{2},\overline{AB}} - (C_1 + \beta) \|r_N\|_{1,S} \|\mu_L\|_{-\frac{1}{2},\overline{AB}}, \end{aligned} \tag{3.6.13}$$

where we have used the bound

$$\int_{\overline{AB}} \mu_L r_N \, d\ell \leq \|\mu_L\|_{-\frac{1}{2},\overline{AB}} \|r_N\|_{\frac{1}{2},\overline{AB}} \leq C_1 \|\mu_L\|_{-\frac{1}{2},\overline{AB}} \|r_N\|_{1,S}. \tag{3.6.14}$$

Since $w_N \in V_N$ converges exponentially to $w \in H_0^1(S)$, there exists an integer $N_0 > 0$ such that for $N \geq N_0$

$$\frac{\|r_N\|_{1,S}}{\|w_N\|_{1,S}} \leq \frac{\beta}{2(C_1 + \beta)}, \tag{3.6.15}$$

where C_1 is the positive constant in eqn. (3.6.14). Then, we obtain from eqns. (3.6.13) and (3.6.15)

$$\int_{\overline{AB}} \mu_L w_N \, d\ell \geq \frac{\beta}{2} \|w_N\|_{1,S} \|\mu_L\|_{-\frac{1}{2},\overline{AB}}.$$

This is the LBB condition, i.e., eqn. (3.6.6) by letting $v_N = w_N$.

Let $\overline{\lambda}_L$ be the L-order polynomial interpolant of λ. Then, we have

$$\begin{aligned} \inf_{\eta \in T_L} \|\lambda - \eta\|_{-\frac{1}{2},\overline{AB}} &\leq \|\lambda - \overline{\lambda}_L|_{-\frac{1}{2},\overline{AB}} \\ &\leq C\|\lambda - \overline{\lambda}_L|_{0,\overline{AB}} \leq Cb^{L+1}, \end{aligned}$$

where $0 < b < 1$. We obtain the following corollary.

Corollary 3.6.1

Let the conditions of Theorem 3.6.1 and Corollary 3.3.1 hold. Then there exist the bounds

$$\|u - \widetilde{u}_N\|_{1,S} \leq C\{\sqrt{N}a^N + b^{L+1}\},$$

$$\|\lambda - \lambda_L\|_{1,S} \leq C\{\sqrt{N}a^N + b^{L+1}\}.$$

Corollary 3.6.1 implies an optimal matching between N and L, obtained as:

$$b^{L+1} = O(\sqrt{N}a^N),$$

which leads approximately to

$$L \approx N \left| \frac{\ln a}{\ln b} \right|.$$

In Kita and Kamiya [249], the CTM was classified as an indirect TM. Below, we briefly give a description of the direct TM [96, 226], which is much closer to that of BEM.

Rewrite Motz's problem in Section 3.1 as

$$\Delta u = 0 \quad \text{in } S,$$

$$u = \overline{u} \quad \text{on } \Gamma_D, \qquad \frac{\partial u}{\partial n} = 0 \quad \text{on } \Gamma_N,$$

where $\Gamma_D = \overline{AB} \cup \overline{OD}$, $\Gamma_N = \overline{BC} \cup \overline{CD} \cup \overline{OA}$, and

$$\overline{u} = \begin{cases} 500 & \text{on } \overline{AB}, \\ 0 & \text{on } \overline{OD}. \end{cases}$$

Applying the penalty plus hybrid TMs with a special case $w = 0$, we obtain

$$\iint_S \nabla u \cdot \nabla v \, ds - \int_{\Gamma_D} \frac{\partial v}{\partial n}(u - \overline{u})\, d\ell - \int_{\Gamma_D} \frac{\partial u}{\partial n} v \, d\ell = 0.$$

Since

$$\iint_S \nabla u \cdot \nabla v \, ds = \int_{\Gamma_D \cup \Gamma_N} u \frac{\partial v}{\partial n}\, d\ell,$$

we have

$$\int_{\Gamma_N} u \frac{\partial v}{\partial n}\, d\ell - \int_{\Gamma_D} \frac{\partial u}{\partial n} v \, d\ell = -\int_{\Gamma_D} \overline{u} \frac{\partial v}{\partial n}\, d\ell. \tag{3.6.16}$$

Denote $q = \frac{\partial u}{\partial n}$, then the eqn. (3.6.16) becomes

$$\int_{\Gamma_N} u \frac{\partial v}{\partial n}\, d\ell - \int_{\Gamma_D} q v \, d\ell = -500 \int_{\overline{AB}} \frac{\partial v}{\partial n}\, d\ell. \tag{3.6.17}$$

In eqn. (3.6.17), u on Γ_N and q on Γ_D are unknowns, we may follow the approaches in the TM, to seek the nodal approximation.

Denote the values

$$q_i = q(Q_i), \quad Q_i \in \Gamma_D,$$

$$u_i = u(Q_i), \quad Q_i \in \Gamma_N.$$

Let V_h and V_H be the finite-dimensional collections of interpolant polynomials of order N and M, based on the nodal values q_i and u_i, respectively, where h and H are the maximal meshspaces of Q_i on Γ_N and Γ_D, respectively. Hence, the direct TM can be written as: To seek $(q_h, u_H) \in (V_h, V_H)$ such that

$$\int_{\Gamma_N} u_H \frac{\partial v}{\partial n}\, d\ell - \int_{\Gamma_D} q_h v \, d\ell = -500 \int_{\overline{AB}} \frac{\partial v}{\partial n}\, d\ell, \quad \forall v \in V_L, \tag{3.6.18}$$

where V_L is spanned by

$$\phi_i = r^{i+\frac{1}{2}} \cos\left(i + \frac{1}{2}\right)\theta, \quad i = 0, 1, 2, \ldots, L. \tag{3.6.19}$$

3.7 Effective condition number

In this section, we turn to study stability by redefining the condition number. The traditional definition of condition number was given in Wilkinson [472], and then used in many books and papers, see Atkinson [9], Golub and van Loan [168], and Quarteroni and Valli [374]. For solving the linear algebraic equation $\mathbf{Fx} = \mathbf{b}$ resulting from elliptic equations, the traditional condition number is defined as $\text{Cond} = \frac{\sigma_1}{\sigma_n}$, where σ_1 and σ_n are the maximal and the minimal singular values of matrix $\mathbf{F}$, respectively. The condition number is used to provide the bounds of the relative errors from the perturbation of both $\mathbf{F}$ and $\mathbf{b}$. However, in real applications, we only deal with a certain vector $\mathbf{b}$, and the real relative errors may be smaller, or even much smaller than the worst Cond. Such a case was first proposed in Rice [383] in 1981 (the condition number for the particular case $\mathbf{b}$ was called the natural condition number), then studied in Chan and Foulser [81], and subsequently applied to a boundary value problem in Christiansen and Hansen [101], who called it the effective condition number. In this section, we will explore the computational formulas to evaluate the effective condition number, denoted by Cond_eff. Moreover, we propose a new simplified effective condition number Cond_E, which is easy to compute, because we only need the eigenvector of the minimal eigenvalue for $\mathbf{F}^T\mathbf{F}$, as additional information.

For smooth solutions of elliptic boundary value problems, the simplified effective condition number Cond_E may be small. For the singular layer solutions, when the FDM with the local refinements of grids near the singular layers are used, the Cond_E may be large, but it is significantly smaller than the huge Cond, defined in Strang and Fix [426]. The rather small Cond_E will benefit the solutions of both smooth and singularity problems. The results are reported in Li, Chien, and Huang [290], Huang and Li [212], Li, Huang, and Huang [294, 295, 296], and Li et al. [297]. The new Cond_E can be applied to the spectral and Trefftz methods, where the solutions obtained are very accurate, while the traditional Cond is huge. Small effective condition number explains well the high accuracy of these solutions, and strengthens the spectral and Trefftz methods reported in Refs. [296, 297].

First, let us derive the computational formulas for the effective condition number. Consider the overdetermined system

$$\mathbf{Fx} = \mathbf{b}, \tag{3.7.1}$$

where matrix $\mathbf{F} \in R^{m\times n}$ and $m \geq n$. Suppose the rank of $\mathbf{F}$ is n. When there exists a perturbation of $\mathbf{F}$ and $\mathbf{b}$, we have

$$\mathbf{F}(\mathbf{x} + \Delta\mathbf{x}) = \mathbf{b} + \Delta\mathbf{b}, \tag{3.7.2}$$

$$(\mathbf{F} + \Delta\mathbf{F})(\mathbf{x} + \Delta\mathbf{x}) = \mathbf{b} + \Delta\mathbf{b}. \tag{3.7.3}$$

Let matrix $\mathbf{F}$ be decomposed by the singular value decomposition

$$\mathbf{F} = \mathbf{U}\Sigma\mathbf{V}^T, \tag{3.7.4}$$

where matrices $\mathbf{U} \in R^{m\times m}$ and $\mathbf{V} \in R^{n\times n}$ are orthogonal, and the matrix $\Sigma \in R^{m\times n}$ is diagonal matrix with the positive singular values σ_i as

$$\sigma_1 \geq \sigma_2 \geq \cdots \geq \sigma_n > 0.$$

The traditional condition number is defined by Golub and van Loan [168], p. 223,

$$\text{Cond} = \frac{\sigma_1}{\sigma_n}. \tag{3.7.5}$$

First consider eqn. (3.7.2). Denote $\mathbf{U} = (\mathbf{u_1}, \ldots, \mathbf{u_m})$, we have the expansions

$$\mathbf{b} = \sum_{i=1}^{m} \beta_i \mathbf{u}_i, \qquad \Delta\mathbf{b} = \sum_{i=1}^{m} \alpha_i \mathbf{u}_i,$$

where

$$\beta_i = \mathbf{u}_i^T \mathbf{b}, \qquad \alpha_i = \mathbf{u}_i^T \Delta\mathbf{b}.$$

Denote the pesudoinverse matrix $\Sigma^+ \in R^{n\times m}$ of Σ to be diagonal with the entries $\frac{1}{\sigma_i}$. Hence, the pesudoinverse matrix of $\mathbf{F}$ is given by

$$\mathbf{F}^+ = \mathbf{V}\Sigma^+\mathbf{U}^\mathbf{T},$$

and the least squares solution is expressed by

$$\mathbf{x} = \mathbf{F}^+\mathbf{b} = \mathbf{V}\boldsymbol{\Sigma}^+\mathbf{U}^\mathbf{T}\mathbf{b}. \tag{3.7.6}$$

Also from eqns. (3.7.1) and (3.7.2)

$$\boldsymbol{\Delta}\mathbf{x} = \mathbf{F}^+\boldsymbol{\Delta}\mathbf{b} = \mathbf{V}\boldsymbol{\Sigma}^+\mathbf{U}^\mathbf{T}\boldsymbol{\Delta}\mathbf{b}.$$

Hence, since $\mathbf{U}$ is orthogonal, we obtain

$$\|\mathbf{x}\| = \|\boldsymbol{\Sigma}^+\mathbf{U}^\mathbf{T}\mathbf{b}\| = \sqrt{\sum_{i=1}^{n} \frac{\beta_i^2}{\sigma_i^2}}, \tag{3.7.7}$$

and

$$\|\boldsymbol{\Delta}\mathbf{x}\| = \|\boldsymbol{\Sigma}^+\mathbf{U}^\mathbf{T}\boldsymbol{\Delta}\mathbf{b}\| = \sqrt{\sum_{i=1}^{n} \frac{\alpha_i^2}{\sigma_i^2}} \tag{3.7.8}$$

$$\leq \frac{1}{\sigma_n}\sqrt{\sum_{i=1}^{n} \alpha_i^2} \leq \frac{\|\boldsymbol{\Delta}\mathbf{b}\|}{\sigma_n}.$$

In practical computation, the worst cases as in eqn. (3.7.8) may or may not happen. Then sometimes, $\|\boldsymbol{\Delta}\mathbf{x}\| < \frac{1}{\sigma_n}\|\boldsymbol{\Delta}\mathbf{b}\|$ may give a better numerical stability than Cond_eff defined in eqn. (3.7.10).

Hence, we obtain

$$\frac{\|\Delta \mathbf{x}\|}{\|\mathbf{x}\|} \leq \frac{\|\Delta \mathbf{b}\|}{\sigma_n} \times \frac{1}{\sqrt{\sum_{i=1}^{n} \frac{\beta_i^2}{\sigma_i^2}}} \leq \text{Cond\_eff} \times \frac{\|\Delta \mathbf{b}\|}{\|\mathbf{b}\|}, \tag{3.7.9}$$

where

$$\text{Cond\_eff} = \frac{\|\mathbf{b}\|}{\sigma_n \|\mathbf{x}\|} = \frac{\|\mathbf{b}\|}{\sigma_n \sqrt{\left(\frac{\beta_1}{\sigma_1}\right)^2 + \cdots + \left(\frac{\beta_n}{\sigma_n}\right)^2}}. \tag{3.7.10}$$

Note that when vector **b** (i.e., **x**) is just parallel to the eigenvector $\mathbf{u_1}$, i.e.,

$$\beta_2 = \cdots = \beta_n = 0, \tag{3.7.11}$$

we have $\|\mathbf{b}\| = |\beta_1|$. Hence, the effective condition number leads to the traditional condition number in eqn. (3.7.5). However, the cases in eqn. (3.7.11) may not happen for the real vector **b**. Hence, the effective condition number may provide a better estimation on the upper bound of relative errors of **x**. Besides, the formula, $\text{Cond\_eff} = \frac{\|\mathbf{b}\|}{\sigma_n \|\mathbf{x}\|}$ in eqn. (3.7.10), can also be found in Rice [383], Christiansen and Saranen [102], and Banoczi et al. [23].

From eqn. (3.7.6),

$$\mathbf{Fx} = \sum_{i=1}^{n} \beta_i \mathbf{u}_i, \qquad \|\mathbf{Fx}\| = \sqrt{\sum_{i=1}^{n} \beta_i^2}.$$

When $m = n$, we have $\|\mathbf{Fx}\| = \|\mathbf{b}\|$. Since

$$\begin{aligned} \sum_{i=1}^{n} \frac{\beta_i^2}{\sigma_i^2} &= \sum_{i=1}^{n-1} \frac{\beta_i^2}{\sigma_i^2} + \frac{\beta_n^2}{\sigma_n^2} \\ &\geq \frac{1}{\sigma_1^2} \sum_{i=1}^{n-1} \beta_i^2 + \frac{\beta_n^2}{\sigma_n^2} = \frac{\|\mathbf{Fx}\|^2 - \beta_n^2}{\sigma_1^2} + \frac{\beta_n^2}{\sigma_n^2}, \end{aligned}$$

the simplified effective condition number is obtained from eqn. (3.7.10)

$$\text{Cond\_E} = \frac{\|\mathbf{b}\|}{\sqrt{\frac{\|\mathbf{Fx}\|^2 - \beta_n^2}{\text{Cond}^2} + \beta_n^2}}. \tag{3.7.12}$$

Moreover, when $\beta_n \neq 0$, we have the simplest effective condition number from an upper bound of eqn. (3.7.10)

$$\text{Cond\_EE} = \frac{\|\mathbf{b}\|}{|\beta_n|}. \tag{3.7.13}$$

Hence, we give the definitions of effective condition number.

Definition 3.7.1

Let $\mathbf{F} \in R^{m\times n}, m \geq n$ *and* $rank(\mathbf{F}) = n$*, the relative errors of the solution* $\mathbf{x}$ *can be bounded by the effective condition number, Cond_eff defined in eqn. (3.7.10), and the simplified effective condition number Cond_E defined in eqn. (3.7.12). Moreover, when* $\beta_n \neq 0$*, and the simplest effective condition number is also defined in eqn. (3.7.13).*

Below, we consider the perturbation in eqn. (3.7.3), and cite a theorem from Lu et al. [317].

Theorem 3.7.1

Let $rank(\mathbf{F}) = rank(\mathbf{F} + \Delta\mathbf{F}) = n$*, and* $\|\mathbf{F}^+\| \, \|\Delta\mathbf{F}\| = \delta < 1$. *There exist the bounds of the relative errors of* $\mathbf{x}$*,*

$$\frac{\|\Delta\mathbf{x}\|}{\|\mathbf{x}\|} \leq \text{Cond\_eff} \times \frac{1}{1-\delta}\left\{\sqrt{2}\delta + \frac{\|\Delta\mathbf{b}\|}{\|\mathbf{b}\|}\right\}. \tag{3.7.14}$$

When $\Delta\mathbf{F} \equiv 0$, then $\delta = 0$, and Theorem 3.7.1 leads to eqn. (3.7.9).

Based on Theorem 3.7.1, Definition 3.7.1 is also valid for eqn. (3.7.3). From eqn. (3.7.4), we have

$$\mathbf{F}^{\mathrm{T}}\mathbf{F} = (\mathbf{U}\boldsymbol{\Sigma}\mathbf{V}^{\mathrm{T}})^{\mathrm{T}}(\mathbf{U}\boldsymbol{\Sigma}\mathbf{V}^{\mathrm{T}}) = \mathbf{V}\boldsymbol{\Sigma}^{\mathrm{T}}\boldsymbol{\Sigma}\mathbf{V}^{\mathrm{T}} = \mathbf{V}\mathbf{D}\mathbf{V}^{\mathrm{T}},$$

where $\mathbf{D} = \boldsymbol{\Sigma}^{\mathrm{T}}\boldsymbol{\Sigma}$ is a diagonal matrix consisting of σ_i^2. Also from eqn. (3.7.4), we have

$$\mathbf{U}\boldsymbol{\Sigma} = \mathbf{F}\mathbf{V},$$

to give (see Atkinson [9], p. 478)

$$\mathbf{u}_i = \frac{1}{\sigma_i}\mathbf{F}\mathbf{v}_i, \quad i = 1, 2, \ldots, n. \tag{3.7.15}$$

Hence, we may first use the power method and the inverse power method (or DEVESF of IMSL subroutines) for matrix $\mathbf{F}^{\mathrm{T}}\mathbf{F}$ to obtain the two eigenpairs $(\sigma_1^2, \mathbf{v_1})$ and $(\sigma_n^2, \mathbf{v_n})$, respectively, and then to obtain vectors $\mathbf{u_1}$ and $\mathbf{u_n}$ from eqn. (3.7.15). Hence, our definition Cond_E in eqn. (3.7.12) can be applied in practical application, since $\|\mathbf{b}\|$, $\|\mathbf{F}\mathbf{x}\|$, and β_n can be easily computed.

3.8 Numerical experiments

In this section, we carry out the numerical experiments for Motz's problem. The particular solutions, i.e., eqn. (3.1.5) are chosen, and the hybrid TM eqn. (3.4.5), the penalty plus hybrid TM eqn. (3.5.11), and the CTM are used, where $P_c = 1$ is selected by trial computation.

For the penalty plus hybrid TM eqn. (3.5.11), we obtain the linear algebraic equations,

$$\mathbf{A_1 x} = \mathbf{b}, \tag{3.8.1}$$

where the unknown $\mathbf{x}$ consists of the leading coefficients D_i, and matrix $\mathbf{A_1}$ is symmetric and positive definite if P_c is chosen to be suitably large, based on Lemma 3.5.5. The condition number is defined by

$$\text{Cond} = \text{Cond}(\mathbf{A_1}) = \frac{\lambda_{\max}(\mathbf{A_1})}{\lambda_{\min}(\mathbf{A_1})}, \tag{3.8.2}$$

where $\lambda_{\max}(\mathbf{A_1})$ and $\lambda_{\min}(\mathbf{A_1})$ are the maximal and minimal eigenvalues of $\mathbf{A_1}$, respectively.

For the hybrid TM, the algebraic equations from eqn. (3.4.5) is given by

$$\mathbf{A_2 x} = \mathbf{b}, \tag{3.8.3}$$

where $\mathbf{A_2}$ is positive definite but non-symmetric. Then, the condition number is given by

$$\text{Cond} = \text{Cond}(\mathbf{A_2}) = \frac{\sqrt{\max_i \lambda_i(\mathbf{A_2^T A_2})}}{\sqrt{\min_i \lambda_i(\mathbf{A_2^T A_2})}}. \tag{3.8.4}$$

As for the CTM, the algorithms are described in the previous chapter. The division number of $\overline{AB}$ is denoted by M, and $h = \frac{1}{M}$. From the boundary conditions, eqns. (3.1.2) and (3.1.3), we establish the linear algebraic equations

$$\mathbf{Fx} = \mathbf{b}, \tag{3.8.5}$$

where $\mathbf{x}$ consists of $N+1$ coefficients D_i, and $\mathbf{F}$ is the $4M \times (N+1)$ matrix. Since we choose $4M \gg (N+1)$, the least squares method (LSM) is used to obtain the solution $\mathbf{x}$, e.g., D_i. Also the condition number is

$$\text{Cond} = \left\{ \frac{\lambda_{\max}(\mathbf{F}^T\mathbf{F})}{\lambda_{\min}(\mathbf{F}^T\mathbf{F})} \right\}^{\frac{1}{2}}. \tag{3.8.6}$$

For the direct TM, the algebraic equations from eqn. (3.6.2) is given by

$$\mathbf{A_3 y} = \mathbf{b_3}, \tag{3.8.7}$$

where $\mathbf{y}$ is the unknown vector consisting of D_i and A_i in eqn. (3.6.3), $\mathbf{b_3}$ is a known vector, and $\mathbf{A_3}$ is non-singular but symmetric. Then, the condition number is given by

$$\text{Cond} = \text{Cond}(\mathbf{A_3}) = \frac{\max_i |\lambda_i(\mathbf{A_3})|}{\min_i |\lambda_i(\mathbf{A_3})|}. \tag{3.8.8}$$

Note that the condition number eqn. (3.8.6) with the square root by the CTM is much smaller than that by the other three TMs.

Denote the error $\epsilon = u - \widetilde{u}$ in the Sobolev norms:

$$|\epsilon|_{0,S} = \left\{ \iint_S (u - \widetilde{u})^2 \, ds \right\}^{\frac{1}{2}}, \tag{3.8.9}$$

$$|\epsilon|_{1,S} = \left\{ \iint_S (|\nabla(u - \widetilde{u})|^2 \, ds \right\}^{\frac{1}{2}} = \left\{ \int_{\Gamma_N^*} (u - \widetilde{u}) \frac{\partial(u - \widetilde{u})}{\partial n} \, d\ell \right\}^{\frac{1}{2}}, \tag{3.8.10}$$

where $\Gamma^* = \overline{AB} \cup \overline{BC} \cup \overline{CD}$, and $\widetilde{u}$ is the approximate solutions by the CTM, the hybrid TM, or the penalty plus hybrid TM, and u is the solutions with more accurate coefficients d_i given in Li and Lu [299]. On ∂S, denote the maximal boundary errors,

$$|\epsilon|_{\infty,\overline{AB}} = \max_{\overline{AB}} |\epsilon|, \tag{3.8.11}$$

$$\left| \frac{\partial \epsilon}{\partial n} \right|_{\infty,\overline{BC}} = \max_{\overline{BC}} \left| \frac{\partial \epsilon}{\partial n} \right|, \tag{3.8.12}$$

$$\left| \frac{\partial \epsilon}{\partial n} \right|_{\infty,\overline{CD}} = \max_{\overline{CD}} \left| \frac{\partial \epsilon}{\partial n} \right|. \tag{3.8.13}$$

The more accurate coefficients D_i can be obtained by the CTM using Mathematica with unlimited significant digits, or by the conformal transformation method using Mathematica. The conformal transformation method was proposed by Whiteman and Papamichael [468], and the series solution in Rosser and Papamichael [393] provided the most accurate coefficients of $D_0 - D_{19}$ under double precision. The coefficients $D_0 - D_{99}$ are obtained by the conformal transformation method using Mathematica with 200 working digits, and listed in Ref. [299], where D_0 and D_{99} have 199 and 89 significant decimal digits, respectively.

The solutions for Motz's problem are obtained from four TMs: (1) the hybrid TM, (2) the penalty plus hybrid TM, (3) the collocation TM (CTM), and (4) the direct TM (i.e., the Lagrange multiplier TM). For the CTM, we choose Gaussian rule with high order. For the integral $\int_{-1}^{1} f(x)\,dx$, the errors by the Gaussian rule with n nodes are found in Atkinson [9]

$$E_n^*(f) = \frac{2^{2n+1}(n!)^4}{(n+1)[(2n)!]^3} f^{(2n)}(\eta), \tag{3.8.14}$$

Table 3.1: The error norms and condition numbers for the three TMs, where $\epsilon = u - \widetilde{u}$.

(a) The hybrid TM

N	$\left\|\frac{\partial\epsilon}{\partial n}\right\|_{\infty,\overline{BC}}$	$\left\|\frac{\partial\epsilon}{\partial n}\right\|_{\infty,\overline{CD}}$	$\|\epsilon\|_{\infty,\overline{AB}}$	$\|\epsilon\|_{0,S}$	$\|\epsilon\|_{1,S}$	Cond	Cond_eff	Cond_EE
10	.400	.551	.397(−1)	.176(−1)	.759 (−1)	.753(3)	26.1	54.9
18	.524(−2)	.675(−2)	.258(−3)	.280(−3)	.844 (−3)	.184(6)	41.7	88.2
26	.719(−4)	.850(−4)	.222(−5)	.759(−5)	.125 (−4)	.464(8)	0.668(4)	0.141(5)
34	.883(−7)	.110(−6)	.196(−7)	.286(−7)	.210 (−6)	.118(11)	0.107(6)	0.226(6)

(b) The penalty plus hybrid TM

N	$\left\|\frac{\partial\epsilon}{\partial n}\right\|_{\infty,\overline{BC}}$	$\left\|\frac{\partial\epsilon}{\partial n}\right\|_{\infty,\overline{CD}}$	$\|\epsilon\|_{\infty,\overline{AB}}$	$\|\epsilon\|_{0,S}$	$\|\epsilon\|_{1,S}$	Cond	Cond_eff	Cond_EE
10	.461	.512	.143(−1)	.175(−1)	.361 (−1)	.104(4)	55.9	113
18	.605(−2)	.604(−2)	.120(−3)	.281 (−3)	.680 (−3)	.251(6)	739	0.148(4)
26	.815(−4)	.749(−4)	.123(−5)	.760 (−5)	.139 (−4)	.628(8)	0.122(5)	0.225(5)
34	.101(−5)	.944(−6)	.141(−7)	.177 (−7)*	.302 (−6)	.159(11)	0.174(6)	0.350(6)

(c) The collocation TM

N	$\left\|\frac{\partial\epsilon}{\partial n}\right\|_{\infty,\overline{BC}}$	$\left\|\frac{\partial\epsilon}{\partial n}\right\|_{\infty,\overline{CD}}$	$\|\epsilon\|_{\infty,\overline{AB}}$	$\|\epsilon\|_{0,S}$	$\|\epsilon\|_{1,S}$	Cond	Cond_eff	Cond_EE
10	.327	.296	.795(−2)	.216 (−1)	.936 (−1)	.955(2)	9.49	20.1
18	.328(−2)	.313(−2)	.658(−4)	.288 (−3)	.901 (−3)	.194(4)	16.4	35.6
26	.354(−4)	.366(−4)	.606(−6)	.761 (−5)	.114 (−4)	.374(5)	23.3	50.6
34	.387(−7)*	.445(−7)*	.596(−8)*	.248 (−7)	.175 (−6)*	.679(6) #	30.2#	65.7#

* Denotes the best results among three TMs.
\# Denotes the significantly better results.

where $\eta \in (-1, 1)$. When $N = 34$, the Gaussian rule of six nodes is used, i.e., $n = 6$. Then, we have

$$E_n^*(f) \approx 2 \times 10^{-11} f^{(12)}(\eta). \tag{3.8.15}$$

The error norms and condition numbers are listed in table 3.1 for the first three TMs. From table 3.1, we can see the exponential rates for the hybrid TM,

$$\begin{aligned} |u-\widetilde{u}|_{\infty,\overline{AB}} &\to 3.5 \times 0.55^N, \\ |\epsilon|_{0,S} &\to 4.9 \times 0.57^N, \qquad |\epsilon|_{1,S} \to 3.5 \times 0.59^N, \\ \text{Cond} &\to 0.7 \times 1.99^N. \end{aligned} \tag{3.8.16}$$

For the penalty plus hybrid TM,

$$\begin{aligned} |u-\widetilde{u}|_{\infty,\overline{AB}} &\to 6.7 \times 0.54^N, \\ |\epsilon|_{0,S} &\to 5.8 \times 0.56^N, \qquad |\epsilon|_{1,S} \to 5.1 \times 0.61^N, \\ \text{Cond} &\to 1.1 \times 1.99^N, \end{aligned} \tag{3.8.17}$$

and for the CTM,

$$
\begin{aligned}
&|u-\widetilde{u}|_{\infty,\overline{AB}} \to 2.8\times 0.55^N, && (3.8.18)\\
&\qquad |\epsilon|_{0,S} \to 5.0\times 0.57^N, \qquad |\epsilon|_{1,S} \to 19.5\times 0.58^N,\\
&\qquad \text{Cond} \to 5.7\times 1.41^N.
\end{aligned}
$$

Let us draw a few conclusions from eqns. (3.8.16)–(3.8.18) and table 3.1.

(a) Global errors. For the error norms, the three TMs give almost the same magnitude, although the CTM yields slightly better accuracy. For $N = 34$, the error norms of the hybrid TM, the penalty plus hybrid TM, and the CTM are given by

$$
\begin{aligned}
&|\epsilon|_{0,S} = 0.286(-7), && |\epsilon|_{1,S} = 0.210(-6), && |u-\widetilde{u}|_{\infty,\overline{AB}} = 0.196(-7),\\
&|\epsilon|_{0,S} = 0.177(-7), && |\epsilon|_{1,S} = 0.302(-6), && |u-\widetilde{u}|_{\infty,\overline{AB}} = 0.141(-7),\\
&|\epsilon|_{0,S} = 0.248(-7), && |\epsilon|_{1,S} = 0.175(-6), && |u-\widetilde{u}|_{\infty,\overline{AB}} = 0.596(-8),
\end{aligned}
$$

respectively. Hence, the CTM is still the best for the global errors in H^1 norms.

(b) Leading coefficients. We are interested in the accuracy of the leading coefficients. For the three TMs, the leading coefficients D_i with $i \le 34$ are listed in Ref. [302] and in table 2.5 in the previous chapter. To provide a clear view of comparisons, the number of significant digits of D_i from all TMs are provided in table 3.3. From table 3.3 we can see that the leading coefficients $D_0 - D_2$ by the CTM are more accurate than those in Ref. [306]; the D_0 by the CTM at $N = 34$ has 17 significant digits.

(c) Stability. For $N = 34$, Cond of the collocation TM, hybrid TM, penalty plus hybrid TM, and the direct TM are

$$
\text{Cond} = 0.679(6), \qquad 0.118(11), \qquad 0.159(11), \qquad 0.207(16), \tag{3.8.19}
$$

respectively. Hence, the Cond of the CTM is significantly smaller than those of the other TMs.

(d) Effective condition number. It is a puzzle that the Cond is large, but the solutions and D_0 are extremely accurate. Such a puzzle can be clarified by the effective condition number given in Section 3.7. Let us evaluate the Cond_eff for the solution given in table 2.5 in Chapter 2. Based on the computation in Ref. [302], we obtain the effective condition numbers, the traditional condition number, and their ratios,

$$
\begin{aligned}
&\text{Cond\_eff} = 30.2, \quad \text{Cond\_EE} = 65.7, \quad \text{Cond} = 0.679(6), && (3.8.20)\\
&\qquad \frac{\text{Cond}}{\text{Cond\_eff}} = 0.225(5), \qquad \frac{\text{Cond}}{\text{Cond\_EE}} = 0.103(5).
\end{aligned}
$$

The fact that the effective condition number is about 30 may explain very well the high accuracy of D_0 with 17 significant digits. In general, the leading

Table 3.2: The error norms, condition number, and errors of leading coefficients from the direct TM for Motz's problem by the Gaussian rule of six nodes rule as $M = 240$, where $|\lambda|_{\infty,\overline{AB}} = 357$.

N	L	$\|\epsilon\|_{\infty,\overline{AB}}$	$\|\lambda-\lambda_L\|_{\infty,\overline{AB}}$	$\|\lambda-\lambda_L\|_{0,\overline{AB}}$	$\|\frac{\partial\epsilon}{\partial n}\|_{\infty,\overline{BC}}$	$\|\frac{\partial\epsilon}{\partial n}\|_{\infty,\overline{CD}}$
10	4	0.166(−1)	0.292	0.135	0.413	0.535
18	5	0.163(−3)	0.103	0.160(−1)	0.789(−2)	0.582(−2)
26	7	0.158(−5)	0.944(−2)	0.112(−2)	0.121(−3)	0.536(−4)
34	9	0.600(−8)	0.168(−3)	0.158(−4)	0.322(−5)	0.211(−5)
40	10	0.200(−7)	0.975(−4)	0.131(−4)	0.591(−5)	0.488(−5)

N	L	$\|\epsilon\|_{0,S}$	$\|\epsilon\|_{1,S}$	$\left\|\frac{\Delta D_0}{D_0}\right\|$	Cond	Cond_eff	Cond_EE
10	4	0.706(−2)	0.175	0.266(−6)	0.650(6)	0.375(3)	0.647(4)
18	5	0.292(−4)	0.493(−2)	0.356(−10)	0.850(9)	0.201(4)	0.235(5)
26	7	0.197(−6)	0.184(−3)	0.467(−12)	0.128(13)	0.121(8)	0.126(6)
34	9	0.463(−8)	0.785(−5)	0.471(−12)	0.207(16)	0.766(5)	0.794(6)
40	10	0.497(−8)	0.693(−6)	0.471(−12)	0.759(17)	0.442(5)	0.428(6)

coefficient D_0 obtained should have 16 significant digits, and occasionally due to the cancellation of rounding errors, D_0 has 17 significant digits, as shown in Chapter 2. A recent study on the effective condition number shows that for Motz's problem by the CTM, there exist the bounds,

$$\text{Cond\_eff} \leq CN, \qquad \text{Cond} \leq CN^{\frac{3}{2}}(\sqrt{2})^N,$$

where C is a constant independent of N.

From table 3.2, we can see for $N = 34$ and $L = 9$

$$\text{Cond\_eff} = 0.766(5), \qquad \text{Cond\_EE} = 0.794(6), \qquad \text{Cond} = 0.207(16), \tag{3.8.21}$$

$$\frac{\text{Cond}}{\text{Cond\_eff}} = 0.270(10), \qquad \frac{\text{Cond}}{\text{Cond\_EE}} = 0.260(9).$$

For the optimal solutions at $N = 34$, the Cond_eff of the hybrid TM, penalty plus hybrid TM, the direct TM, and the CTM are given by

$$\text{Cond\_eff} = 0.107(6), \quad 0.174(6), \quad 0.766(5), \quad 30.2, \tag{3.8.22}$$

respectively. The effective condition number of the CTM is also significantly smaller than that of the other TMs. Interestingly, Cond_eff of the direct TM

Table 3.3: The number of significant decimal digits of leading coefficients for the three BAMs at $N = 34$, as well as those in Georgiou, Olson, and Smyrlis [161] at $N = 74$, and $N_\lambda = 33$, where N_λ is the number of the Lagrange multiplier used.

i	Original TM	P-H TM	Direct TM	[161]	Collocation TM
0	16	15	16	13	17
1	13	13	14	12	15
2	14	12	14	12	15
3	13	13	13	11	12
4	13	11	13	11	12
5	12	11	12	10	12
6	11	11	10	10	11
7	9	9	9	9	9
8	9	9	9	9	9
9	9	9	8	9	9
10	8	8	7	8	7
11	7	7	7	8	7
12	7	7	6	8	7
13	6	6	6	7	6
14	6	6	6	7	6
15	5	5	6	6	5
16	5	5	5	5	4
17	5	5	4	5	4
18	4	5	4	5	5

is not the largest, although its Cond is, indeed, huge. A new stability analysis of the penalty plus hybrid and the direct TMs is made in Ref. [297], based on effective condition number.

(e) Complexity of the algorithms. Let us compare the number of nodes used in the Gaussian rule of six nodes. For the best leading coefficients in Ref. [302] and table 2.5 in Chapter 2,

$$N_p = 7680, 1920, 7680, \text{ and } 30$$

by the original TM, the penalty plus hybrid TM, the direct TM, and the CTM, respectively. Note that $N_p = 30$ by the CTM is significantly smaller than that by the other TMs. Hence, the CTM needs less CPU time. The small number N_p is sufficient for the CTM, based on the analysis in Section 2.3 of Chapter 2, and the large N_p needed by the other TMs based on the analysis in Theorem 3.4.2. The algorithms of the CTM are simple; the algorithms of the Lagrange multiplier TM are the most complicated, because extra Lagrange multipliers as unknowns are needed, also see Refs. [161, 280].

(f) Limitation of applications. There exist some limitations for the hybrid TM and the direct TM to solve Laplace's equation. For example, when the solution domain S is divided into several subdomains, the interior subdomain $S_i \cap \Gamma = \emptyset$ is not allowed. However, such a limitation does not exist for the CTM and the penalty plus hybrid TM. Analysis of hybrid and other coupling techniques using piecewise particular solutions is explored in Li and Huang [293].

In summary, in this chapter we explore three new efficient TMs: (1) the hybrid TM, (2) the penalty plus hybrid TM, and (3) the direct TM, beyond the CTM in the Chapter 2. The drawback of these three TMs is the worse stability; details are reported in Refs. [294, 297]. Based on the theoretical analysis in Sections 3.2–3.6 and the numerical experiments in this section, we can conclude that the hybrid TM, the penalty plus hybrid TM, and the Lagrange multiplier TM in Ref. [161] are all efficient, but the CTM is the best. They all form a family of GTMs, accompanied by the CTM described in Chapters 1 and 2. The TM and GTMs are highly efficient for singularity problems, to compete with the BEM. Details of comparisons for the TMs are made in Li et al. [302].

4 Biharmonic equations with singularities

In this chapter, we give an extension of the collocation Trefftz method (CTM or collocation TM) in Chapter 2 for the biharmonic equations with crack singularities. First, this chapter derives the Green formulas for biharmonic equations on bounded domains with a non-smooth boundary, and corner terms are developed. The Green formulas are important to provide all the exterior and interior boundary conditions, which will be used in the CTM. Second, this chapter proposes three singularity models (called Models I, II, and III), and the CTM provides their most accurate solutions. In fact, Models I and II resemble Motz's problem in Chapter 2, and Model III with all the clamped boundary conditions originated from Schiff, Fishelov, and Whiteman [404]. Moreover, a brief analysis of error bounds for the CTM is made. Since the accuracy of the solutions obtained in this chapter is very high, they can be used as the typical models in testing numerical methods. The computed results in Ref. [300] show that as the singularity models, Models I and II are superior to Model III, because more accurate solutions can be obtained by the CTM.

4.1 Introduction

When an interior crack occurs within a thin elastic plate, determination of a stress intensity factor at the crack front is significant in fracture mechanics. Such a mechanical problem can be described as the biharmonic equations with the crack singularity, and the stress intensity factor is given by $K=\sqrt{2\pi}d_1$, where d_1 is the leading coefficient of singular particular solutions.

The singular problems have drawn much attention in the last several decades, and reported in many papers. Most of them deal with the second-order partial differential equations (PDEs); there exist a few books and papers for the fourth-order PDEs. Examples of textbooks and papers for biharmonic equations by the finite element method (FEM), finite difference method (FDM), and the boundary element method (BEM) include Chien [98], Carey and Oden [75], Birkhoff and

Lynch [45], Arad, Yakhot, and Ben-Dor [4], and Brebbia and Dominguez [61]. Besides, the methods using the series expansion solutions can be found in Whiteman [467], Lefeber [270], and Elliotis, Georgiou, and Xenophontos [135].

In this chapter, we pursue better models with series expansion solutions of very high convergent rates. Three singularity models are investigated: Models I and II are mimic Motz's problem in Chapter 2, and Model III results from Schiff, Fishelov, and Whiteman [404]. The TM and CTM introduced in Chapter 2 were studied in Li [280], Li and Mathon [304, 305], and Li, Mathon, and Sermer [306] as the boundary approximation method. In Chapter 2, by means of central and Gaussian rules, more accurate leading coefficient has also been obtained from the CTM. More detailed discussions about the model of Schiff et al. are provided by Hsu [204]. In this chapter, the CTM is developed to yield very accurate solutions for the singularity models of biharmonic equations. The materials of this chapter are adapted from Li, Lu, and Hu [300]. Besides, an error bound in L_2 norm for biharmonic equations is given in Comodi and Mathon [105].

This chapter is organized as follows. In the next section, we derive the Green formulas for rectangular and polygonal domains, and provide different types of boundary conditions. In Section 4.3, three singularity models are developed, and the CTM is described. In Section 4.4, a brief analysis for the CTM is made, and the CTM may be applied to the interior boundary (see Ref. [300]).

4.2 The Green formulas of $\Delta^2 u$

4.2.1 On rectangular domains

First, consider the rectangular domain $S = \{(x, y) \mid 0 < x < a, 0 < y < b\}$. We will derive the following Green formulas for $\Delta^2 u$,

$$\iint_S \Lambda_\mu(u, v) = \iint_S v\Delta^2 u - \int_{\partial S} m(u) v_n - \int_{\partial S} p(u)v + 2(1-\mu)([u_{xy}v]_3^4 - [u_{xy}v]_1^2), \qquad (4.2.1)$$

where $\Delta u = \frac{\partial^2 u}{\partial x^2} + \frac{\partial^2 u}{\partial y^2}$, $[v]_1^2 = v_2 - v_1$, and 1, 2, 3, 4 are four corners of S, see fig. 4.1. The notations are

$$\begin{aligned}\Lambda_\mu(u, v) &= \Delta u \Delta v + (1-\mu)(2u_{xy}v_{xy} - u_{xx}v_{yy} - u_{yy}v_{xx}) \\ &= u_{xx}v_{xx} + u_{yy}v_{yy} + \mu(u_{xx}v_{yy} + u_{yy}v_{xx}) + 2(1-\mu)u_{xy}v_{xy}, \\ m(u) &= -u_{nn} - \mu u_{ss}, \qquad p(u) = u_{nnn} + (2-\mu)u_{ssn},\end{aligned}$$

where $0 \le \mu < 1$ and n and s are normal and tangent directions along the boundary ∂S of S, respectively, and $u_{xy} = \frac{\partial^2 u}{\partial x \partial y}$, $u_{xx} = \frac{\partial^2 u}{\partial x^2}$, $u_{nn} = \frac{\partial^2 u}{\partial n^2}$, $u_{ss} = \frac{\partial^2 u}{\partial s^2}$, etc. In eqn. (4.2.1), we assume that all integrands therein are continuous.

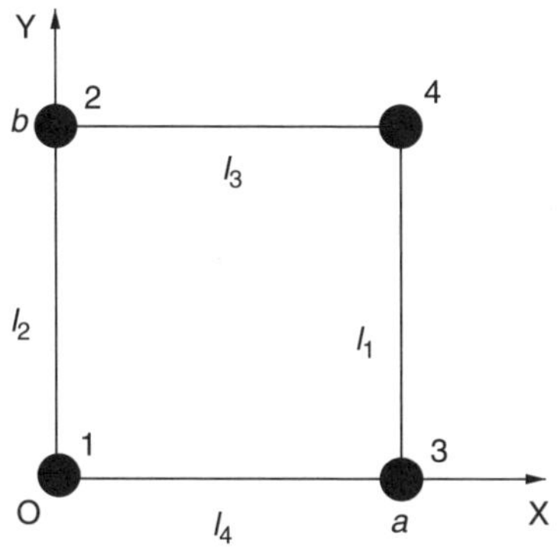

Figure 4.1: A rectangle.

Let us prove eqn. (4.2.1). In fig. 4.1, let ℓ_i with $i = 1, 2, 3, 4$ denote four edges of S, we obtain from integration by parts,

$$\iint_S u_{xx} v_{xx} = \left(\int_{\ell_1} - \int_{\ell_2}\right) v_x u_{xx} - \iint_S v_x u_{xxx}$$
$$= \left(\int_{\ell_1} - \int_{\ell_2}\right) v_x u_{xx} - \left(\int_{\ell_1} - \int_{\ell_2}\right) v u_{xxx} + \iint_S v u_{xxxx}, \quad (4.2.2)$$

and

$$\iint_S u_{yy} v_{yy} = \left(\int_{\ell_3} - \int_{\ell_4}\right) v_y u_{yy} - \left(\int_{\ell_3} - \int_{\ell_4}\right) v u_{yyy} + \iint_S v u_{yyyy}. \quad (4.2.3)$$

Similarly, we have

$$\iint_S u_{yy} v_{xx} = \left(\int_{\ell_1} - \int_{\ell_2}\right) v_x u_{yy} - \left(\int_{\ell_1} - \int_{\ell_2}\right) v u_{xyy} + \iint_S v u_{xxyy}, \quad (4.2.4)$$

and

$$\iint_S u_{xx} v_{yy} = \left(\int_{\ell_3} - \int_{\ell_4}\right) v_y u_{xx} - \left(\int_{\ell_3} - \int_{\ell_4}\right) v u_{xxy} + \iint_S v u_{xxyy}. \quad (4.2.5)$$

Also, from integration by parts again we have

$$\iint_S u_{xy} v_{xy} = \left(\int_{\ell_1} - \int_{\ell_2}\right) v_y u_{xy} - \iint_S v_y u_{xxy}$$
$$= \left(\int_{\ell_1} - \int_{\ell_2}\right) v_y u_{xy} - \left(\int_{\ell_3} - \int_{\ell_4}\right) v u_{xxy} + \iint_S v u_{xxyy}, \quad (4.2.6)$$

where

$$\left(\int_{\ell_1} - \int_{\ell_2}\right) v_y u_{xy} = [u_{xy}v]_3^4 - [u_{xy}v]_1^2 - \left(\int_{\ell_1} - \int_{\ell_2}\right) v u_{xyy}. \tag{4.2.7}$$

After some manipulations, we obtain from eqns. (4.2.2)–(4.2.7)

$$\begin{aligned}
\iint_S \Lambda_\mu(u,v) &= \iint_S \{u_{xx}v_{xx} + u_{yy}v_{yy} + \mu(u_{xx}v_{yy} + u_{yy}v_{xx}) + 2(1-\mu)u_{xy}v_{xy}\} \\
&= \iint_S v\Delta^2 u + \left(\int_{\ell_1} - \int_{\ell_2}\right)(u_{xx} + \mu u_{yy})v_x \\
&\quad + \left(\int_{\ell_3} - \int_{\ell_4}\right)(u_{yy} + \mu u_{xx})v_y - \left(\int_{\ell_1} - \int_{\ell_2}\right)(u_{xxx} + (2-\mu)u_{xyy})v \\
&\quad - \left(\int_{\ell_3} - \int_{\ell_4}\right)(u_{yyy} + (2-\mu)u_{xxy})v + 2(1-\mu)([u_{xy}v]_3^4 - [u_{xy}v]_1^2) \\
&= \iint_S v\Delta^2 u - \int_{\partial S} m(u)v_n - \int_{\partial S} p(u)v + 2(1-\mu)([u_{xy}v]_3^4 - [u_{xy}v]_1^2).
\end{aligned}$$

■

4.2.2 Corner effects on polygons

For the rectangular domains in fig. 4.1, there do exist the corner terms

$$2(1-\mu)([u_{xy}v]_3^4 - [u_{xy}v]_1^2) \tag{4.2.8}$$

in the Green formulas, which are different from those in Courant and Hilbert [108], p. 252 and Carey and Oden [75], p. 250. In fact, the formulas in Ref. [108] are valid only for the smooth boundary ∂S. There are many papers on Green formulas, see Herrera [192], Gourgeon and Herrera [171], and Russo [399], but only a few reports (e.g., Chien [98]) mention corner effects for biharmonic equations. Below, we will derive the Green formulas by different approaches from Section 4.2.1 and Chien [98].

Consider a polygon S in fig. 4.2, where the boundary $\partial S = \bigcup_{i=1}^m \Gamma_i$ and Γ_i are straight line segments. The corners are denoted by $P_1, P_2, \ldots, P_m$. We have from calculus,

$$\iint_S \Delta u\, \Delta v = \iint_S (\Delta^2 u)v + \int_{\partial S} (\Delta u)v_n - \int_{\partial S} \frac{\partial(\Delta u)}{\partial n} v, \tag{4.2.9}$$

and

$$\begin{aligned}
&\iint_S (u_{xx}v_{yy} + u_{yy}v_{xx}) \\
&= \int_{\partial S} (u_{xx}v_y y_n + u_{yy}v_x x_n) - \iint_S (u_{xxy}v_y + u_{xyy}v_x) \\
&= \int_{\partial S} (u_{xx}v_y y_n + u_{yy}v_x x_n) - \int_{\partial S} u_{xy}(v_y x_n + v_x y_n) + 2\iint_S u_{xy}v_{xy},
\end{aligned}$$

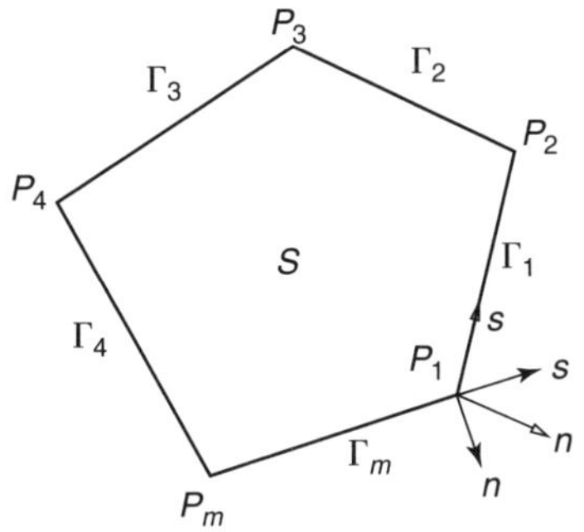

Figure 4.2: A polygon with corners.

where x_n, y_n and x_s, y_s are the directional cosine of the outward normal and the tangent vectors, respectively. Since $x_n = y_s$ and $y_n = -x_s$, we obtain

$$\iint_S (2u_{xy}v_{xy} - u_{xx}v_{yy} - u_{yy}v_{xx})$$
$$= -\int_{\partial S} (u_{xx}v_y y_n + u_{yy}v_x x_n) + \int_{\partial S} u_{xy}(v_y y_s - v_x x_s). \tag{4.2.10}$$

Since x_s, x_n, y_s, and y_n on the straight segments Γ_i are constant, we have the following derivative relations

$$\begin{aligned} v_y &= v_n y_n + v_s y_s, \qquad v_x = v_n x_n + v_s x_s, \\ u_n &= u_x x_n + u_y y_n, \qquad u_s = u_x x_s + u_y y_s, \\ u_{ss} &= u_{xx} x_s^2 + 2u_{xy} x_s y_s + u_{yy} y_s^2, \\ u_{ns} &= u_{xx} x_n x_s + u_{xy}(x_n y_s + x_s y_n) + u_{yy} y_n y_s. \end{aligned} \tag{4.2.11}$$

Equation (4.2.10) is then reduced to

$$\begin{aligned} &\iint_S (2u_{xy}v_{xy} - u_{xx}v_{yy} - u_{yy}v_{xx}) \\ &= -\int_{\partial S} (u_{xx}x_s^2 + 2u_{xy}x_s y_s + u_{yy}y_s^2)v_n \\ &\quad + \int_{\partial S} \{u_{xx}x_n x_s + u_{xy}(x_n y_s + x_s y_n) + u_{yy}y_n y_s\}v_s \\ &= -\int_{\partial S} u_{ss}v_n + \int_{\partial S} u_{ns}v_s. \end{aligned} \tag{4.2.12}$$

Next, we have from integration by parts,

$$\int_{\partial S} u_{ns} v_s = \sum_{i=1}^{m} \int_{\Gamma_i} u_{ns} v_s = -\sum_{i=1}^{m} \delta[u_{ns}]_i v_i - \sum_{i=1}^{m} \int_{\Gamma_i} \left(\frac{\partial}{\partial s} u_{ns} \right) v, \tag{4.2.13}$$

where $v_i = v(P_i)$ and $\delta[u_{ns}]_i$ denote the jumps of u_{ns} at corner P_i counter-clockwise,

$$\delta[u_{ns}]_i = u_{ns}(P_i) \,|\, _{\Gamma_i} - u_{ns}(P_i) \,|\, _{\Gamma_{i-1}},$$

and $\Gamma_m = \Gamma_0$. From eqns. (4.2.9), (4.2.12), and (4.2.13), we obtain the Green formulas,

$$\begin{aligned} \iint_S \Lambda_\mu(u, v) &= \iint_S \Delta u \Delta v + (1-\mu) \int_{\partial S} (2u_{xy} v_{xy} - u_{xx} v_{yy} - u_{yy} v_{xx}) \\ &= \iint_S (\Delta^2 u) v + \int_{\partial S} (\Delta u) v_n - \int_{\partial S} \frac{\partial (\Delta u)}{\partial n} v \\ &\quad + (1-\mu) \left\{ - \int_{\partial S} u_{ss} v_n - \int_{\partial S} \left(\frac{\partial}{\partial s} u_{ns} \right) v - \sum_{i=1}^{m} \delta[u_{ns}]_i v_i \right\} \\ &= \iint_S v\, \Delta u^2 - \int_{\partial S} (m(v) v_n + p(v) v) - (1-\mu) \sum_{i=1}^{m} \delta[u_{ns}]_i v_i, \end{aligned} \tag{4.2.14}$$

where the general forms of $m(v)$ and $p(v)$ are denoted by (see Ref. [108])

$$\begin{aligned} m(u) &= -\Delta u + (1-\mu) u_{ss} = -(u_{nn} + \mu u_{ss}), \\ p(u) &= \frac{\partial}{\partial n} \Delta u + (1-\mu) u_{nss} = u_{nnn} + (2-\mu) u_{nss}. \end{aligned} \tag{4.2.15}$$

There also exist the corner terms in the Green formulas, i.e., eqn. (4.2.14),

$$-(1-\mu) \sum_{i=1}^{m} \delta[u_{ns}]_i v_i. \tag{4.2.16}$$

When the corner angles at P_i in fig. 4.2 are just $\frac{\pi}{2}$, and when the second-order derivatives are also continuous:

$$u_{ns}(P_i) \,|\, _{\Gamma_i} = -u_{ns}(P_i) \,|\, _{\Gamma_{i-1}},$$

the corner conditions, i.e., eqn. (4.2.16) are reduced to

$$-(1-\mu) \delta[u_{ns}]_i v_i = 2(1-\mu) u_{ns}(P_i) \,|\, _{\Gamma_{i-1}}. \tag{4.2.17}$$

Note that eqn. (4.2.17) coincides well with eqn. (4.2.8). Obviously, for the smooth boundary ∂S, the corner terms disappear, and the Green formulas in Courant and Hilbert [108] are obtained. This shows that the Green formulas in Ref. [108] are the special cases of eqn. (4.2.14).

To close this subsection, let us consider the piecewise curved boundary Γ_i. Denote by $\alpha = \alpha(s)$ the angle between the tangent direction of Γ_i and the x axis, then

$$\begin{aligned} x_s &= \cos\alpha, \qquad y_n = -\cos\alpha, \qquad x_n = y_s = \sin\alpha, \\ u_n &= u_x x_n + u_y y_n = (\sin\alpha)u_x - (\cos\alpha)u_y, \\ u_s &= u_x x_s + u_y y_s = (\cos\alpha)u_x + (\sin\alpha)u_y. \end{aligned} \tag{4.2.18}$$

There exist the derivatives of α with respect to s: $\frac{\partial\alpha}{\partial s} = \frac{1}{\rho}$, where ρ denotes the curvature radius of Γ_i, ρ is positive if the curvature center is within S, or negative otherwise. We have from eqn. (4.2.18) by calculus,

$$\begin{aligned} u_{ss} &= u_{xx}x_s^2 + 2u_{xy}x_s y_s + u_{yy}y_s^2 - \frac{u_n}{\rho}, \\ u_{ns} &= u_{xx}x_n x_s + u_{xy}(x_n y_s + x_s y_n) + u_{yy}y_n y_s + \frac{u_s}{\rho}. \end{aligned} \tag{4.2.19}$$

Note that there are two additional terms, $-\frac{u_n}{\rho}$ and $\frac{u_s}{\rho}$ in eqn. (4.2.19), compared with those in eqn. (4.2.11).

Similarly, eqn. (4.2.10) is reduced to

$$\begin{aligned} &\iint_S (2u_{xy}v_{xy} - u_{xx}v_{yy} - u_{yy}v_{xx}) \\ &= -\int_{\partial S} (u_{xx}x_s^2 + 2u_{xy}x_s y_s + u_{yy}y_s^2)v_n \\ &\quad + \int_{\partial S} \{u_{xx}x_n x_s + u_{xy}(x_n y_s + x_s y_n) + u_{yy}y_n y_s\}v_s \\ &= -\int_{\partial S} \left(u_{ss} + \frac{u_n}{\rho}\right) v_n + \int_{\partial S} \left(u_{ns} - \frac{u_s}{\rho}\right) v_s \\ &= -\int_{\partial S} \left(u_{ss} + \frac{u_n}{\rho}\right) v_n - \sum_{i=1}^{n} \delta\left[u_{ns} - \frac{u_s}{\rho}\right]_i v_i - \int_{\partial S} \frac{\partial}{\partial s}\left(u_{ns} - \frac{u_s}{\rho}\right) v. \end{aligned} \tag{4.2.20}$$

From eqns. (4.2.9) and (4.2.20), we obtain the Green formulas,

$$\begin{aligned} &\iint_S \Lambda_\mu(u, v) \\ &= \iint_S \Delta u\, \Delta v + (1-\mu)\int_{\partial S} (2u_{xy}v_{xy} - u_{xx}v_{yy} - u_{yy}v_{xx}) \\ &= \iint_S v\, \Delta u^2 - \int_{\partial S} \{m^*(v)v_n + p^*(v)v\} - (1-\mu)\sum_{i=1}^{m} \delta\left[u_{ns} - \frac{u_s}{\rho}\right]_i v_i, \end{aligned} \tag{4.2.21}$$

where

$$
\begin{aligned}
m^*(u) &= -\Delta u + (1-\mu)\left(u_{ss} + \frac{u_n}{\rho}\right) = m(u) + (1-\mu)\frac{u_n}{\rho},\\
p^*(u) &= \frac{\partial}{\partial n}\Delta u + (1-\mu)\left(u_{nss} - \frac{\partial}{\partial s}\left(\frac{u_s}{\rho}\right)\right)\\
&= p(u) - (1-\mu)\frac{\partial}{\partial s}\left(\frac{u_s}{\rho}\right),
\end{aligned}
$$

and $m(u)$ and $p(u)$ are defined in eqn. (4.2.15). When natural boundary conditions are subjected to Γ_{i-1} and Γ_i, we obtain the boundary equations

$$
m^*(u) = 0, \qquad p^*(u) = 0 \quad \text{on } \Gamma_{i-1} \cup \Gamma_i, \tag{4.2.22}
$$

and the corner conditions

$$
\delta\left[u_{ns} - \frac{u_s}{\rho}\right]_i = 0 \quad \text{at } P_i. \tag{4.2.23}
$$

Equations (4.2.21), (4.2.22), and (4.2.23) are given in Chien [98], p. 245, p. 238, and p. 59, respectively. Where Γ_i are straight lines, $\rho = \infty$, eqn. (4.2.21) leads to eqn. (4.2.14), and the corner terms $-(1-\mu)\sum_{i=1}^{m} \delta[u_{ns} - \frac{u_s}{\rho}]_i v_i$ in eqns. (4.2.21)–(4.2.16).

4.2.3 Boundary conditions for biharmonic equations on polygons

Consider a polygon S, and the exterior boundary conditions on ∂S can be easily derived from the Green formulas. Let the solution $u \in H^2(S)$, where $H^2(S)$ is the Sobolev space defined in Ref. [417]. In $H^2(S)$, the biharmonic solution and its derivatives are continuous on the entire S, i.e., $u \in C^1(S)$. Then the generalized solution of the biharmonic equation, $\Delta^2 u + f = 0$ in S, can be expressed in a weak form: To seek $u \in H^2(S)$ such that

$$
\iint_S \Lambda_\mu(u, v) + \iint_S f v = 0, \quad v \in H_0^2(S),
$$

where $H_0^2(S)$ is a subspace of $H^2(S)$ satisfying suitable homogeneous boundary conditions for v. Based on the Green formulas in Section 4.2.1,

$$
0 = \iint_S (\Delta^2 u + f)v - \int_{\partial S} (m(u)v_n + p(u)v) + 2(1-\mu)([u_{xy}v]_3^4 - [u_{xy}v]_1^2),
$$

we obtain three equations from an arbitrary function v,

$$\iint_S (\Delta^2 u + f)v = 0, \tag{4.2.24}$$

$$\int_{\partial S} (m(u)v_n + p(u)v) = 0, \tag{4.2.25}$$

$$2(1-\mu)([u_{xy}v]_3^4 - [u_{xy}v]_1^2) = 0. \tag{4.2.26}$$

Since v in S is arbitrary, we obtain the biharmonic equation, $\Delta^2 u + f = 0$ in S from eqn. (4.2.24), where f also represents the exterior surface force. Next, let us consider different exterior boundary conditions. For the clamped condition: $u = g_1$ and $u_n = g_2$ on ∂S, in view of $v = 0$ and $v_n = 0$ on ∂S, we can see that the boundary integrals satisfy eqn. (4.2.25) automatically. Next, for the simply supported condition $u = g_1$, since v_n on ∂S is arbitrary, then the additional condition, the boundary blending moment $m(u) = 0$ on ∂S, is obtained from eqn. (4.2.25). For the special case: $u = constant$, we have $u_{ss} = 0$, which gives $m(u) = -u_{nn} - \mu u_{ss} = -u_{nn} = 0$. Hence, we obtain a concise form of the simply supported conditions: $u = constant$ and $u_{nn} = 0$ on ∂S.

For symmetric conditions, $u_n = 0$ on ∂S, we have the additional condition: $p(u) = u_{nnn} + (2-\mu)u_{nss} = 0$ from eqn. (4.2.25). Since $u_{nss} = 0$ on ∂S, $p(u) = 0$ is also simplified to $u_{nnn} = 0$ on ∂S. So, we obtain the symmetric conditions: $u_n = u_{nnn} = 0$ on ∂S. For the simple natural boundary condition, e.g., no constraints are given on ∂S, then $m(u) = 0$ and $p(u) = 0$ from eqn. (4.2.25). Suppose that the natural boundary conditions are given by the exterior boundary force $p(u) = g_3$ and the bending moment $m(u) = g_4$ on Γ_N, where Γ_N is part of ∂S, and the clamped boundary condition is subjected on $\partial S \backslash \Gamma_N$. The unique solution of biharmonic equations is then expressed as: To seek $u \in H^2(S)$ such that

$$\iint_S \Lambda_\mu(u, v) + \iint_S f v + \int_{\Gamma_N} (g_3 v + g_4 v_n) = 0, \quad v \in H_0^2(S).$$

We may consider the mixed types of different boundary conditions, where different conditions are given on different edges of ∂S. In this case, the corner terms must be considered for the natural corners, where two adjacent edges are *all* subjected to the natural conditions. Since v is arbitrary, we obtain the corner condition $u_{xy} = 0$ from eqn. (4.2.26), which is important for the CTM, see Section 4.3.3, because all boundary conditions including the corner conditions must be satisfied as best as possible. Moreover, the corner condition $u_{xy}v = 0$ is satisfied automatically, if one adjacent edge of the corner is subjected to one of the following cases: (1) the clamped condition, (2) the symmetric condition, or (3) the simply supported condition with $u_n = constant$. It is easy to see that either $v = 0$ or $u_{ns} = u_{xy} = 0$ on one edge yields $u_{xy}v = 0$ automatically at the corner.

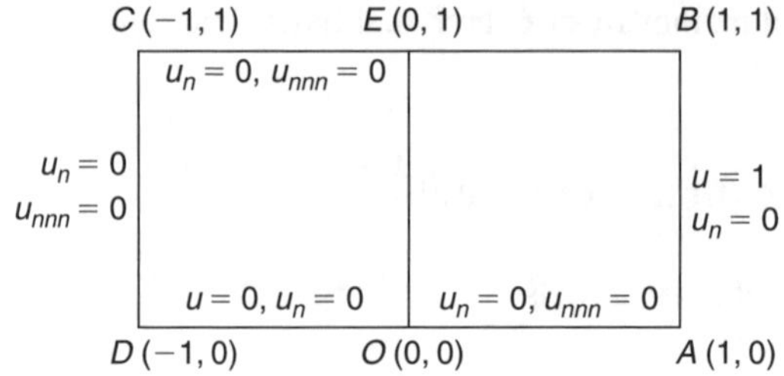

Figure 4.3: The boundary conditions of biharmonic equations for Model I.

We may discuss the uniqueness of the solutions by considering the homogeneous biharmonic equation $\Delta^2 u = 0$. A solution u can be also described as the minimal energy:

$$E(u) = \min_{v \in H_0^2(S)} E(v),$$

where

$$\begin{aligned} E(v) &= \frac{1}{2} \iint_S \Lambda_\mu(v, v) \\ &= \frac{1}{2} \iint_S \mu(v_{xx}^2 + v_{yy}^2) + (1-\mu)(v_{xx}^2 + v_{yy}^2 + 2v_{xy}^2). \end{aligned}$$

When $0 \le \mu < 1$, condition $E(v) = 0$ leads to that $v_{xx} = v_{yy} = v_{xy} = 0$, then the linear functions are obtained, i.e., $v = a + bx + cy$ with constants a, b, and c. To guarantee the unique solutions, the linear functions must be zero: $v = a + bx + cy \equiv 0$. For the mixed types of boundary conditions, there exist unique solutions for one edge, e.g., $\overline{AB}$ in fig. 4.3, subjected to the clamped boundary condition. Since v_n (e.g., v_x) $= v = 0$ on $\overline{AB}$, then the constants $b = 0$ and $(a + cy)\big|_{x=1} = 0$. So, $a = c = 0$ and then $v \equiv 0$. This confirms the unique solutions if one edge of a corner is subjected to the clamped boundary condition.

In summary, we have derived five typical boundary conditions for biharmonic equations on polygonal domains:

1. The symmetric condition: $u_n = 0$, $u_{nnn} = 0$.
2. The clamped condition: $u = c_0$, $u_n = c_1$.
3. The simply supported condition: $u = c_0$, $u_{nn} = c_2$.
4. The natural condition:

$$m(u) = -u_{nn} - \mu u_{ss} = c_3, \qquad p(u) = u_{nnn} + (2-\mu)u_{nss} = c_4.$$

5. The natural corner condition: $u_{xy} = c_5$.

Here, c_i are the given constants, which are dependent on the problems to be solved. Note that for biharmonic equations, the interior and exterior boundary conditions and the corner conditions are important not only to the CTM in this chapter but also to the collocation methods using the radial basis functions, or the Sinc functions etc., see Chapter 7.

4.3 The collocation Trefftz methods

4.3.1 Three singularity models

Consider the homogeneous biharmonic equation

$$\Delta^2 u = 0 \quad \text{in } S, \tag{4.3.1}$$

where the solution domain is the rectangle: $S = \{(x,y) \mid -1 < x < 1,\ 0 < y < 1\}$. In this chapter, we study three models of singularity problems, shown in figs. 4.3–4.5. The section $\overline{OD}$ represents an interior crack under the clamped condition $u = u_n = 0$. From Section 4.2.3, the symmetric conditions, $u_n = u_{nnn} = 0$ on $\overline{OA} \cup \overline{BC} \cup \overline{CD}$, are required. Here, n is the outward normal direction to the boundary ∂S. On $\overline{AB}$, when the clamped conditions are provided, we propose the biharmonic boundary value problem with the following conditions, called Model I in this chapter, see fig. 4.3:

$$u \mid_{\overline{OD}} = 0, \qquad u_y \mid_{\overline{OD}} = 0, \tag{4.3.2}$$

$$u_y \mid_{\overline{OA}} = 0, \qquad u_{yyy} \mid_{\overline{OA}} = 0, \tag{4.3.3}$$

$$u \mid_{\overline{AB}} = 1, \qquad u_x \mid_{\overline{AB}} = 0, \tag{4.3.4}$$

$$u_y \mid_{\overline{BC}} = 0, \qquad u_{yyy} \mid_{\overline{BC}} = 0, \tag{4.3.5}$$

$$u_x \mid_{\overline{CD}} = 0, \qquad u_{xxx} \mid_{\overline{CD}} = 0. \tag{4.3.6}$$

We may replace the clamped condition on $\overline{AB}$ by the simply supported condition,

$$u \mid_{\overline{AB}} = 1, \qquad u_{xx} \mid_{\overline{AB}} = 0,$$

but the other boundary conditions remain the same as those in Model I. Such a model is called Model II, see fig. 4.4. Note that Models I and II resemble Motz's problem in Chapter 2.

Next, we choose the models in Schiff, Fishelov, and Whiteman [404] with all the clamped conditions on ∂S except $\overline{OA}$, see fig. 4.5,

$$\Delta^2 u = 0, \quad \text{on } S, \tag{4.3.7}$$

$$u \mid_{\overline{OD}} = 0, \qquad u_y \mid_{\overline{OD}} = 0, \tag{4.3.8}$$

$$u_y \mid_{\overline{OA}} = 0, \qquad u_{yyy} \mid_{\overline{OA}} = 0, \tag{4.3.9}$$

$$u \mid_{\overline{AB}} = 2a^2, \qquad u_x \mid_{\overline{AB}} = 2a, \tag{4.3.10}$$

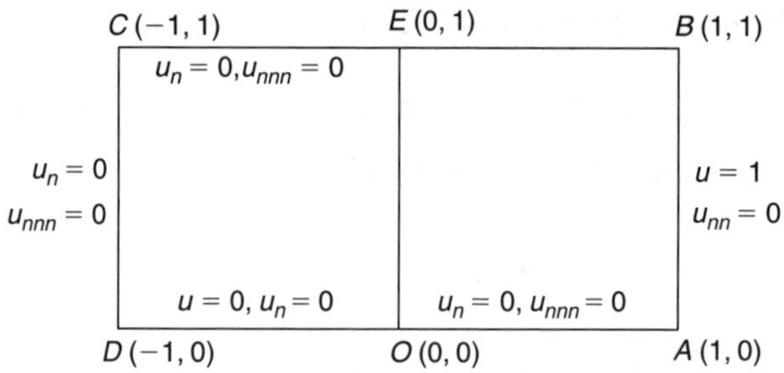

Figure 4.4: The boundary conditions of biharmonic equations for Model II.

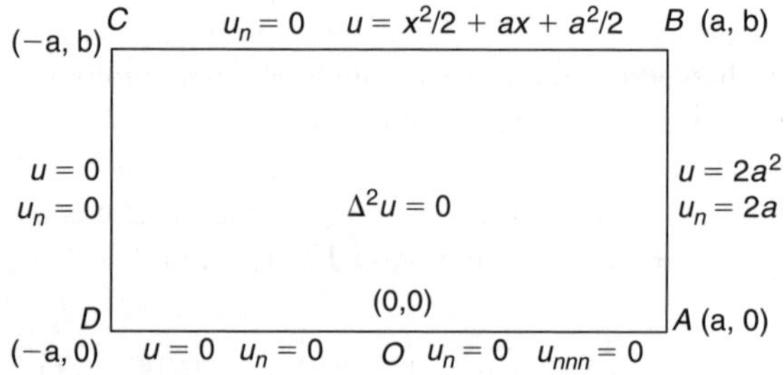

Figure 4.5: The boundary conditions of biharmonic equations for Model III.

$$u\,|_{\overline{BC}} = \frac{x^2}{2} + ax + \frac{a^2}{2}, \qquad u_y\,|_{\overline{BC}} = 0, \tag{4.3.11}$$

$$u\,|_{\overline{CD}} = 0, \qquad u_x\,|_{\overline{CD}} = 0, \tag{4.3.12}$$

where $S = \{(x,y)| -a < x < a, 0 < y < b\}$. When the parameters $a = b = 1$, the model, i.e., eqns. (4.3.7)–(4.3.12) is called Model III in this chapter.

For the clamped crack on $\overline{OD}$, the particular solutions of these models are known and given in Schiff, Fishelov, and Whiteman [404]:

$$u = \sum_{i=1}^{\infty} (\overline{d}_i \phi_i(r,\theta) + \overline{c}_i f_i(r,\theta)),$$

where $\overline{d}_i$ and $\overline{c}_i$ are expansion coefficients, and the singular particular solutions are

$$\phi_i(r,\theta) = r^{i+\frac{1}{2}} \left\{ \cos\left(i - \frac{3}{2}\right)\theta - \frac{i - \frac{3}{2}}{i + \frac{1}{2}} \cos\left(i + \frac{1}{2}\right)\theta \right\}, \tag{4.3.13}$$

and the analytic particular solutions

$$f_i(r,\theta) = r^{i+1}\{\cos(i-1)\theta - \cos(i+1)\theta\}. \tag{4.3.14}$$

4.3.2 Description of the method

We shall take Model I as an example to describe the CTM, since the algorithms of the CTM for other models may be similarly described. Choose finite terms of particular solutions

$$u_N = \sum_{i=1}^{N} (d_i\phi_i(r,\theta) + c_i f_i(r,\theta)), \tag{4.3.15}$$

where d_i and c_i are approximate coefficients to be sought. Since the particular solutions, i.e., eqns. (4.3.13) and (4.3.14) satisfy the biharmonic eqn. (4.3.1) in S and the boundary conditions on $\overline{OD}$ and $\overline{OA}$ already, the unknown coefficients c_i and d_i can be obtained by satisfying the rest of boundary conditions as best as possible.

For Model I, there exists the crack tip at O. We split S into S^+ and S^-, where $S^+ = S \cap (x \ge 0)$ and $S^- = S \cap (x \le 0)$. Then the Green formulas are applied to S^+ and S^- in fig. 4.3, to give

$$\begin{aligned}\iint_S \Lambda_\mu(u,v) &= \iint_{S^+} \Lambda_\mu(u,v) + \iint_{S^-} \Lambda_\mu(u,v)\\ &= \iint_S v\,\Delta^2 u - \int_{\partial S} \{m(u)v_n + p(u)v\} + 2(1-\mu)\\ &\quad \times ([u_{xy}v]_A^B - [u_{xy}v]_{O^+}^E + [u_{xy}v]_{O^-}^E - [u_{xy}v]_D^C),\end{aligned} \tag{4.3.16}$$

where point $E = (0, 1)$. From eqns. (4.3.2)–(4.3.6), we obtain

$$v(A) = v(B) = v(D) = v(O^-) = 0, \qquad u_{xy}(C) = u_{xy}(E) = u_{xy}(O^+) = 0.$$

Hence, the last term on the right-hand side of eqn. (4.3.16) is zero automatically. In this case, the corner conditions at E and O may not be needed in the CTM, either. We confirm again the boundary conditions given in eqns. (4.3.2)–(4.3.6).

For Model I with the boundary conditions, i.e., eqns. (4.3.2)–(4.3.6), define an energy on the boundary by

$$\begin{aligned}I(v) = I(d_i,c_i) &= \int_{\overline{AB}} \{(v-1)^2 + w_1^2 v_x^2\} + \int_{\overline{BC}} (w_1^2 v_y^2 + w_3^2 v_{yyy}^2)\\ &\quad + \int_{\overline{CD}} (w_1^2 v_x^2 + w_3^2 v_{xxx}^2),\end{aligned} \tag{4.3.17}$$

where the weights $w_i = \frac{1}{(N+1)^i}$, based on the analysis in Chapter 2. The approximate coefficients $\widetilde{d}_i$ and $\widetilde{c}_i$ can be found by

$$I(u_N) = I(\widetilde{d}_i, \widetilde{c}_i) = \min_{d_i,c_i} I(d_i, c_i).$$

To be more precise, we use the central rule to discretize the integrals in eqn. (4.3.17) with the uniform partitions chosen for $\overline{AB} \cup \overline{BC} \cup \overline{CD}$, and the division number of

$\overline{AB}$ is denoted by M. This is equivalent to the direct collocation method to establish the linear algebraic equations,

$$\mathbf{F}\mathbf{x} = \mathbf{b},$$

where $\mathbf{x}$ consists of $2N$ coefficients $\widetilde{d}_i$ and $\widetilde{c}_i$, and $\mathbf{F} \in R^{8M \times 2N}$ is the matrix, where $8M \gg 2N$. Its least squares solution is just the desired solution $\mathbf{x}$, e.g., the coefficients $\widetilde{d}_i$ and $\widetilde{c}_i$.

In computation, we use the error norm for accuracy,

$$E_2 = \|\epsilon\|_{\mathrm{B}} = \|u - v\|_{\mathrm{B}} = \sqrt{I(v)}, \tag{4.3.18}$$

and the condition number for stability

$$\mathrm{Cond} = \left\{ \frac{\lambda_{\max}(\mathbf{F}^T\mathbf{F})}{\lambda_{\min}(\mathbf{F}^T\mathbf{F})} \right\}^{\frac{1}{2}}, \tag{4.3.19}$$

where $\lambda_{\max}(\mathbf{A})$ and $\lambda_{\min}(\mathbf{A})$ are the maximal and minimal eigenvalues, respectively, of the matrix $\mathbf{A} = \mathbf{F}^T\mathbf{F}$.

4.3.3 The collocation Trefftz method with natural corners

Let the clamped and the natural conditions be given on $\overline{AB}$ and $\overline{BC} \cup \overline{CD}$ in fig. 4.3, respectively:

$$u = 1, \qquad u_n = 0, \quad \text{on } \overline{AB}; \qquad m(u) = p(u) = 0 \quad \text{on } \overline{BC} \cup \overline{CD}.$$

Since point C is a natural corner, the corner condition is needed: $u_{xy}(C) = 0$. The CTM involving the natural corner C is given by

$$I^*(u_N) = I^*(\widetilde{d}_i, \widetilde{c}_i) = \min_{d_i, c_i} I^*(d_i, c_i),$$

where

$$\begin{aligned} I^*(v) = & \int_{\overline{AB}} ((v-1)^2 + w_1^2 v_x^2) \\ & + \int_{\overline{BC} \cup \overline{CD}} (w_2^2 m^2(v) + w_3^2 p^2(v)) + 2(1-\mu) w_2^2 v_{xy}^2(C), \end{aligned} \tag{4.3.20}$$

and the weights are $w_i = \frac{1}{(N+1)^i}$. Note that the corner term involving $v_{xy}^2(C)$ is *necessary* to the CTM for the biharmonic equations with natural corners. When the corner angle at $\angle DCB$ is not just $\pi/2$, the corner contribution in eqn. (4.3.20) is replaced by $(1-\mu)w_2^2(v_{ns}^+(C) - v_{ns}^-(C))^2$ based on the analysis in Section 4.2.2.

4.3.4 Formulas of partial derivatives

In this section, we provide useful formulas for partial derivatives $u_x, u_y, u_{xx}, u_{yy}, u_{xy}$, $u_{xxx}, u_{yyy}, u_{xxy}$, and u_{xyy}, which are required by the CTM. Since the particular solutions, i.e., eqns. (4.3.13) and (4.3.14) in S are given in polar coordinates, we will find their explicit formulas of partial derivatives with respect to x and y. Let the origins of the Cartesian and polar coordinates be the same, and we obtain

$$u_x = \cos\theta\frac{\partial u}{\partial r} - \sin\theta\frac{\partial u}{r\,\partial\theta}, \qquad u_y = \sin\theta\frac{\partial u}{\partial r} + \cos\theta\frac{\partial u}{r\,\partial\theta}. \tag{4.3.21}$$

Based on eqn. (4.3.21), we have

$$\begin{aligned} u_{xx} &= \left(\cos\theta\frac{\partial}{\partial r} - \sin\theta\frac{\partial}{r\,\partial\theta}\right)\left(\cos\theta\frac{\partial}{\partial r} - \sin\theta\frac{\partial}{r\,\partial\theta}\right)u \\ &= \cos^2\theta\frac{\partial^2 u}{\partial r^2} - \sin2\theta\left(\frac{\partial}{r\,\partial r}\frac{\partial u}{\partial\theta} - \frac{\partial u}{r^2\,\partial\theta}\right) + \frac{\sin^2\theta}{r}\frac{\partial u}{\partial r} + \frac{\sin^2\theta}{r^2}\frac{\partial^2 u}{\partial\theta^2}. \end{aligned} \tag{4.3.22}$$

Similarly, we have

$$u_{yy} = \sin^2\theta\frac{\partial^2 u}{\partial r^2} + \sin2\theta\left(\frac{\partial}{r\,\partial r}\frac{\partial u}{\partial\theta} - \frac{\partial u}{r^2\,\partial\theta}\right) + \frac{\cos^2\theta}{r}\frac{\partial u}{\partial r} + \frac{\cos^2\theta}{r^2}\frac{\partial^2 u}{\partial\theta^2}, \tag{4.3.23}$$

and

$$\begin{aligned} u_{xy} &= \cos\theta\sin\theta\frac{\partial^2 u}{\partial r^2} + \frac{\cos^2\theta}{r}\frac{\partial}{\partial r}\frac{\partial u}{\partial\theta} - \frac{\cos^2\theta}{r^2}\frac{\partial u}{\partial\theta} - \frac{\sin^2\theta}{r}\frac{\partial}{\partial r}\frac{\partial u}{\partial\theta} \\ &\quad - \frac{\cos\theta\sin\theta}{r}\frac{\partial u}{\partial r} - \frac{\cos\theta\sin\theta}{r^2}\frac{\partial^2 u}{\partial\theta^2} + \frac{\sin^2\theta}{r^2}\frac{\partial u}{\partial\theta}. \end{aligned} \tag{4.3.24}$$

After some manipulation, we can also obtain

$$\begin{aligned} u_{xxx} &= \left(\cos\theta\frac{\partial}{\partial r} - \sin\theta\frac{\partial}{r\,\partial\theta}\right) \\ &\quad \times\left[\cos^2\theta\frac{\partial^2}{\partial r^2} - \sin2\theta\left(\frac{\partial}{r\,\partial r}\frac{\partial}{\partial\theta} - \frac{\partial}{r^2\,\partial\theta}\right) + \frac{\sin^2\theta}{r}\frac{\partial}{\partial r} + \frac{\sin^2\theta}{r^2}\frac{\partial^2}{\partial\theta^2}\right]u \\ &= \cos^3\theta\frac{\partial^3 u}{\partial r^3} - \frac{3\cos\theta\sin2\theta}{2r}\frac{\partial^2}{\partial r^2}\frac{\partial u}{\partial\theta} + \frac{3\sin\theta\sin2\theta}{2r^2}\frac{\partial}{\partial r}\frac{\partial^2 u}{\partial\theta^2} \\ &\quad - \frac{\sin^3\theta}{r^3}\frac{\partial^3 u}{\partial\theta^3} + \frac{3\sin\theta\sin2\theta}{2r}\frac{\partial^2 u}{\partial r^2} - \frac{3(\sin\theta - 3\sin3\theta)}{4r^2}\frac{\partial}{\partial r}\frac{\partial u}{\partial\theta} \\ &\quad - \frac{3\sin\theta\sin2\theta}{r^3}\frac{\partial^2 u}{\partial\theta^2} - \frac{3\sin\theta\sin2\theta}{2r^2}\frac{\partial u}{\partial r} - \frac{2\sin3\theta}{r^3}\frac{\partial u}{\partial\theta}. \end{aligned} \tag{4.3.25}$$

Other derivatives of third order are given by

$$\begin{aligned} u_{yyy} &= \sin^3\theta \frac{\partial^3 u}{\partial r^3} + \frac{3\sin\theta\sin 2\theta}{2r}\frac{\partial^2}{\partial r^2}\frac{\partial u}{\partial\theta} + \frac{3\cos\theta\sin 2\theta}{2r^2}\frac{\partial}{\partial r}\frac{\partial^2 u}{\partial\theta^2} \\ &+ \frac{\cos^3\theta}{r^3}\frac{\partial^3 u}{\partial\theta^3} + \frac{3\cos\theta\sin 2\theta}{2r}\frac{\partial^2 u}{\partial r^2} + \frac{3(\cos\theta + 3\cos 3\theta)}{4r^2}\frac{\partial}{\partial r}\frac{\partial u}{\partial\theta} \\ &- \frac{3\cos\theta\sin 2\theta}{r^3}\frac{\partial^2 u}{\partial\theta^2} - \frac{3\cos\theta\sin 2\theta}{2r^2}\frac{\partial u}{\partial r} - \frac{2\cos 3\theta}{r^3}\frac{\partial u}{\partial\theta}, \end{aligned} \tag{4.3.26}$$

$$\begin{aligned} u_{xxy} &= \sin\theta\cos^2\theta \frac{\partial^3 u}{\partial r^3} + \frac{\cos\theta + 3\cos 3\theta}{4r}\frac{\partial^2}{\partial r^2}\frac{\partial u}{\partial\theta} + \frac{\sin\theta - 3\sin 3\theta}{4r^2}\frac{\partial}{\partial r}\frac{\partial^2 u}{\partial\theta^2} \\ &+ \frac{\sin\theta\sin 2\theta}{2r^3}\frac{\partial^3 u}{\partial\theta^3} + \frac{\sin\theta - 3\sin 3\theta}{4r}\frac{\partial^2 u}{\partial r^2} + \frac{\cos\theta - 9\cos 3\theta}{4r^2}\frac{\partial}{\partial r}\frac{\partial u}{\partial\theta} \\ &- \frac{2\sin\theta - 3\sin 3\theta}{2r^3}\frac{\partial^2 u}{\partial\theta^2} - \frac{\sin\theta - 3\sin 3\theta}{4r^2}\frac{\partial u}{\partial r} + \frac{2\cos 3\theta}{r^3}\frac{\partial u}{\partial\theta}, \end{aligned} \tag{4.3.27}$$

and

$$\begin{aligned} u_{xyy} &= \cos\theta\sin^2\theta \frac{\partial^3 u}{\partial r^3} + \frac{3\sin 3\theta - \sin\theta}{4r}\frac{\partial^2}{\partial r^2}\frac{\partial u}{\partial\theta} + \frac{\cos\theta + 3\cos 3\theta}{4r^2}\frac{\partial}{\partial r}\frac{\partial^2 u}{\partial\theta^2} \\ &- \frac{\sin\theta\cos^2\theta}{r^3}\frac{\partial^3 u}{\partial\theta^3} + \frac{\cos\theta + 3\cos 3\theta}{4r}\frac{\partial^2 u}{\partial r^2} - \frac{\sin\theta + 9\sin 3\theta}{4r^2}\frac{\partial}{\partial r}\frac{\partial u}{\partial\theta} \\ &\frac{\cos\theta + 3\cos 3\theta}{2r^3}\frac{\partial^2 u}{\partial\theta^2} - \frac{\cos\theta + 3\cos 3\theta}{4r^2}\frac{\partial u}{\partial r} + \frac{2\sin 3\theta}{r^3}\frac{\partial u}{\partial\theta}. \end{aligned} \tag{4.3.28}$$

In the computation of eqns. (4.3.21)–(4.3.28), the derivatives of u with respect to r and θ can be easily obtained directly from eqns. (4.3.13) and (4.3.14). For the analytical particular solutions $f_i(r,\theta)$ in eqn. (4.3.14), the explicit derivatives, $u_x, u_y, \ldots, u_{xyy}$, can be derived straightforward. However, for the singular particular solutions, $\phi_i(r,\theta)$ in eqn. (4.3.13), the above computational formulas are *essential* in computations.

4.4 Error bounds

For simplicity, we consider the CTM for Model I in fig. 4.3. Let $S = S^+ \cup S^-$, where $S^- = S \cap (x < 0)$ and $S^+ = S \cap (x > 0)$. We have the Green formulas, i.e., eqn. (4.3.16) for $v = u_N$,

$$\begin{aligned} \iint_S \Lambda_\mu(v,v) = \iint_S v\,\Delta^2 v - \int_{\partial S} m(v)v_n - \int_{\partial S} p(v)v \\ + 2(1-\mu)\{[v_{xy}v]_A^B - [v_{xy}v]_{O^+}^E + [v_{xy}v]_{O^-}^E - [v_{xy}v]_D^C\}, \end{aligned}$$

where $E=(0,1)$, $m(v)=-v_{nn}$ and $p(v)=v_{nnn}$ on $\overline{AB}\cup\overline{BC}\cup\overline{CD}$ for Model I. Hence, for the particular solutions $v=v_N$ of biharmonic equations, since $v\,|\,_D=v\,|\,_{O^-}=0$ and $v_{xy}\,|\,_A=v_{xy}\,|\,_{O^+}=0$, the Green formulas are simplified to

$$\iint_S \Lambda_\mu(v,v)=\int_{\overline{AB}\cup\overline{BC}\cup\overline{CD}}(v_{nn}v_n-v_{nnn}v)+2(1-\mu)[v_{xy}v]_C^B. \tag{4.4.1}$$

Moreover, the boundary norm eqn. (4.3.18) is expressed as

$$\begin{aligned}\|\epsilon\|_{\mathrm{B}}&=\|u-v\|_{\mathrm{B}}=\sqrt{I(v)}\\&=\{\|v-1\|^2_{0,\overline{AB}}+w_1^2\|v_n\|^2_{0,\overline{AB}}+w_1^2\|v_n\|^2_{0,\overline{BC}}\\&\quad+w_3^2\|v_{nnn}\|^2_{0,\overline{BC}}+w_1^2\|v_n\|^2_{0,\overline{CD}}+w_3^2\|v_{nnn}\|^2_{0,\overline{CD}}\}^{\frac{1}{2}},\end{aligned} \tag{4.4.2}$$

where $w_i=1/(N+1)^i$. Then, we have the following theorem.

Theorem 4.4.1

Let $v=u_N$ in eqn. (4.3.15) be chosen for Model I. Suppose that the inverse inequalities hold:

$$\|\epsilon_{nnn}\|_{0,\overline{AB}}\le K_N\|\epsilon\|_{2,S},\qquad \|\epsilon_{nn}\|_{0,\overline{AB}\cup\overline{BC}\cup\overline{CD}}\le K_N^*\|\epsilon\|_{2,S}, \tag{4.4.3}$$

$$|v_{xy}(B)|\le K_N^{**}\|v_n\|_{0,\overline{BC}},\qquad |v_{xy}(C)|\le K_N^{**}\|v_n\|_{0,\overline{BC}}, \tag{4.4.4}$$

where K_N, K_N^*, and K_N^{**} may be unbounded as $N\to\infty$. Then there exists the error bound,

$$\begin{aligned}\|\epsilon\|_{2,S}&=\|u-v\|_{2,S}\\&\le C\left\{K_N+\frac{(K_N^{**}+K_N^*)}{w_1}+\frac{1}{w_3}\right\}\|u-v\|_{\mathrm{B}},\end{aligned}$$

where C is a bounded constant independent of N.

Proof.

For $\mu\in[0,1)$ and the clamped boundary conditions on $\overline{OD}\cup\overline{AB}$, from Marti [327] and eqn. (4.4.1) we have

$$\begin{aligned}\|\epsilon\|^2_{2,S}&\le C|\epsilon|^2_{2,S}\le C\iint_S\Lambda_\mu(\epsilon,\epsilon)\\&\le C\{\|\epsilon_{nn}\|_{0,\overline{AB}}\|v_n\|_{0,\overline{AB}}+\|\epsilon_{nnn}\|_{0,\overline{AB}}\|v-1\|_{0,\overline{AB}}+\|\epsilon_{nn}\|_{0,\overline{BC}}\|v_n\|_{0,\overline{BC}}\\&\quad+\|v_{nnn}\|_{0,\overline{BC}}\|\epsilon\|_{0,\overline{BC}}+\|\epsilon_{nn}\|_{0,\overline{CD}}\|v_n\|_{0,\overline{CD}}\\&\quad+\|v_{nnn}\|_{0,\overline{CD}}\|\epsilon\|_{0,\overline{CD}}+|v_{xy}(B)|\,|\epsilon(B)|+|v_{xy}(C)|\,|\epsilon(C)|\}.\end{aligned} \tag{4.4.5}$$

From the Sobolev imbedding theorem [417], there exist the bounds,

$$\|\epsilon\|_{0,\overline{BC}\cup\overline{CD}} \le C\|\epsilon\|_{2,S}, \qquad |\epsilon(B)| \le C\|\epsilon\|_{2,S}, \qquad |\epsilon(C)| \le C\|\epsilon\|_{2,S}. \quad (4.4.6)$$

Hence, from eqns. (4.4.3), (4.4.5), and (4.4.6) we have

$$\begin{aligned}\|\epsilon\|_{2,S}^2 \le C\{&K_N^*\|v_n\|_{0,\overline{AB}} + K_N\|v-1\|_{0,\overline{AB}} + K_N^*\|v_n\|_{0,\overline{BC}} + \|v_{nnn}\|_{0,\overline{BC}} \\ &+ K_N^*\|v_n\|_{0,\overline{CD}} + \|v_{nnn}\|_{0,\overline{CD}} + |v_{xy}(B)| + |v_{xy}(C)|\}\|\epsilon\|_{2,S}. \quad (4.4.7)\end{aligned}$$

Moreover, from eqn. (4.4.4)

$$|v_{xy}(B)| + |v_{xy}(C)| \le 2K_N^{**}\|v_n\|_{0,\overline{BC}} \le 2\frac{K_N^{**}}{w_1}\|\epsilon\|_{\mathrm{B}}. \quad (4.4.8)$$

Combining eqns. (4.4.7) and (4.4.8) leads to

$$\begin{aligned}\|\epsilon\|_{2,S} &\le C\{K_N^*\|v_n\|_{0,\overline{AB}} + K_N\|v-1\|_{0,\overline{AB}} + K_N^*\|v_n\|_{0,\overline{BC}} + \|v_{nnn}\|_{0,\overline{BC}} \\ &\quad + K_N^*\|v_n\|_{0,\overline{CD}} + \|v_{nnn}\|_{0,\overline{CD}}\} + \frac{2K_N^{**}}{w_1}\|\epsilon\|_{\mathrm{B}} \\ &\le C\left\{K_N + \frac{K_N^* + K_N^{**}}{w_1} + \frac{1}{w_3}\right\}\|\epsilon\|_{\mathrm{B}},\end{aligned}$$

where we have used eqn. (4.4.2). ■

Note that although the corner condition $v_{xy}(B) = v_{xy}(C)$ is not imposed explicitly on v, it is also satisfied approximately based on Theorem 4.4.1. The analysis for the CTM with interior boundary conditions can be made similarly by following Chapter 2 and this chapter. In fact, the assumption, i.e., eqn. (4.4.4) can be obtained by

$$\begin{aligned}|v_{xy}(B)| &\le |v_{xy}|_{\infty,\overline{BC}} \le K_1\|v_n\|_{\frac{3}{2},\overline{BC}} \\ &\le K_1K_2\|v_n\|_{0,\overline{BC}} = K_N^{**}\|v_n\|_{0,\overline{BC}},\end{aligned}$$

where $K_N^{**} = K_1K_2$. When S is a sector, we can show that $K_N = O(N^2)$, $K_N^* = O(N)$, and $K_N^{**} = O(N^{\frac{3}{2}})$.

What is reasonable is real;
that which is real is reasonable.
——— *Philosophy of Right (1821)* ———

Georg Wilhelm Friedrich Hegel
(1770–1831)

Part II
Collocation methods

In fact, the collocation method (CM) can be classified into the category of the domain-type solution procedures. Compared with other domain-type methods such as finite element method (FEM), finite difference method (FDM), and finite volume method (FVM), CM is much easier to perform due to its simplicity in numerical algorithms. It is known that many methods have been developed to approximate the solutions of partial differential equations (PDEs). Since the FEM has been developed in both wide applications and deep theoretical analysis, we employ the FEM theory in Ciarlet [103] and Oden and Reddy [348], to explore the theoretical framework of the CM and its combinations with other methods. Because the solution domains of problems will not be confined in polygons, domain decomposition approach can be taken into account.

If the admissible functions are chosen to be analytical functions, e.g., trigonometric or other orthogonal functions, we may enforce them to satisfy exactly the PDEs at certain collocation nodes, by simply letting the residuals to be zero. This leads to the CM. The traditional CM has been described in a number of books: Bernardi and Maday [35], Canuto et al. [72], Gottlieb and Orszag [170], Li [280], Quarteroni and Valli [374], and Mercier [335]. Here, we also mention several important studies of CM. Bernardi, Debit, and Maday [34] provided a coupling FEM and spectral method with two kinds of matching conditions on interface. Shen [409, 410, 411] gave a series of research study on spectral-Galerkin methods for elliptic equations. Haidvogel [181] applied double Chebyshev polynomials to Poisson's equation. Yin [477] used the Sinc-collocation method to singular Poisson-like problems. Other reports on CM are given by Arnold and Wendland [6], Canuto, Hariharan, and Lustman [71], Pathria and Karniadakis [359], and Sneddon [416].

In this part, a new unified framework of combination of CM with other methods such as FEM, etc., will be analyzed. For the smooth solutions of problems, polynomials of different degree can be chosen to approximate the solutions properly. Besides, different kinds of admissible functions can be chosen such as orthogonal polynomials [35, 72], trigonometric functions [170], radial basis functions (RBFs) [236, 237], the Sinc functions [427], and particular solutions [280]. Using different admissible functions is more flexible for CM to apply the practical problems and to easily fit those on rather arbitrary domains.

This part consists of three chapters:

Chapter 5: Collocation Methods.
Chapter 6: Combinations of Collocation and Finite Element Methods.
Chapter 7: Radial Basis Function Collocation Methods.

The contents of this part are adopted mainly from Refs. [205, 207, 208, 209]. A brief description is given as follows.

Chapter 5 presents the generalized collocation methods (GCM), in which the piecewise admissible functions are used. Besides, the CM for the Robin boundary conditions is discussed. Three typical boundary conditions, Dirichlet, Neumann, and Robin, can be handled well in the CM. Moreover, the collocation Trefftz methods (CTM) in Part I is, in fact, the special cases of the CM.

Chapter 6 provides a framework of combinations of CM with the FEM. The optimal convergence rates can be achieved, where the quadrature formulas play a role only in satisfying the uniformly V_h-elliptic inequality, because the higher order quadrature formulas do not improve much accuracy.

Chapter 7 is a sequential study for previous two chapters. We apply the RBFs (radial basis functions) to CM, simply denoted by RBCM (radial basis collocation method, or called Kansa's method in engineering literature). One important development is to apply RBFs for PDEs, while most of the existing literatures of RBFs deal with surface fitting and functional approximation. The optimal error bounds are derived; the important inverse inequalities are derived, although the entire approaches of proofs are similar to those in Chapters 5 and 6. Some numerical examples are given to display the effectiveness of the RBCM as well.

5 Collocation methods

In Chapters 1–4, the collocation method (CM) has been employed only on the boundary conditions. In Chapters 5–7, we employ the CM both in the solution domain and on its boundaries. Hence, orthogonal polynomials, trigonometric functions, radial basis functions (RBFs), special functions, etc., can also be considered as the admissible functions that do not satisfy the governing equations. The CM in this part can be viewed as the collocation spectral method, or the collocation Ritz–Galerkin method, see Li [280]. In this chapter, we provide an analysis on the CM, which uses a broad range of admissible functions such as orthogonal polynomials, trigonometric functions, RBFs, and particular solutions. The admissible functions can be chosen to be *piecewise*, i.e., different functions are used in different subdomains. The key idea is that the CM can be regarded as the least squares method (LSM) involving integration approximation, and optimal convergence rates can be easily achieved from the traditional analysis of the finite element method (FEM). The key analysis is to prove the uniformly V_h-elliptic inequality and some inverse inequalities used. This chapter explores the interesting fact that for the CM, the integration rules only affect the uniformly V_h-elliptic inequality, but not the solution accuracy. The advantage of the CM is to formulate easily the collocation equations as the associated algebraic equations, which can be solved directly by the LSM, thus to greatly reduce the condition number of the stiffness matrix. Note that the boundary approximation method (BAM) in Ref. [280] and the CTM in Part I are a special case of the CM, where the admissible functions satisfy the governing equations exactly. Numerical experiments in Hu and Li [208] are also carried for Poisson's problem to support the analysis made.

5.1 Introduction

In this chapter, we present a new analysis of CM by following the ideas in Ref. [280] that every numerical method can be regarded as a special kind of the Ritz–Galerkin method. The CM is, indeed, the LSM, which is treated as the Ritz–Galerkin method involving integration approximation. The advantages of the CM are twofold: (1) flexibility of application to different geometric shapes and different elliptic equations, and (2) simplicity of computer programming. The optimal error bounds can be easily derived, based on the uniformly V_h-elliptic inequality, which is proved in detail in this chapter.

This chapter is organized as follows. In the next section, the CM with an interior interface is described, and in Section 5.3 an analysis is given. In Section 5.4, the CM for the Robin boundary conditions is discussed, and in Section 5.5, some inverse inequalities are proven. In the last section, some remarks are made.

5.2 Description of collocation methods

Consider Poisson's equation on domain S with the mixed type of the Dirichlet and Neumann conditions,

$$-\Delta u = -\left(\frac{\partial^2 u}{\partial x^2} + \frac{\partial^2 u}{\partial y^2}\right) = f(x, y) \quad \text{in } S, \tag{5.2.1}$$

$$u \mid_{\Gamma_D} = g_1 \quad \text{on } \Gamma_D, \tag{5.2.2}$$

$$u_\nu \mid_{\Gamma_N} = g_2 \quad \text{on } \Gamma_N, \tag{5.2.3}$$

where S is a polygon, $\partial S = \Gamma = \Gamma_D \cup \Gamma_N$ is its boundary, $u_\nu = \frac{\partial u}{\partial \nu}$, and ν is the unit outnormal to ∂S. Let S be divided by Γ_0 into two disjoint subregions, S_1 and S_2 (see fig. 5.1): $S = S_1 \cup S_2 \cup \Gamma_0$ and $S_1 \cap S_2 = \emptyset$. We give a few assumptions.

(A1) The solutions in S_1 and S_2 can be expanded as

$$v = \begin{cases} v^- = \sum_{i,j=1}^{\infty} a_{ij} \Phi_i(x)\Phi_j(y) & \text{in } S_1, \\ v^+ = \sum_{i,j=1}^{\infty} b_{ij} \Psi_i(x)\Psi_j(y) & \text{in } S_2, \end{cases} \tag{5.2.4}$$

where $\{\Phi_i(x)\Phi_j(y)\}$ and $\{\Psi_i(x)\Psi_j(y)\}$ are complete and independent bases in S_1 and S_2, respectively, and a_{ij} and b_{ij} are the expansion coefficients.

(A2) The basis functions

$$\Phi_i(x)\Phi_j(y) \in C^2(S_1) \cap C^1(\partial S_1), \qquad \Psi_i(x)\Psi_j(y) \in C^2(S_2) \cap C^1(\partial S_2).$$

(A3) The expansions in eqn. (5.2.4) converge exponentially to the true solutions $u^\pm$. Let

$$u^- = u_m^- + R_m^-, \qquad u^+ = u_n^+ + R_n^+,$$

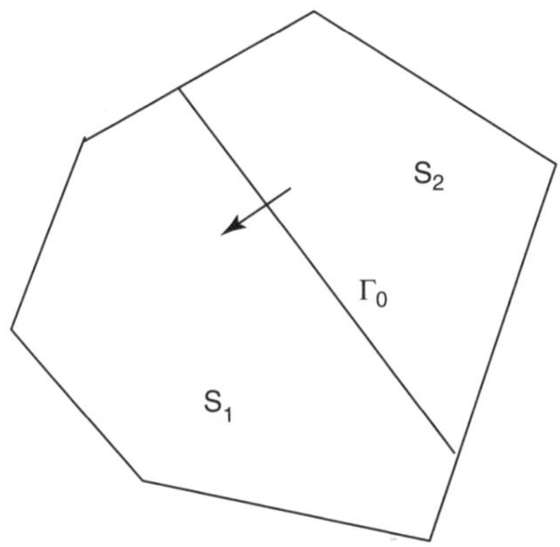

Figure 5.1: Partition of a convex polygon.

where $u^- = u\,|_{S_1}$ and $u^+ = u\,|_{S_2}$, and

$$u_m^- = \sum_{i,j=1}^{m} a_{ij}\Phi_i(x)\Phi_j(y), \qquad u_n^+ = \sum_{i,j=1}^{n} b_{ij}\Psi_i(x)\Psi_j(y),$$

R_m^- and R_n^+ are the remainders, and a_{ij} and b_{ij} are the true expansion coefficients. Then

$$\max_{S_1} |R_m^-| = O(\mathrm{e}^{-\bar{c}m}), \qquad \max_{S_2} |R_n^+| = O(\mathrm{e}^{-\bar{c}n}), \tag{5.2.5}$$

where $\bar{c} > 0$, $m > 1$, and $n > 1$.

Based on **A1**–**A3** we may choose the *piecewise* admissible functions,

$$v = \begin{cases} v^- = \sum_{i,j=1}^{m} \widetilde{a}_{ij}\Phi_i(x)\Phi_j(y) & \text{in } \overline{S}_1, \\ v^+ = \sum_{i,j=1}^{n} \widetilde{b}_{ij}\Psi_i(x)\Psi_j(y) & \text{in } \overline{S}_2, \end{cases} \tag{5.2.6}$$

where $\widetilde{a}_{ij}$ and $\widetilde{b}_{ij}$ are unknown coefficients to be sought, and $\overline{S}_i = S_i \cup \partial S_i$. Since v on Γ_0 are not continuous, $v^\pm$ have to satisfy the interior continuity conditions

$$u^+ = u^-, \qquad u_\nu^+ = u_\nu^-, \quad \text{on } \Gamma_0, \tag{5.2.7}$$

where $u_\nu = \frac{\partial u}{\partial \nu}$, and ν is the outward unit normal of ∂S_2.

Based on **A2** we may seek the coefficients $\widetilde{a}_{ij}$ and $\widetilde{b}_{ij}$ by satisfying eqns. (5.2.1), (5.2.2), (5.2.3), and (5.2.7) directly at nodes $Q_{ij}^\pm$ and Q_i,

$$(\Delta v^\pm + f)(Q_{ij}^\pm) = 0, \quad Q_{ij}^\pm \in S^\pm, \tag{5.2.8}$$

$$(v - g_1)(Q_i) = 0, \quad Q_i \in \Gamma_D, \tag{5.2.9}$$

$$(v_\nu - g_2)(Q_i) = 0, \quad Q_i \in \Gamma_N, \tag{5.2.10}$$

$$(v^+ - v^-)(Q_i) = 0, \quad Q_i \in \Gamma_0, \tag{5.2.11}$$

$$(v_\nu^+ - v_\nu^-)(Q_i) = 0, \quad Q_i \in \Gamma_0, \tag{5.2.12}$$

where $S^- = S_1$ and $S^+ = S_2$. The eqns. (5.2.8)–(5.2.12) can be expressed by the linear algebraic equations,

$$\mathbf{Fx} = \mathbf{b}, \tag{5.2.13}$$

where $\mathbf{x}$ is the unknown vector consisting of $\widetilde{a}_{ij}$ and $\widetilde{b}_{ij}$, $\mathbf{b}$ is the known vector, and $\mathbf{F} \in R^{M\times(m^2+n^2)}$, where $M(\geq m^2+n^2)$ is the total number of collocation nodes $Q_{ij}^{\pm}$ in $S^{\pm}$, and Q_i on $\Gamma_D \cup \Gamma_N \cup \Gamma_0$. In this chapter, we always choose $M > (m^2+n^2)$ and even $M \gg (m^2+n^2)$. Then eqn. (5.2.13) is an overdetermined system of linear algebraic equations. Hence, we may use the LSM (i.e., the QR method or the singular value decomposition (SVD) method) for solving eqn. (5.2.13), see Golub and van Loan [168].

Below, let us view the CM as the LSM involving integration approximation. Denote V_h the finite dimensional collection of the admissible functions, i.e., eqn. (5.2.6). We give one more assumption.

(A4) Suppose that there exists a positive constant $\mu(>0)$ such that

$$\|v_\nu^\pm\|_{0,\Gamma_D\cap \overline{S}^\pm} \leq CL^\mu \|v^\pm\|_{1,S^\pm}, \quad v \in V_h, \tag{5.2.14}$$

$$\|v_\nu^+\|_{0,\Gamma_0\cap \overline{S}^+} \leq CL^\mu \|v^+\|_{1,S^+}, \quad v \in V_h. \tag{5.2.15}$$

For polynomials of degree L, we will prove eqns. (5.2.14) and (5.2.15) with $\mu = 2$ in Section 5.5.

Then the approximate coefficients $\widetilde{a}_{ij}$ and $\widetilde{b}_{ij}$ can be obtained by the LSMs: To seek the approximation solution $u_{m,n} \in V_h$ such that

$$E(u_{m,n}) = \min_{v\in V_h} E(v),$$

where

$$E(v) = \frac{1}{2}\left\{\iint_{S_1} (\Delta v^- + f)^2 + \iint_{S_2} (\Delta v^+ + f)^2 + L^{2\mu}\int_{\Gamma_0} (v^+ - v^-)^2 \right.$$
$$\left. + \int_{\Gamma_0} (v_\nu^+ - v_\nu^-)^2 + L^{2\mu}\int_{\Gamma_D} (v-g_1)^2 + \int_{\Gamma_N} (v_\nu - g_2)^2 \right\}, \tag{5.2.16}$$

where $L = \max\{m, n\}$, and ν is also the outward unit normal to ∂S_2. The eqn. (5.2.16) can be described equivalently

$$a(u_{m,n}, v) = f(v), \quad \forall v \in V_h, \tag{5.2.17}$$

where

$$
\begin{aligned}
a(u,v) = {} & \iint_{S_1} \Delta u\, \Delta v + \iint_{S_2} \Delta u\, \Delta v + L^{2\mu} \int_{\Gamma_0} (u^+ - u^-)(v^+ - v^-) \\
& + \int_{\Gamma_0} (u_\nu^+ - u_\nu^-)(v_\nu^+ - v_\nu^-) + L^{2\mu} \int_{\Gamma_D} uv + \int_{\Gamma_N} u_\nu v_\nu, \\
f(v) = {} & - \iint_{S_1} f\, \Delta v - \iint_{S_2} f\, \Delta v + L^{2\mu} \int_{\Gamma_D} g_1 v + \int_{\Gamma_N} g_2 v_\nu. \qquad (5.2.18)
\end{aligned}
$$

The integrals in eqn. (5.2.18) can be approximated by some rules of integration:

$$
\begin{aligned}
\widehat{\iint_{S^\pm}} g &= \sum_{ij} \alpha_{ij}^\pm g(Q_{ij}^\pm), && Q_{ij}^\pm \in S^\pm, \\
\widehat{\int_{\Gamma_0}} g &= \sum_i \alpha_i g(Q_i), && Q_i \in \Gamma_0, \\
\widehat{\int_{\Gamma_D}} g &= \sum_i \alpha_i^D g(Q_i), && Q_i \in \Gamma_D, \qquad (5.2.19) \\
\widehat{\int_{\Gamma_N}} g &= \sum_i \alpha_i^N g(Q_i), && Q_i \in \Gamma_N,
\end{aligned}
$$

where $\alpha_{ij}^\pm$, α_i, α_i^D, and α_i^N are positive weights, and $Q_{ij}^\pm$ and Q_i are integration nodes. The LSM, i.e., eqn. (5.2.17) is then reduced to

$$
\hat{a}(\hat{u}_{m,n}, v) = \hat{f}(v), \quad \forall v \in V_h, \qquad (5.2.20)
$$

where

$$
\begin{aligned}
\hat{a}(u,v) = {} & \widehat{\iint_{S_1}} \Delta u\, \Delta v + \widehat{\iint_{S_2}} \Delta u\, \Delta v + L^{2\mu} \widehat{\int_{\Gamma_0}} (u^+ - u^-)(v^+ - v^-) \\
& + \widehat{\int_{\Gamma_0}} (u_\nu^+ - u_\nu^-)(v_\nu^+ - v_\nu^-) + L^{2\mu} \widehat{\int_{\Gamma_D}} uv + \widehat{\int_{\Gamma_N}} u_\nu v_\nu, \\
\hat{f}(v) = {} & - \widehat{\iint_{S_1}} f\, \Delta v - \widehat{\iint_{S_2}} f\, \Delta v + L^{2\mu} \widehat{\int_{\Gamma_D}} g_1 v + \widehat{\int_{\Gamma_N}} g_2 v_\nu.
\end{aligned}
$$

It is easy to see that by the rules, i.e., eqn. (5.2.19), the following algebraic equations can be obtained from eqn. (5.2.20) directly,

$$
\sqrt{\alpha_{ij}^\pm}(\Delta v^\pm + f)(Q_{ij}^\pm) = 0, \quad Q_{ij}^\pm \in S^\pm, \qquad (5.2.21)
$$

$$
\sqrt{\alpha_i} L^\mu (v^+ - v^-)(Q_i) = 0, \quad Q_i \in \Gamma_0, \qquad (5.2.22)
$$

$$\sqrt{\alpha_i}(v_\nu^+ - v_\nu^-)(Q_i) = 0, \quad Q_i \in \Gamma_0, \tag{5.2.23}$$

$$\sqrt{\alpha_i^D} L^\mu (v - g_1)(Q_i) = 0, \quad Q_i \in \Gamma_D, \tag{5.2.24}$$

$$\sqrt{\alpha_i^N}(v_\nu - g_2)(Q_i) = 0, \quad Q_i \in \Gamma_N. \tag{5.2.25}$$

Compared with eqns. (5.2.8)–(5.2.12), the eqns. (5.2.21)–(5.2.25) can be denoted by

$$\mathbf{WFx} = \mathbf{Wb},$$

where $\mathbf{F}$ is given in eqn. (5.2.13), and $\mathbf{W} \in R^{M\times M}$ is the diagonal weight matrix, consisting of the weights, $\sqrt{\alpha_{ij}^\pm}$, $L^\mu\sqrt{\alpha_i}$, $\sqrt{\alpha_i}$, $L^\mu\sqrt{\alpha_i^D}$, and $\sqrt{\alpha_i^N}$. We may also obtain the coefficients (i.e., $\mathbf{x}$) by solving the normal equations:

$$\mathbf{Ax} = \mathbf{b}^*,$$

where matrix $\mathbf{A} = \mathbf{F}^T\mathbf{W}^T\mathbf{WF}$ is symmetric and positive definite, and the known vector $\mathbf{b}^* = \mathbf{F}^T\mathbf{W}^T\mathbf{Wb}$.

5.3 Error analysis

We will provide the error bounds for the solutions from eqns. (5.2.17) and (5.2.20). Denote the space

$$H^* = \{v \mid v \in L^2(S),\ v^\pm \in H^1(S^\pm), \Delta v^\pm \in L^2(S^\pm)\},$$

accompanied with the norm

$$\|v\|_H = \{\|v\|_1^2 + \|\Delta v\|_{0,S_1}^2 + \|\Delta v\|_{0,S_2}^2 + L^{2\mu}\|v^+ - v^-\|_{0,\Gamma_0}^2 + \|v_\nu^+ - v_\nu^-\|_{0,\Gamma_0}^2 + L^{2\mu}\|v\|_{0,\Gamma_D}^2 + \|v_\nu\|_{0,\Gamma_N}^2\}^{\frac{1}{2}},$$

where

$$\|v\|_1 = \{\|v\|_{1,S_1}^2 + \|v\|_{1,S_2}^2\}^{\frac{1}{2}}, \qquad |v|_1 = \{|v|_{1,S_1}^2 + |v|_{1,S_2}^2\}^{\frac{1}{2}},$$

and $\|v\|_{1,S_1}$, $\|v\|_{0,\Gamma_0}$, etc., are the Sobolev norms [417]. Obviously, $V_h \subset H^*$. Then

$$\|v\|_H^2 = \|v\|_1^2 + a(v,v).$$

Now, we have a theorem.

Theorem 5.3.1

Suppose that there exist two inequalities,

$$a(u,v) \le C\|u\|_H \times \|v\|_H, \quad \forall v \in V_h, \tag{5.3.1}$$

$$a(v,v) \ge C_0\|v\|_H^2, \quad \forall v \in V_h, \tag{5.3.2}$$

where $C_0(>0)$ and C are two constants independent of m and n. Then, the solution of the LSM, i.e., eqn. (5.2.17) has the error bound,

$$\|u - u_{m,n}\|_H = C \inf_{v \in V_h} \|u - v\|_H \tag{5.3.3}$$

$$\leq \varepsilon_1 = \|R_m^-\|_{2,S_1} + \|R_n^+\|_{2,S_2} + L^\mu \|R_L\|_{0,\Gamma_D \cup \Gamma_0} + \|(R_L)_\nu\|_{0,\Gamma_N \cup \Gamma_0},$$

where $|R_L| = |R_m^-| + |R_n^+|$.

Proof.

For the true solution, we have $a(u, v) = f(v), \forall v \in V_h$. Then

$$a(u - u_{m,n}, v) = 0, \quad \forall v \in V_h. \tag{5.3.4}$$

Denote the projection solution on V_h

$$u_I = \begin{cases} u_I^- = \sum_{i,j=1}^{m} a_{ij} \Phi_i(x) \Phi_j(y) & \text{in } \overline{S}_1, \\ u_I^+ = \sum_{i,j=1}^{n} b_{ij} \Psi_i(x) \Psi_j(y) & \text{in } \overline{S}_2, \end{cases}$$

where a_{ij} and b_{ij} are the true expansion coefficients. Then, $u_I \in V_h$. Let $v \in V_h$, and $w = u_{m,n} - v \in V_h$. We have from eqns. (5.3.2), (5.3.4), and (5.3.1)

$$C_0 \|w\|_H^2 \leq a(u_{m,n} - v, w) = a(u - v, w)$$
$$\leq C\|u - v\|_H \|w\|_H.$$

This leads to

$$\|u_{m,n} - v\|_H = \|w\|_H \leq C\|u - v\|_H.$$

Then, we obtain

$$\|u_{m,n} - u\|_H \leq \|u_{m,n} - v\|_H + \|u - v\|_H \leq C\|u - v\|_H,$$

and

$$\|u_{m,n} - u\|_H \leq C \inf_{v \in V_h} \|u - v\|_H.$$

Let $v = u_I$, we have

$$\|u_{m,n} - u\|_H \leq C \inf_{v \in V_h} \|u - v\|_H \leq C\|u - u_I\|_H$$

$$\leq C\{\|R_m^-\|_{2,S_1} + \|R_n^+\|_{2,S_2} + L^\mu \|R_L\|_{0,\Gamma_D \cup \Gamma_0} + \|(R_L)_\nu\|_{0,\Gamma_N \cup \Gamma_0}\}.$$

■

Since the solution u of eqns. (5.2.1)–(5.2.3) satisfies eqns. (5.2.21)–(5.2.25) exactly, then

$$\hat{a}(u,v) = \hat{f}(v), \quad \forall v \in V_h. \tag{5.3.5}$$

We can also prove the following theorem similarly, see Ciarlet [103] and Strang and Fix [426].

Theorem 5.3.2

Suppose that there exist two inequalities

$$\begin{aligned} \hat{a}(u,v) &\le C\|u\|_H \times \|v\|_H, && \forall v \in V_h, \\ \hat{a}(v,v) &\ge C_0\|v\|_H^2, && \forall v \in V_h, \end{aligned} \tag{5.3.6}$$

where $C_0(>0)$ and C are two constants independent of m and n. Then, the solution of the CM, i.e., eqn. (5.2.20) has the error bound,

$$\|u - \hat{u}_{m,n}\|_H = C \inf_{v \in V_h} \|u - v\|_H \le \varepsilon_1,$$

where ε_1 is given in eqn. (5.3.3).

Note that for the FEM, finite difference method (FDM), etc., the true solution does not satisfy eqn. (5.3.5); then Theorem 5.3.2 may not hold. Also, the analysis in this chapter is different from the traditional analysis in the CM in Refs. [35, 72, 170, 335, 374] where only the zeros of polynomials are used as the collocation nodes.

Below, we prove the uniformly V_h-elliptic inequalities in eqns. (5.3.2) and (5.3.6). We cite Lemma 1.1.1 of Chapter 1 as a lemma.

Lemma 5.3.1

Let $\Gamma_D \cap S_1 \neq \emptyset$. If $v \in H^*$, then there exists a positive constant C independent of v such that

$$\|v\|_1 \le C\{|v|_1 + \|v\|_{0,\Gamma_D} + \|v^+ - v^-\|_{0,\Gamma_0}\}.$$

Lemma 5.3.2

Let **A4** *be given, and $\Gamma_D \cap S_1 \neq \emptyset$. There exists the bound for all $v \in V_h$,*

$$C_0\|v\|_1^2 \le a(v,v), \tag{5.3.7}$$

where $C_0(>0)$ is a constant independent of m and n.

Proof.

We have

$$|v|_1^2 = \iint_{S_1} |\nabla v|^2 + \iint_{S_2} |\nabla v|^2 \tag{5.3.8}$$
$$= -\iint_{S_1} v\,\Delta v - \iint_{S_2} v\,\Delta v + \int_{\partial S_1} v_\nu^- v^- + \int_{\partial S_2} v_\nu^+ v^+$$
$$= -\iint_{S_1} v\,\Delta v - \iint_{S_2} v\,\Delta v + \int_{\Gamma_0} (v_\nu^+ v^+ - v_\nu^- v^-) + \int_{\Gamma_D} v_\nu v + \int_{\Gamma_N} v_\nu v,$$

where ν is the unit outnormal to ∂S or ∂S_2. Below, we give the bounds of all terms on the right-hand side in the above equation.

First we have from eqn. (5.2.15)

$$\left|\int_{\Gamma_0} (v_\nu^+ v^+ - v_\nu^- v^-)\right| \le \left|\int_{\Gamma_0} (v^+ - v^-) v_\nu^+\right| + \left|\int_{\Gamma_0} (v_\nu^+ - v_\nu^-) v^-\right| \tag{5.3.9}$$
$$\le \|v^+ - v^-\|_{0,\Gamma_0} \|v_\nu^+\|_{0,\Gamma_0} + \|v_\nu^+ - v_\nu^-\|_{0,\Gamma_0} \|v^-\|_{0,\Gamma_0}$$
$$\le C\{L^\mu \|v^+ - v^-\|_{0,\Gamma_0} + \|v_\nu^+ - v_\nu^-\|_{0,\Gamma_0}\} \|v\|_1,$$

where we have used the bounds,

$$\|v^-\|_{0,\Gamma_0} \le \|v^-\|_{1,S^-}, \qquad \|v^\pm\|_{1,S^\pm} \le \|v\|_1.$$

Next, we obtain from eqn. (5.2.14)

$$\left|\int_{\Gamma_D} v_\nu v\right| \le \|v\|_{0,\Gamma_D} \|v_\nu\|_{0,\Gamma_D} \le CL^\mu \|v\|_{0,\Gamma_D} \|v\|_1, \tag{5.3.10}$$
$$\left|\int_{\Gamma_N} v_\nu v\right| \le \|v_\nu\|_{0,\Gamma_N} \|v\|_{0,\Gamma_N} \le C\|v_\nu\|_{0,\Gamma_N} \|v\|_1. \tag{5.3.11}$$

Moreover, there exist the bounds,

$$\left|\iint_{S_1} v\,\Delta v\right| \le \|\Delta v\|_{0,S_1} \|v\|_{0,S_1} \le \|\Delta v\|_{0,S_1} \|v\|_1, \tag{5.3.12}$$
$$\left|\iint_{S_2} v\,\Delta v\right| \le \|\Delta v\|_{0,S_2} \|v\|_1. \tag{5.3.13}$$

From eqns. (5.3.8)–(5.3.13),

$$|v|_1^2 \le \{\|\Delta v\|_{0,S_1} + \|\Delta v\|_{0,S_2} + CL^\mu(\|v^+ - v^-\|_{0,\Gamma_0} + \|v\|_{0,\Gamma_D})$$
$$+ \|v_\nu^+ - v_\nu^-\|_{0,\Gamma_0} + \|v_\nu\|_{0,\Gamma_N}\} \|v\|_1. \tag{5.3.14}$$

Hence, we have from Lemma 5.3.1

$$\begin{aligned}\|v\|_1^2 &\le C\{|v|_1^2 + \|v\|_{0,\Gamma_D}^2 + \|v^+ - v^-\|_{0,\Gamma_0}^2\} \\ &\le C\{|v|_1^2 + (\|v\|_{0,\Gamma_D} + \|v^+ - v^-\|_{0,\Gamma_0})\|v\|_1\}.\end{aligned} \tag{5.3.15}$$

Combining eqns. (5.3.14) and (5.3.15) gives

$$\begin{aligned}\|v\|_1^2 \le C\{&\|\Delta v\|_{0,S_1} + \|\Delta v\|_{0,S_2} + L^{\mu}(\|v^+ - v^-\|_{0,\Gamma_0} + \|v\|_{0,\Gamma_D}) \\ &+ \|v_\nu^+ - v_\nu^-\|_{0,\Gamma_0} + \|v_\nu\|_{0,\Gamma_N}\}\|v\|_1.\end{aligned}$$

This leads to

$$\begin{aligned}\|v\|_1 \le C\{&\|\Delta v\|_{0,S_1} + \|\Delta v\|_{0,S_2} + L^{\mu}(\|v^+ - v^-\|_{0,\Gamma_0} + \|v\|_{0,\Gamma_D}) \\ &+ \|v_\nu^+ - v_\nu^-\|_{0,\Gamma_0} + \|v_\nu\|_{0,\Gamma_N}\},\end{aligned}$$

and then

$$\begin{aligned}\|v\|_1^2 \le C\{&\|\Delta v\|_{0,S_1}^2 + \|\Delta v\|_{0,S_2}^2 + L^{2\mu}(\|v^+ - v^-\|_{0,\Gamma_0}^2 + \|v\|_{0,\Gamma_D}^2) \\ &+ \|v_\nu^+ - v_\nu^-\|_{0,\Gamma_0}^2 + \|v_\nu\|_{0,\Gamma_N}^2\} = Ca(v,v).\end{aligned}$$

This is the desired result, i.e., eqn. (5.3.7). ■

Theorem 5.3.3

Let **A4** *and* $\Gamma_D \cap \partial S_1 \neq \emptyset$ *hold. Then there exists the uniformly* V_h*-elliptic inequality eqn. (5.3.2).*

Proof.

From Lemma 5.3.2, we have the bound,

$$\begin{aligned}a(v,v) &= \frac{1}{2}a(v,v) + \frac{1}{2}a(v,v) \\ &\ge C_0\|v\|_1^2 + \frac{1}{2}\{\|\Delta v\|_{0,S_1}^2 + \|\Delta v\|_{0,S_2}^2 + L^{2\mu}(\|v^+ - v^-\|_{0,\Gamma_0}^2 + \|v\|_{0,\Gamma_D}^2) \\ &\qquad + \|v_\nu^+ - v_\nu^-\|_{0,\Gamma_0}^2 + \|v_\nu\|_{0,\Gamma_N}^2\} \\ &\ge \overline{C}_0\|v\|_H^2,\end{aligned}$$

where $\overline{C}_0 = \min\{\frac{1}{2}, C_0\}$. ■

Next, we derive the uniformly V_h-elliptic inequality eqn. (5.3.6). We need a stronger assumption than **A4**.

(A5) Suppose that there exists a positive constant $\mu(>0)$ such that for $v \in V_h$

$$
\begin{aligned}
\|v^{\pm}\|_{k,\Gamma_D\cap\overline{S}^{\pm}} &\leq CL^{k\mu}\|v^{\pm}\|_{0,\Gamma_D\cap S^{\pm}},\\
\|v^{\pm}\|_{k,\Gamma_0} &\leq CL^{k\mu}\|v^{\pm}\|_{0,\Gamma_0},\\
\|v_\nu^{\pm}\|_{k,\Gamma_N\cap\overline{S}^{\pm}} &\leq CL^{(k+1)\mu}\|v^{\pm}\|_{1,S^{\pm}},\\
\|v_\nu^{\pm}\|_{k,\Gamma_0\cap\overline{S}^{\pm}} &\leq CL^{(k+1)\mu}\|v^{\pm}\|_{1,S^{\pm}},
\end{aligned}
\tag{5.3.16}
$$

where $k=0,1,\ldots$.

We will give an analysis for the integration approximation. Take $\widehat{\int}_{\Gamma_0}(v_\nu^+ - v_\nu^-)^2$ as example. Choose the integral rule of order r,

$$\widehat{\int_{\Gamma_0}} g = \int_{\Gamma_0} \hat{g}, \tag{5.3.17}$$

where $\hat{g}$ is the interpolant polynomial of g with order r on the partition of Γ_0 with the maximal meshspacing h. Denote

$$\overline{\|v\|}^2_{0,\Gamma_0} = \widehat{\int_{\Gamma_0}} v^2.$$

We have the following lemma.

Lemma 5.3.3

Let eqn. (5.3.16) be given. For rule, i.e., eqn. (5.3.17) with order r, there exists the bound for $v \in V_h$,

$$\left|\overline{\|v_\nu^+ - v_\nu^-\|}^2_{0,\Gamma_0} - \|v_\nu^+ - v_\nu^-\|^2_{0,\Gamma_0}\right| \leq Ch^{r+1}L^{(r+3)\mu}\|v\|_1^2. \tag{5.3.18}$$

Proof.

Let $g=(v_\nu^+ - v_\nu^-)^2$. We have

$$
\begin{aligned}
\left|\overline{\|v_\nu^+ - v_\nu^-\|}^2_{0,\Gamma_0} - \|v_\nu^+ - v_\nu^-\|^2_{0,\Gamma_0}\right| &= \left|\int_{\Gamma_0}(\hat{g}-g)\right|\\
&\leq Ch^{r+1}|g|_{r+1,\Gamma_0},
\end{aligned}
\tag{5.3.19}
$$

where

$$
\begin{aligned}
|g|_{r+1,\Gamma_0} &= |(v_\nu^+ - v_\nu^-)^2|_{r+1,\Gamma_0}\\
&\leq 2|(v_\nu^+)^2|_{r+1,\Gamma_0} + 2|(v_\nu^-)^2|_{r+1,\Gamma_0}.
\end{aligned}
\tag{5.3.20}
$$

From eqn. (5.3.16),

$$
\begin{aligned}
|(v_\nu^+)^2|_{r+1,\Gamma_0} &\le C\sum_{i=0}^{r+1} |v_\nu^+|_{r+1-i,\Gamma_0}|v_\nu^+|_{i,\Gamma_0} \\
&\le C\sum_{i=0}^{r+1}(L^{(r-i+2)\mu}\|v\|_{1,S_2})\times(L^{(i+1)\mu}\|v\|_{1,S_2}) \\
&\le CL^{(r+3)\mu}\|v\|_{1,S_2}^2.
\end{aligned}
\tag{5.3.21}
$$

Similarly,

$$
|(v_\nu^-)^2|_{r+1,\Gamma_0} \le CL^{(r+3)\mu}\|v\|_{1,S_1}^2. \tag{5.3.22}
$$

Combining eqns. (5.3.19)–(5.3.22) gives the desired result, i.e., eqn. (5.3.18). ■

Similarly, we can prove the following lemma.

Lemma 5.3.4

Let **A5** *be given. For the rule, i.e., eqn. (5.3.17) with order r, there exist the bounds for $v \in V_h$,*

$$
\begin{aligned}
\left|\overline{\|v^+ - v^-\|_{0,\Gamma_0}^2} - \|v^+ - v^-\|_{0,\Gamma_0}^2\right| &\le Ch^{r+1}L^{(r+1)\mu}\|v\|_1^2, \\
\left|\overline{\|v\|_{0,\Gamma_D}^2} - \|v\|_{0,\Gamma_D}^2\right| &\le Ch^{r+1}L^{(r+1)\mu}\|v\|_1^2, \\
\left|\overline{\|v_\nu\|_{0,\Gamma_N}^2} - \|v_\nu\|_{0,\Gamma_N}^2\right| &\le Ch^{r+1}L^{(r+3)\mu}\|v\|_1^2.
\end{aligned}
$$

Now, we give an essential assumption.

(A6) Suppose that

$$
\|v^\pm\|_{k,S^\pm} \le CL^{(k-1)\mu}\|v\|_{1,S^\pm}, \quad k \ge 1, \ \ v \in V_h,
$$

where μ is a constant independent of m and n. Choose the integral rule of order r in S,

$$
\widehat{\iint_S} g = \iint_S \hat{g}, \tag{5.3.23}
$$

where $\hat{g}$ is the interpolant of polynomials of order r. We can also prove the following lemma easily.

Lemma 5.3.5

Let **A6** *be given and the rule, i.e., eqn. (5.3.23) be chosen with order r. There exists the bound,*

$$\left| \left(\widehat{\iint_{S^\pm}} - \iint_{S^\pm} \right) (\Delta v)^2 \right| \leq C h^{r+1} L^{(r+3)\mu} \|v\|_{1,S^\pm}^2 .$$

Theorem 5.3.4

Let **A5–A6** *and $\Gamma_D \cap \partial S_1 \neq \emptyset$ hold. We choose h to satisfy*

$$L^{(r+3)\mu} h^{r+1} = o(1). \tag{5.3.24}$$

Then there exists the uniformly V_h-elliptic inequality eqn. (5.3.6).

Proof.

We have from Theorem 5.3.3 and Lemmas 5.3.3–5.3.5,

$$\begin{aligned}
\hat{a}(v, v) &\geq a(v, v) - C L^{(r+3)\mu} h^{r+1} \|v\|_1^2 \\
&\geq C_0 \|v\|_H^2 - C L^{(r+3)\mu} h^{r+1} \|v\|_1^2 \\
&\geq C_0 \left\{ \left(1 - \frac{C}{C_0} L^{(r+3)\mu} h^{r+1} \right) \|v\|_1^2 + \|\Delta v\|_{0,S_1}^2 + \|\Delta v\|_{0,S_2}^2 \right. \\
&\qquad \left. + L^{2\mu} \|v^+ - v^-\|_{0,\Gamma_0}^2 + \|v_\nu^+ - v_\nu^-\|_{0,\Gamma_0}^2 + L^{2\mu} \|v\|_{0,\Gamma_D}^2 + \|v_\nu\|_{0,\Gamma_N}^2 \right\} \\
&\geq \frac{C_0}{2} \|v\|_H^2,
\end{aligned}$$

provided that

$$\frac{C}{C_0} L^{(r+3)\mu} h^{r+1} \leq \frac{1}{2},$$

which is valid due to eqn. (5.3.24). ■

5.4 Robin boundary conditions

In the above sections, only the Dirichlet and Neumann boundary conditions are discussed. In this section, we consider Poisson's equation involving the

Robin boundary condition

$$-\triangle u = -\left(\frac{\partial^2 u}{\partial x^2} + \frac{\partial^2 u}{\partial y^2}\right) = f(x, y) \quad \text{in } S, \tag{5.4.1}$$

$$u_\nu \mid_{\Gamma_N} = g_1 \quad \text{on } \Gamma_N, \tag{5.4.2}$$

$$(u_\nu + \beta u) \mid_{\Gamma_R} = g_2 \quad \text{on } \Gamma_R, \tag{5.4.3}$$

where $\beta \geq \beta_0 > 0$, $\partial S = \Gamma = \Gamma_N \cup \Gamma_R$. Assume Meas$(\Gamma_R) > 0$ for the unique solution. For simplicity, let $\Gamma_0 = \emptyset$. (When $\Gamma_0 \neq \emptyset$, a similar analysis can be made easily by following Sections 5.2 and 5.3.) We also give two more assumptions.

(A7) The solutions in S can be expanded as

$$v = \sum_{i=1}^{\infty} a_i \Phi_i \quad \text{in } S, \tag{5.4.4}$$

where $\Phi_i (\in C^2(S) \cap C^1(\partial S))$ are complete and independent bases in S, and a_i are the exact expansion coefficients.

(A8) The expansions in eqn. (5.4.4) converge exponentially to the true solution u. Let

$$u = u_m + R_m,$$

where

$$u_m = \sum_{i=1}^{m} a_i \Phi_i, \qquad R_m = \sum_{i=m+1}^{\infty} a_i \Phi_i,$$

and a_i are the true expansion coefficients. Then

$$\max_S |R_m| = O(\mathrm{e}^{-\overline{c}m}),$$

where $\overline{c} > 0$ and $m > 1$.

Based on **A7**–**A8** we may choose the uniform admissible functions,

$$v = \sum_{i=1}^{m} \widetilde{a}_i \Phi_i \quad \text{in } S, \tag{5.4.5}$$

where $\widetilde{a}_i$ are unknown coefficients to be sought. Denote by V_h the collection of the functions, i.e., eqn. (5.4.5).

Choose the integral rules:

$$\widehat{\iint_S} g = \sum_{ij} \alpha_{ij} g(Q_{ij}), \quad Q_{ij} \in S, \tag{5.4.6}$$

$$\widehat{\int_{\Gamma_N}} g = \sum_i \alpha_i^N g(Q_i), \quad Q_i \in \Gamma_N, \tag{5.4.7}$$

$$\widehat{\int_{\Gamma_R}} g = \sum_i \alpha_i^R g(Q_i), \quad Q_i \in \Gamma_R. \tag{5.4.8}$$

We may seek the coefficients $\widetilde{a}_i$ by satisfying the eqns. (5.4.1)–(5.4.3) directly at Q_{ij} and Q_i,

$$\sqrt{\alpha_{ij}}(\Delta v + f)(Q_{ij}) = 0, \quad Q_{ij} \in S, \tag{5.4.9}$$

$$\sqrt{\alpha_i^N}(v_\nu - g_1)(Q_i) = 0, \quad Q_i \in \Gamma_N, \tag{5.4.10}$$

$$\sqrt{\alpha_i^R}(v_\nu + \beta v - g_2)(Q_i) = 0, \quad Q_i \in \Gamma_R. \tag{5.4.11}$$

The CM described in eqns. (5.4.9)–(5.4.11) can be written as

$$\hat{b}(\hat{u}_m, v) = \hat{f}(v), \quad \forall v \in V_h, \tag{5.4.12}$$

where

$$\hat{b}(u, v) = \widehat{\iint_S} \Delta u \, \Delta v + \widehat{\int_{\Gamma_N}} u_\nu v_\nu + \widehat{\int_{\Gamma_R}} (u_\nu + \beta u)(v_\nu + \beta v),$$

$$\hat{f}(v) = -\widehat{\iint_S} f \, \Delta v + \widehat{\int_{\Gamma_N}} g_1 v_\nu + \widehat{\int_{\Gamma_R}} g_2(v_\nu + \beta v).$$

The corresponding LSMs are then denoted by

$$b(u_m, v) = f(v), \quad \forall v \in V_h,$$

where

$$b(u, v) = \iint_S \Delta u \, \Delta v + \int_{\Gamma_N} u_\nu v_\nu + \int_{\Gamma_R} (u_\nu + \beta u)(v_\nu + \beta v),$$

$$f(v) = -\iint_S f \, \Delta v + \int_{\Gamma_N} g_1 v_\nu + \int_{\Gamma_R} g_2(v_\nu + \beta v).$$

Denote the norm

$$\|v\|_h = \{\|v\|_{1,S}^2 + \|\Delta v\|_{0,S}^2 + \|v_\nu\|_{0,\Gamma_N}^2 + \|(v_\nu + \beta v)\|_{0,\Gamma_R}^2\}^{\frac{1}{2}}.$$

Now, we have a lemma.

Lemma 5.4.1

Let $Meas(\Gamma_R) > 0$. There exists the uniformly V_h-elliptic inequality

$$b(v, v) \geq C_0 \|v\|_{1,S}^2, \quad \forall v \in V_h. \tag{5.4.13}$$

Proof.

We have

$$
\begin{aligned}
|v|_{1,S}^2 &= \iint_S |\nabla v|^2 = -\iint_S v\,\Delta v + \int_{\partial S} v_\nu v \\
&\le \|\Delta v\|_{0,S}\|v\|_{0,S} + \int_{\Gamma_N} v_\nu v + \int_{\Gamma_R} (v_\nu + \beta v)v - \int_{\Gamma_R} \beta v^2 \\
&\le \|\Delta v\|_{0,S}\|v\|_{0,S} + \|v_\nu\|_{0,\Gamma_N}\|v\|_{0,\Gamma_N} + \|v_\nu + \beta v\|_{0,\Gamma_R}\|v\|_{0,\Gamma_R} - \int_{\Gamma_R} \beta v^2 \\
&\le \{\|\Delta v\|_{0,S} + \|v_\nu\|_{0,\Gamma_N} + \|v_\nu + \beta v\|_{0,\Gamma_R}\}\|v\|_{1,S} - \int_{\Gamma_R} \beta v^2,
\end{aligned}
$$

where we have used the bounds

$$
\|v\|_{0,\Gamma_N} \le C\|v\|_{1,S}, \qquad \|v\|_{0,\Gamma_R} \le C\|v\|_{1,S}.
$$

This leads to

$$
|v|_{1,S}^2 + \int_{\Gamma_R} \beta v^2 \le \{\|\Delta v\|_{0,S} + \|v_\nu\|_{0,\Gamma_N} + \|v_\nu + \beta v\|_{0,\Gamma_R}\}\|v\|_{1,S}. \tag{5.4.14}
$$

On the other hand, for $\text{Meas}(\Gamma_R) > 0$,

$$
\|v\|_{1,S}^2 \le C(|v|_{1,S}^2 + \beta\|v\|_{0,\Gamma_R}^2). \tag{5.4.15}
$$

Combining eqns. (5.4.14) and (5.4.15) gives

$$
\begin{aligned}
\|v\|_{1,S}^2 &\le C\left(|v|_{1,S}^2 + \int_{\Gamma_R} \beta v^2\right) \\
&\le C\{\|\Delta v\|_{0,S} + \|v_\nu\|_{0,\Gamma_N} + \|v_\nu + \beta v\|_{0,\Gamma_R}\}\|v\|_{1,S}.
\end{aligned}
$$

This leads to

$$
\|v\|_{1,S} \le C\{\|\Delta v\|_{0,S} + \|v_\nu\|_{0,\Gamma_N} + \|v_\nu + \beta v\|_{0,\Gamma_R}\},
$$

and then

$$
\|v\|_{1,S}^2 \le C\{\|\Delta v\|_{0,S}^2 + \|v_\nu\|_{0,\Gamma_N}^2 + \|v_\nu + \beta v\|_{0,\Gamma_R}^2\} = Cb(v, v).
$$

This is eqn. (5.4.13). ■

Note that the true solution u also satisfies eqn. (5.4.12) exactly, $\hat{b}(\hat{u}_m, v) = \hat{f}(v)$, $\forall v \in V_h$. A similar argument as in Theorem 5.3.4 can be given for eqn. (5.4.16): $\hat{b}(v, v) \ge C_0\|v\|_h^2$, $\forall v \in V_h$. We can obtain the following theorem by following Sections 5.2 and 5.3.

Theorem 5.4.1

Suppose that there exist two inequalities,

$$\begin{aligned} \hat{b}(u, v) &\leq C\|u\|_h \times \|v\|_h, && \forall v \in V_h, \\ \hat{b}(v, v) &\geq C_0\|v\|_h^2, && \forall v \in V_h, \end{aligned} \tag{5.4.16}$$

where $C_0(> 0)$ and C are two constants independent of m. Then the solution of the CM, i.e., eqn. (5.4.12) has the error bound,

$$\begin{aligned} \|u - \hat{u}_m\|_h &= C \inf_{v \in V_h} \|u - v\|_h \\ &\leq C\{\|R_m\|_{2,S} + \|(R_m)_\nu\|_{0,\Gamma_N} + \|(R_m)_\nu\|_{0,\Gamma_R}\}. \end{aligned}$$

Note that when the admissible functions are chosen to satisfy Poisson's equation, the CTM, i.e., the BAM, is then obtained from the CM. Hence, the CTM in Chapter 2 is a special case of the CM in this chapter. Moreover, in traditional CM, some difficulties are encountered for the Neumann boundary conditions, see Ref. [374]. In this chapter, the techniques given can handle very well for both the Neumann and the Robin boundary conditions.

5.5 Inverse inequalities

In the above analysis, we need the inverse estimates in **A4**, **A5**, and **A6**. In fact, the inverse estimates in **A6** is essential. Take the norms on Γ_0 as example. We have from assumption **A6**

$$\|v^+\|_{k,\Gamma_0} \leq C\|v^+\|_{k+1,S^+} \leq CL^{k\mu}\|v^+\|_{1,S^+}, \tag{5.5.1}$$

$$\|v_\nu^+\|_{k,\Gamma_0} \leq C\|v^+\|_{k+2,S^+} \leq CL^{(k+1)\mu}\|v^+\|_{1,S^+}. \tag{5.5.2}$$

Hence, **A5** can be replaced by eqns. (5.5.1), (5.5.2), etc., and the proof for Lemma 5.3.4 and Theorem 5.3.4 is similar.

To prove the inverse inequalities, in this chapter we confine ourselves to the smooth solution of eqns. (5.2.1)–(5.2.3), and choose admissible functions $\Phi_i(x)$ and $\Psi_i(x)$ in eqn. (5.2.4), and Φ_i in eqn. (5.4.4) as polynomials of order i. Theorem 5.5.1 yields the essential inverse inequality. As to other admissible functions, such as RBFs, the inverse inequality will be proven in Chapter 7. As long as the inverse inequalities hold, the uniformly V_h-elliptic inequality holds and then the optimal error estimates can be achieved easily.

First, we cite the results in Li [280], pp. 161–163, as two lemmas.

Lemma 5.5.1

Let $\rho_L = \rho_L(x)$ be an L-order polynomial on $[-1,1]$. Then there exists a constant C independent of L such that

$$\|\rho_L'\|_{0,[-1,1]} \le CL^2 \|\rho_L\|_{0,[-1,1]}.$$

Lemma 5.5.2

Suppose that Γ_0 is made up of finite sections of straight lines, and that the admissible function w_h in S_2 is an L-order polynomial. Then there exists a constant C independent of L such that

$$\sup_{w_h \in V_h} \frac{|w_h^+|_{k+1,\Gamma_0}}{\|w_h\|_1} \le CL^{2(k+1)}.$$

Lemma 5.5.3

Let $\square = \{(x, y) \mid -1 \le x \le 1, -1 \le y \le 1\}$, and choose

$$w_L = \sum_{i,j=0}^{L} a_{ij} T_i(x) T_j(y), \quad (x, y) \in \square, \tag{5.5.3}$$

where a_{ij} are expansion coefficients, and $T_i(x)$ are the Chebyshev polynomials of order i. Then there exist the inverse inequalities,

$$\left\| \frac{\partial}{\partial x} w_L \right\|_{0,\square} \le CL^2 \|w_L\|_{0,\square}, \tag{5.5.4}$$

$$\left\| \frac{\partial}{\partial y} w_L \right\|_{0,\square} \le CL^2 \|w_L\|_{0,\square}, \tag{5.5.5}$$

where C is a constant independent of L.

Proof.

We prove eqn. (5.5.4) only, since the proof for eqn. (5.5.5) is similar. We may express w_L by the Legendre polynomials

$$w_L = \sum_{i,j=0}^{L} b_{ij} P_i(x) P_j(y), \quad (x, y) \in \square,$$

where the coefficients b_{ij} from eqn. (5.5.3) are uniquely determined. We have from the orthogonality of the Legendre polynomials,

$$\begin{aligned}\|w_L\|_{0,\square}^2 &= \iint_\square \left(\sum_{i,j=0}^{L} b_{ij}P_i(x)P_j(y)\right)^2 \\ &= \sum_{i,j=0}^{L} \frac{4b_{ij}^2}{(2i+1)(2j+1)} = \sum_{j=0}^{L} \frac{2}{2j+1} \sum_{i=0}^{L} \frac{2b_{ij}^2}{2i+1} \\ &= \sum_{j=0}^{L} \frac{2}{2j+1} \|z_j\|_{0,[-1,1]}^2,\end{aligned} \tag{5.5.6}$$

where z_j are polynomials of order L,

$$z_j = z_j(x) = \sum_{i=0}^{L} b_{ij}P_i(x), \quad x \in [-1,1]. \tag{5.5.7}$$

On the other hand, we have

$$\begin{aligned}\left\|\frac{\partial}{\partial x}w_L\right\|_{0,\square}^2 &= \iint_\square \left(\sum_{i,j=0}^{L} b_{ij}^2 P_i'(x)P_j(y)\right)^2 \\ &= \sum_{j=0}^{L} \frac{2}{2j+1} \int_{-1}^{1} \sum_{i,\bar{i}=0}^{L} b_{ij}b_{\bar{i}j}P_i'(x)P_{\bar{i}}'(x)\,dx \\ &= \sum_{j=0}^{L} \frac{2}{2j+1} \|z_j'(x)\|_{0,[-1,1]}^2,\end{aligned} \tag{5.5.8}$$

where the polynomials $z_j(x)$ are given in eqn. (5.5.7). Based on Lemma 5.5.1, we have from eqns. (5.5.6) and (5.5.8)

$$\begin{aligned}\left\|\frac{\partial}{\partial x}w_L\right\|_{0,\square}^2 &= \sum_{j=0}^{L} \frac{2}{2j+1} \|z_j'(x)\|_{0,[-1,1]}^2 \\ &\le CL^4 \sum_{j=0}^{L} \frac{2}{2j+1} \|z_j(x)\|_{0,[-1,1]}^2 \\ &= CL^4 \sum_{i,j=0}^{L} \frac{4b_{ij}^2}{(2i+1)(2j+1)} = CL^4 \|w_L\|_{0,\square}^2.\end{aligned}$$

This is the desired result, i.e., eqn. (5.5.4). ■

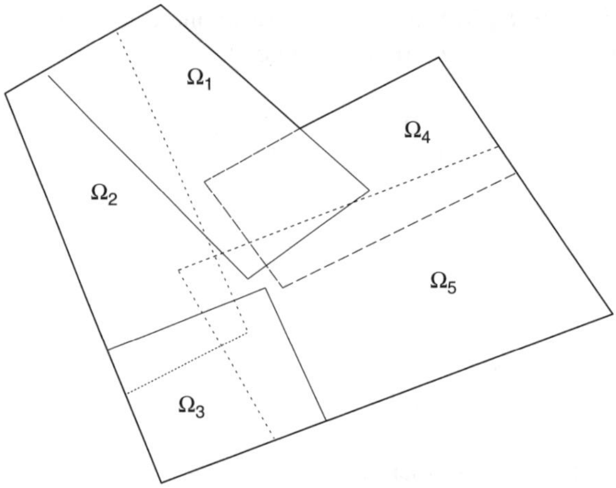

Figure 5.2: A polygon decomposed into finite number of parallelograms Ω_i.

(A9) Let S be a polygon shown in fig. 5.2. Then S can be decomposed into finite number of quasi-uniform parallelograms Ω_i: $S = \cup_i \Omega_i$, where overlap of Ω_i is allowed, see Babuska and Guo [17]. By the diagonal line, Ω_i is split into two triangles, $\triangle_i^+$ and $\triangle_i^-$. Suppose that all $\triangle_i^\pm$ are quasi-uniform, e.g., $c_0 \le \rho_i^\pm$, where $\rho_i^\pm$ denotes the radius of the largest inscribed sphere of $\triangle_i^\pm$, and c_0 is a positive constant. We have the following theorem.

Theorem 5.5.1

Let **A9** *be given. Then for the polynomial w_L in eqn. (5.5.3) exists a constant C independent of L such that*

$$\|w_L\|_{k,S} \le CL^{2k}\|w_L\|_{0,S}.$$

Proof.

Consider the parallelograms Ω_i in fig. 5.3, where a_i and b_i are the lengths of two edges, and α_i are the angles between Ω_i and the Y-axis. From the quasi-uniform parallelograms, there exist the bounds,

$$0 < a_i, \quad b_i < C, \quad \frac{\max\{a_i, b_i\}}{\min\{a_i, b_i\}} \le C, \quad 0 \le \alpha_i \le \alpha_M < \frac{\pi}{2}. \tag{5.5.9}$$

The parallelograms Ω_i can be transformed to $\square$ by the linear transformation T: $(x, y) \to (\hat{x}, \hat{y})$, where

$$\hat{x} = \frac{2}{a_i}[x - (\tan\alpha_i)y] - 1,$$

$$\hat{y} = \frac{2}{b_i}y - 1.$$

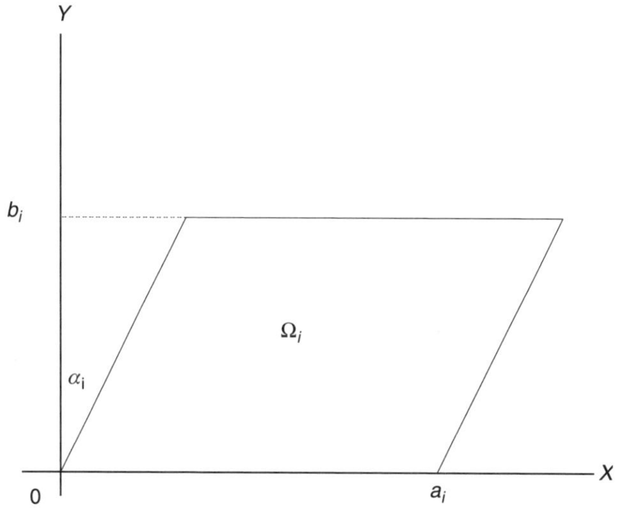

Figure 5.3: A parallelogram Ω_i.

Denote $\hat{w}(\hat{x}, \hat{y}) = \hat{w}(T(x, y)) = w(x, y)$. We have

$$\frac{\partial w}{\partial x} = \frac{2}{a_i}\frac{\partial \hat{w}}{\partial \hat{x}},$$
$$\frac{\partial w}{\partial y} = -\frac{2}{a_i}(\tan\alpha_i)\frac{\partial \hat{w}}{\partial \hat{x}} + \frac{2}{b_i}\frac{\partial \hat{w}}{\partial \hat{y}}.$$

Through the linear transformations T, we obtain

$$\begin{aligned} |w|^2_{1,\Omega_i} &= \iint_{\Omega_i} (w_x^2 + w_y^2)\,dx\,dy \qquad (5.5.10)\\ &= \frac{a_i b_i}{4}\iint_{\square}\left\{\left(\frac{2}{a_i}\frac{\partial \hat{w}}{\partial \hat{x}}\right)^2 + \left(-\frac{2}{a_i}(\tan\alpha_i)\frac{\partial \hat{w}}{\partial \hat{x}} + \frac{2}{b_i}\frac{\partial \hat{w}}{\partial \hat{y}}\right)^2\right\} d\hat{x}\,d\hat{y}\\ &\le \frac{a_i b_i}{4}\iint_{\square}\left\{\frac{4}{a_i^2}(1 + 2\tan^2\alpha_i)\left(\frac{\partial \hat{w}}{\partial \hat{x}}\right)^2 + 2\frac{4}{b_i^2}\left(\frac{\partial \hat{w}}{\partial \hat{y}}\right)^2\right\} d\hat{x}\,d\hat{y}\\ &\le C\iint_{\square}\left\{\left(\frac{\partial \hat{w}}{\partial \hat{x}}\right)^2 + \left(\frac{\partial \hat{w}}{\partial \hat{y}}\right)^2\right\} d\hat{x}\,d\hat{y}, \end{aligned}$$

where the constant

$$C = \max_i\left\{\frac{b_i}{a_i}(1 + 2\tan^2\alpha_i) + \frac{2a_i}{b_i}\right\}.$$

The constant C is independent of i due to assumption, i.e., eqn. (5.5.9). Under the linear transformation T, the polynomials of order L remain as well. Based on Lemma 5.5.3, we have for $w = w_L$,

$$\iint_{\square} \left\{ \left(\frac{\partial \hat{w}}{\partial \hat{x}} \right)^2 + \left(\frac{\partial \hat{w}}{\partial \hat{y}} \right)^2 \right\} d\hat{x}\, d\hat{y} \leq CL^4 \iint_{\square} (\hat{w})^2 d\hat{x}\, d\hat{y}. \tag{5.5.11}$$

Moreover, through the inverse transformation $\hat{T}$, we obtain for $w = w_L$,

$$\iint_{\square} (\hat{w})^2\, d\hat{x}\, d\hat{y} = \frac{4}{a_i b_i} \iint_{\Omega_i} w^2\, dx\, dy \leq C\|w\|_{0,\Omega_i}^2. \tag{5.5.12}$$

Combining eqns. (5.5.10), (5.5.11), and (5.5.12) gives

$$|w|_{1,\Omega_i} \leq CL^2 \|w\|_{0,\Omega_i},$$

and

$$\|w\|_{1,\Omega_i} \leq CL^2 \|w\|_{0,\Omega_i}.$$

Consequently, for parallelograms Ω_i we have

$$\begin{aligned} \|w_L\|_{k,\Omega_i} &\leq C(L-k)^2 \|w_L\|_{k-1,\Omega_i} \\ &\leq CL^2 \|w_L\|_{k-1,\Omega_i} \leq CL^{2k} \|w_L\|_{0,\Omega_i}. \end{aligned}$$

From **A9**, we obtain

$$\|w_L\|_{k,S}^2 \leq \sum_i \|w_L\|_{k,\Omega_i}^2 \leq CL^{4k} \sum_i \|w_L\|_{0,\Omega_i}^2 \leq CL^{4k} \|w_L\|_{0,S}^2,$$

by noting finite overlaps of Ω_i. ∎

5.6 Final remarks

1. In this chapter, the CM is treated as the LSM involving integration approximation. We employ the FEM theory to develop the theoretical analysis of CMs, in which the key analysis for the CMs is to prove the new V_h-elliptic inequality.
2. Three typical boundary conditions, Dirichlet, Neumann, and Robin, can be handled well by the techniques of CMs in this chapter. The number of collocation nodes may be chosen to be much larger than the number of RBFs (i.e., source points). The collocation nodes are, indeed, the integration nodes of the rules used. Based on integration approximation, not only the collocation nodes can be easily located, but also the error analysis has been developed, see Sections 5.2 and 5.3.

3. Piecewise admissible functions can be used in the CM; both the analysis in this chapter and the computation in Ref. [208] are provided, to enable the CM to be more flexible to complicated geometric domains for general PDEs, because different admissible functions can be chosen in different subdomains. Such an idea is also similar to the p-version FEM of Babuška and Guo [17], and the analysis of the CM may also be extended to singularity problems by following Ref. [17].
4. Usually, the solution of eqns. (5.2.1)–(5.2.3) is highly smooth inside S, but less smooth on ∂S. In particular, for concave corners of polygons or the intersection points of the Dirichlet and Neumann boundary conditions, the solution near the boundary nodes is singular with infinite derivatives. In this case, some special treatments should be solicited. For Motz's problem with the singular origin, the collocation equations may be established at the nodes Q_i far from the origin. Hence, assumptions **A2** can be relaxed to

$$v^{\pm} \in H^{1+\delta}(S) \cap C^2(D), \quad 0 < \delta < 1, \tag{5.6.1}$$

where the subdomain $D \subset S$ is far from the singularity. Moreover, when the singular functions or singular particular solution $v^{\pm}$ are chosen, assumption **A3** may also hold. Hence, the CMs may also be applied to singularity problems if suitable treatments are used.

6 Combinations of collocation and finite element methods

The collocation method (CM) in Chapter 5 can be combined with other numerical methods, such as the finite element method (FEM), the finite difference method (FDM), or the finite volume method (FVM) to solve complicated problems, e.g., those with singularities. Since the CM is, indeed, the Ritz–Galerkin method (i.e., the spectral method) involving integration approximation, the combined methods in this chapter can also be classified into those in Ref. [280]. However, new algorithms and error analysis are explored below. We only choose the FEM for exposition of the combinations of CM; other kinds of combinations of CM can be easily developed. In this chapter, we provide a framework of combinations of CM with the FEM. The key idea is to link the Galerkin method to the least squares method which is then approximated by integration rule and led to the CM. The new important uniformly V_h^0-elliptic inequality is proved. Interestingly, the integration approximation plays a role only in satisfying the uniformly V_h^0-elliptic inequality. For the combinations of the finite element and collocation methods (FEM–CM), the optimal convergence rates can be achieved. The advantage of the CM is to easily formulate the linear algebraic equations, where the stiffness matrices are positive definite but non-symmetric. We may also solve the algebraic equations of FEM and the collocation equations directly by the LSM, thus to greatly improve numerical stability. Numerical experiments in Refs. [205, 207] are also carried for Poisson's problem to support the analysis. Note that the analysis in this chapter is distinct from the existing literature, and it covers a large class of the CM using various admissible functions, such as the radial basis functions (RBFs), the Sinc functions, etc.

6.1 Introduction

In this chapter, we follow the ideas of the combined methods in Ref. [280] and provide the combination of the FEM–CM. The advantages of this combination are threefold: (1) flexibility of applications to different geometric shapes and different elliptic equations, (2) simplicity of computer programming by straightforward mimicking the partial differential equations (PDEs) and the boundary conditions, and (3) varieties of CMs using particular solutions, orthogonal polynomials, RBFs, the Sinc functions, etc. Moreover, optimal error bounds are derived, mainly based on the uniformly V_h^0-elliptic inequalities, which are also proved in Sections 6.4 and 6.5. Note that the analysis of the CM in this chapter is distinct from the existing literature of CMs.

This chapter is organized as follows. In the next section, the combinations of FEM–CM are described, and in Section 6.3, linear algebraic equations are formulated, and the solution methods are provided. In Section 6.4, the important uniformly V_h^0-elliptic inequality is derived, and in Section 6.5, the CM is expressed by approximation of integrals, and error bounds are derived. Numerical experiments to support the analysis made are provided in Refs. [205, 207].

6.2 Combinations of FEMs

Consider Poisson's equation with the Dirichlet boundary condition,

$$-\Delta u = -\left(\frac{\partial^2 u}{\partial x^2} + \frac{\partial^2 u}{\partial y^2}\right) = f(x,y) \quad \text{in } S, \tag{6.2.1}$$

$$u\,|_\Gamma = 0 \quad \text{on } \Gamma,$$

where S is a polygon, and Γ is its boundary. Let S be divided by Γ_0 into two disjoint subregions, S_1 and S_2 (see fig. 6.1): $S = S_1 \cup S_2 \cup \Gamma_0$, $S_1 \cap S_2 = \emptyset$ and $\partial\overline{S}_1 \cap \partial\overline{S}_2 = \Gamma_0$.

On the interior boundary Γ_0 there exist the interior continuity conditions:

$$u^+ = u^-, \qquad u_n^+ = u_n^- \quad \text{on } \Gamma_0, \tag{6.2.2}$$

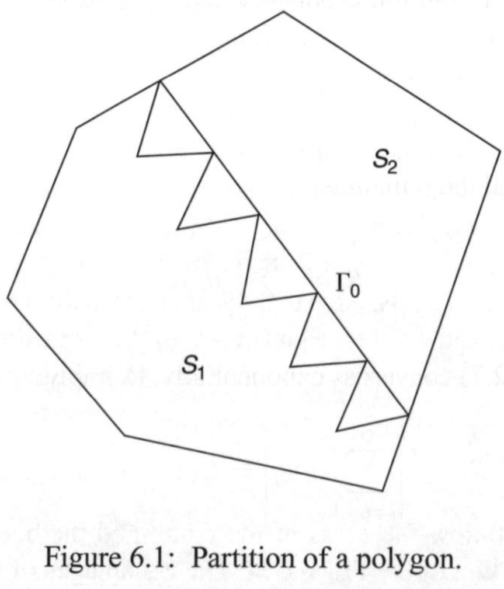

Figure 6.1: Partition of a polygon.

where $u_n = \frac{\partial u}{\partial n}$, $u^+ = u$ on $\Gamma_0 \cup S_2$ and $u^- = u$ on $\Gamma_0 \cup S_1$. Assume that the solution u in S_2 is smoother than u in S_1. We choose the FEM in S_1 and the LSM in S_2, whose discrete forms lead to the CM (see Section 6.3). Let S_1 be partitioned into small triangles: $\triangle_{ij}$, i.e., $S_1 = \cup_{ij} \triangle_{ij}$. Denote h_{ij} the boundary length of $\triangle_{ij}$. The $\triangle_{ij}$ are said to be quasi-uniform if $\frac{h}{\min\{h_{ij}\}} \le C$, where $h = \max\{h_{ij}\}$, and C is a constant independent of h. Then the admissible functions may be expressed by

$$v = \begin{cases} v^- = v_k & \text{in } S_1, \\ v^+ = \sum_{i=1}^{L} \widetilde{a}_i \Psi_i & \text{in } S_2, \end{cases} \tag{6.2.3}$$

where $\widetilde{a}_i$ are unknown coefficients, and v_k are piecewise k-order Lagrange polynomials in S_1. Assume that $\Psi_i \in C^2(S_2 \cup \partial S_2)$ so that $v^+ \in C^2(S_2 \cup \partial S_2)$. Therefore, we may evaluate eqn. (6.2.1) directly

$$(\Delta v^+ + f)(P_i) = 0 \quad \text{for } P_i \in S_2, \tag{6.2.4}$$

at certain collocation nodes $P_i \in S_2$. Note that v in eqn. (6.2.3) is not continuous on the interior interface Γ_0. Hence, to satisfy eqn. (6.2.2) the interior collocation equations are needed:

$$v^+(P_i) = v^-(P_i) \quad \text{for } P_i \in \Gamma_0, \tag{6.2.5}$$

$$v_n^+(P_i) = v_n^-(P_i) \quad \text{for } P_i \in \Gamma_0. \tag{6.2.6}$$

The eqns. (6.2.4)–(6.2.6) are straightforward and easy to be formulated. In this chapter, we choose the total number of collocation nodes (e.g., P_i) to be larger (or much larger) than the number of unknown coefficients $\widetilde{a}_i$. Hence, we may seek the solutions of the entire CM by the LSM in Golub and van Loan [168], see comments in Remark 6.3.1 below.

We assume that the solution expansion: $u = \sum_{i=1}^{\infty} a_i \Psi_i$ in S_2 where a_i are the true coefficients. Denote

$$u_L = \sum_{i=1}^{L} a_i \Psi_i \quad \text{in } S_2. \tag{6.2.7}$$

Then $u = u_L + R_L$, and the remainder

$$R_L = \sum_{i=L+1}^{\infty} a_i \Psi_i.$$

Assume that eqn. (6.2.7) converges exponentially, to imply:

$$|R_L| = \left| \sum_{i=L+1}^{\infty} a_i \Psi_i \right| = O(e^{-\bar{c}L}) \quad \text{in } S_2,$$

where $\bar{c} > 0$ and $L > 1$.

Denote by V_h^0 the finite-dimensional collections of eqn. (6.2.3) satisfying $v\,|_{\Gamma}=0$, where we simply assume $\Psi_i\,|_{\partial S_2\cap\Gamma}=0$. If such a condition does not hold, the corresponding collocation equations on $\partial S_2\cap\Gamma$ are also needed, and the arguments of error analysis can be provided similarly. Then the combination of the FEM–CM is designed to seek the approximate solution $u_h\in V_h^0$ such that

$$a(u_h, v) = f(v), \quad \forall v \in V_h^0, \tag{6.2.8}$$

where

$$\begin{aligned} a(u,v) = & \iint_{S_1} \nabla u\cdot\nabla v + \int_{\Gamma_0} u_n^- v^- + P_c \iint_{S_2} \Delta u\,\Delta v \\ & + \frac{P_c}{h}\int_{\Gamma_0}(u^+ - u^-)(v^+ - v^-) + P_c\int_{\Gamma_0}(u_n^+ - u_n^-)(v_n^+ - v_n^-), \\ f(v) = & \iint_{S_1} fv - P_c\iint_{S_2} f\,\Delta v, \end{aligned}$$

where $\nabla u = u_x\mathbf{i} + u_y\mathbf{j}$, $u_x=\frac{\partial u}{\partial x}$, $u_y=\frac{\partial u}{\partial y}$, $u_n=\frac{\partial u}{\partial n}$, and n is the unit outward normal to ∂S_2. h is the maximal boundary length of Δ_{ij} or $\square_{ij}$ in S_1, and constant $P_c>0$ is chosen to be suitably large but still independent of h.

Denote the space

$$H^* = \{v\,|\,v\in L^2(S),\ \ v\in H^1(S_1),\ \ v\in H^1(S_2),\ \ \Delta v\in L^2(S_2),\ \text{ and }\ v\,|_{\Gamma}=0\},$$

accompanied with the norm

$$\begin{aligned} |||v||| = \Bigg(& \|v\|_{1,S_1}^2 + P_c\|v\|_{1,S_2}^2 + P_c\|\Delta v\|_{0,S_2}^2 \\ & + \frac{P_c}{h}\|v^+ - v^-\|_{0,\Gamma_0}^2 + P_c\|v_n^+ - v_n^-\|_{0,\Gamma_0}^2\Bigg)^{\frac{1}{2}}, \end{aligned} \tag{6.2.9}$$

where $\|v\|_{1,S_1}$ and $\|v\|_{1,S_2}$ are the Sobolev norms. Obviously, $V_h^0\subset H^*$. For the true solution u to eqn. (6.2.1), we have $a(u-u_h, v)=0$, $\forall v\in V_h^0$. By means of a traditional argument in Refs. [103, 426], we have the following theorem.

Theorem 6.2.1

Suppose that there exist two inequalities,

$$a(u,v)\le C|||u|||\times|||v|||, \quad \forall v\in V_h^0, \tag{6.2.10}$$

$$a(v,v)\ge C_0|||v|||^2, \quad \forall v\in V_h^0, \tag{6.2.11}$$

where $C_0(>0)$ and C are two constants independent of h and L. Then, the solution of combination, i.e., eqn. (6.2.8) has the error bound,

$$|||u-u_h||| = C\inf_{v\in V_h^0}|||u-v|||.$$

The proof for eqn. (6.2.11) is non-trivial and complicated, which is deferred to Section 6.4.

Choose an auxiliary function:

$$u_{I,L} = \begin{cases} u_I & \text{in } S_1, \\ \sum_{i=1}^{L} a_i \Psi_i & \text{in } S_2, \end{cases} \tag{6.2.12}$$

where u_I is the piecewise k-order Lagrange interpolant of the true solution u, and a_i are the true coefficients. Then $u = \sum_{i=1}^{L} a_i \Psi_i + R_L$ in S_2. By means of the auxiliary function, i.e., eqn. (6.2.12), we obtain the following corollary.

Corollary 6.2.1

Let all conditions in Theorem 6.2.1 hold. Suppose that

$$u \in H^{k+1}(S_1) \quad \textit{and} \quad u \in H^{k+1}(\Gamma_0). \tag{6.2.13}$$

Then there exists the error bound,

$$|||u - u_h||| \le C \left\{ h^k |u|_{k+1,S_1} + \sqrt{P_c}\, \|R_L\|_{2,S_2} + \sqrt{P_c} \left(h^{k+\frac{1}{2}} |u|_{k+1,\Gamma_0} + \frac{1}{\sqrt{h}} \|R_L\|_{0,\Gamma_0} + \|(R_L)_n\|_{0,\Gamma_0} \right) \right\}.$$

Furthermore, suppose that the number L of v^+ in eqn. (6.2.3) is chosen such that

$$\|R_L\|_{2,S_2} = O(h^k), \qquad \|R_L\|_{0,\Gamma_0} = O(h^{k+\frac{1}{2}}), \qquad \|(R_L)_n\|_{0,\Gamma_0} = O(h^k). \tag{6.2.14}$$

Then, there exists the optimal convergence rate,

$$|||u - u_h||| = O(h^k). \tag{6.2.15}$$

Remark 6.2.1

The combination, i.e., eqn. (6.2.8) is nothing new (cf. [280]), although the proof of eqn. (6.2.11) is challenging. Particularly, the CM is used in S_2, which can be obtained from eqn. (6.2.8) involving approximation of integration. Hence, the combination of Ritz–Galerkin–FEM is a backbone for the study, but more justification will be provided below.

6.3 Linear algebraic equations of combination of FEM and CM

Let $\widehat{\iint_{S_2}}$ and $\widehat{\int_{\Gamma_0}}$ denote the approximations of $\iint_{S_2}$ and $\int_{\Gamma_0}$ by some integration rules, respectively. The combination of FEM–CM of eqn. (6.2.8) involving integration approximation is given by: To seek the approximation solution $\hat{u}_h \in V_h^0$ such that

$$\hat{a}(\hat{u}_h, v) = \hat{f}(v), \quad \forall v \in V_h^0, \tag{6.3.1}$$

where

$$\hat{a}(u,v) = \iint_{S_1} \nabla u \cdot \nabla v + \int_{\Gamma_0} u_n^- v^- + P_c \widehat{\iint_{S_2}} \Delta u \, \Delta v$$
$$+ \frac{P_c}{h} \widehat{\int_{\Gamma_0}} (u^+ - u^-)(v^+ - v^-) + P_c \int_{\Gamma_0} (u_n^+ - u_n^-)(v_n^+ - v_n^-),$$
$$\hat{f}(v) = \iint_{S_1} f v - P_c \widehat{\iint_{S_2}} f \, \Delta v.$$

The eqn. (6.3.1) can be described equivalently as:

$$\hat{a}^*(\hat{u}_h, v) = f_1(v), \quad \forall v \in V_h^0, \tag{6.3.2}$$

where

$$\hat{a}^*(u,v) = \iint_{S_1} \nabla u \cdot \nabla v + \int_{\Gamma_0} u_n^- v^- + P_c \widehat{\iint_{S_2}} (\Delta u + f)(\Delta v + f)$$
$$+ \frac{P_c}{h} \widehat{\int_{\Gamma_0}} (u^+ - u^-)(v^+ - v^-) + P_c \int_{\Gamma_0} (u_n^+ - u_n^-)(v_n^+ - v_n^-),$$
$$f_1(v) = \iint_{S_1} f v.$$

In S_2, we choose the integration rules

$$\widehat{\iint_{S_2}} g^2 = \sum_{ij} \alpha_{ij} g^2(P_{ij}), \quad P_{ij} \in S_2, \tag{6.3.3}$$
$$\widehat{\int_{\Gamma_0}} g^2 = \sum_{j} \alpha_j g^2(P_j), \quad P_j \in \Gamma_0,$$

where α_{ij} and α_j are positive weights. In fact, we may formulate the collocation equations at $P_{ij} \in S_2$, and $P_j \in \Gamma_0$ directly. The collocation equations at P_{ij} and P_j

are given by

$$(\Delta v^+ + f)(P_{ij}) = 0, \quad P_{ij} \in S_2, \tag{6.3.4}$$

$$(v^+ - v^-)(P_j) = 0, \quad P_j \in \Gamma_0, \tag{6.3.5}$$

$$(v_n^+ - v_n^-)(P_j) = 0, \quad P_j \in \Gamma_0, \tag{6.3.6}$$

where $v_n^-(P_j) = \frac{v_{1j} - v_{0j}}{h}$, $v_{0j} = v(P_j)$ and v_{1j} are the nodal variables in S_1 normal to Γ_0. By introducing suitable weight functions, we rewrite the eqns. (6.3.4)–(6.3.6) as

$$\sqrt{P_c\alpha_{ij}}(\Delta v^+ + f)(P_{ij}) = 0, \quad P_{ij} \in S_2, \tag{6.3.7}$$

$$\sqrt{\frac{P_c\alpha_j}{2h}}(v^+ - v^-)(P_j) = 0, \quad P_j \in \Gamma_0, \tag{6.3.8}$$

$$\sqrt{\frac{P_c\alpha_j}{2}}(v_n^+ - v_n^-)(P_j) = 0, \quad P_j \in \Gamma_0, \tag{6.3.9}$$

where P_{ij} are the interior element nodes of Γ_0.

We give some rules of integration with explicit weights α_{ij} in eqn. (6.3.7). First, choose the trapezoidal rule,

$$\widehat{\int_{\overline{P_1P_2}}} g^2 = \frac{\overline{P_1P_2}}{2}(g^2(P_1) + g^2(P_2)) = \frac{H}{2}(g^2(P_1) + g^2(P_2)).$$

Let S_2 be a rectangle shown in fig. 6.2, and be divided into uniform difference grids with the meshspacing H, where P_{ij} denote the collocation nodes (i, j). Hence, the weights α_{ij} in eqn. (6.3.7) have the following values,

$$\alpha_{ij} = \begin{cases} H^2, & (i,j) \in S_2, \\ \frac{1}{2}H^2, & (i,j) \in \partial S_2 \text{ excluding corners of } \partial S_2, \\ \frac{1}{4}H^2, & (i,j) \in \quad \text{corners of } \partial S_2. \end{cases}$$

We may choose more efficient rules, such as the Legendre–Gauss rule with two boundary nodes fixed in Refs. [9, 374],

$$\widehat{\int_{-1}^{1}} g^2(x)\, dx = \sum_{j=1}^{n} w_j g^2(x_j), \tag{6.3.10}$$

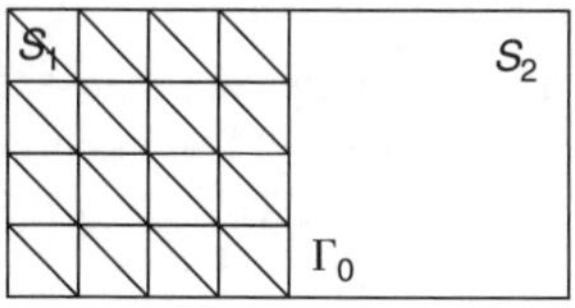

Figure 6.2: Uniform partition of a rectangular solution domain with $M = 4$, where M denotes the numbers of partitions along the Y-direction in S_1.

where x_j is the jth zero of $P_n(x)$, and $P_n(x)$ are the Legendre polynomials defined by

$$P_n(x) = \frac{(-1)^n}{2^n n!} \cdot \frac{d^n}{dx^n}[(1-x^2)^n], \quad n \geq 1. \tag{6.3.11}$$

The weights are given by

$$w_j = \frac{2}{(1-x_j)[P_n'(x_j)]^2}. \tag{6.3.12}$$

Then at the collocation nodes $P_{ij} = (x_i, y_j)$, the weights in eqn. (6.3.7) are obtained by

$$\alpha_{ij} = w_i w_j, \quad (i,j) \in S_2. \tag{6.3.13}$$

Let $f = g^2$ and $f \in C^{2n}[-1,1]$, then the remainder of eqn. (6.3.10) is given by

$$E(f) = \frac{2^{2n+1}[n!]^4}{(2n+1)[(2n)!]^3} f^{(2n)}(\xi), \quad f = g^2, \ -1 < \xi < 1.$$

Let g in $[-1,1]$ be polynomials of order L. Then $f (= g^2)$ are polynomials of order $2L$. Choose $n = L+1$, then the derivatives $f^{(2n)}(\xi) = f^{(2L+2)} \equiv 0$ and $E(f) \equiv 0$. Therefore, when S_2 is a rectangle, the functions v^+ in S_2 are chosen to be polynomials of order L. Since functions Δv^+ are polynomials of order $L-2$, the Legendre–Gauss rule with $n = L-1$ in eqns. (6.3.3) and (6.3.13) offers no error for $\widehat{\iint_{S_2}} (\Delta v^+)^2$, i.e.,

$$\widehat{\iint_{S_2}} (\Delta v^+)^2 = \iint_{S_2} (\Delta v^+)^2.$$

Now let us establish the linear algebraic equations of combination, i.e., eqn. (6.3.2) of FEM–CM. First, consider the entire FEM in S_1 only,

$$a_1(\hat{u}_h, v) = f_1(v), \quad \forall v \in V_h,$$

where

$$a_1(u,v) = \iint_{S_1} \nabla u \nabla v + \int_{\Gamma_0} u_n^- v^-, \qquad f_1(v) = \iint_{S_1} fv.$$

We obtain the linear algebraic equations,

$$\mathbf{A_1 x_1} = \mathbf{b_1}, \tag{6.3.14}$$

where $\mathbf{x_1}$ is a vector consisting of v_{ij} only, and matrix $\mathbf{A_1}$ is non-symmetric.

Next, the eqns. (6.3.7)–(6.3.9) in $S_2 \cup \Gamma_0$ are denoted by

$$\mathbf{A_2 x_2} = \mathbf{b_2}, \tag{6.3.15}$$

where $\mathbf{x_2}$ is a vector consisting of $\tilde{a}_i$, v_{1j}, and v_{0j}, and v_{0j} and v_{1j} are the unknowns on the two boundary layer nodes in S_1 close to Γ_0. Denote by M_1 the number of all collocation nodes in S_2 and ∂S_2, and by N_1 the number of v_{1j} and v_{0j}. Then matrix $\mathbf{A_2} \in R^{M_1 \times (L+N_1)}$. Therefore, we can see

$$\frac{1}{2}\mathbf{x}_2^T\mathbf{A}_2^T\mathbf{A}_2\mathbf{x}_2 - \mathbf{A}_2^T\mathbf{b}_2\mathbf{x}_2 + \mathbf{c}$$

$$= \frac{P_c}{2}\widehat{\iint_{S_2}}(\Delta v + f)^2 + \frac{P_c}{2h}\widehat{\int_{\Gamma_0}}(v^+ - v^-)^2 + \frac{P_c}{2}\widehat{\int_{\Gamma_0}}(v_n^+ - v_n^-)^2. \tag{6.3.16}$$

Combining eqns. (6.3.14) and (6.3.16) yields explicitly:

$$\mathbf{Ax} = \mathbf{b}, \tag{6.3.17}$$

$$\mathbf{A} = \mathbf{A}_1 + \mathbf{A}_2^T\mathbf{A}_2, \qquad \mathbf{b} = \mathbf{b}_1 + \mathbf{A}_2^T\mathbf{b}_2, \tag{6.3.18}$$

where $\mathbf{x}$ is a vector consisting of the coefficients $\widetilde{a}_i$ and v_{ij} in $S_1 \cup \Gamma_0$. Denote by N the number of nodes on $S_1 \cup \Gamma_0$, then the vector $\mathbf{x}$ in eqn. (6.3.17) has $N + L$ dimensions. The matrix $\mathbf{A}$ is non-symmetric, but positive definite, based on Theorem 6.5.1 given later. Note that the matrix and vector operations in eqn. (6.3.18) may be performed after extending the dimensions of $\mathbf{A}$ and $\mathbf{b}$ by filling up zero entries.

Strictly speaking, the dimensions of matrices and vectors in eqn. (6.3.18) are inconsistent. Hence, the equalities in eqns. (6.3.18), (6.3.22), and (6.3.27) read as: the smaller dimensions are extended to the larger dimensions by filling out zero entries.

Let us briefly address the solution methods for eqn. (6.3.17). When P_c is chosen large enough, matrix $\mathbf{A} \in R^{(L+N)\times(L+N)}$ in eqn. (6.3.17) is positive definite and non-symmetric and sparse when $N \gg L$. When $L + N$ is not huge, to solve eqn. (6.3.17) we may choose the Gaussian elimination without pivoting, see Golub and van Loan [168].

Again since the true solution u satisfies exactly $\hat{a}(u, v) = \hat{f}(v)$, $\forall v \in V_h^0$, we have $\hat{a}(u - \hat{u}_h, v) = 0$, $\forall v \in V_h^0$. We obtain the following theorem.

Theorem 6.3.1

Suppose that there exist two inequalities,

$$\hat{a}(u, v) \le C|||u||| \times |||v|||, \quad \forall v \in V_h^0, \tag{6.3.19}$$

$$\hat{a}(v, v) \ge C_0|||v|||^2, \qquad \forall v \in V_h^0, \tag{6.3.20}$$

where $C_0(> 0)$ and C are two constants independent of h and L. Then, the solution of combination, i.e., eqn. (6.3.1) has the error bound,

$$|||u - \hat{u}_h||| \le C \inf_{v \in V_h} |||u - v|||.$$

Moreover, the optimal convergence rate, i.e., eqn. (6.2.15) holds if the conditions, i.e., eqns. (6.2.13) and (6.2.14) are satisfied.

The proof of inequality eqn. (6.3.20) is deferred to Section 6.5.

Remark 6.3.1

Note that eqn. (6.3.17), called Method I, presents exactly the combination, i.e., eqn. (6.3.2). There arises a question. Since eqn. (6.3.17) results from eqns. (6.3.14) and (6.3.15), should we solve both eqns. (6.3.14) and (6.3.15) directly by the LSM? The following arguments give a positive justification.

Method I. *We rewrite eqns. (6.3.17) and (6.3.18) as*

$$\mathbf{A}\mathbf{y} = \mathbf{b}, \tag{6.3.21}$$

$$\mathbf{A}_1\mathbf{y} + \mathbf{A}_2^{\mathbf{T}}\mathbf{A}_2\mathbf{y} = \mathbf{b}_1 + \mathbf{A}_2^{\mathbf{T}}\mathbf{b}_2. \tag{6.3.22}$$

Method II. *The direct LSM. Solve*

$$\mathbf{A}_1\mathbf{x} = \mathbf{b}_1, \qquad \mathbf{A}_2\mathbf{x} = \mathbf{b}_2, \tag{6.3.23}$$

by

$$I(\mathbf{x}) = \min_{\mathbf{z}} I(\mathbf{z}), \tag{6.3.24}$$

where

$$I(\mathbf{z}) = \|\mathbf{A}_1\mathbf{z} - \mathbf{b}_1\|^2 + \|\mathbf{A}_2\mathbf{z} - \mathbf{b}_2\|^2,$$

and $\|\cdot\|$ *is the Euclidean norm.*

Proposition 6.3.1

Let $\mathbf{x}$ *and* $\mathbf{y}$ *be the solutions from eqns. (6.3.23) and (6.3.21), respectively, then* $\mathbf{x} \approx \mathbf{y}$ *with the relative error bound,*

$$\frac{\|\mathbf{x}-\mathbf{y}\|}{\|\mathbf{y}\|} \le \mathrm{Cond}\frac{\|\mathbf{r}\|}{\|\mathbf{b}\|} \le \mathrm{Cond}(1 + \|\mathbf{A}_2\|)\frac{\varepsilon}{\|\mathbf{b}\|}, \tag{6.3.25}$$

where the error of Method II is

$$\varepsilon = (\|\mathbf{A}_1\mathbf{x} - \mathbf{b}_1\|^2 + \|\mathbf{A}_2\mathbf{x} - \mathbf{b}_2\|^2)^{\frac{1}{2}}.$$

Cond denotes the condition number

$$\mathrm{Cond} = \sqrt{\frac{\lambda_{\max}(\mathbf{A}^T\mathbf{A})}{\lambda_{\min}(\mathbf{A}^T\mathbf{A})}},$$

and $\lambda_{\max}(\mathbf{A}^T\mathbf{A})$ *and* $\lambda_{\min}(\mathbf{A}^T\mathbf{A})$ *are the maximal and minimal eigenvalues of* $\mathbf{A}^T\mathbf{A}$*, respectively.*

Proof.

Let $I(\mathbf{x}) = \varepsilon^2$, we then obtain eqn. (6.3.24)

$$\|\mathbf{A}_1\mathbf{x} - \mathbf{b}_1\| \le \varepsilon, \qquad \|\mathbf{A}_2\mathbf{x} - \mathbf{b}_2\| \le \varepsilon. \tag{6.3.26}$$

Consider the remainder of eqn. (6.3.22) when $\mathbf{x}$ replaces $\mathbf{y}$:

$$\mathbf{r} = \mathbf{A}\mathbf{x} - \mathbf{b} = \mathbf{A}_1\mathbf{x} - \mathbf{b}_1 + \mathbf{A}_2^T\mathbf{A}_2\mathbf{x} - \mathbf{A}_2^T\mathbf{b}_2. \tag{6.3.27}$$

We have from eqn. (6.3.26)

$$\|\mathbf{r}\| \le \|\mathbf{A}_1\mathbf{x} - \mathbf{b}_1\| + \|\mathbf{A}_2\|\|\mathbf{A}_2\mathbf{x} - \mathbf{b}_2\| \le (1 + \|\mathbf{A}_2\|)\varepsilon.$$

Moreover, we have from eqns. (6.3.21) and (6.3.27)

$$\mathbf{A}(\mathbf{x} - \mathbf{y}) = \mathbf{r}.$$

Since $\|\mathbf{y}\| \ge \|\mathbf{b}\|/\|\mathbf{A}\|$, the desired result, i.e., eqn. (6.3.25) is obtained by following Refs. [9, 168]. Since ε is very small, the solution $\mathbf{x}$ of Method II is very approximated to the solution $\mathbf{y}$ of Method I. ■

6.4 Uniformly V_h^0-elliptic inequality

The key analysis of combinations, i.e., eqns. (6.2.8) and (6.3.1) is to prove the uniformly V_h^0-elliptic inequalities, i.e., eqns. (6.2.11) and (6.3.20), since the proof for eqns. (6.2.10) and (6.3.19) is much simpler. We shall prove eqn. (6.2.11) in this section and then eqn. (6.3.20) in the next section.

First, we consider $a(v, v)$ without the term $\int_{\Gamma_0} v_n^- v^-$. Define the norms

$$\|v\|_E = \left(|v|_{1,S_1}^2 + P_c\|\Delta v\|_{0,S_2}^2 + \frac{P_c}{h}\|v^+ - v^-\|_{0,\Gamma_0}^2 + P_c\|v_n^+ - v_n^-\|_{0,\Gamma_0}^2\right)^{\frac{1}{2}}, \tag{6.4.1}$$

and

$$\overline{\|v^+ - v^-\|_{\ell,\Gamma_0}} = \|\hat{v}^+ - v^-\|_{\ell,\Gamma_0},$$

where $\ell = 0, \frac{1}{2}$, and $\hat{v}^+$ is the piecewise k-order polynomial interpolant of v^+ in S_2. Then we have the following lemma.

Lemma 6.4.1

Suppose that there exists a positive constant $\nu(> 0)$ *such that*

$$\|v^+\|_{\ell,\Gamma_0} \le CL^{\ell\nu}\|v^+\|_{0,\Gamma_0}, \quad \ell = 1, 2, \ldots \tag{6.4.2}$$

Then there exists the bound for $v \in V_h^0$,

$$\|v^+ - v^-\|_{\frac{1}{2},\Gamma_0} \le \frac{C}{\sqrt{h}}\|v^+ - v^-\|_{0,\Gamma_0} + Ch^{\frac{3}{2}}L^{2\nu}\|v^+\|_{1,S_2}, \tag{6.4.3}$$

where the norms

$$\|v\|_{\frac{1}{2},\Gamma_0} = \left\{ \|v\|_{0,\Gamma_0}^2 + \int_{\Gamma_0}\int_{\Gamma_0} \frac{[v(P)-v(Q)]^2}{\|P-Q\|^2}\, d\ell(P)\, d\ell(Q) \right\}^{\frac{1}{2}},$$

$$\|u\|_{-\frac{1}{2},\Gamma_0} = \sup_{v\neq 0} \frac{\left|\int_{\Gamma_0} uv\, d\ell\right|}{\|v\|_{\frac{1}{2},\Gamma_0}}.$$

Proof.

We have from triangle inequalities,

$$\|v^+ - v^-\|_{\frac{1}{2},\Gamma_0} \le \overline{\|v^+ - v^-\|}_{\frac{1}{2},\Gamma_0} + \|\hat{v}^+ - v^+\|_{\frac{1}{2},\Gamma_0},$$
$$\overline{\|v^+ - v^-\|}_{0,\Gamma_0} \le \|v^+ - v^-\|_{0,\Gamma_0} + \|\hat{v}^+ - v^+\|_{0,\Gamma_0}.$$

Then from the inverse inequality for piecewise k-order polynomials, there exists the bound,

$$\begin{aligned}\|v^+ - v^-\|_{\frac{1}{2},\Gamma_0} &\le \overline{\|v^+ - v^-\|}_{\frac{1}{2},\Gamma_0} + \|\hat{v}^+ - v^+\|_{\frac{1}{2},\Gamma_0} \\ &\le \frac{C}{\sqrt{h}}\overline{\|v^+ - v^-\|}_{0,\Gamma_0} + \|\hat{v}^+ - v^+\|_{\frac{1}{2},\Gamma_0} \\ &\le \frac{C}{\sqrt{h}}\|v^+ - v^-\|_{0,\Gamma_0} + \frac{C}{\sqrt{h}}\|\hat{v}^+ - v^+\|_{0,\Gamma_0} + \|\hat{v}^+ - v^+\|_{\frac{1}{2},\Gamma_0}.\end{aligned} \tag{6.4.4}$$

Moreover, from eqn. (6.4.2) we have

$$\begin{aligned}&h^{-\frac{1}{2}}\|\hat{v}^+ - v^+\|_{0,\Gamma_0} + \|\hat{v}^+ - v^+\|_{\frac{1}{2},\Gamma_0} \\ &\le Ch^{\frac{3}{2}}\|v^+\|_{2,\Gamma_0} \le Ch^{\frac{3}{2}}L^{2\nu}\|v^+\|_{0,\Gamma_0} \le Ch^{\frac{3}{2}}L^{2\nu}\|v^+\|_{1,S_2}.\end{aligned} \tag{6.4.5}$$

Combining eqns. (6.4.4) and (6.4.5) yields the desired result, i.e., eqn. (6.4.3). ■

Lemma 6.4.2

There exist the bounds for $v \in V_h^0$,

$$\|v^+\|_{1,S_2} \le C\{\|\Delta v^+\|_{-1,S_2} + \|v^+\|_{\frac{1}{2},\partial S_2}\}, \tag{6.4.6}$$

$$\|v_n^+\|_{-\frac{1}{2},\Gamma_0} \le C\{\|\Delta v^+\|_{-1,S_2} + \|v^+\|_{\frac{1}{2},\partial S_2}\}, \tag{6.4.7}$$

where C is a constant independent of h and L, and the negative norm is defined by $\|u\|_{-1,S} = \sup_{v\in H_0^1(S)} \frac{\left|\iint_S uv\, ds\right|}{\|v\|_{1,S}}$.

Proof.

We cite the bounds from Oden and Reddy [348], p. 189–192,

$$\|u\|^2_{\widetilde{H}^{s,2m}(\Omega)} \leq C\left\{\|Au\|^2_{H^{s-2m}(\Omega)} + \sum_{k=0}^{m-1}\|B_k u\|^2_{H^{s-g_k-\frac{1}{2}}(\partial\Omega)}\right\}, \tag{6.4.8}$$

where $s < 2m$, and m is a positive integer. The notations are: $Au = \Delta u$, $B_0 u = u$, $B_1 u = u_n$, $g_0 = 0$ and $g_1 = 1$. The norm on the left-hand side in eqn. (6.4.8) is defined in Ref. [348], p. 183,

$$\|u\|^2_{\widetilde{H}^{s,r}(\Omega)} = \|u\|^2_{H^s(\Omega)} + \sum_{k=0}^{r-1}\|D_n^k u\|^2_{H^{s-k-\frac{1}{2}}(\partial\Omega)}, \tag{6.4.9}$$

where r is a positive integer, s is an integer, and $D_n^k = \frac{\partial^k}{\partial n^k}$ is the kth normal derivatives. In eqn. (6.4.8), choosing $s = 1$, $m = 1$, and $\Omega = S_2$, we obtain

$$\|u\|^2_{\widetilde{H}^{1,2}(S_2)} \leq C\left\{\|\Delta u\|^2_{H^{-1}(S_2)} + \|u\|^2_{H^{\frac{1}{2}}(\partial S_2)}\right\}, \tag{6.4.10}$$

where the norm in the left-hand side is given in eqn. (6.4.9) with $s = 1$, $r = 2$, and $\Omega = S_2$,

$$\|u\|^2_{\widetilde{H}^{1,2}(S_2)} = \|u\|^2_{H^1(S_2)} + \|u\|^2_{H^{\frac{1}{2}}(\partial S_2)} + \|u_n\|^2_{H^{-\frac{1}{2}}(\partial S_2)}. \tag{6.4.11}$$

Combining eqns. (6.4.10) and (6.4.11) gives the following bound,

$$\|u\|^2_{1,S_2} + \|u\|^2_{\frac{1}{2},\partial S_2} + \|u_n\|^2_{-\frac{1}{2},\partial S_2} \leq C\{\|\Delta u\|^2_{-1,S_2} + \|u\|^2_{\frac{1}{2},\partial S_2}\}. \tag{6.4.12}$$

The desired results, i.e., eqns. (6.4.6) and (6.4.7) are obtained directly from eqn. (6.4.12). ■

Lemma 6.4.3

Let eqn. (6.4.2) and the following bound hold,

$$h^{\frac{3}{2}} L^{2\nu} = o(1). \tag{6.4.13}$$

Then for $v \in V_h^0$ there exists the bound,

$$\|v^+\|_{1,S_2} \leq C\left\{\|v^-\|_{\frac{1}{2},\Gamma_0} + \frac{1}{\sqrt{h}}\|v^+ - v^-\|_{0,\Gamma_0} + \|\Delta v^+\|_{0,S_2}\right\}, \tag{6.4.14}$$

where C is a constant independent of h and L.

Proof.

From Lemma 6.4.1, we have

$$\|v^+\|_{\frac{1}{2},\Gamma_0} \le \|v^-\|_{\frac{1}{2},\Gamma_0} + \|v^+ - v^-\|_{\frac{1}{2},\Gamma_0} \tag{6.4.15}$$
$$\le \|v^-\|_{\frac{1}{2},\Gamma_0} + \frac{C}{\sqrt{h}}\|v^+ - v^-\|_{0,\Gamma_0} + Ch^{\frac{3}{2}}L^{2\nu}\|v^+\|_{1,S_2}.$$

From Lemma 6.4.2 and eqn. (6.4.15)

$$\|v^+\|_{1,S_2} \le C\{\|v^+\|_{\frac{1}{2},\Gamma_0} + \|\Delta v^+\|_{-1,S_2}\}$$
$$\le C\left\{\|v^-\|_{\frac{1}{2},\Gamma_0} + \frac{1}{\sqrt{h}}\|v^+ - v^-\|_{0,\Gamma_0} + h^{\frac{3}{2}}L^{2\nu}\|v^+\|_{1,S_2} + \|\Delta v^+\|_{0,S_2}\right\}.$$

This leads to

$$\|v^+\|_{1,S_2} \le \frac{C}{1 - Ch^{\frac{3}{2}}L^{2\nu}}\left\{\|v^-\|_{\frac{1}{2},\Gamma_0} + \frac{1}{\sqrt{h}}\|v^+ - v^-\|_{0,\Gamma_0} + \|\Delta v^+\|_{0,S_2}\right\}.$$

The desired result, i.e., eqn. (6.4.14) follows from $Ch^{\frac{3}{2}}L^{2\nu} \le \frac{1}{2}$ by assumption, i.e., eqn. (6.4.13). ■

Lemma 6.4.4

Let $\Gamma \cap \partial S_1 \ne \emptyset$, eqns. (6.4.2) and (6.4.13) hold; then there exists an inequality

$$C_0 |||v||| \le \|v\|_E, \quad \forall v \in V_h^0, \tag{6.4.16}$$

where $|||v|||$ and $\|v\|_E$ are defined in eqns. (6.2.9) and (6.4.1), respectively, and $C_0 > 0$ is a lower bound independent of h and L.

Proof.

We prove by the contradiction. Suppose that we can find a sequence $\{v_\ell\} \subseteq H^*$ such that

$$|||v_\ell||| = 1, \qquad \|v_\ell\|_E \to 0, \quad \text{as } \ell \to \infty. \tag{6.4.17}$$

First, $\|v_\ell\|_E \to 0$ implies that for large ℓ, $|v_\ell^-|_{1,S_1} \le 1$ and $v_\ell^- \mid_{\partial S_1 \cap \Gamma} = 0$, and then $\|v_\ell^-\|_{1,S_1}$ is bounded. Based on the Kandrosov or Rellich theorem [103], there exists a subsequence $\{v_\ell^-\}$ in $L^2(S_1)$ (also written as $\{v_\ell^-\}$) such that $v_\ell^- \to \overline{v}^- \in L^2(S_1)$. Then $\overline{v}^- \in H^1(S_1)$, since $|\overline{v}^-|_{1,S_1}$ are bounded due to $|v_\ell^-|_{1,S_1} \le 1$. Moreover, $\|v_\ell\|_E \to 0$ gives $|v_\ell^-|_{1,S_1} \to 0$ as $\ell \to \infty$. Since $H^1(S_1)$ is complete, we conclude that $|\overline{v}^-|_{1,S_1} = \lim_{\ell\to\infty} |v_\ell^-|_{1,S_1} = 0$. Hence, $\overline{v}^-$ is a constant, and $\overline{v}^- \equiv 0$ in S_1 due to $\overline{v}^- \mid_{\partial S_1 \cap \Gamma} = 0$.

From the trace theorem in Ref. [417]

$$\|v_\ell^-\|_{\frac{1}{2},\Gamma_0} \le C\|v_\ell^-\|_{1,S_1},$$

$\|v_\ell^-\|_{\frac{1}{2},\Gamma_0}$ is also bounded, and

$$\lim_{\ell\to\infty} \|v_\ell^-\|_{\frac{1}{2},\Gamma_0} = \|\overline{v}^-\|_{\frac{1}{2},\Gamma_0} = 0. \tag{6.4.18}$$

Next, consider the sequence v_ℓ^+ in S_2. We have from Lemma 6.4.3,

$$\begin{aligned}\|v_\ell^+\|_{\frac{1}{2},\Gamma_0} &\le C\|v_\ell^+\|_{1,S_2} \\ &\le C\left\{\|v_\ell^-\|_{\frac{1}{2},\Gamma_0} + \frac{1}{\sqrt{h}}\|v_\ell^+ - v_\ell^-\|_{0,\Gamma_0} + \|\Delta v_\ell^+\|_{0,S_2}\right\}.\end{aligned} \tag{6.4.19}$$

We conclude that $\|v_\ell^+\|_{\frac{1}{2},\Gamma_0}$ is bounded, and that $\lim_{\ell\to\infty}\|v_\ell^+\|_{\frac{1}{2},\Gamma_0}=0$ from eqns. (6.4.19), (6.4.18) and $\|v_\ell\|_E \to 0$. Then, based on Lemma 6.4.2,

$$\begin{aligned}\|v_\ell^+\|_{1,S_2} &\le C\{\|\Delta v_\ell^+\|_{-1,S_2} + \|v_\ell^+\|_{\frac{1}{2},\partial S_2}\} \\ &\le C\{\|\Delta v_\ell^+\|_{0,S_2} + \|v_\ell^+\|_{\frac{1}{2},\Gamma_0}\}.\end{aligned}$$

Hence, $\|v_\ell^+\|_{1,S_2}$ is also bounded from $\|v_\ell\|_E \to 0$, and then $\lim_{\ell\to\infty}\|v_\ell^+\|_{1,S_2}=0$. By repeating the above arguments, there also exists a subsequence v_ℓ^+ to converge $\overline{v}^+ \in H^1(S_2)$. Moreover, we have $\|\overline{v}^+\|_{1,S_2} = \lim_{\ell\to\infty}\|v_\ell^+\|_{1,S_2}=0$, and then $\overline{v}^+\equiv 0$ in S_2. Hence, $\overline{v}\equiv 0$ in the entire S and $|||\overline{v}|||=0$. This contradicts the assumption $|||\overline{v}||| = \lim_{\ell\to\infty}|||v_\ell|||=1$ in eqn. (6.4.17). ■

Now, we give the main theorem.

Theorem 6.4.1

Let $\Gamma\cap\partial S_1 \ne \emptyset$, eqns. (6.4.2) and (6.4.13) hold, and P_c be chosen to be suitably large but still independent of h. Then the uniformly V_h^0-elliptic inequality holds,

$$C_0|||v|||^2 \le a(v,v), \quad \forall v \in V_h^0, \tag{6.4.20}$$

where $C_0 > 0$ is a constant independent of h and L.

Proof.

From Lemma 6.4.4, we obtain the bound,

$$\begin{aligned}a(v,v) &\ge \|v\|_E^2 - \int_{\Gamma_0} v_n^- v^- \\ &\ge C_1|||v|||^2 - \int_{\Gamma_0} v_n^- v^- \\ &= C_1\left(\|v\|_{1,S_1}^2 + P_c\|v\|_{1,S_2}^2 + P_c\|\Delta v\|_{0,S_2}^2 + \frac{P_c}{h}\|v^+ - v^-\|_{0,\Gamma_0}^2\right. \\ &\qquad \left. + P_c\|v_n^+ - v_n^-\|_{0,\Gamma_0}^2\right) - \int_{\Gamma_0} v_n^- v^-,\end{aligned} \tag{6.4.21}$$

where $C_1 > 0$ is a lower bound independent of h and L. Next, we have

$$\left|\int_{\Gamma_0} v_n^- v^-\right| \le \|v_n^-\|_{-\frac{1}{2},\Gamma_0}\|v^-\|_{\frac{1}{2},\Gamma_0}.$$

Moreover, there exist the bounds for $v \in V_h^0$,

$$\|v^-\|_{\frac{1}{2},\Gamma_0} \le C\|v^-\|_{1,S_1}, \tag{6.4.22}$$

$$\begin{aligned} \|v_n^-\|_{-\frac{1}{2},\Gamma_0} &\le \|v_n^+\|_{-\frac{1}{2},\Gamma_0} + \|v_n^+ - v_n^-\|_{-\frac{1}{2},\Gamma_0} \\ &\le C\{\|v^+\|_{1,S_2} + \|\Delta v^+\|_{0,S_2} + \|v_n^+ - v_n^-\|_{0,\Gamma_0}\}, \end{aligned} \tag{6.4.23}$$

where we have used the bound from Lemma 6.4.2,

$$\begin{aligned} \|v_n^+\|_{-\frac{1}{2},\Gamma_0} &\le C\{\|\Delta v^+\|_{-1,S_2} + \|v^+\|_{\frac{1}{2},\partial S_2}\} \\ &\le C\{\|\Delta v^+\|_{0,S_2} + \|v^+\|_{1,S_2}\}. \end{aligned}$$

Since $Cab \le \epsilon a^2 + \frac{C^2}{4\epsilon}b^2$ for any $\epsilon > 0$, we obtain from eqns. (6.4.22) and (6.4.23)

$$\begin{aligned} \left|\int_{\Gamma_0} v_n^- v^-\right| &\le C\|v^-\|_{1,S_1}\{\|v^+\|_{1,S_2} + \|\Delta v^+\|_{0,S_2} + \|v_n^+ - v_n^-\|_{0,\Gamma_0}\} \\ &\le \frac{C_1}{2}\|v^-\|_{1,S_1}^2 + \frac{C^2}{2C_1}\{\|v^+\|_{1,S_2} + \|\Delta v^+\|_{0,S_2} + \|v_n^+ - v_n^-\|_{0,\Gamma_0}\}^2 \\ &\le \frac{C_1}{2}\|v^-\|_{1,S_1}^2 + \frac{3C^2}{2C_1}\{\|v^+\|_{1,S_2}^2 + \|\Delta v^+\|_{0,S_2}^2 + \|v_n^+ - v_n^-\|_{0,\Gamma_0}^2\}, \end{aligned} \tag{6.4.24}$$

where C_1 is given in eqn. (6.4.21). Combining eqns. (6.4.21) and (6.4.24) gives

$$\begin{aligned} a(v,v) \ge{}& \frac{C_1}{2}\|v\|_{1,S_1}^2 + \left(C_1P_c - \frac{3C^2}{2C_1}\right)(\|v\|_{1,S_2}^2 + \|\Delta v\|_{0,S_2}^2 + \|v_n^+ - v_n^-\|_{0,\Gamma_0}^2) \\ &+ C_1\frac{P_c}{h}\|v^+ - v^-\|_{0,\Gamma_0}^2 \ge \frac{C_1}{2}|||v|||^2, \end{aligned}$$

provided that $C_1P_c - \frac{3C^2}{2C_1} \ge \frac{1}{2}C_1P_c$. This leads to $P_c \ge 3\frac{C^2}{C_1^2}$, which is suitably large but still independent of h and L. Then the uniformly V_h^0-elliptic inequality eqn. (6.4.20) holds with $C_0 = \frac{C_1}{2}$. ∎

6.5 Uniformly V_h^0-elliptic inequality involving integration approximation

In this section, we prove the uniformly V_h^0-elliptic inequality, eqn. (6.3.20). Choose the integration rule

$$\widehat{\int_{\Gamma_0}} v^2 = \int_{\Gamma_0} \hat{v}^2 = \overline{\|v\|}_{0,\Gamma_0}^2,$$

where $\hat{v}$ is the k-order polynomial interpolant of v. First, we give a few lemmas.

Lemma 6.5.1

Let eqn. (6.4.2) and

$$\|v_n^+\|_{1,\Gamma_0} \leq CL^{2\nu}\|v^+\|_{1,S_2}, \quad \forall v \in V_h^0 \tag{6.5.1}$$

hold, where $\nu(> 0)$ is a positive constant. There exist the bounds for $v \in V_h^0$,

$$\overline{\|v^+ - v^-\|}_{0,\Gamma_0} \geq \|v^+ - v^-\|_{0,\Gamma_0} - Ch^2L^{2\nu}\|v^+\|_{1,S_2}, \tag{6.5.2}$$

$$\overline{\|v_n^+ - v_n^-\|}_{0,\Gamma_0} \geq \|v_n^+ - v_n^-\|_{0,\Gamma_0} - ChL^{2\nu}\|v^+\|_{1,S_2}, \tag{6.5.3}$$

where C is a constant independent of h and L.

Proof.

We have

$$\|v^+ - v^-\|_{0,\Gamma_0} \leq \overline{\|v^+ - v^-\|}_{0,\Gamma_0} + \|\hat{v}^+ - v^+\|_{0,\Gamma_0}, \tag{6.5.4}$$

and from eqn. (6.4.2)

$$\|\hat{v}^+ - v^+\|_{0,\Gamma_0} \leq Ch^2|v^+|_{2,\Gamma_0} \leq Ch^2L^{2\nu}\|v\|_{0,\Gamma_0} \leq Ch^2L^{2\nu}\|v\|_{1,S_2}. \tag{6.5.5}$$

Then combining eqns. (6.5.4) and (6.5.5) gives the first desired bound, i.e., eqn. (6.5.2),

$$\overline{\|v^+ - v^-\|}_{0,\Gamma_0} \geq \|v^+ - v^-\|_{0,\Gamma_0} - Ch^2L^{2\nu}\|v\|_{1,S_2}.$$

Similarly, we obtain from eqn. (6.5.1)

$$\begin{aligned}\overline{\|v_n^+ - v_n^-\|}_{0,\Gamma_0} &\geq \|v_n^+ - v_n^-\|_{0,\Gamma_0} - \|v_n^+ - \hat{v}_n^+\|_{0,\Gamma_0} \\ &\geq \|v_n^+ - v_n^-\|_{0,\Gamma_0} - Ch\|v_n^+\|_{1,\Gamma_0} \\ &\geq \|v_n^+ - v_n^-\|_{0,\Gamma_0} - ChL^{2\nu}\|v^+\|_{1,S_2}.\end{aligned}$$

This is the second desired bound, i.e., eqn. (6.5.3). ■

Lemma 6.5.2

Let all conditions in Lemma 6.5.1 hold. Then

$$\overline{\|v^+ - v^-\|}^2_{0,\Gamma_0} \geq \frac{1}{2}\|v^+ - v^-\|^2_{0,\Gamma_0} - Ch^4L^{4\nu}\|v^+\|^2_{1,S_2}, \tag{6.5.6}$$

$$\overline{\|v_n^+ - v_n^-\|}^2_{0,\Gamma_0} \geq \frac{1}{2}\|v_n^+ - v_n^-\|^2_{0,\Gamma_0} - Ch^2L^{4\nu}\|v^+\|^2_{1,S_2}, \tag{6.5.7}$$

where C is a constant independent of h and L.

Proof.

Denote

$$x = \overline{\|v^+ - v^-\|}_{0,\Gamma_0}, \qquad y = \|v^+ - v^-\|_{0,\Gamma_0},$$
$$z = \|v^+\|_{1,S_2}, \qquad w = Ch^2L^{2\nu}. \tag{6.5.8}$$

The eqn. (6.5.2) is written simply as $x \geq y - wz \geq 0$ for small h. We have

$$x^2 \geq (y - wz)^2 = y^2 - 2wyz + w^2z^2.$$

Since $2wyz \leq \frac{y^2}{2} + 2w^2z^2$, we obtain

$$x^2 \geq y^2 - \left(\frac{y^2}{2} + 2w^2z^2\right) + w^2z^2 = \frac{y^2}{2} - w^2z^2.$$

This is the desired result, i.e., eqn. (6.5.6) by noting eqn. (6.5.8). The proof for eqn. (6.5.7) is similar. ■

Second, let us consider integration approximation for $\iint_{S_2} t$, where $t = t(x,y) = (\Delta u + f)(\Delta v + f)$. Let S_2 be divided into small triangles Δ_{ij} and small rectangles $\square_{ij}$,

$$S_2 = (\cup_{ij}\Delta_{ij}) \cup (\cup_{ij}\square_{ij}). \tag{6.5.9}$$

Denote by $\hat{t}_r$ the piecewise r-order interpolant of t on S_2, i.e.,

$$\hat{t}_r = P_r(x,y) = \sum_{i+j=0}^{r} a_{ij}x^iy^j, \quad (x, y) \in \Delta_{ij},$$

or

$$\hat{t}_r = Q_r(x,y) = \sum_{i,j=0}^{r} a_{ij}x^iy^j, \quad (x, y) \in \square_{ij},$$

with the coefficients a_{ij}. Then, the integration rule in eqn. (6.3.3) can be viewed as

$$\begin{aligned}\sum_{ij} \alpha_{ij} g^2(P_{ij}) &= \widehat{\iint_{S_2}} g^2 = \widehat{\iint_{S_2}} t = \iint_{S_2} \hat{t}_r \\ &= \sum_{ij} \iint_{\Delta_{ij}} \hat{t}_r + \sum_{ij} \iint_{\Box_{ij}} \hat{t}_r.\end{aligned} \tag{6.5.10}$$

The partition in eqn. (6.5.9) is regular if $\max_{ij} \frac{H_{ij}}{\rho_{ij}} \le C$, where H_{ij} is the maximal boundary length of Δ_{ij} and $\Box_{ij}$, ρ_{ij} is the diameter of the inscribed circle of Δ_{ij} and $\Box_{ij}$, and C is constant independent of $H(=\max_{ij} H_{ij})$. The partition eqn. (6.5.9) is quasi-uniform if $\frac{H}{\min_{ij} H_{ij}} \le C$. Then, we have the following lemma from the Bramble–Hilbert lemma [103].

Lemma 6.5.3

Let the partition eqn. (6.5.9) be regular and quasi-uniform. Then the integration rule, i.e., eqn. (6.5.10) has the error bound,

$$\left| \iint_{S_2} t - \widehat{\iint_{S_2}} t \right| = \left| \iint_{S_2} (t - \hat{t}_r) \right| \le C H^{r+1} |t|_{r+1,S_2},$$

where C is a constant independent of H.

The integration rule on Δ_{ij} can be found in Strang and Fix [426], and the rule on $\Box_{ij}$ can be formulated by the tensor product of the rule in one dimension, such as the Newton–Cotes rule or the Gaussian rule. The Legendre–Gauss rule given in eqns. (6.3.10)–(6.3.13) is just one kind of Gaussian rules with two boundary nodes fixed. For the Newton–Cotes rule with order r, we may choose the uniform integration nodes. When $r=1$ and 2, the popular trapezoidal and Simpson's rules are given. When v^+ in S_2 are polynomials of order L and choose $r=2L$, the exact integration holds,

$$\widehat{\iint_{S_2}} (\Delta v^+)^2 = \iint_{S_2} (\Delta v^+)^2. \tag{6.5.11}$$

Below, we consider the approximate integration

$$\widehat{\iint_{S_2}} (\Delta v^+)^2 \approx \iint_{S_2} (\Delta v^+)^2,$$

by the rule with integration orders $r \le 2L-1$. We have the following lemma.

Lemma 6.5.4

Let v^+ in S_2 be polynomials of order L, and the rule, i.e., eqn. (6.5.10) with order $r \le 2L-1$ be used for $\int\int_{S_2}(\Delta u+f)(\Delta v+f)$. Also assume

$$\|v^+\|_{\ell,S_2} \le CL^{(\ell-1)\nu}\|v^+\|_{1,S_2}, \quad \ell \ge 1, \ \forall v \in V_h^0, \tag{6.5.12}$$

where $\nu > 0$ is a constant independent of L. Then there exists the bound,

$$\left|\left(\iint_{S_2} - \widehat{\iint}_{S_2}\right)(\Delta v)^2\right| \le CH^{r+1}L^{(r+3)\nu}\|v\|^2_{1,S_2},$$

where H is the meshspacing of uniform integration nodes in S_2, and C is a constant independent of H and L.

Proof.

For the rule of order $r \le 2L-1$, we have from Lemma 6.5.3 and eqn. (6.5.12),

$$\begin{aligned}\left|\left(\iint_{S_2} - \widehat{\iint}_{S_2}\right)(\Delta v)^2\right| &\le CH^{r+1}|(\Delta v)^2|_{r+1,S_2} \\ &\le CH^{r+1}\sum_{i=0}^{r+1}|\Delta v|_{i,S_2}|\Delta v|_{r+1-i,S_2} \\ &\le CH^{r+1}\sum_{i=0}^{r+1}\|v\|_{i+2,S_2}\|v\|_{r+3-i,S_2} \\ &\le CH^{r+1}\sum_{i=0}^{r+1}(L^{(i+1)\nu}\|v\|_{1,S_2})(L^{(r+2-i)\nu}\|v\|_{1,S_2}) \\ &\le CH^{r+1}L^{(r+3)\nu}\|v\|^2_{1,S_2}.\end{aligned}$$

■

Theorem 6.5.1

Let eqn. (6.5.1) and all conditions of Theorem 6.4.1 and Lemma 6.5.4 hold. Suppose

$$hL^{2\nu} = o(1), \tag{6.5.13}$$

$$HL^{(1+\frac{2}{r+1})\nu} = o(1). \tag{6.5.14}$$

Then the uniformly V_h^0-elliptic inequality eqn. (6.3.20) holds.

Proof.

From Lemmas 6.5.2 and 6.5.4 and Theorem 6.4.1, we have

$$
\begin{aligned}
\hat{a}(v,v) &= \iint_{S_1} |\nabla v^-|^2 + \int_{\Gamma_0} v_n^- v^- + P_c \widehat{\iint_{S_2}} (\Delta v^+)^2 \\
&\quad + \frac{P_c}{h}\overline{\|v^+ - v^-\|_{0,\Gamma_0}^2} + P_c \overline{\|v_n^+ - v_n^-\|_{0,\Gamma_0}^2} \\
&\geq \iint_{S_1} |\nabla v^-|^2 + \int_{\Gamma_0} v_n^- v^- + P_c \iint_{S_2} (\Delta v^+)^2 \\
&\quad + \frac{P_c}{2h}\|v^+ - v^-\|_{0,\Gamma_0}^2 + \frac{P_c}{2}\|v_n^+ - v_n^-\|_{0,\Gamma_0}^2 \\
&\quad - CP_c(h^3L^{4\nu} + h^2L^{4\nu})\|v\|_{1,S_2}^2 - CH^{r+1}L^{(r+3)\nu}\|v\|_{1,S_2}^2 \\
&\geq \frac{1}{2}a(v,v) - C\{P_c(h^3L^{4\nu} + h^2L^{4\nu}) + H^{r+1}L^{(r+3)\nu}\}\|v\|_{1,S_2}^2 \\
&\geq \frac{C_0}{2}|||v|||^2 - C\{P_c(h^3L^{4\nu} + h^2L^{4\nu}) + H^{r+1}L^{(r+3)\nu}\}\|v\|_{1,S_2}^2 \\
&\geq \frac{C_0}{2}\{\|v\|_{1,S_1}^2 + \left\{P_c - 2\frac{C}{C_0}[P_c(h^3L^{4\nu} + h^2L^{4\nu}) + H^{r+1}L^{(r+3)\nu}]\right\}\|v\|_{1,S_2}^2 \\
&\quad + P_c\|\Delta v\|_{0,S_2}^2 + \frac{P_c}{h}\|v^+ - v^-\|_{0,\Gamma_0}^2 + P_c\|v_n^+ - v_n^-\|_{0,\Gamma_0}^2\} \\
&\geq \frac{C_0}{4}|||v|||^2,
\end{aligned}
$$

provided that

$$
2\frac{C}{C_0}[P_c(h^3L^{4\nu} + h^2L^{4\nu}) + H^{r+1}L^{(r+3)\nu}] \leq \frac{P_c}{2},
$$

which is satisfied by eqns. (6.5.13) and (6.5.14). ∎

When there is no approximation for $\iint_{S_2} \Delta u^+ \Delta v^+$, we have the following corollary.

Corollary 6.5.1

Let eqn. (6.5.1) and all conditions of Theorem 6.4.1 hold. Also let the numerical integration eqn. (6.5.11) in S_2 be exact. Suppose

$$
hL^{2\nu} = o(1).
$$

Then, the uniformly V_h^0-elliptic inequality eqn. (6.3.20) holds.

Corollary 6.5.1 holds for the case that v^+ in S_2 are polynomials of order $L(>k)$, and that the Legendre–Gauss rule in eqn. (6.3.10) with $n=L-1$ is used for $\widehat{\iint}_{S_2}(\Delta v^+)^2$. Next, let us consider a special case: The functions v^+ in S_2 are chosen to be the particular solutions satisfying $-\Delta v^+=f$ in S_2 exactly. The combination of FEM–CM in eqn. (6.3.2) is given by

$$\hat{a}^*(\hat{u}_h, v) = f_1(v), \quad \forall v \in V_h^0,$$

where

$$\hat{a}^*(u,v) = \iint_{S_1} \nabla u \cdot \nabla v + \int_{\Gamma_0} u_n^- v^- + \frac{P_c}{h}\widehat{\int}_{\Gamma_0} (u^+ - u^-)(v^+ - v^-) + P_c \widehat{\int}_{\Gamma_0} (u_n^+ - u_n^-)(v_n^+ - v_n^-),$$

$$f_1(v) = \iint_{S_1} f v.$$

In this case, we have

$$\hat{a}(v,v) = \iint_{S_1} |\nabla v|^2 + \int_{\Gamma_0} v_n^- v^- + P_c \iint_{S_2} (\Delta v)^2 + \frac{P_c}{h}\widehat{\int}_{\Gamma_0} (v^+ - v^-)^2 + P_c \widehat{\int}_{\Gamma_0} (v_n^+ - v_n^-)^2.$$

Note that the term, $P_c \iint_{S_2} (\Delta v)^2$, disappears in computation.

Remark 6.5.1

Different integration rules for $\iint_{S_2}(\Delta u+f)(\Delta v+f)$ do not influence upon errors of the solutions by combinations of FEM–CM, but guarantee the uniformly V_h^0-elliptic inequality eqn. (6.3.20), as long as H is chosen so small to satisfy eqn. (6.5.14), e.g., as long as the number of collocation nodes P_{ij} in quasi-uniform distribution is large enough. This conclusion is a distinctive feature from that in the conventional analysis of FEMs.

Remark 6.5.2

For Theorems 6.4.1 and 6.5.1, three inverse inequalities, the eqns. (6.4.2), (6.5.1), and (6.5.12), are needed for a polygon S_2. For polynomials v^+ of order L, the eqn. (6.4.2) holds for $\nu=2$ in Ref. [280]. The proof of eqns. (6.5.1) and (6.5.12) has been given in Section 5.5 of Chapter 5.

Remark 6.5.3

The eqns. (6.3.7)–(6.3.9) represent the generalized collocation equations using other admissible functions, such as RBFs, the Sinc functions, etc. The analysis of this chapter holds provided that the inverse inequalities, i.e., eqns. (6.4.2), (6.5.1), and (6.5.12) are satisfied. In fact, these inequalities can be proved for RBFs, the Sinc functions, etc. Details of analysis and numerical examples will be given in Chapter 7.

6.6 Final remarks

1. This chapter provides a theoretical framework of combinations of CM with other methods. The basic idea is to interpret CM as a special FEM, i.e., the LSM involving integration approximation. The eqns. (6.3.4)–(6.3.6) in CM are straightforward, and easily incorporated in the combined methods, see eqns. (6.3.14) and (6.3.15). The combination of CM in this chapter is also an important development from Li [280].
2. The key analysis for combinations of CMs is to prove the new uniformly V_h^0- elliptic inequalities, i.e., eqns. (6.2.11) and (6.3.20). The non-trivial proofs in Section 6.3 are new and intriguing, which consists of two steps: Step I for the simple one, i.e., eqn. (6.4.16) without $\int_{\Gamma_0} v_n^- v^-$; Step II for Theorem 6.4.1. Note that both eqns. (6.2.5) and (6.2.6) are required in combinations, i.e., eqn. (6.2.8) because the integral $P_c \iint_{S_2} \Delta u \, \Delta v$ worked like the biharmonic equation in S_2 in the traditional FEMs, see Ciarlet [103], where the essential continuity conditions $u^+ = u^-$ and $u_n^+ = u_n^-$ should be imposed on the interior boundary Γ_0.
3. In numerical algorithms, the integration approximation leads the LSM to the CM. In error analysis, the integration approximation plays a role only for satisfying the uniformly V_h^0-elliptic inequality, but not for improving accuracy of the solutions. The algorithms and analysis in this chapter are distinctive from the existing literature in CMs.
4. In S_2, Poisson's equation and the interior and exterior boundary conditions are copied straightforward into the collocation equations. This simple approach covers a large class of the CMs using various admissible functions, such as particular solutions, orthogonal polynomials, the RBFs (see Chapter 7), the Sinc functions, etc.

7 Radial basis function collocation methods

In this chapter, we choose the radial basis functions (RBFs) as the admissible functions, and use the collocation method (CM) in Chapter 5 and the combination of CM and finite element method (FEM) in Chapter 6. The algorithms are called the radial basis function collocation method (RBFCM), and the error analysis can be made for other kinds of continuous admissible functions. This chapter also displays the importance of CMs given in previous two chapters. The RBFs are introduced into the CMs and the combined methods for elliptic boundary value problems. First, for Poisson's equation the Ritz–Galerkin method (RGM) is chosen using the RBFs, and the integration approximation leads to the CM of RBFs. The combinations of RBFs with the FEM, the finite difference method (FDM), etc., can be easily formulated by following Li [280] and Chapters 5 and 6. More analysis of inverse estimates is explored in this chapter. Since the RBFs have the exponential convergence rates, and since the collocation nodes may be scattered in rather arbitrary fashions in various applications, the RBFs become the popular approximate tools for smooth solutions. Moreover, for singular solutions, we may use some singular functions and RBFs together. Numerical examples for smooth and singular problems are provided, to demonstrate the effectiveness of the methods and to support the analysis.

7.1 Introduction

In recent years, there have been many new developments for RBFs. The RBFs can be used for the interpolatory tool for smooth solutions, $u \in C^{\infty}(S)$. The convergence of their interpolants to a given continuous function has been discussed in the following work: Kansa [236] provided the surface approximations and partial derivative estimates, Madych [322] established several types of error bounds for multiquadric and related interpolators, Wu and Schaback [473] focussed on local errors of scattered data interpolated by RBFs in suitable variational formulation, and Yoon [478] regarded the convergence of RBFs in an arbitrary Sobolev space. All of those reports show exponential convergence rates. Moreover, the applications of RBFs have been given as follows: Kansa [236, 237] presented a series of applications in computational fluid dynamics. Franke and Schaback [150] gave some theoretical foundation for methods solving partial differential equations (PDEs). Wendland [463] derived error estimates for the solutions of smooth problems. Cheng et al. [94] introduced the $h-c$ meshless scheme for smooth problems, where numerical experiments were also provided. Hon and Schaback [201] used unsymmetrical collocation by RBFs, Mai-Duy and Tran-Cong [324, 325] used the RBF network methods for Poisson's equations, Hu, Chen, and Hu [206] used the weighted RBCM for elasticity problems, and Chen, Hon, and Schaback [85] introduced systematically the method of RBFs.

We may classify the CM as a special kind of spectral methods, whose numerical solutions have high accuracy, but with high instability due to the large condition numbers. In fact, the effective condition numbers given in Section 3.7 for the practical applications may be much smaller. Fortunately, in practice, only a few terms of RBFs are needed so that the condition number and the effective condition number will not be huge, and useful numerical computation can be carried out even in double precision. Since by using Mathematica, unlimited number of significant digits are available, the CM and the spectral methods using RBFs are very promising for numerical PDEs.

In this chapter, we consider the CM using RBFs, simply called the radial basis function collocation method (RBFCM). We derive inverse estimates and new error analysis. The CM is treated as the Ritz–Galerkin method involving integration approximation. However, the integration quadrature is used in analysis only to satisfy the V_h-elliptic inequality, but not to reach the exponentially high convergence rates, also see Chapters 2, 5, and 6. More explanations are given in Section 7.3. The advantages of the RBCM are twofold: (1) Source points of RBFs and collocation nodes may be scattered in rather arbitrary fashions in various applications, in which the solution domain is not confined to a rectangle. We need only a dense set of collocation nodes in any irregular domain. (2) Simplicity of the computing codes. A drawback of the RBCM is its high instability with large condition number.

In some literature (e.g., Chen, Hon, and Schaback [85] and Cho et al. [99]), the RBFCM for elliptic equations is called Kansa's method. The broad approaches of Kansa's method are provided in this chapter, by combining it with other numerical methods, such as FEM, and by applying it to singularity problems. More importantly, the convergence and the error analysis are given in the chapter.

This chapter is organized as follows. In the next section, the RBFs are described, and in Section 7.3, the CMs for different boundary conditions and combinations of FEM are discussed. In Section 7.4, the inverse estimates are derived. In Section 7.5, numerical experiments including smooth and singular problems are carried out to demonstrate the effectiveness of the methods proposed, and to support the analysis made.

7.2 Radial basis functions

For surface fitting on scattered points, using the RBFs shows remarkable advantages, see Hardy [186], Franke [149], Franke and Schaback [150], Schaback [403], Golberg [164], Kansa [236, 237], Madych and Nelson [323], Wu and Schaback [473], and Yoon [478]. Based on the theory of the FEM in Ciarlet [103], the errors of numerical solutions for elliptic equations are, basically, those of optimal approximations of admissible functions to the true solutions. Hence, the RBFs can be definitely applied to solve elliptic equations.

Let us describe the RBFs. The multiquadrics, the thin-plate splines, and the Gaussian functions are defined by, see Cheng et al. [94],

$$g_i(\mathbf{x}) = (r_i^2 + c^2)^{n-\frac{3}{2}}, \qquad n = 1, 2, \ldots,$$

$$g_i(\mathbf{x}) = \begin{cases} r_i^{2n} \ln r_i, & n = 1, 2, \ldots, \text{ in } R^2, \\ r_i^{2n-1}, & n = 1, 2, \ldots, \text{ in } R^3, \end{cases}$$

$$g_i(\mathbf{x}) = \exp\left(-\frac{r_i^2}{c^2}\right),$$

$$g_i(\mathbf{x}) = (r_i^2 + c^2)^{n-\frac{3}{2}} \exp\left(-\frac{r_i^2}{a^2}\right), \quad n = 1, 2, \ldots,$$

where $\mathbf{x} = (x, y)$ in R^2, and $\mathbf{x} = (x, y, z)$ in R^3. The radius $r_i = \{(x - x_i)^2 + (y - y_i)^2\}^{\frac{1}{2}}$ in R^2 and $r_i = \{(x - x_i)^2 + (y - y_i)^2 + (z - z_i)^2\}^{\frac{1}{2}}$ in R^3, where (x_i, y_i) and (x_i, y_i, z_i) are called source points of RBFs. The constants c and a are the shape parameters to be chosen later. When parameters c and a become larger, the RBFs become flatter.

Choose a linear combination of RBFs,

$$v = \sum_{i=1}^{n} a_i g_i(\mathbf{x}), \tag{7.2.1}$$

where the coefficients a_i are sought by the RGM or the CMs. For surface fitting, the coefficients are sought by satisfying

$$\sum_{i=1}^{n} a_i g_i(\mathbf{x}_j) + \sum_{i=1}^{m} b_i P_i(\mathbf{x}_j) = f_j, \quad j = 1, 2, \ldots, n,$$

$$\sum_{j=1}^{n} a_j P_i(\mathbf{x}_j) = 0, \quad i = 1, 2, \ldots, m,$$

where $P_i(\mathbf{x})$ is a nominal function of order i.

Based on the same ideas, we may add some singular functions $\psi_i(\mathbf{x})$ for fitting singular surfaces, or for solving singular problems of elliptic equations,

$$v = \sum_{i=1}^{n} a_i g_i(\mathbf{x}) + \sum_{i=0}^{m} b_i \psi_i(\mathbf{x}),$$

where

$$\sum_{i=1}^{n} a_i g_i(\mathbf{x}_j) + \sum_{i=0}^{m} b_i \psi_i(\mathbf{x}_j) = f_j, \quad j = 1, 2, \ldots, n. \tag{7.2.2}$$

For Motz's problem given in Chapter 2, a benchmark of singularity problems, we may add a few leading singular functions as

$$\psi_i(\mathbf{x}) = r^{i+\frac{1}{2}} \cos\left(i + \frac{1}{2}\right)\theta, \quad i = 0, 1, \ldots, L,$$

where (r, θ) are the polar coordinates with the origin (0,0), $r = \sqrt{x^2 + y^2}$ and $\tan\theta = \frac{y}{x}$.

7.3 Description of radial basis function collocation methods

By following the approaches in Chapters 5 and 6, the RBFs are chosen as the admissible functions in the RGMs. Since the RBFs do not satisfy the PDE and the boundary conditions, the residuals have to be enforced to zero at collocation points both in the solution domain and on its boundary. We obtain the CMs of RBFs, simply denoted by RBFCM, and the combined methods of RBFCM with other numerical methods. The optimal error bounds are provided, and the proofs are given in the following subsection, or can be done by following the analysis in Chapters 5 and 6 straightforwardly. The crucial inverse estimates for RBFs will be proven in the next section.

7.3.1 RBFCM for different boundary conditions

Consider Poisson's equation with Robin boundary condition:

$$-\Delta u = -\left(\frac{\partial^2 u}{\partial x^2} + \frac{\partial^2 u}{\partial y^2}\right) = f(x, y) \quad \text{in } S, \tag{7.3.1}$$

$$u_\nu \mid_{\Gamma_N} = q_1 \quad \text{on } \Gamma_N, \tag{7.3.2}$$

$$(u_\nu + \beta u) \mid_{\Gamma_R} = q_2 \quad \text{on } \Gamma_R, \tag{7.3.3}$$

where $\beta \geq \beta_0 > 0$, β_0 is constant, S is a polygon, $\partial S = \Gamma = \Gamma_N \cup \Gamma_R$, and u_ν is the unit outer normal derivative to ∂S. Assume Meas$(\Gamma_R) > 0$ for guaranteeing the unique solution. We make two assumptions.

(A1) The solutions in S can be expanded as

$$u = \sum_{i=1}^{\infty} a_i g_i(x, y) \quad \text{in } S, \tag{7.3.4}$$

where $g_i(x, y) \in C^2(S)$ and $g_i(x,y) \in C(\partial S)$. For the boundary singularity problems, since we may choose the particular solutions $g_i(x,y)$, the collocation nodes may be far from the singular points, so assumption, i.e., eqn. (7.3.4) may be relaxed by $g_i(x, y) \in C^2(D) \cap C^1(\partial S^*)$, where $D \subset S$ and $\partial S^* \subset \partial S$, and a_i are the expansion coefficients.

(A2) The expansions in eqn. (7.3.4) converge exponentially to the true solutions u,

$$u = u_L + R_L,$$

where $u_L = \sum_{i=1}^{L} a_i g_i(x, y)$ and $R_L = \sum_{i=L+1}^{\infty} a_i g_i(x, y)$. Then

$$\max_S |R_L| = O(\lambda^{c/\delta}),$$

where $c > 0$, $L > 1$, $0 < \lambda < 1$, and δ is the radial distance defined as

$$\delta = \sup_{(x,y)\in S} \{\inf_i [(x - x_i)^2 + (y - y_i)^2]^{\frac{1}{2}}\}, \tag{7.3.5}$$

where (x_i, y_i) are the source points of RBFs.

Based on **A1**–**A2** we may choose the admissible functions,

$$v = \sum_{i=1}^{L} \widetilde{a}_i g_i(x, y) \quad \text{in } S, \tag{7.3.6}$$

where $\widetilde{a}_i$ are unknown coefficients to be sought. Denote by V_h the finite-dimensional collection of the admissible functions, i.e., eqn. (7.3.6). To solve eqns. (7.3.1)–(7.3.3), the RGM can be written as follows: To seek solution $u_L \in V_h$ such that

$$b(u_L, v) = f(v), \quad \forall v \in V_h,$$

where

$$b(u, v) = \iint_S \Delta u\, \Delta v + \int_{\Gamma_N} u_\nu v_\nu + \int_{\Gamma_R} (u_\nu + \beta u)(v_\nu + \beta v), \tag{7.3.7}$$

$$f(v) = -\iint_S f\, \Delta v + \int_{\Gamma_N} q_1 v_\nu + \int_{\Gamma_R} q_2(v_\nu + \beta v).$$

When we view the CM as a RGM involving approximate quadratures, the RBFCM can be written as: To seek solution $\hat{u}_L \in V_h$ such that

$$\hat{b}(\hat{u}_L, v) = \hat{f}(v), \quad \forall v \in V_h, \tag{7.3.8}$$

where

$$\hat{b}(u, v) = \widehat{\iint_S} \Delta u\, \Delta v + \widehat{\int_{\Gamma_N}} u_\nu v_\nu + \widehat{\int_{\Gamma_R}} (u_\nu + \beta u)(v_\nu + \beta v),$$

$$\hat{f}(v) = -\widehat{\iint_S} f\, \Delta v + \widehat{\int_{\Gamma_N}} q_1 v_\nu + \widehat{\int_{\Gamma_R}} q_2(v_\nu + \beta v),$$

where $\widehat{\iint_S}$, $\widehat{\int_{\Gamma_N}}$, and $\widehat{\int_{\Gamma_R}}$ denote the approximations of $\iint_S$, $\int_{\Gamma_N}$, and $\int_{\Gamma_R}$ by some integration rules, respectively. We may choose the Newton–Cotes rules or the Legendre–Gauss rules:

$$\widehat{\iint_S} F = \sum_{ij} \alpha_{ij} F(Q_{ij}), \quad Q_{ij} \in S, \tag{7.3.9}$$

$$\widehat{\int_\Gamma} F = \sum_{i} \alpha_{i} F(Q_{i}), \quad Q_{i} \in \Gamma, \tag{7.3.10}$$

where α_{ij} and α_i are positive weights. We can then formulate the collocation equations directly at Q_{ij} and Q_i:

$$\sqrt{\alpha_{ij}}(\Delta v + f)(Q_{ij}) = 0, \quad Q_{ij} \in S, \tag{7.3.11}$$

$$\sqrt{\alpha_i^N}(v_\nu - q_1)(Q_i) = 0, \quad Q_i \in \Gamma_N, \tag{7.3.12}$$

$$\sqrt{\alpha_i^R}(v_\nu + \beta v - q_2)(Q_i) = 0, \quad Q_i \in \Gamma_R. \tag{7.3.13}$$

The eqns. (7.3.11)–(7.3.13) in $S \cup \Gamma_N \cup \Gamma_R$ are written simply as

$$\mathbf{F}\mathbf{x} = \mathbf{b}, \tag{7.3.14}$$

where $\mathbf{x}$ is a vector consisting of $\widetilde{a}_i$, $\mathbf{b}$ is a known vector, matrix $\mathbf{F} \in R^{N_c \times L}$, and N_c is the number of collocation nodes in $S \cup \Gamma_N \cup \Gamma_R$. In this chapter, let $N_c \geq L$, and we may seek the solutions of the entire RBFCM by the least squares method (LSM) in Golub and van Loan [168].

Now, we will provide the error estimates for solution $\hat{u}_L$ in eqn. (7.3.8) by following the FEM theory, see Ref. [103] for more details. Denote the space

$$H^* = \{v \mid v \in L^2(S),\ v \in H^1(S),\ \Delta v \in L^2(S)\},$$

accompanied by the norm

$$\|v\|_h = \{\|v\|_{1,S}^2 + \|\Delta v\|_{0,S}^2 + \|v_\nu\|_{0,\Gamma_N}^2 + \|v_\nu + \beta v\|_{0,\Gamma_R}^2\}^{\frac{1}{2}},$$

where $\|v\|_{1,S}$ is the Sobolev norm. In order to derive our main theorem (Theorem 7.3.1) given later, the following lemmas are needed.

Lemma 7.3.1

There exist two inequalities

$$b(u, v) \leq C\|u\|_h \times \|v\|_h, \quad \forall v \in V_h, \tag{7.3.15}$$

$$b(v, v) \geq C_0\|v\|_h^2, \quad \forall v \in V_h, \tag{7.3.16}$$

where C_0 and C are two positive constants independent of L.

Proof.

From the eqn. (7.3.7), we have

$$
\begin{aligned}
b(u, v) &\leq C_1 \sqrt{\iint_S (\Delta u)^2} \sqrt{\iint_S (\Delta v)^2} + C_2 \sqrt{\int_{\Gamma_N} (u_\nu)^2} \sqrt{\int_{\Gamma_N} (v_\nu)^2} \\
&\quad + C_3 \sqrt{\int_{\Gamma_R} (u_\nu + \beta u)^2} \sqrt{\int_{\Gamma_R} (v_\nu + \beta v)^2} \\
&\leq C\{\|\Delta u\|_{0,S} \|\Delta v\|_{0,S} + \|u_\nu\|_{0,\Gamma_N} \|v_\nu\|_{0,\Gamma_N} \\
&\qquad + \|u_\nu + \beta u\|_{0,\Gamma_R} \|v_\nu + \beta v\|_{0,\Gamma_R} + \|u\|_{1,S} \|v\|_{1,S}\} \\
&\leq C\{\|\Delta u\|_{0,S}^2 + \|u_\nu\|_{0,\Gamma_N}^2 + \|u_\nu + \beta u\|_{0,\Gamma_R}^2 + \|u\|_{1,S}^2\}^{\frac{1}{2}} \\
&\quad \times \{\|\Delta v\|_{0,S}^2 + \|v_\nu\|_{0,\Gamma_N}^2 + \|v_\nu + \beta v\|_{0,\Gamma_R}^2 + \|v\|_{1,S}^2\}^{\frac{1}{2}} \\
&\leq C\|u\|_h \times \|v\|_h,
\end{aligned}
$$

where C_i and C are generic constants, and their values may be different in different contexts. The first desired result, i.e., eqn. (7.3.15) is obtained.

Next, from the Green formula, we have

$$
\begin{aligned}
|v|_{1,S}^2 &= \iint_S \nabla v \cdot \nabla v = -\iint_S v\,\Delta v + \int_{\partial S} v_\nu v \\
&\leq \|\Delta v\|_{0,S} \|v\|_{0,S} + \int_{\Gamma_N} v_\nu v + \int_{\Gamma_R} (v_\nu + \beta v) v - \int_{\Gamma_R} \beta v^2 \\
&\leq \{\|\Delta v\|_{0,S} + \|v_\nu\|_{0,\Gamma_N} + \|v_\nu + \beta v\|_{0,\Gamma_R}\} \|v\|_{1,S} - \int_{\Gamma_R} \beta v^2,
\end{aligned}
\tag{7.3.17}
$$

where $\partial S = \Gamma = \Gamma_N \cup \Gamma_R$, and two bounds are used:

$$
\|v\|_{0,\Gamma_N} \leq C\|v\|_{1,S}, \qquad \|v\|_{0,\Gamma_R} \leq C\|v\|_{1,S}.
$$

Besides, from eqn. (7.3.17) we obtain

$$
\begin{aligned}
\|v\|_{1,S}^2 &\leq C\left(|v|_{1,S}^2 + \int_{\Gamma_R} \beta v^2\right) \\
&\leq C\{\|\Delta v\|_{0,S} + \|v_\nu\|_{0,\Gamma_N} + \|v_\nu + \beta v\|_{0,\Gamma_R}\} \|v\|_{1,S}.
\end{aligned}
$$

This leads to

$$
\|v\|_{1,S} \leq C\{\|\Delta v\|_{0,S} + \|v_\nu\|_{0,\Gamma_N} + \|v_\nu + \beta v\|_{0,\Gamma_R}\},
$$

and then

$$
\|v\|_{1,S}^2 \leq C\{\|\Delta v\|_{0,S}^2 + \|v_\nu\|_{0,\Gamma_N}^2 + \|v_\nu + \beta v\|_{0,\Gamma_R}^2\} = Cb(v, v). \tag{7.3.18}
$$

Moreover, from eqns. (7.3.18) and (7.3.7) we obtain

$$b(v,v) = \frac{1}{2}b(v,v) + \frac{1}{2}b(v,v)$$
$$\geq C_0\|v\|_{1,S}^2 + \frac{1}{2}\{\|\Delta v\|_{0,S}^2 + \|v_\nu\|_{0,\Gamma_N}^2 + \|v_\nu + \beta v\|_{0,\Gamma_R}^2\} \geq \overline{C}_0\|v\|_h^2,$$

where $\overline{C}_0 = \min\{\frac{1}{2}, C_0\}$. The second desired result, i.e., eqn. (7.3.16) is obtained. ■

We make more assumptions.

(A3) Suppose that for v in eqn. (7.3.6) there exists positive constant C independent of L, k, and v such that

$$\|v\|_{k,S} \leq CL^k\|v\|_{0,S}, \qquad v \in V_h, \tag{7.3.19}$$
$$\|v\|_{k,\Gamma} \leq CL^k\|v\|_{0,\Gamma}, \qquad v \in V_h, \tag{7.3.20}$$
$$\|v_\nu\|_{k,\Gamma} \leq CL^{k+1}\|v\|_{1,\Gamma}, \quad v \in V_h, \tag{7.3.21}$$

where ν is unit outward normal to Γ.

The inverse inequalities in eqns. (7.3.19)–(7.3.21) are crucial to following lemmas and theorem. Their proofs are new to Chapters 5 and 6, and are deferred to Section 7.4.

Lemma 7.3.2

For the rules, i.e., eqns. (7.3.9) and (7.3.10) with order r, there exist the bounds for $v \in V_h$,

$$\left|\left(\iint_S - \widehat{\iint_S}\right)(\Delta v)^2\right| \leq CH^{r+1}L^{r+3}\|v\|_{1,S}^2, \tag{7.3.22}$$
$$\left|\left(\int_{\Gamma_N} - \widehat{\int_{\Gamma_N}}\right)(v_\nu)^2\right| \leq CH^{r+1}L^{r+3}\|v\|_{1,S}^2, \tag{7.3.23}$$
$$\left|\left(\int_{\Gamma_R} - \widehat{\int_{\Gamma_R}}\right)(v_\nu + \beta v)^2\right| \leq CH^{r+1}L^{r+3}\|v\|_{1,S}^2, \tag{7.3.24}$$

where H denotes the maximal spacing between the consecutive integration nodes (i.e., collocation points), and L denotes the number of RBFs.

Proof.

Choose the integration rule of order r

$$\widehat{\int_\Gamma} g = \int_\Gamma \hat{g},$$

where $\hat{g}$ is the polynomial interpolant of order r on the partition of Γ with the maximal meshspacing H. Then, we obtain

$$\left|\left(\int_\Gamma - \widehat{\int_\Gamma}\right) g\right| = \left|\int_\Gamma g - \int_\Gamma \hat{g}\right| = \left|\int_\Gamma (g - \hat{g})\right| \leq CH^{r+1}|g|_{r+1,\Gamma}.$$

Based on **A3**, letting $g = (v_\nu)^2$ and $\Gamma = \Gamma_N$ we have

$$\begin{aligned} |g|_{r+1,\Gamma} &= |(v_\nu)^2|_{r+1,\Gamma_N} \leq C\sum_{i=0}^{r+1} |v_\nu|_{r+1-i,\Gamma_N}|v_\nu|_{i,\Gamma_N} \\ &\leq C\sum_{i=0}^{r+1} (L^{r+2-i}\|v\|_{1,S} \times L^{i+1}\|v\|_{1,S}) \leq CL^{r+3}\|v\|^2_{1,S}. \end{aligned}$$

Combining above two inequalities gives the desired result, i.e., eqn. (7.3.23). The eqns. (7.3.22) and (7.3.24) can be similarly proven. ■

Lemma 7.3.3

Let Lemmas 7.3.1 and 7.3.2 hold. We choose H to satisfy

$$H^{r+1}L^{r+3} = o(1), \tag{7.3.25}$$

then there exists the uniformly V_h-elliptic inequality

$$\hat{b}(v,v) \geq C\|v\|_h^2, \quad \forall v \in V_h. \tag{7.3.26}$$

Proof.

From eqn. (7.3.16) and Lemma 7.3.2, we have

$$\begin{aligned} \hat{b}(v,v) &\geq b(v,v) - CH^{r+1}L^{r+3}\|v\|^2_{1,S} \\ &\geq C_0\|v\|_h^2 - CH^{r+1}L^{r+3}\|v\|^2_{1,S} \\ &\geq C_0\left\{\left(1 - \frac{C}{C_0}H^{r+1}L^{r+3}\right)\|v\|^2_{1,S} + \|\Delta v\|^2_{0,S} + \|v_\nu\|^2_{0,\Gamma_N} + \|v_\nu + \beta v\|^2_{0,\Gamma_R}\right\} \\ &\geq \frac{C_0}{2}\|v\|_h^2. \end{aligned}$$

This is eqn. (7.3.26) with $C = C_0/2$. ■

We obtain an important theorem as follows.

Theorem 7.3.1

Suppose that there exist two inequalities

$$\begin{aligned} \hat{b}(u,v) &\le C\|u\|_h \times \|v\|_h, \quad &&\forall v \in V_h, \\ \hat{b}(v,v) &\ge C_0\|v\|_h^2, &&\forall v \in V_h, \end{aligned} \tag{7.3.27}$$

where C_0 and C are two positive constants independent of L. Then, when choosing H and L as eqn. (7.3.25), the solution of the RBFCM in eqn. (7.3.8) has the error bound,

$$\begin{aligned} \|u-\hat{u}_L\|_h &= C \inf_{v\in V_h} \|u-v\|_h \\ &\le C\{\|R_L\|_{2,S} + \|(R_L)_\nu\|_{0,\Gamma_N} + \|(R_L)_\nu\|_{0,\Gamma_R} + \|R_L\|_{0,\Gamma_R}\}, \end{aligned}$$

where C is a constant independent of L.

Proof.

Since the true solution u satisfies the collocation equations exactly, we have $\hat{b}(u,v)=\hat{f}(v)$. By using Lemma 7.3.2, we obtain

$$\hat{b}(u,v) \le \hat{f}(v) + CH^{r+1}L^{r+3}\|v\|_{1,S}^2, \quad \forall v \in V_h.$$

Since $\hat{u}_L$ is the solution of RBFCM, we have

$$\hat{b}(\hat{u}_L,v) = \hat{f}(v), \quad \forall v \in V_h,$$

and then

$$\hat{b}(u-\hat{u}_L,v) \le CH^{r+1}L^{r+3}\|v\|_{1,S}^2, \quad \forall v \in V_h.$$

Let $w=\hat{u}_L - v \in V_h$, from the above bound we obtain

$$\begin{aligned} C_0\|w\|_h^2 &\le \hat{b}(\hat{u}_L - v, w) = \hat{b}(\hat{u}_L - u + u - v, w) \\ &\le C(\|u-v\|_h\|w\|_h + H^{r+1}L^{r+3}\|w\|_{1,S}^2) \\ &\le C(\|u-v\|_h\|w\|_h + H^{r+1}L^{r+3}\|w\|_h^2). \end{aligned}$$

This leads to

$$\{C_0 - CH^{r+1}L^{r+3}\}\|w\|_h^2 \le \|u-v\|_h\|w\|_h.$$

Moreover, from eqn. (7.3.25) we have

$$\|\hat{u}_L - v\|_h = \|w\|_h \le \frac{1}{(C_0 - C \times o(1))}\|u-v\|_h \le C_1\|u-v\|_h.$$

From the triangle inequality,

$$\|u-\hat{u}_L\|_h \le \|u-v\|_h + \|\hat{u}_L - v\|_h \le C\|u-v\|_h,$$

and then

$$\begin{aligned}\|u-\hat{u}_L\|_h &\le C \inf_{\forall v \in V_h} \|u-v\|_h \\ &\le C\{\|R_L\|_{1,S} + \|\Delta R_L\|_{0,S} + \|(R_L)_\nu\|_{0,\Gamma_N} + \|(R_L)_\nu\|_{0,\Gamma_R} + \|R_L\|_{0,\Gamma_R}\}.\end{aligned}$$

■

Remark 7.3.1

*Theorem 7.3.1 implies that the errors of the solutions for Poisson's equation using RBFCM are bounded by the truncation errors of RBFs multiplied by a constant. From assumption **A2**, we assure that the errors of solution $\hat{u}_L$ have exponential convergence. Moreover, the eqn. (7.3.25) has the following relation*

$$HL^{\left(1+\frac{2}{r+1}\right)} = o(1),$$

where H and L are given in Lemma 7.3.2. Then we take $L = N^2$, and assume that the radial distance δ defined in eqn. (7.3.5) satisfies the relation $\delta \approx O(\frac{1}{N})$. Furthermore, we obtain an important relation

$$H = o\left(\delta^{\left(2+\frac{4}{r+1}\right)}\right). \tag{7.3.28}$$

From eqn. (7.3.28), we know how to balance the collocation nodes and the source points of RBFs.

Remark 7.3.2

In this chapter, the Newton–Cotes rules of integration are chosen for simplicity in exposition. From eqn. (7.3.28), we need a dense set of integration nodes in order to satisfy the V_h-elliptic inequality, and then to obtain the exponential convergence rates. The solution domain may not be confined to rectangles (or boxes for three dimensions). When the CMs using the RBFs are applied to solving PDEs in three dimensions, the simplest central rule may also be used. An integral in a closed region of three dimensions is approximated by the value of the integrand at the center of gravity (or roughly at any point) of the region multiplied by the volume of the region. Hence, the integration quadrature is not a severe problem in the CMs described in this chapter.

7.3.2 Combination of FEM and RBFCM

Consider Poisson's equation with the Dirichlet boundary condition,

$$\begin{aligned} -\Delta u &= f(x,y) \quad \text{in } S, \\ u &= 0 \qquad\quad\; \text{on } \Gamma, \end{aligned} \tag{7.3.29}$$

where S is a polygon, and Γ is its boundary. Let S be divided by Γ_0 into two disjoint subregions, S_1 and S_2: $S = S_1 \cup S_2 \cup \Gamma_0$ and $\partial S_1 \cap \partial S_2 = \Gamma_0$. On the interior boundary Γ_0, there hold the interior continuity conditions:

$$u^+ = u^-, \qquad u_\nu^+ = u_\nu^- \quad \text{on } \Gamma_0, \tag{7.3.30}$$

where $u_\nu = \frac{\partial u}{\partial \nu}$, $u^+ = u$ on $\Gamma_0 \cup S_2$, and $u^- = u$ on $\Gamma_0 \cup S_1$. Assume that the solution u in S_2 is smoother than u in S_1. We choose the FEM in S_1 and the RGM in S_2, whose discrete forms lead to the CM. Let S_1 be partitioned into small triangles: $\triangle_{ij}$, i.e., $S_1 = \cup_{ij} \triangle_{ij}$. Denote by h_{ij} the boundary length of $\triangle_{ij}$. The $\triangle_{ij}$ are said to be quasi-uniform if $\frac{h}{\min\{h_{ij}\}} \le C$, $h = \max_{\triangle_{ij} \subset S_1} \{h_{ij}\}$, and C is a constant independent of h. Then the admissible functions may be expressed by

$$v = \begin{cases} v^- = v_k & \text{in } S_1, \\ v^+ = \displaystyle\sum_{i=1}^{L} \widetilde{a}_i g_i(x, y) & \text{in } S_2, \end{cases} \tag{7.3.31}$$

where $\widetilde{a}_i$ are unknown coefficients, and v_k are piecewise Lagrange polynomials of power k in S_1 in the FEM. Assume that **A1** and **A2** hold in S_2, and $g_i(x,y) \in C^2(S_2 \cup \partial S_2)$ are the RBFs so that $v^+ \in C^2(S_2 \cup \partial S_2)$. Therefore, we may evaluate eqn. (7.3.29) directly from

$$(\Delta v^+ + f)(Q_{ij}) = 0, \tag{7.3.32}$$

at certain collocation nodes $Q_{ij} \in S_2$. Note that v in eqn. (7.3.31) is not continuous on the interior boundary Γ_0. Hence, to satisfy eqn. (7.3.30), the interior collocation equations are obtained:

$$v^+(Q_i) = v^-(Q_i) \quad \text{for } Q_i \in \Gamma_0, \tag{7.3.33}$$

$$v_\nu^+(Q_i) = v_\nu^-(Q_i) \quad \text{for } Q_i \in \Gamma_0. \tag{7.3.34}$$

Note that the eqns. (7.3.32)–(7.3.34) are straightforward and easy to be formulated.

Denote by V_h^0 the finite-dimensional collection of eqn. (7.3.31) satisfying $v|_\Gamma = 0$, where we simply assume $g_i(x,y)|_{\partial S_2 \cap \Gamma} = 0$. If such a condition does not hold, the corresponding collocation equations on $\partial S_2 \cap \Gamma$ are also needed, and the arguments can be provided similarly. The combination of the FEM–RBFCM involving integration approximation is given by: To seek the approximation solution $\hat{u}_h \in V_h^0$ such that

$$\hat{a}^*(\hat{u}_h, v) = f_1(v), \quad \forall v \in V_h^0, \tag{7.3.35}$$

where

$$\hat{a}^*(u,v) = \iint_{S_1} \nabla u \cdot \nabla v + \int_{\Gamma_0} u_\nu^- v^- + P_c \widehat{\iint_{S_2}} (\Delta u + f)(\Delta v + f)$$
$$+ \frac{P_c}{h} \widehat{\int_{\Gamma_0}} (u^+ - u^-)(v^+ - v^-) + P_c \widehat{\int_{\Gamma_0}} (u_\nu^+ - u_\nu^-)(v_\nu^+ - v_\nu^-),$$
$$f_1(v) = \iint_{S_1} f v,$$

where $\nabla u = u_x \mathbf{i} + u_y \mathbf{j}$, $u_x = \frac{\partial u}{\partial x}$, $u_y = \frac{\partial u}{\partial y}$, $u_\nu = \frac{\partial u}{\partial \nu}$, and ν is the unit outward normal to ∂S_2. Also h is the maximal boundary length of $\triangle_{ij}$ or $\square_{ij}$ in S_1, and $P_c > 0$ is chosen to be suitably large but still independent of h.

Now, let us establish the linear algebraic equations of combinations, i.e., eqn. (7.3.35) of FEM–RBFCM. First, considering the FEM in S_1 only, we have

$$a_1(\hat{u}_h, v) = f_1(v), \quad \forall v \in V_h,$$

where

$$a_1(u,v) = \iint_{S_1} \nabla u \cdot \nabla v + \int_{\Gamma_0} u_\nu^- v^-, \qquad f_1(v) = \iint_{S_1} f v.$$

By means of the traditional procedure of FEM [103], we obtain the linear algebraic equations,

$$\mathbf{A_1 x_1} = \mathbf{b_1}, \tag{7.3.36}$$

where $\mathbf{x_1}$ is a vector consisting of v_{ij} only, and matrix $\mathbf{A_1}$ is non-symmetric.

Next, we choose the integration rules (Newton–Cotes rules or Legendre–Gauss rules) in S_2. We may formulate collocation equations at $Q_{ij} \in S_2$, and $Q_i \in \Gamma_0$ directly. The collocation equations at Q_{ij} and Q_i are given by:

$$\sqrt{P_c \alpha_{ij}}(\Delta v^+ + f)(Q_{ij}) = 0, \quad Q_{ij} \in S_2, \tag{7.3.37}$$

$$\sqrt{\frac{P_c \alpha_i}{h}}(v^+ - v^-)(Q_i) = 0, \quad Q_i \in \Gamma_0, \tag{7.3.38}$$

$$\sqrt{\frac{P_c \alpha_i}{h}}(v_\nu^+ - v_\nu^-)(Q_i) = 0, \quad Q_i \in \Gamma_0, \tag{7.3.39}$$

where $v_\nu^-(Q_i) = \frac{v_{1i} - v_{0i}}{h}$, $v_{0i} = v(Q_i)$ and v_{1i} are the nodal variables in S_1 normal to Γ_0.

The eqns. (7.3.37)–(7.3.39) in $S_2 \cup \Gamma_0$ are denoted by

$$\mathbf{A_2 x_2} = \mathbf{b_2},$$

where $\mathbf{x_2}$ is a vector consisting of $\widetilde{a}_i$, v_{1i} and v_{0i}, and v_{0i} and v_{1i} are the unknowns on the two boundary layers nodes in S_1 close to Γ_0. Denote by M_C the number of all collocation nodes in S_2 and ∂S_2, and by N_B the number of v_{1i} and v_{0i}. The matrix $\mathbf{A_2} \in R^{M_C \times (L+N_B)}$. Therefore, we can see

$$\frac{P_c}{2}\widehat{\iint_{S_2}}(\Delta v+f)^2 + \frac{P_c}{2h}\widehat{\int_{\Gamma_0}}(v^+ - v^-)^2 + \frac{P_c}{2}\widehat{\int_{\Gamma_0}}(v_\nu^+ - v_\nu^-)^2$$
$$= \frac{1}{2}\mathbf{x_2^T A_2^T A_2 x_2} - \mathbf{x_2^T A_2^T b_2} + \mathbf{c}. \tag{7.3.40}$$

Combining eqns. (7.3.36) and (7.3.40) yields explicitly:

$$\mathbf{Ax} = \mathbf{b},$$
$$\mathbf{A} = \mathbf{A_1} + \mathbf{A_2^T A_2}, \qquad \mathbf{b} = \mathbf{b_1} + \mathbf{A_2^T b_2}, \tag{7.3.41}$$

where $\mathbf{x}$ is a vector consisting of the coefficients $\widetilde{a}_i$ and v_{ij} in $S_1 \cup \Gamma_0$. Denote by N_E the number of nodes on $S_1 \cup \Gamma_0$. Then vector $\mathbf{x}$ in eqn. (7.3.41) has $N_E + L$ dimensions. When P_c is chosen large enough, matrix $\mathbf{A} \in R^{(L+N_E)\times(L+N_E)}$ in eqn. (7.3.41) is positive definite but non-symmetric, and sparse when $N_E \gg L$. When $L + N_E$ is not huge, we may choose the Gaussian elimination without pivoting to obtain $\mathbf{x}$ from eqn. (7.3.41), see Golub and van Loan [168].

Now, we give error bounds for the solution $\hat{u}_h$ from eqn. (7.3.35) whose proofs have been reported in Chapter 6. The combination, i.e., eqn. (7.3.35) can be described equivalently as follows: To seek $\hat{u}_h \in V_h^0$ such that

$$\hat{a}(\hat{u}_h, v) = \hat{f}(v), \quad \forall v \in V_h^0,$$

where

$$\hat{a}(u,v) = \iint_{S_1} \nabla u \cdot \nabla v + \int_{\Gamma_0} u_\nu^- v^- + P_c \widehat{\iint_{S_2}} \Delta u\, \Delta v$$
$$+ \frac{P_c}{h}\widehat{\int_{\Gamma_0}}(u^+ - u^-)(v^+ - v^-) + P_c \int_{\Gamma_0}(u_\nu^+ - u_\nu^-)(v_\nu^+ - v_\nu^-),$$
$$\hat{f}(v) = \iint_{S_1} fv - P_c \widehat{\iint_{S_2}} f\,\Delta v. \tag{7.3.42}$$

Denote the space

$$H^{**} = \left\{ v \,|\, v \in L^2(S),\ v \in H^1(S_1),\ v \in H^1(S_2),\ \Delta v \in L^2(S_2), \text{ and } v\,|_\Gamma = 0 \right\},$$

accompanied with the norm

$$|||v||| = \left\{ \|v\|_{1,S_1}^2 + P_c\|v\|_{1,S_2}^2 + P_c\|\Delta v\|_{0,S_2}^2 + \frac{P_c}{h}\|v^+ - v^-\|_{0,\Gamma_0}^2 \right.$$
$$\left. + P_c\|v_\nu^+ - v_\nu^-\|_{0,\Gamma_0}^2 \right\}^{\frac{1}{2}},$$

where

$$\|v\|_1 = \{\|v\|_{1,S_1}^2 + \|v\|_{1,S_2}^2\}^{\frac{1}{2}}, \qquad |v|_1 = \{|v|_{1,S_1}^2 + |v|_{1,S_2}^2\}^{\frac{1}{2}},$$

and $\|v\|_{1,S_1}$ and $\|v\|_{1,S_2}$ are the Sobolev norms. Obviously, $V_h^0 \subset H^{**}$. We obtain the following theorem.

Theorem 7.3.2

Suppose that there exist two inequalities,

$$\begin{aligned} \hat{a}(u,v) &\le C|||u||| \times |||v|||, \quad && \forall v \in V_h^0, \\ \hat{a}(v,v) &\ge C_0|||v|||^2, && \forall v \in V_h^0, \end{aligned} \tag{7.3.43}$$

where $C_0(>0)$ and C are two constants independent of h and L. Then the solution of combination, i.e., eqn. (7.3.35) has the error bound,

$$|||u - \hat{u}_h||| \le C \inf_{v \in V_h} |||u - v|||.$$

We can obtain the following corollary easily from Theorem 7.3.2, also see Chapter 6.

Corollary 7.3.1

Let all conditions in Theorem 7.3.2 hold. Suppose that

$$u \in H^{k+1}(S_1) \quad \text{and} \quad u \in H^{k+1}(\Gamma_0).$$

Then there exists the error bound,

$$\begin{aligned} |||u - \hat{u}_h||| \le C \Bigg\{ & h^k |u|_{k+1,S_1} + \sqrt{P_c}\|R_L\|_{2,S_2} \\ & + \sqrt{P_c}\left(h^{k+\frac{1}{2}}|u|_{k+1,\Gamma_0} + \frac{1}{\sqrt{h}}\|R_L\|_{0,\Gamma_0} + \|(R_L)_\nu\|_{0,\Gamma_0} \right) \Bigg\}. \end{aligned}$$

Furthermore, suppose that the number L in eqn. (7.3.31) is chosen such that

$$\|R_L\|_{2,S_2} = O(h^k), \qquad \|R_L\|_{0,\Gamma_0} = O(h^{k+\frac{1}{2}}), \qquad \|(R_L)_\nu\|_{0,\Gamma_0} = O(h^k).$$

Then the optimal convergence rate is given by

$$|||u - \hat{u}_h||| = O(h^k).$$

Remark 7.3.3

The detailed proof for the V_h^0-elliptic inequality, i.e., eqn. (7.3.43) of Theorem 7.3.2 may follow Sections 6.4 and 6.5. Here we only give an outline of the proof, which *consists of two steps: Step I for the original equation of eqn. (7.3.42) without integration approximation, we need to prove $a(v,v) \geq C_0 |||v|||^2$, $\forall v \in V_h^0$; Step II for the Newton–Cotes integration rules of order r. Assuming that the inverse inequalities in* **A3** *hold, we obtain*

$$hL^2 = o(1), \qquad HL^{\left(1+\frac{2}{r+1}\right)} = o(1),$$

where $h = \max_{ij}\{h_{ij}\}$ in S_1, H denotes the maximal meshspacing between consecutive integration nodes (i.e., collocation nodes), and L denotes the number of RBFs in S_2. When we take $L = N^2$, and when the radial distance δ defined in eqn. (7.3.5) satisfies that $\delta \approx O(\frac{1}{N})$, we have the relations

$$h = o(\delta^4), \quad H = o\left(\delta^{\left(2+\frac{4}{r+1}\right)}\right). \tag{7.3.44}$$

From eqn. (7.3.44), we also know how to balance the mesh of FEM and the source points of RBFs.

Remark 7.3.4

From the above error analysis, we discover that the integration quadrature plays a role only for satisfying the uniformly V_h^0-elliptic inequalities, i.e., eqns. (7.3.27) and (7.3.43), but not for improving the accuracy of the solutions. As long as the maximal spacing H between consecutive collocation nodes is small enough, there always exist the optimal orders of solution errors from the RBCMs and their combinations.

7.4 Inverse estimates for radial basis functions

Although there exist many papers on RBFs, only a few of them are related to solutions of PDEs, see Kansa [237], Franke and Schaback [150], and Golberg [164]. In the Ritz–Galerkin method (RGM) [280], orthogonal polynomials and particular solutions are chosen, which have been replaced by RBFs in Section 7.3.

Note that the theoretical framework for the CMs has been established in Chapters 5 and 6, which can be easily applied to RBFCMs and their combinations, except that the crucial inverse estimates, i.e., eqns. (7.3.19)–(7.3.21) of RBFs need to be proven. Also note that our results in Chapter 5 and 6 and in this chapter are more comprehensive than those in Refs. [150, 164], because different boundary conditions are also involved, and because the combined methods are developed. Below, let us explore the inverse estimates, i.e., eqns. (7.3.19)–(7.3.21) needed for the RBFs. Take the multiquadric functions for example,

$$Q_L(x,y) = a_0 + \sum_{i=1}^{L} a_i g_i(x,y),$$

where $g_i(x,y)=(r_i^2+c^2)^{\frac{1}{2}}$. To approximate function $f(x,y)$, the collocation equations are given by

$$Q_L(x_i,y_i)=f(x_i,y_i),\quad i=1,2,\ldots,L,\qquad \sum_{i=1}^{L}a_i=0. \tag{7.4.1}$$

The eqn. (7.3.19) is essential, because the eqns. (7.3.20) and (7.3.21) can be easily derived from (or replaced by) eqn. (7.3.19), see Chapter 5. Hence, we focus on the proof of eqn. (7.3.19) for the multiquadric functions. We have the following lemma.

Lemma 7.4.1

Let $f(x,y)$ be defined on a rectangle and $c\geq 1$. There exists $\lambda\in(0,1)$ independent of f, c, and δ such that for sufficiently small δ,

$$\|f-Q_L\|_{0,S}\leq C\exp(\alpha c^2)\lambda^{c/\delta}\|f\|_{0,S}, \tag{7.4.2}$$

$$\|f-Q_L\|_{k,S}\leq C\delta^c\|f\|_{0,S}, \tag{7.4.3}$$

where $\alpha=\frac{\epsilon_h^2\sigma}{2}$, ϵ_h, and σ are two real constants in the Fourier transform, and C is a constant independent of δ and α.

Proof.

Based on Madych [322], Theorem 1 (cf. The eqn. (10) in p. 124), there exists a bound,

$$\|f-Q_L\|_{0,\infty,S}\leq C\exp(\alpha c^2)\lambda^{c/\delta}\|f\|_{0,S},\quad \alpha=\frac{\epsilon_h^2\sigma}{2}.$$

Since S is a bounded domain, we have

$$\|f-Q_L\|_{0,S}\leq C\|f-Q_L\|_{0,\infty,S}\leq C\exp(\alpha c^2)\lambda^{c/\delta}\|f\|_{0,S}.$$

This is the first result, i.e., eqn. (7.4.2).

Next, we also obtain from Ref. [322], Theorem 4, p. 127,

$$\|f-Q_L\|_{k,S}\leq C\delta^c\|f\|_{C_{hc}}, \tag{7.4.4}$$

where the norm is defined by (cf. The eqn. (6) in p. 124 of Ref. [322])

$$\|f\|_{C_{hc}}^2=\sum_{i=1}^{2}\iint_S|\xi_i\hat{f}(\xi)|^2(|\xi|^2h_c(\xi))^{-1}\,d\xi,$$

where $h_c = \sqrt{r^2+c^2}$, and the Fourier transform

$$\hat{f}(x) = \iint_S f(x)\exp(-i(x,\xi))\,dx.$$

Since $h_c \geq h_1$ for $c \geq 1$ we have

$$\|f\|_{C_{h_c}} \leq \|f\|_{C_{h_1}}, \quad c \geq 1. \tag{7.4.5}$$

Moreover, there is an estimate in Ref. [322], p. 124, that for $g(x)=f(cx)$,

$$\|g\|^2_{C_{h_1}} \leq C\exp(2\epsilon_h\sigma c)\|f\|^2_{0,S}, \tag{7.4.6}$$

where ϵ_h and σ are constants. Hence, if $c=1$ and $g(x)=f(x)$, we have from eqn. (7.4.6)

$$\|f\|^2_{C_{h_1}} \leq C\exp(2\epsilon_h\sigma)\|f\|^2_{0,S}. \tag{7.4.7}$$

Combining eqns. (7.4.4), (7.4.5), and (7.4.7) yields

$$\|f - Q_L\|_{k,S} \leq C\delta^c\|f\|_{C_{h_1}} \leq C\exp(\epsilon_h\sigma)\delta^c\|f\|_{0,S} \leq C_1\delta^c\|f\|_{0,S},$$

where $C_1 = C\exp(\epsilon_h\sigma)$. This is the second result, i.e., eqn. (7.4.3). ■

Lemma 7.4.1 and Madych [322] imply that the approximate solutions have the exponential convergence rate $O(\lambda^{c/\delta})$ with respect to δ, but the derivatives of low orders have only polynomially convergence rates $O(\delta^c)$. Although eqn. (7.4.3) is very conservative, the linear convergence rate $O(\delta)$, which is used as reasonable assumptions, i.e., eqn. (7.4.9), leads to the inverse estimates shown below. Suppose that there exist the bounds,

$$\|f - Q_L\|_{0,S} \leq o(1)\|f\|_{0,S}, \tag{7.4.8}$$

$$\|f - Q_L\|_{k,S} \leq C\|f\|_{0,S}, \quad k = 1,2,\ldots, \tag{7.4.9}$$

where $0 < o(1) \ll 1$, and C is a bounded constant as $\delta \to 0$. Denote polynomials f by

$$f = f_N(x,y) = \sum_{i,j=0}^{N} b_{ij}x^iy^j. \tag{7.4.10}$$

We cite Theorem 5.5.1 in Chapter 5 as a lemma.

Lemma 7.4.2

For the polynomials f of order N defined in S, there exists a constant independent of N such that

$$\|f_N\|_{k,S} \leq CN^{2k}\|f_N\|_{0,S}. \tag{7.4.11}$$

Let eqn. (7.4.1) be given. For uniquely determining $Q_L(x,y)$, we choose

$$N^2 < L \leq (N+1)^2, \tag{7.4.12}$$

and employ more collocation equations

$$Q_L(x_i,y_i) = f(x_i,y_i), \quad i = 1,2,\ldots,(N+1)^2, \qquad \sum_{i=1}^{(N+1)^2} a_i = 0, \tag{7.4.13}$$

where the first L collocation equations are the same as in eqn. (7.4.1). Hence, the errors of the solutions and their derivatives would not decrease, and Lemma 7.4.1 also holds. Then we may still assume eqns. (7.4.8) and (7.4.9), and prove the following theorem.

Theorem 7.4.1

Let eqns. (7.4.8) and (7.4.9) hold for f_N satisfying eqns. (7.4.10) and (7.4.13), where S is a rectangle. Then there exists the bound,

$$\|Q_L\|_{k,S} \leq CL^k \|Q_L\|_{0,S}. \tag{7.4.14}$$

Proof.

From the triangle inequality, we have

$$\|Q_L\|_{k,S} \leq \|f_N\|_{k,S} + \|Q_L - f_N\|_{k,S}.$$

From Lemma 7.4.2 and eqn. (7.4.9),

$$\|Q_L\|_{k,S} \leq CN^{2k}\|f_N\|_{0,S} + C\|f_N\|_{0,S} \leq CN^{2k}\|f_N\|_{0,S}. \tag{7.4.15}$$

Similarly, from eqn. (7.4.8) we have

$$\|f_N\|_{0,S} \leq \|Q_L\|_{0,S} + \|Q_L - f_N\|_{0,S} \leq \|Q_L\|_{0,S} + o(1)\|f_N\|_{0,S}.$$

This leads to

$$\|f_N\|_{0,S} \leq \frac{1}{(1-o(1))}\|Q_L\|_{0,S} \leq \frac{1}{2}\|Q_L\|_{0,S}. \tag{7.4.16}$$

Since $N \leq \sqrt{L}$ from eqn. (7.4.12), combining eqns. (7.4.15) and (7.4.16) yields

$$\|Q_L\|_{k,S} \leq CN^{2k}\|Q_L\|_{0,S} \leq CL^k\|Q_L\|_{0,S}.$$

This is the desired result, i.e., eqn. (7.4.14). ■

For polynomials, the inverse estimates, i.e., eqn. (7.4.11) are proven in Chapter 5, and the similar inverse estimates, i.e., eqn. (7.4.14) hold for RBFs, based on their approximation properties. Theorem 7.4.1 enables us to extend the combined methods and CMs in Chapters 5 and 6 to those using the RBFs. Similar arguments may be used for the CMs and their combinations using the Sinc functions [427], or the fundamental functions.

Remark 7.4.1

The eqn. (7.4.8) implies the approximation $f \approx Q_L$, *with small relative errors. Note that the bound of eqn. (7.4.9) can also be derived from Wendland [463]. In fact, we have, from Ref. [463], Theorem 5.3,*

$$\|f - Q_L\|_{k,S} \le Ch^{m-k}\|f\|_{m,S}, \quad m \ge k,$$

where $h = \delta$. *Choosing* $m = 2k$, *we obtain from eqn. (7.4.11),*

$$\|f_N - Q_L\|_{k,S} \le Ch^k\|f_N\|_{2k,S} \le Ch^k N^{4k}\|f_N\|_{0,S} \le C_1\|f_N\|_{0,S},$$

provided that h *can be chosen so small that* $h^k N^{4k} \le C_1$. *This also leads to assumption, i.e., eqn. (7.4.9). For* $c \ge 1$, *we may choose small* h *such that* $\left(\frac{h}{c}\right)^k N^{4k} \le C$.

Remark 7.4.2

By following Chapter 5, the solution domain can be extended to a polygon. The reason is that a polygon can be decomposed into a finite number of parallelograms with overlaps. A parallelogram may be transformed to a rectangle by linear transformations. Then the eqn. (7.4.14) also holds for the polygonal domain.

7.5 Numerical experiments

In this section, first we carry out a computational procedure for smooth problems to support the theoretical analysis in Section 7.3. Then we carry out two methods for solving Motz's problem.

7.5.1 Different boundary conditions

Consider Poisson's equation,

$$-\Delta u = -\left(\frac{\partial^2 u}{\partial x^2} + \frac{\partial^2 u}{\partial y^2}\right) = f(x,y) \quad \text{in } S, \tag{7.5.1}$$

where $S = \{(x,y) \mid 0 < x < 1,\ 0 < y < 1\}$, with the mixed type of different boundary conditions:

$$\begin{aligned} u\mid_{x=0} &= 0, \qquad u\mid_{y=0} = 0, \\ u_\nu \mid_{x=1} &= g_N, \\ u_\nu + \alpha u \mid_{y=1} &= g_R, \end{aligned} \tag{7.5.2}$$

where $\alpha > 0$, (e.g., $\alpha = 2$). The exact solution is chosen to be exactly the same as in Ref. [94]

$$u = \sin\left(\frac{\pi x}{6}\right) \sin\left(\frac{7\pi x}{4}\right) \sin\left(\frac{3\pi y}{4}\right) \sin\left(\frac{5\pi y}{4}\right). \tag{7.5.3}$$

The functions f, g_N, and g_R are then given explicitly.

The admissible functions are chosen as follows:

$$v = \sum_{i=1}^{L} \widetilde{a}_i g_i(x,y) \quad \text{in } S, \tag{7.5.4}$$

where $\widetilde{a}_i$ are unknown coefficients to be determined, and $g_i(x,y)$ are the RBFs. First, we use the inverse multiquadric radial basis functions (IMQRB)

$$g_i(x,y) = \frac{1}{\sqrt{r_i^2 + c^2}}, \tag{7.5.5}$$

where c is a parameter constant of shapeness, $r_i = \sqrt{(x-x_i)^2 + (y-y_i)^2}$, and $(x_i,\ y_i)$ are the source points which should also be chosen as the collocation nodes. Suitable additional functions may be added into eqn. (7.5.4), such as some polynomials or singular functions if necessary. Next, we may use the Gaussian radial basis functions (GRB),

$$g_i(x,y) = \exp\left(-\frac{r_i^2}{c^2}\right). \tag{7.5.6}$$

In Section 7.4, we address that the number of collocation nodes may be larger than the number of terms of RBFs. Let L be the number of RBFs, and N be the number of source points in one direction. We use the collocation eqns. (7.3.11)–(7.3.13) on the uniform interior and boundary nodes with $L = N^2$. The distribution of source points in this chapter is chosen to be uniform for easy test, but it may, of course, be chosen rather arbitrarily. Then the radial distance $\delta = O(\frac{1}{N})$. The error norms by using IMQRB with shape parameter $c = 2.0$ are listed in table 7.1, where the collocation nodes are also chosen to be uniform with the total number 196.

Table 7.1: The error norms and condition number by the IMQRB collocation method with parameter $c=2.0$.

$L=N^2$	7^2	9^2	11^2	13^2
$\|u-v\|_{0,\infty,S}$	9.30(−3)	5.92(−5)	4.32(−6)	1.10(−6)
$\|u-v\|_{0,S}$	1.60(−3)	1.51(−5)	8.19(−7)	3.10(−7)
$\|u-v\|_{1,S}$	2.28(−2)	1.32(−4)	5.22(−6)	1.67(−6)
Cond(**A**)	1.02(7)	2.28(8)	1.21(9)	2.01(9)

From table 7.1, we can see the following asymptotic relations,

$$\|u-v\|_{0,\infty,S} = O((0.22)^N), \tag{7.5.7}$$

$$\|u-v\|_{0,S} = O((0.24)^N), \tag{7.5.8}$$

$$\|u-v\|_{1,S} = O((0.20)^N). \tag{7.5.9}$$

The eqns. (7.5.7)–(7.5.9) indicate that the numerical solutions have the exponential convergence rates. In table 7.1, the errors in the Sobolev norm and the infinite norm can be achieved at the order of 10^{-6}, when the number of RBFs is given by $L=11^2=121$. Moreover, the shape parameter c of RBFs can also be chosen larger, e.g., $c=2.5$ or 3.0. It seems that those numerical results are better than those in Cheng et al. [94].

7.5.2 Adding method of singular functions

Consider Motz's problem, its exact solution is given in Ref. [299], and the leading six coefficients for eqn. (7.2.3) are

$$d_0 = 401.1624, \quad d_1 = 87.6559, \quad d_2 = 17.2379,$$
$$d_3 = -8.0712, \quad d_4 = 1.44027, \quad d_5 = 0.33105.$$

Because of singularity, some singular functions may be added into the RBFs. The admissible functions are chosen as

$$v = \sum_{i=1}^{L} \widetilde{a}_i g_i(x,y) + \sum_{n=0}^{M} \widetilde{d}_n \varphi_n(r,\theta) \quad \text{in } S, \tag{7.5.10}$$

where $\widetilde{a}_i$ and $\widetilde{d}_n$ are unknown coefficients to be determined, and $g_i(x,y)$ are the RBFs. In eqn. (7.5.10), the singular particular solutions $\varphi_n(r,\theta)=r^{n+\frac{1}{2}}\cos(n+\frac{1}{2})\theta$. We choose IMQRB and GRB. Let the distribution of source points be uniform in our computation, and use the collocation equations on uniform interior and

Table 7.2: The error norms and condition number by the IMQRB collocation method adding singular functions with parameter $c=2.0$.

$L=N^2,\ M$	$4^2,\ 2$	$6^2,\ 3$	$8^2,\ 4$	$10^2,\ 5$
$\|u-v\|_{0,\infty,S}$	11.34	1.07	7.46(−2)	5.50(−3)
$\|u-v\|_{0,S}$	3.00	2.85(−1)	1.27(−2)	1.70(−3)
$\|u-v\|_{1,S}$	12.0	1.68	1.03(−1)	1.45(−2)
$\tilde{d}_0$	423.8219	403.9276	401.0821	401.1446
$\tilde{d}_1$	95.9485	111.1686	89.9612	87.7384
$\tilde{d}_2$	12.7307	26.5319	8.6550	16.5567
$\tilde{d}_3$	/	−12.2469	−10.7710	−3.4625
$\tilde{d}_4$	/	/	1.8867	−3.1864
$\tilde{d}_5$	/	/	/	−0.8645
Cond(**A**)	2.58(3)	8.60(5)	5.38(8)	3.21(9)

Table 7.3: The error norms and condition number by the GRB collocation method adding singular functions with parameter $c=2.0$.

$L=N^2,\ M$	$4^2,\ 2$	$6^2,\ 3$	$8^2,\ 4$	$10^2,\ 5$
$\|u-v\|_{0,\infty,S}$	8.07	7.32(−1)	9.40(−2)	5.90(−3)
$\|u-v\|_{0,S}$	1.86	1.95(−1)	1.39(−2)	1.60(−3)
$\|u-v\|_{1,S}$	8.13	1.07	1.45(−1)	1.78(−2)
$\tilde{d}_0$	416.4048	403.6369	401.0648	401.1865
$\tilde{d}_1$	72.9120	103.3334	88.9340	87.3105
$\tilde{d}_2$	16.4691	45.5082	24.7180	17.2233
$\tilde{d}_3$	/	−19.0671	−11.6812	−13.7762
$\tilde{d}_4$	/	/	−1.7538	7.8214
$\tilde{d}_5$	/	/	/	−0.4210
Cond(**A**)	7.16(3)	2.73(8)	1.36(9)	5.79(9)

boundary collocation nodes. The error norms are listed in tables 7.2 and 7.3 for IMQRB($c=2.0$) and GRB($c=2.0$), respectively. The L denotes the number of RBFs, $L=N^2$, and M denotes the number of singular functions. The coupling techniques for L and M may refer to Ref. [280]. From table 7.2, we can see the following asymptotic relations,

$$\|u - v\|_{0,\infty,S} = O((0.28)^N), \tag{7.5.11}$$
$$\|u - v\|_{0,S} = O((0.29)^N), \tag{7.5.12}$$
$$\|u - v\|_{1,S} = O((0.33)^N). \tag{7.5.13}$$

Also, from table 7.3, we can see for GRB,

$$\|u - v\|_{0,\infty,S} = O((0.30)^N), \tag{7.5.14}$$
$$\|u - v\|_{0,S} = O((0.31)^N), \tag{7.5.15}$$
$$\|u - v\|_{1,S} = O((0.36)^N). \tag{7.5.16}$$

The eqns. (7.5.11)–(7.5.16) indicate that the numerical solutions obtained also have the exponential convergence rates, which verify the error bounds obtained.

7.5.3 Subtracting method of singular functions

We also consider Motz's problem here, but which is solved by slightly different method. First, choose purely the RBFs,

$$\bar{u} = \sum_{i=1}^{L} a_i g_i(x, y). \tag{7.5.17}$$

The Motz's solutions are known as

$$u(r, \theta) = \sum_{i=0}^{\infty} d_i r^{i+\frac{1}{2}} \cos\left(i + \frac{1}{2}\right)\theta, \tag{7.5.18}$$

where the coefficients can also be obtained from

$$d_i = \frac{2}{\pi} r^{-(i+\frac{1}{2})} \int_0^{\pi} u(r, \theta) \cos\left(i + \frac{1}{2}\right)\theta \, d\theta. \tag{7.5.19}$$

Usually, the solutions from the CM using eqn. (7.5.17) are poor only near the origin due to the singularity. Hence, choosing $r \geq \frac{1}{2}$, we may evaluate the approximate coefficients $\widetilde{d}_i$ from eqn. (7.5.19) very well, and obtain the singular solutions,

$$\overline{w} = \sum_{i=0}^{M} \widetilde{d}_i r^{i+\frac{1}{2}} \cos\left(i + \frac{1}{2}\right)\theta. \tag{7.5.20}$$

Table 7.4: The error norms and condition number by the IMQRB collocation method subtracting singular functions with parameter $c=2.0$.

$L=N^2,\ M$	$4^2,\ 2$	$6^2,\ 3$	$8^2,\ 4$	$10^2,\ 5$
$\|u-(\overline{u}+\overline{w})\|_{0,\infty,S}$	6.99	6.10(−1)	7.40(−2)	7.00(−3)
$\|u-(\overline{u}+\overline{w})\|_{0,S}$	2.24	2.64(−1)	8.90(−3)	1.40(−3)
$\|u-(\overline{u}+\overline{w})\|_{1,S}$	7.82	1.45	1.11(−1)	1.63(−2)
$\widetilde{d}_0$	399.8414	401.0809	401.1633	401.1627
$\widetilde{d}_1$	86.0459	87.6951	87.6526	87.6556
$\widetilde{d}_2$	18.0743	17.0752	17.2360	17.2378
$\widetilde{d}_3$	/	7.8863	−8.0718	−8.0714
$\widetilde{d}_4$	/	/	1.4386	1.4403
$\widetilde{d}_5$	/	/	/	0.3310
Cond(**A**)	1.14(3)	4.21(5)	6.16(8)	3.21(9)
Num. of iter.	10	10	9	9

We may subtract the singular part, i.e., eqn. (7.5.20) from u, and then obtain a rather smooth problem for $\overline{u}=u-\overline{w}$, with the following equations,

$$\Delta\overline{u} = 0 \qquad \text{in } S, \tag{7.5.21}$$

$$\begin{aligned} \overline{u}_x &= -\overline{w}_x && \text{on } x=-1 \wedge 0 \le y \le 1, \\ \overline{u} &= 500-\overline{w} && \text{on } x=1 \wedge 0 \le y \le 1, \\ \overline{u}_y &= -\overline{w}_y && \text{on } y=1 \wedge -1 \le x \le 1, \\ \overline{u} &= 0 && \text{on } y=0 \wedge -1 \le x < 0, \\ \overline{u}_y &= 0 && \text{on } y=0 \wedge 0 < x \le 1. \end{aligned} \tag{7.5.22}$$

Again, we choose eqn. (7.5.17) and use the RBCM in Section 7.3 to seek the rather smooth problem eqns. (7.5.21) and (7.5.22). We add singular part $\overline{w}$ to the smooth part $\overline{u}$ to obtain a better approximation to u. Repeat the evaluation, i.e., eqn. (7.5.19) of coefficients d_i, and then subtract $\overline{w}$ of eqn. (7.5.20) again. The above iteration repeats until a convergent solution is obtained.

The error norms are listed in tables 7.4 and 7.5 for IMQRB($c=2.0$) and GRB($c=2.0$), respectively. We choose $r=1$ to evaluate the approximate coefficients $\widetilde{d}_n,\ n=0,1,\ldots,5$. The termination of iterations occurs when the absolute error becomes less than 10^{-6}, and the number of iteration needed is about 10. From

Table 7.5: The error norms and condition number by the GRB collocation method subtracting singular functions with parameter $c=2.0$.

$L=N^2,\ M$	$4^2,\ 2$	$6^2,\ 3$	$8^2,\ 4$	$10^2,\ 5$
$\|u-(\bar{u}+\bar{w})\|_{0,\infty,S}$	5.24	5.99(−1)	1.01(−1)	1.18(−2)
$\|u-(\bar{u}+\bar{w})\|_{0,S}$	1.62	2.69(−1)	2.10(−2)	2.10(−3)
$\|u-(\bar{u}+\bar{w})\|_{1,S}$	6.14	1.48	1.64(−1)	1.87(−2)
$\tilde{d}_0$	399.7470	401.0842	401.1625	401.1617
$\tilde{d}_1$	87.2125	87.7272	87.6501	87.6571
$\tilde{d}_2$	17.7888	17.1083	17.2329	17.2373
$\tilde{d}_3$	/	−7.8454	−8.0616	−8.0708
$\tilde{d}_4$	/	/	1.4393	1.4403
$\tilde{d}_5$	/	/	/	0.3310
Cond(**A**)	6.06(3)	1.89(8)	1.01(9)	2.70(10)
Num. of iter.	11	10	10	9

table 7.4, we can also observe the exponential convergence rates,

$$\|u-(\bar{u}+\bar{w})\|_{0,\infty,S} = O((0.34)^N), \tag{7.5.23}$$

$$\|u-(\bar{u}+\bar{w})\|_{0,S} = O((0.30)^N), \tag{7.5.24}$$

$$\|u-(\bar{u}+\bar{w})\|_{1,S} = O((0.33)^N). \tag{7.5.25}$$

And from table 7.5,

$$\|u-(\bar{u}+\bar{w})\|_{0,\infty,S} = O((0.39)^N), \tag{7.5.26}$$

$$\|u-(\bar{u}+\bar{w})\|_{0,S} = O((0.34)^N), \tag{7.5.27}$$

$$\|u-(\bar{u}+\bar{w})\|_{1,S} = O((0.38)^N). \tag{7.5.28}$$

The exponential convergence rates, eqns. (7.5.23)–(7.5.28), also support the theoretical analysis made.

The adding method of singular solutions was reported in Fix, Gulati, and Wakoff [146], and the subtracting method of singular solutions in Wigley [469, 470], also see Ref. [280].

7.6 Comparisons and conclusions

1. To solve Poisson's problem, this chapter provides the theoretical framework of RBFCM, also called Kansa's method, and its combinations with other methods.

This chapter is also an important extension of RBFs from the approximation theory of smooth functions to the solutions of PDEs. The RBFCM is also an efficient tool for solving singular PDEs, in which a combination of RBFs and singular functions is used.

2. From table 7.1, we can observe the exponential convergence rates for smooth problems, which may be competitive to orthogonal polynomials. From tables 7.2–7.5, we can find that the exponential convergence rates also exist for singularity problems, when some singular functions are applied as basis functions. From these tables, the following asymptotic relations are also observed,

$$\|u - v\|_{k,S} = O(\lambda^N) = O(\lambda^{\sqrt{L}}), \quad k = 0, 1,$$

where $0 < \lambda < 1$.

3. For Motz's problem, we find that the approximate solutions by RBFCM adding method of singular solutions in Section 7.5.2 have much better convergence rates than those by RBFCM subtracting method of singular solutions in Section 7.5.3. On the contrary, the leading coefficients of singular functions, $\widetilde{d}_n$, $n=0,1,\ldots,5$, obtained by RBFCM subtracting method are more accurate than those by RBFCM adding method. We also observe that the IMQRB is superior to the GRB, in both accuracy and stability.
4. From the numerical results, we see that the RBFs have large condition numbers which imply high instability. This is the drawback of the RBCM. In practical computation, since only a few terms of RBFs are needed, such a drawback is not serious. Moreover, its effective condition number may be much smaller, see Section 3.7 of Chapter 3. In spite of this drawback, the RBFCM is still a competitive method for PDEs due to high accuracy and a very low computational cost.
5. From tables 7.6 and 7.7 and fig. 7.1 we can see that there exists convergence by increasing parameter c from 1.0 to 3.0. The following asymptotic relations are also observed,

$$\|u - v\|_{k,S} = O(\lambda^c), \quad k = 0, 1,$$

where $0 < \lambda < 1$.

6. All the numerical experiments indicate that the RBFCM have exponential convergence rates,

$$\|u - v\|_{k,S} = O(\lambda^{c/\delta}) \approx O(\lambda^{cN}) \approx O(\lambda^{c\sqrt{L}}), \quad k = 0, 1,$$

where $0 < \lambda < 1$. From the viewpoint of accuracy, the errors tend to zero as $c/\delta \to \infty$ (i.e., $L \to \infty$ and $c \to \infty$). However, from the viewpoint of stability, we cannot increase L too much due to large condition number. Also, we cannot increase c too much due to the flatness of RBFs, causing the ill-conditioned $\mathbf{F}$ in eqn. (7.3.14). It seems that accuracy and instability are twins. Such a statement is also called the *Uncertainty Principle* in Schaback [403]. Either one goes for a small error and gets a bad sensitivity, or one wants a stable algorithm and has to take a comparably large error. In practical computation, we should keep some

Table 7.6: The error norms and condition number by the IMQRB collocation method adding singular functions with $L = 8^2$ and $M = 4$.

c	$\|u-v\|_{0,\infty,S}$	$\|u-v\|_{0,S}$	$\|u-v\|_{1,S}$	Cond(**A**)
1.0	6.71(−2)	1.96(−2)	1.17(−1)	2.34(6)
1.2	7.93(−2)	2.16(−2)	1.39(−1)	5.56(6)
1.4	9.36(−2)	2.12(−2)	1.44(−1)	1.66(7)
1.6	9.36(−2)	1.94(−2)	1.37(−1)	5.49(7)
1.8	8.59(−2)	1.63(−2)	1.21(−1)	1.94(8)
2.0	7.46(−2)	1.27(−2)	1.03(−1)	1.70(9)
2.2	6.29(−2)	9.70(−3)	8.74(−2)	8.27(8)
2.4	5.25(−2)	7.60(−3)	7.60(−2)	1.83(9)
2.6	4.38(−2)	6.10(−3)	6.81(−2)	1.92(9)
2.8	3.65(−2)	5.20(−3)	6.28(−2)	3.35(9)
3.0	3.30(−2)	4.80(−3)	5.96(−2)	2.01(9)

Table 7.7: The error norms and condition number by the IMQRB collocation method subtracting singular functions with $L = 8^2$ and $M = 4$.

c	$\|u-(\overline{u}+\overline{w})\|_{0,\infty,S}$	$\|u-(\overline{u}+\overline{w})\|_{0,S}$	$\|u-(\overline{u}+\overline{w})\|_{1,S}$	Cond(**A**)
1.0	5.28(−2)	8.70(−3)	9.19(−2)	1.35(6)
1.2	8.02(−2)	9.80(−3)	1.14(−1)	2.97(6)
1.4	9.78(−2)	1.03(−2)	1.27(−1)	7.81(6)
1.6	9.63(−2)	1.01(−2)	1.27(−1)	2.32(7)
1.8	8.66(−2)	9.60(−3)	1.20(−1)	7.35(7)
2.0	7.40(−2)	8.90(−3)	1.11(−1)	1.95(8)
2.2	6.16(−2)	8.20(−3)	1.00(−1)	3.18(8)
2.4	5.06(−2)	7.50(−3)	9.06(−2)	4.75(8)
2.6	4.13(−2)	6.80(−3)	8.20(−2)	5.59(8)
2.8	3.36(−2)	6.20(−3)	7.48(−2)	9.11(8)
3.0	2.73(−2)	5.70(−3)	6.89(−2)	1.11(9)

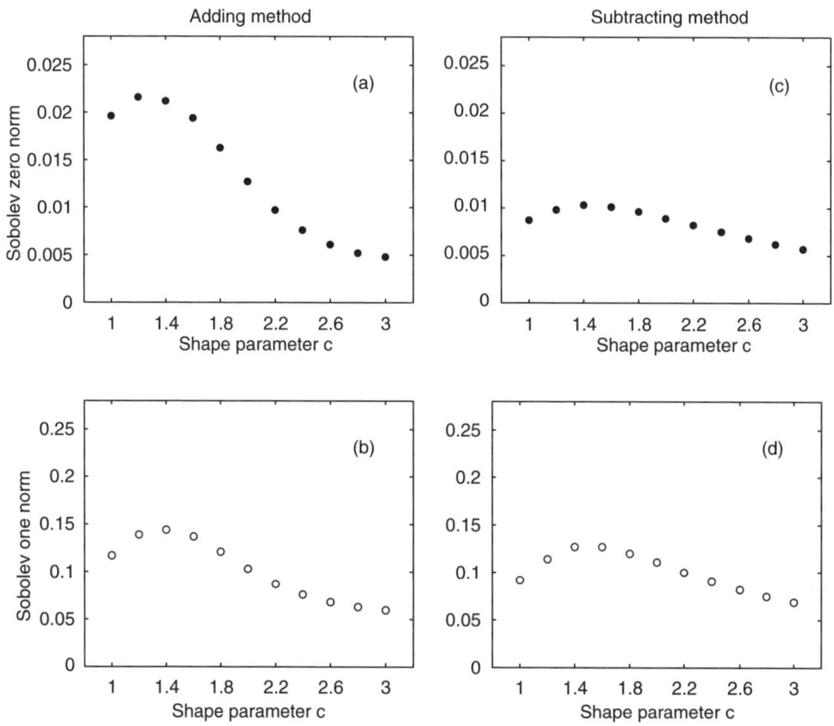

Figure 7.1: The solution errors versus parameter c, with $L = 8^2$ and $M = 4$.

balance between them. In summary, Theorem 7.3.1 and the numerical examples in this section display that the errors of the solutions of Poisson's equation by RBFCM have the same exponential convergence rates as those of surface fitting by RBFs.

The firm, the enduring, the simple,
and the modest are near to virtue.
——— *The Confucian Analects* ———

Confucius
(551–479 B.C.)

Part III
Advanced topics

The remainder of this book is devoted to advanced topics, such as the combinations with high-order FEMs, the eigenvalue problems, and the Helmholtz equation. In the last chapter of this part, explicit harmonic solutions are provided for the Laplace equation on sectors, and new models involving the discontinuity and mild singularity of the solutions are designed. The advanced topics in this part are related more to the collocation Trefftz method (CTM) in Part I.

Let the solution domain S be split into S_1 and S_2. Suppose that the CTM is used in S_2 to deal with the solution singularity, and the FEM in S_1 where the solution is smooth enough. The first topic is to combine the CTM with high-order FEMs. The penalty plus hybrid techniques of Chapter 3 are used, and global superconvergence can be achieved. The second topic is to use the CTM for eigenvalue problems and the Helmholtz equation. By an iteration process, when there occurs a degeneracy of the Helmholtz equation, an eigenvalue is obtained. High accuracy of the eigenvalues and eigenfunctions can also be achieved. This is just an example to extend the Trefftz method (TM) from elliptic equations in Ref. [280] to other problems. The TM can be applied to parabolic and hyperbolic equations, see Cho et al. [99] and Chen, Hon, and Schaback [85].

This book is confined to elliptic and eigenvalue problems. In Chen, Hon, and Schaback [85] and Cho et al. [99], the TM is extended to time-dependent partial differential equations (PDEs). By the time differencing or the Laplace transform or the Laguerre transform, the initial boundary value problems can be converted to the modified inhomogeneous Helmholtz equation $-\Delta u \pm k^2 u = f$. Once one of its particular solutions is found by means of the techniques in Chen, Hon, and Schaback [85], the homogeneous equation $-\Delta u \pm k^2 u = 0$ is reduced. Then, the Trefftz techniques for the Helmholtz equation in Chapter 10 and for the Debye–Huckel equation in Chapter 1 can be employed.

When the series solutions satisfy the PDEs exactly, the Ritz–Galerkin method (RGM) and the spectral methods are simplified to the TM. An efficient implementation of TM is developed in this book, called the CTM. The CTM becomes a very competitive method for solving PDEs, due to several merits: (1) simplicity of the algorithms, (2) high accuracy of the solutions, (3) explicit forms of the solutions, (4) the ease in dealing with singularity, (5) facile analysis, and (6) savings in CPU time and computer storage. Evidently, the CTM is advantageous over the FEM, the finite difference method (FDM), and the finite volume method (FVM). Both the TM and the boundary element method (BEM) may reduce the solved problems by one dimension. However, the CTM is more efficient and effective.

The particular solutions of PDEs are essential to the TM. Our efforts are focused on the explicit harmonic solutions of the Laplace equation on a polygon. Although these particular solutions can be found in Volkov [452], the formulas of the harmonic functions presented in Chapter 11 are easier to expose the mild singularity at the domain corners.

This part consists of four chapters for the advanced topics of Parts I and II.

Chapter 8: Combinations with High-Order FEMs.
Chapter 9: Eigenvalue Problems.

Chapter 10: The Helmholtz Equation.
Chapter 11: Explicit Harmonic Solutions of Laplace's Equation.

A brief description of Part III is given as follows.

Chapter 8 solves the Poisson equation on a polygonal domain wherein the bi-Lagrange p-order FEMs are chosen. We call these methods the p-rectangles. Let S be split into S_1 and S_2, where only S_2 includes a singular point. Suppose that the solution in S_1 is highly smooth. Then the CTM and the p-rectangles may be used in S_2 and S_1, respectively. In this chapter, we invoke the penalty plus hybrid techniques to couple the CTM and the p-rectangles, and derive *almost* the *best* global superconvergence $O(h^{p+2-\delta})$, where h denotes the meshspacing and $0 < \delta \ll 1$. When Adini's elements are used in S_1 instead of the p-rectangles, the *best* global superconvergence $O(h^{3.5})$ can be achieved.

Chapter 9 deals with eigenvalue problems, by the new algorithms of the TM, based on the degeneracy of the Helmholtz solutions. The degeneracy means that as the parameter k^2 used in the Helmholtz equation approaches to an eigenvalue of the Laplace operator, the solution matrix becomes singular. Piecewise particular solutions are obtained for a sample of eigenvalue problems. Using piecewise particular solutions is advantageous for solving problems with complicated geometry, because uniform particular solutions may not always exist in the entire solution domain. Numerical experiments are reported only for the eigenvalue problems of a basic model in this chapter, and the model involving singularity of eigenfunctions in Ref. [301].

Chapter 10 employs the TM for finding the solution of the Helmoholtz problem, where k^2 is very close to eigenvalues λ_l of the Laplace operator, but not exactly equal to. Denote the relative distance $\delta = \frac{k^2 - \lambda_l}{\lambda_l}$. Error analysis is made to derive the error bounds of the solutions with respect to δ, thus linking to the analysis of Part I and Chapter 9.

Chapter 11 derives explicitly the harmonic functions, i.e., the solutions of Laplace's equations, for the Dirichlet and Neumann boundary conditions on sectors. These harmonic functions clearly expose strong or mild singularity properties of the Laplace solutions at the domain corners. We also explore in detail the singularities of the sectors with the boundary angles, $\Theta = \frac{\pi}{2}, \frac{3\pi}{2}, \pi$, and 2π, which often occur in many testing models. Two new rectangular models with singularities are provided. One model involves the discontinuity of the solution, and the other the mild singularities. The CTM, the Schwarz alternating method (SAM), and their combinations may be chosen to seek the solutions with high accuracy.

8 Combinations with high-order FEMs

To solve Poisson's equation on a polygonal domain, the bi-Lagrange $p(\geq 2)$-order finite element methods (FEMs) (called the p-rectangles) are chosen. The global superconvergence $O(h^{p+2})$ in the H^1 norm was first derived in Lin and Yan [311] on the entire solution domain for smooth solutions, by means of an *a posteriori* polynomial interpolant of higher order. In this chapter, the *high* global superconvergence is applied to Poisson's equation with singularities. Let the solution domain S be split into S_1 and S_2, where only S_2 includes a singular point. Suppose that the solution in S_1 is highly smooth. Then the Trefftz method (TM) using singular particular solutions and the $p(\geq 2)$-rectangles may be used in S_2 and S_1, respectively. In this chapter, we invoke the penalty plus hybrid techniques in Section 3.5 to couple the TM and the $p(\geq 2)$-rectangles, and derive almost the *best* global superconvergence $O(h^{p+2-\delta})$, $0 < \delta \ll 1$. When Adini's elements are used in S_1 instead of the p-rectangles, the *best* global superconvergence $O(h^{3.5})$ can be achieved. Numerical experiments are carried out by the combination of the TM and Adini's elements in Huang and Li [212], to verify the superconvergence $O(h^{3.5})$.

8.1 Introduction

In Li [280] only the linear and bilinear elements were discussed; in this chapter we intend to achieve *high* global superconvergence by using high-order Lagrange FEMs and Adini's elements. There exist two kinds of superconvergence: *global* and *local interior pointwise*. The global superconvergence was reported in Křížek and Neittaanmäki [256] and Lin and Yan [311], and *local interior pointwise* superconvergence in Křížek and Neittaanmäki [255], MacKinnon and Carey [321], Nakao [345], Pehlivanov et al. [360], Wheeler and Whiteman [466], and Wahlbin [459, 460].

To solve singularity problems, let the solution domain S be split into two disjoint subdomains S_1 and S_2. The $p(\geq 2)$-order Lagrange rectangles (or simply

$p(\geq 2)$-rectangles) are used in S_1 where the solution is highly smooth, and the singular particular solutions are used in S_2 where the solution has a singularity. Coupling techniques play an important role in combining different numerical methods for solving elliptic equations. In a recent study in Refs. [211, 292], the simplified hybrid techniques were chosen, to couple the $p(\geq 2)$-order Lagrange rectangles in S_1 and the singular particular solutions in S_2. The superclose $\|u_h - u_I\|_1 = O(h^{p+\frac{3}{2}})$ was obtained, where u_h is the approximate solution and u_I is the piecewise bi-p-order Lagrange polynomial interpolant of the true solution u. There is a loss of $O(h^{\frac{1}{2}})$ in superclose, compared with the best superclose $O(h^{p+2})$ given in Ref. [311]. Can we find better couple techniques to retain the best superclose $O(h^{p+2})$? To answer this question is an aim of this chapter. In this chapter, we employ the penalty plus hybrid techniques in Section 3.5 to couple the different methods, and *almost* the *best* superclose estimate, $\|u_h - u_I\|_1 = O(h^{p+2-\delta})$, $0 < \delta \ll 1$, can be reached. Hence, by means of an *a posteriori* polynomial interpolant of order $p+2$ in S_1, almost the *best* global superconvergence $O(h^{p+2-\delta})$ over the entire subdomains can also be achieved. The local superconvergence is also developed when the highly smooth solutions exist only in partial of S_1. Moreover, when Adini's elements are used in S_1 instead of p-rectangles, the superconvergence $O(h^{3.5})$ can be obtained and verified by numerical experiments presented [212]. Here, we briefly mention the references of the penalty plus hybrid techniques, which are first proposed in Nitsche [346], and then developed in Arnold [5], Baker [20], Barrett and Elliott [25], Fairweather [142], and Li [280]. The materials in this chapter are adapted from Huang [210] and Huang and Li [211, 212].

Since combining different numerical methods is an important tool to solve complicated PDEs, combinations of the TM with other methods, in particular with FEMs, is significant for advanced numerical analysis (see Li [280]). This chapter is also a further development of the book [280], where only linear or bilinear FEMs in the TM combinations are used. The high-order FEMs, such as the bi-p-order FEM and Adini's elements, are chosen in this chapter, in order to achieve the highly accurate solutions.

This chapter is organized as follows. The combined algorithm of the two methods is described in the next section. In Section 8.3, the $p(\geq 2)$-rectangles are chosen in S_1, to derive the global superconvergence $O(h^{p+2-\delta})$ of the solution derivatives. In Section 8.4, Adini's elements are chosen in S_1, to yield the superconvergence $O(h^{3.5})$ of the solution derivatives.

8.2 Combinations of TM and Lagrange FEMs

Consider Poisson's equation with the Dirichlet boundary condition:

$$-\Delta u = -\left(\frac{\partial^2 u}{\partial x^2} + \frac{\partial^2 u}{\partial y^2}\right) = f(x, y) \quad \text{in } S, \tag{8.2.1}$$

$$u\,|_\Gamma = 0 \quad \text{on } \Gamma,$$

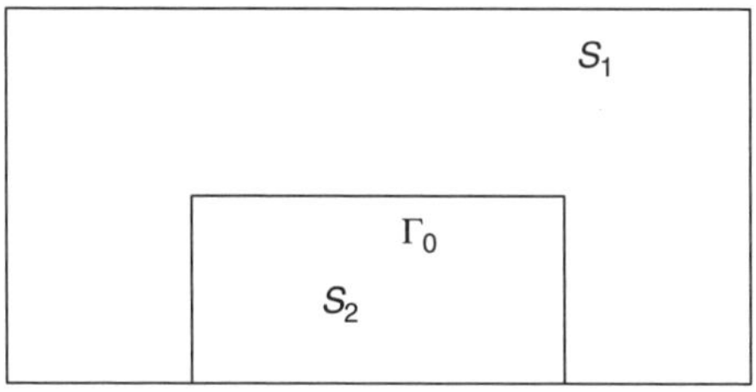

Figure 8.1: Partition of a rectangular domain.

where S is a polygon and Γ is its boundary. Let S be divided by Γ_0 into two disjoint subregions, S_1 and S_2, $S = S_1 \cup S_2 \cup \Gamma_0$ and $S_1 \cap S_2 = \emptyset$, see fig. 8.1. In this chapter we assume the following conditions.

(A1) There exists a boundary singularity in S_2, where $u \in H^{1+\alpha}(S_2)$ for $0 < \alpha < 1$.

(A2) In S_2, the true solution u can be spanned by particular solutions, $\{\Psi_i\}$, singular or analytical,

$$u = \Psi_0 + \sum_{i=1}^{\infty} a_i \Psi_i \quad \text{in } \overline{S}_2, \tag{8.2.2}$$

where $\overline{S}_2 = S_2 \cup \partial S_2$, a_i are the expansion coefficients, and

$$-\Delta\Psi_0 = f, \qquad \Delta\Psi_i = 0, \quad i \geq 1. \tag{8.2.3}$$

Also the particular solutions $\{\Psi_i\}$, $i = 1, 2, \ldots$, are assumed complete and linearly independent.

(A3) The expansion, i.e., eqn. (8.2.2) converges exponentially. Denote

$$u_L = g_L(a_1, \ldots, a_L) = \Psi_0 + \sum_{i=1}^{L} a_i \Psi_i,$$

then $u = u_L + R_L$, where the remainder

$$R_L = \sum_{i=L+1}^{\infty} a_i \Psi_i.$$

The exponential convergence rates imply

$$\max |R_L| = \max \left| \sum_{i=L+1}^{\infty} a_i \Psi_i \right| = O(e^{-\overline{c}L}) \quad \text{in } S_2,$$

where $\overline{c} > 0$ and $L \geq 1$.

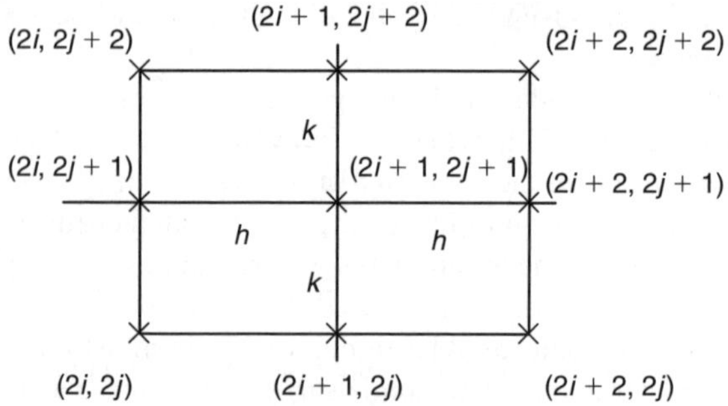

Figure 8.2: $\square^{2\times 2}_{2i+1,2j+1}$ in the 2×2 fashion of partition.

(A4) In S_1, the solution u is highly smooth such that $u \in H^k(S_1), k \geq 5$. Later, we may relax this assumption to $u \in H^k(D_1), k \geq 5$, where D_1 is a subdomain of S_1.

(A5) S_1 is partitioned again into quasi-uniform rectangular elements in 2×2 fashion shown in fig. 8.2, denoted by $S_1^h = \bigcup_{ij} \square_{ij}$, and h denotes the maximal boundary length of all $\square_{ij}$. The $\square_{ij}$ are said to be *quasi-uniform* if the following ratios are bounded,

$$\frac{h}{\min_{ij}\{h_i, k_j\}} \leq C,$$

where C is a constant independent of h.

When S_1 is not a rectangle, the triangulation is used, and the combinations with k-order FEMs are similar. However, only the superconvergence $O(h^{k+1})$ can be achieved.

Based on **A1–A3**, we may choose the finite expansions

$$v^+ = g_L(\widetilde{a}_1, \ldots, \widetilde{a}_L) = \Psi_0 + \sum_{i=0}^{L} \widetilde{a}_i \Psi_i \tag{8.2.4}$$

as admissible functions, where Ψ_i are known and satisfy eqn. (8.2.3); but the coefficients $\widetilde{a}_i$ are unknown, to be sought by the combined algorithms given below.

Based on **A4–A5**, we will choose the piecewise Lagrange interpolation polynomial v_p on S_1, and follow the superconvergence techniques of Lin and Yan [311] and Křížek and Neittaanmäki [256], to achieve the global superconvergence on the entire domain. The effort in this chapter is to combine the TM using the singular particular solutions and the high-order FEMs, to solve effectively the singularity problems.

The admissible functions in S should be chosen as

$$v = \begin{cases} v^- = v_p & \text{in } \overline{S}_1, \\ v^+ = g_L(\widetilde{a}_1, \ldots, \widetilde{a}_L) & \text{in } \overline{S}_2, \end{cases}$$

where $g_L(\widetilde{a}_1,\ldots,\widetilde{a}_L)$ are given in eqn. (8.2.4), and v_p are the piecewise bi-p-order Lagrange polynomials.

In this chapter we consider the Lagrange elements with order $p(\geq 2)$ on rectangles. Based on Refs. [214, 311], the polynomial interpolant are designed sophisticatedly by means of the solution u at the vertices of $\square_{ij}$, the integrals of u along edges $\partial\square_{ij}$, and the integrals of u on the $\square_{ij}$. We call them the point–line–area elements (or variables). In the traditional $p(\geq 2)$-rectangles in Ref. [103], only the nodal elements (or point variables) are used. We may obtain the point–line–area elements from the pure point variables through a matrix transformation.

The piecewise interpolation polynomials u_I^p are formulated as follows.

$$u_I^p(Z_i) = u(Z_i), \quad i = 1,2,3,4.$$

When $p \geq 2$, more equations are satisfied:

$$\int_{\ell_r} (u_I^p - u)v\, d\ell = 0, \quad \forall v \in P_{p-2}(\ell_r), \;\; r = 1,2,3,4, \tag{8.2.5}$$

$$\iint_{\square_{ij}} (u_I^p - u)v\, ds = 0, \quad \forall v \in Q_{p-2}(\square_{ij}), \tag{8.2.6}$$

where ℓ_r are the edges of $\square_{ij}$, and $P_p(x)$ and $Q_p(x, y)$ are the polynomials of order p defined by

$$P_p(x) = \sum_{i=0}^{p} a_i x^i, \qquad Q_p(x, y) = \sum_{i,j=0}^{p} a_{ij} x^i y^j.$$

Construct the interpolant of u of eqn. (8.2.1):

$$u_{I,L} = \begin{cases} u_I^p, & \text{in } \overline{S}_1, \\ u_L = g_L(a_1,\ldots,a_L), & \text{in } \overline{S}_2. \end{cases} \tag{8.2.7}$$

We choose the specific rules of integration,

$$\widehat{\int_0^1} uv = \int_0^1 uv, \quad \forall u, v \in P_p. \tag{8.2.8}$$

Also denote by V_h^* and V_h^0 the finite collections of the functions satisfying $v\,|\,_\Gamma = 0$ of

$$v_h = \begin{cases} v_p, & \text{in } \overline{S}_1, \\ v_L = \Psi_0 + \sum_{i=1}^{L} \widetilde{a}_i \Psi_i, & \text{in } \overline{S}_2, \end{cases} \tag{8.2.9}$$

and

$$v_h = \begin{cases} v_p, & \text{in } \overline{S}_1, \\ v_L = \sum_{i=1}^{L} \widetilde{a}_i \Psi_i, & \text{in } \overline{S}_2, \end{cases} \tag{8.2.10}$$

respectively.

In Refs. [211, 292], the simplified hybrid techniques are used to couple the TM and the $p(\geq 2)$-rectangles. Unfortunately, there is a loss of $O(h^{\frac{1}{2}})$ in superclose [211, 292]. In this chapter, we invoke the penalty plus hybrid techniques in Section 3.5: To seek $u_h \in V_h^*$ such that

$$\widehat{A}_h(u_h, v) = \widehat{f}(v), \quad \forall v \in V_h^0, \tag{8.2.11}$$

where

$$\widehat{A}_h(u, v) = \iint_{S_1} \nabla u \cdot \nabla v + \iint_{S_2} \nabla u \cdot \nabla v + \widehat{D}_h(u, v) + \widehat{E}_h(u, v), \tag{8.2.12}$$

$$\widehat{f}(v) = \widehat{\iint_{S_1}} f v + \iint_{S_2} f v, \tag{8.2.13}$$

$$\widehat{D}_h(u, v) = \frac{P_c}{h^{2\sigma}} \widehat{\int_{\Gamma_0}} (u^+ - u^-)(v^+ - v^-), \tag{8.2.14}$$

$$\widehat{E}_h(u, v) = -\widehat{\int_{\Gamma_0}} \frac{\partial u^+}{\partial n}(v^+ - v^-) - \widehat{\int_{\Gamma_0}} \frac{\partial v^+}{\partial n}(u^+ - u^-). \tag{8.2.15}$$

$\widehat{D}_h(u, v)$ and $\widehat{E}_h(u, v)$ are the penalty and hybrid integrals, respectively, and $P_c(> 0)$, $\sigma(> 0)$ are the parameters. In eqns. (8.2.13)–(8.2.15) we choose the rules of integration:

$$\widehat{\int_{\Gamma_0}} uv = \int_{\Gamma_0} \hat{u}\hat{v}, \qquad \widehat{\iint_{S_1}} f v = \iint_{S_1} \hat{f}\hat{v},$$

where $\hat{u}$ is the piecewise p-order Lagrange polynomial interpolant of u. We define the appropriate norm

$$\overline{\|v\|}_H = \left\{ \|v\|_{1,S_1}^2 + \|v\|_{1,S_2}^2 + \frac{1}{h^{2\sigma}} \overline{\|v^+ - v^-\|}_{0,\Gamma_0}^2 \right\}^{\frac{1}{2}}, \tag{8.2.16}$$

where

$$\overline{\|v\|}_{0,\Gamma_0}^2 = \int_{\Gamma_0} \hat{v}^2. \tag{8.2.17}$$

First, let us give a basic theorem.

Theorem 8.2.1 (Basic Theorem)

Let $\square_{ij}$ be quasi-uniform. Assume that the uniformly V_h^0-elliptic inequality holds,

$$C_0 \overline{\|v\|}_H^2 \leq \widehat{A}_h(v, v), \quad \forall v \in V_h^0, \tag{8.2.18}$$

where $C_0(>0)$ is a constant independent of h. Then there exists a constant C independent of h such that

$$\overline{\|u_h - u_{I,L}\|_H}$$
$$\leq C \sup_{w \in V_h^0} \frac{1}{\overline{\|w\|_H}} \left\{ \left| \iint_{S_1} \nabla(u - u_I) \cdot \nabla w \right| + \left| \iint_{S_2} \nabla(u - u_L) \cdot \nabla w \right| \right.$$
$$+ \left| \left(\iint_{S_1} - \widehat{\iint_{S_1}} \right) fw \right| + \left| \left(\int_{\Gamma_0} - \widehat{\int_{\Gamma_0}} \right) \frac{\partial u}{\partial n}(w^+ - w^-) \right|$$
$$\left. + |\widehat{D}_h(u - u_{I,L}, w)| + |\widehat{E}_h(u - u_{I,L}, w)| \right\}, \tag{8.2.19}$$

where u_h and u are the solutions of eqns. (8.2.11) *and* (8.2.1)*, respectively, and $u_{I,L}$ is the interpolant eqn.* (8.2.7) *of u.*

Proof.

For the true solution, we have

$$\widehat{A}_h(u, v) = \left(\int_{\Gamma_0} - \widehat{\int_{\Gamma_0}} \right) \frac{\partial u}{\partial n}(v^+ - v^-) + \iint_{S_1} f v + \iint_{S_2} f v,$$

which leads to

$$\widehat{A}_h(u - u_h, v) = \left(\int_{\Gamma_0} - \widehat{\int_{\Gamma_0}} \right) \frac{\partial u}{\partial n}(v^+ - v^-) + \left(\iint_{S_1} - \widehat{\iint_{S_1}} \right) f v.$$

Let $w = u_h - u_{I,L} \in V_h^0$, then we obtain from the assumption, i.e., eqn. (8.2.18)

$$C_0 \overline{\|w\|}_H^2 \leq \widehat{A}_h(u_h - u_{I,L}, w)$$
$$= \widehat{A}_h(u - u_{I,L}, w) - \left(\int_{\Gamma_0} - \widehat{\int_{\Gamma_0}} \right) \frac{\partial u}{\partial n}(w^+ - w^-) - \left(\iint_{S_1} - \widehat{\iint_{S_1}} \right) fw$$
$$\leq |\widehat{A}_h(u - u_{I,L}, w)| + \left| \left(\int_{\Gamma_0} - \widehat{\int_{\Gamma_0}} \right) \frac{\partial u}{\partial n}(w^+ - w^-) \right|$$
$$+ \left| \left(\iint_{S_1} - \widehat{\iint_{S_1}} \right) fw \right|.$$

Hence, we have

$$\overline{\|u_h - u_{I,L}\|_H} \leq C \frac{1}{\overline{\|w\|_H}} \left\{ |\widehat{A}_h(u - u_{I,L}, w)| + \left| \left(\int_{\Gamma_0} - \widehat{\int_{\Gamma_0}} \right) \frac{\partial u}{\partial n}(w^+ - w^-) \right| \right.$$
$$\left. + \left| \left(\iint_{S_1} - \widehat{\iint_{S_1}} \right) fw \right| \right\}. \tag{8.2.20}$$

Also from eqn. (8.2.12),

$$|\widehat{A}_h(u-u_{I,L},w)| \le \left|\iint_{S_1} \nabla(u-u_I)\cdot\nabla w\right| + \left|\iint_{S_2} \nabla(u-u_L)\cdot\nabla w\right|$$
$$+|\widehat{D}_h(u-u_{I,L},w)| + |\widehat{E}_h(u-u_{I,L},w)|. \qquad (8.2.21)$$

The desired result, i.e., eqn. (8.2.19) follows from eqns. (8.2.20) and (8.2.21). ■

8.3 Global superconvergence

We will derive the bounds of all terms on the right-hand side of eqn. (8.2.19) in Theorem 8.2.1. First define a semi-norm

$$\overline{|v^+ - v^-|_{1,\Gamma_0}} = |\hat{v}^+ - \hat{v}^-|_{1,\Gamma_0},$$

where $\hat{v}$ is the piecewise p-order polynomial interpolant of v. We have the following lemma.

Lemma 8.3.1

Suppose that there exists a constant $\nu(>0)$ *such that*

$$\|v^+\|_{k,\Gamma_0} \le CL^{k\nu}\|v^+\|_{0,\Gamma_0}, \quad k=1,2, \ \ v \in V_h^0. \qquad (8.3.1)$$

Then there exists the bound,

$$\|v^-\|_{1,\Gamma_0} \le C\{L^\nu + hL^{2\nu} + h^{\sigma-1}\}\overline{\|v\|}_H. \qquad (8.3.2)$$

Proof.

We have from eqns. (8.2.17) and (8.3.1)

$$\|v^-\|_{1,\Gamma_0} \le \|v^+\|_{1,\Gamma_0} + \|v^+ - v^-\|_{1,\Gamma_0} \qquad (8.3.3)$$
$$\le CL^\nu\|v^+\|_{0,\Gamma_0} + \overline{\|v^+ - v^-\|_{1,\Gamma_0}} + \|v^+ - \hat{v}^+\|_{1,\Gamma_0} + \|v^- - \hat{v}^-\|_{1,\Gamma_0}.$$

Noting that $v^- = \hat{v}^-$ in S_1, we have $\|v^- - \hat{v}^-\|_{1,\Gamma_0} = 0$. Moreover, since $\hat{v}^+$ are piecewise p-order polynomials, there exists the inverse inequality:

$$\overline{\|v^+ - v^-\|_{1,\Gamma_0}} = \|\hat{v}^+ - v^-\|_{1,\Gamma_0} \le Ch^{-1}\|\hat{v}^+ - v^-\|_{0,\Gamma_0}$$
$$= Ch^{-1}\overline{\|v^+ - v^-\|_{0,\Gamma_0}} \le Ch^{\sigma-1}\overline{\|v\|}_H. \qquad (8.3.4)$$

Also we have from eqn. (8.3.1)

$$\|v^+ - \hat{v}^+\|_{1,\Gamma_0} \le Ch|v^+|_{2,\Gamma_0} \le ChL^{2\nu}\|v^+\|_{0,\Gamma_0} \qquad (8.3.5)$$
$$\le ChL^{2\nu}\|v^+\|_{1,S_2} \le ChL^{2\nu}\overline{\|v\|}_H.$$

Combining eqns. (8.3.3)–(8.3.5) yields the desired result, i.e., eqn. (8.3.2). ■

Lemma 8.3.2

Let eqn. (8.3.1) *be given. Then there exists the bound,*

$$\left|\iint_{S_1} \nabla(u-u_I)\cdot\nabla w\right| \le Ch^{p+2}\|u\|_{p+3,S_1}(1+L^\nu+hL^{2\nu}+h^{\sigma-1})\overline{\|w\|}_H. \tag{8.3.6}$$

Proof.

From Refs. [214, 311], we have

$$\left|\iint_{S_1} \nabla(u-u_I)\,\nabla w\right| \le Ch^{p+2}\left\{\|u\|_{p+3,S_1}\|w^-\|_{1,S_1} + \left|\frac{\partial u}{\partial n}\right|_{p+1,\Gamma_0}\|w^-\|_{1,\Gamma_0}\right\}.$$

The desired result, i.e., eqn. (8.3.6) follows directly from Lemma 8.3.1 and $\left|\frac{\partial u}{\partial n}\right|_{p+1,\Gamma_0} \le C\|u\|_{p+3,S_1}$. ■

Lemma 8.3.3

Suppose there exists a constant $\nu(>0)$ such that

$$\left|\frac{\partial w^+}{\partial n}\right|_{k,\Gamma_0} \le CL^{(k+1)\nu}\|w^+\|_{1,S_2},\quad k=0,1,\quad w\in V_h^0. \tag{8.3.7}$$

Then there exist the bounds,

$$\left|\widehat{D}_h(u-u_{I,L},w)\right| \le Ch^{-\sigma}\overline{\|R_L\|}_{0,\Gamma_0}\overline{\|w\|}_H, \tag{8.3.8}$$

and

$$|\widehat{E}_h(u-u_{I,L},w)| \le C\left\{h^\sigma\overline{\left\|\frac{\partial R_L}{\partial n}\right\|}_{0,\Gamma_0} + (L^\nu+hL^{2\nu})\overline{\|R_L\|}_{0,\Gamma_0}\right\}\overline{\|w\|}_H, \tag{8.3.9}$$

where the remainder $R_L = u-u_L$ in S_2.

Proof.

Since $\widehat{\int}_{\Gamma_0}(u-u_I)(w^+-w^-)=0$ in view of eqn. (8.2.8), we have from eqns. (8.2.14) and (8.2.17),

$$\begin{aligned}|\widehat{D}_h(u-u_{I,L},w)| &= \frac{P_c}{h^{2\sigma}}\left|\widehat{\int}_{\Gamma_0}((u_L^+-u)+(u-u_I^-))(w^+-w^-)\right| \\ &= \frac{P_c}{h^{2\sigma}}\left|\widehat{\int}_{\Gamma_0}(u_L^+-u)(w^+-w^-)\right| \le \frac{P_c}{h^{2\sigma}}\|\widehat{R}_L\|_{0,\Gamma_0}\overline{\|w^+-w^-\|}_{0,\Gamma_0} \\ &\le Ch^{-\sigma}\|\widehat{R}_L\|_{0,\Gamma_0}\overline{\|w\|}_H = Ch^{-\sigma}\overline{\|R_L\|}_{0,\Gamma_0}\overline{\|w\|}_H.\end{aligned}$$

This is the first estimate, i.e., eqn. (8.3.8). Next we show eqn. (8.3.9). We have from eqn. (8.2.15)

$$\widehat{E}_h(u-u_{I,L},w) = -\widehat{\int_{\Gamma_0}} \left(\frac{\partial u^+}{\partial n} - \frac{\partial u_L^+}{\partial n}\right)(w^+ - w^-) + \widehat{\int_{\Gamma_0}} \frac{\partial w^+}{\partial n}(u_L^+ - u + u - u_I^-). \tag{8.3.10}$$

There exists the bound,

$$\begin{aligned} \left|\widehat{\int_{\Gamma_0}} \left(\frac{\partial u^+}{\partial n} - \frac{\partial u_L^+}{\partial n}\right)(w^+ - w^-)\right| &\le \left\|\frac{\partial \widehat{R_L}}{\partial n}\right\|_{0,\Gamma_0} \overline{\|w^+ - w^-\|_{0,\Gamma_0}} \\ &\le Ch^\sigma \left\|\frac{\partial \widehat{R_L}}{\partial n}\right\|_{0,\Gamma_0} \overline{\|w\|_H} \\ &= Ch^\sigma \overline{\left\|\frac{\partial R_L}{\partial n}\right\|_{0,\Gamma_0}} \overline{\|w\|_H}. \end{aligned} \tag{8.3.11}$$

Moreover, we obtain

$$\begin{aligned} \left|\widehat{\int_{\Gamma_0}} \frac{\partial w^+}{\partial n}((u_L^+ - u) + (u - u_I^-))\right| &= \left|\widehat{\int_{\Gamma_0}} \frac{\partial w^+}{\partial n}(u_L^+ - u)\right| \le \|\widehat{R}_L\|_{0,\Gamma_0} \left\|\frac{\partial \widehat{w}^+}{\partial n}\right\|_{0,\Gamma_0} \\ &= \overline{\|R_L\|_{0,\Gamma_0}} \overline{\left\|\frac{\partial w^+}{\partial n}\right\|_{0,\Gamma_0}}. \end{aligned} \tag{8.3.12}$$

From assumption, i.e., eqn. (8.3.7), we have

$$\begin{aligned} \left\|\frac{\partial \widehat{w}^+}{\partial n}\right\|_{0,\Gamma_0} &\le \left\|\frac{\partial w^+}{\partial n}\right\|_{0,\Gamma_0} + \left\|\frac{\partial w^+}{\partial n} - \frac{\partial \widehat{w}^+}{\partial n}\right\|_{0,\Gamma_0} \\ &\le CL^\nu\|w^+\|_{1,S_2} + Ch\left|\frac{\partial w^+}{\partial n}\right|_{1,\Gamma_0} \\ &\le C(L^\nu + hL^{2\nu})\|w\|_{1,S_2} \le C(L^\nu + hL^{2\nu})\overline{\|w\|_H}. \end{aligned} \tag{8.3.13}$$

Combining eqns. (8.3.10)–(8.3.13) yields

$$|\widehat{E}_h(u-u_{I,L},w)| \le C\left\{h^\sigma \overline{\left\|\frac{\partial R_L}{\partial n}\right\|_{0,\Gamma_0}} + (L^\nu + hL^{2\nu})\overline{\|R_L\|_{0,\Gamma_0}}\right\}\overline{\|w\|_H}.$$

This is eqn. (8.3.9). ∎

Lemma 8.3.4

Let eqn. (8.3.1) *hold for* $k \le p+1$. *There exists the bound,*

$$\left|\left(\int_{\Gamma_0} - \widehat{\int_{\Gamma_0}}\right) \frac{\partial u}{\partial n}(w^+ - w^-)\right|$$
$$\le C\left\{h^{p+2}L^{(p+1)\nu}\left|\frac{\partial u}{\partial n}\right|_{1,\Gamma_0} + h^{p+1+\sigma}\left|\frac{\partial u}{\partial n}\right|_{p+1,\Gamma_0}\right\}\overline{\|w\|_H}. \quad (8.3.14)$$

Proof.

Since $xy - \hat{x}\hat{y} = x(y-\hat{y}) + (x-\hat{x})\hat{y}$, we have

$$\left(\int_{\Gamma_0} - \widehat{\int_{\Gamma_0}}\right) \frac{\partial u}{\partial n}(w^+ - w^-)$$
$$= \int_{\Gamma_0} \frac{\partial u}{\partial n}(w^+ - w^- - \hat{w}^+ + \hat{w}^-) + \int_{\Gamma_0}\left(\frac{\partial u}{\partial n} - \widehat{\frac{\partial u}{\partial n}}\right)(\hat{w}^+ - \hat{w}^-). \quad (8.3.15)$$

Then, we have

$$\left|\int_{\Gamma_0}\left(\frac{\partial u}{\partial n} - \widehat{\frac{\partial u}{\partial n}}\right)(\hat{w}^+ - \hat{w}^-)\right| \le C\left\|\frac{\partial u}{\partial n} - \widehat{\frac{\partial u}{\partial n}}\right\|_{0,\Gamma_0}\overline{\|w^+ - w^-\|_{0,\Gamma_0}} \quad (8.3.16)$$
$$\le Ch^{p+1}\left|\frac{\partial u}{\partial n}\right|_{p+1,\Gamma_0}\overline{\|w^+ - w^-\|_{0,\Gamma_0}}$$
$$\le Ch^{p+1+\sigma}\left|\frac{\partial u}{\partial n}\right|_{p+1,\Gamma_0}\overline{\|w\|_H}.$$

For $p \ge 2$, there exists the equality from eqn. (8.2.5)

$$\int_{\Gamma_0}(u - u_I)v = \sum_{ij}\int_{\ell_{ij}}(u - u_I)v = 0, \quad \forall v \in P_{p-2}, \quad (8.3.17)$$

where P_{p-2} are the piecewise polynomials of order $p-2$ on edges $\ell_{ij} = \Gamma_0 \cap \square_{ij}$. Choose the piecewise constant with the mean $\overline{\frac{\partial u}{\partial n}}\big|_{\ell_{ij}} = \int_{\ell_{ij}} \frac{\partial u}{\partial n}/|\ell_{ij}|$, then $\int_{\Gamma_0} \overline{\frac{\partial u}{\partial n}}(w^+ - \hat{w}^+) = 0$. We have from eqns. (8.3.17) and (8.3.1) for $k \le p+1$

$$\left|\int_{\Gamma_0}\frac{\partial u}{\partial n}(w^+ - w^- - \hat{w}^+ + \hat{w}^-)\right| = \left|\int_{\Gamma_0}\frac{\partial u}{\partial n}(w^+ - \hat{w}^+)\right| \quad (8.3.18)$$
$$\le \left|\int_{\Gamma_0}\left(\frac{\partial u}{\partial n} - \overline{\frac{\partial u}{\partial n}}\right)(w^+ - \hat{w}^+)\right| + \left|\int_{\Gamma_0}\overline{\frac{\partial u}{\partial n}}(w^+ - \hat{w}^+)\right|$$

$$= \left| \int_{\Gamma_0} \left(\frac{\partial u}{\partial n} - \overline{\frac{\partial u}{\partial n}} \right) (w^+ - \hat{w}^+) \right| \leq Ch \left| \frac{\partial u}{\partial n} \right|_{1,\Gamma_0} h^{p+1} |w^+|_{p+1,\Gamma_0}$$

$$\leq Ch^{p+2} \left| \frac{\partial u}{\partial n} \right|_{1,\Gamma_0} L^{(p+1)\nu} \|w^+\|_{0,\Gamma_0} \leq Ch^{p+2} L^{(p+1)\nu} \left| \frac{\partial u}{\partial n} \right|_{1,\Gamma_0} \|w\|_{1,S_2}$$

$$\leq Ch^{p+2} L^{(p+1)\nu} \left| \frac{\partial u}{\partial n} \right|_{1,\Gamma_0} \overline{\|w\|}_H.$$

Combining eqns. (8.3.15), (8.3.16), and (8.3.18) yields the desired result, i.e., eqn. (8.3.14). ■

Moreover, we have

$$\left| \iint_{S_2} \nabla(u - u_L) \cdot \nabla w \right| \leq C |R_L|_{1,S_2} \overline{\|w\|}_H. \tag{8.3.19}$$

Denote the piecewise constant interpolant $\overline{w}$ in S_1 with the mean $\iint_{\square_{ij}} w / |\square_{ij}|$. We obtain from eqn. (8.2.6) that $\iint_{S_1} (f - f_I)\overline{w} = 0$ for $p \geq 2$. Hence, we have

$$\left| \left(\iint_{S_1} - \widehat{\iint_{S_1}} \right) fw \right| = \left| \iint_{S_1} (f - f_I) w \right|$$

$$= \left| \iint_{S_1} (f - f_I)(w - \overline{w}) \right| \leq \|f - f_I\|_{0,S_1} \|w - \overline{w}\|_{0,S_1} \tag{8.3.20}$$

$$\leq Ch^{p+1} |f|_{p+1,S_1} \times h |w|_{1,S_1} \leq Ch^{p+2} |f|_{p+1,S_1} \overline{\|w\|}_H.$$

Based on Theorem 8.2.1, Lemmas 8.3.2–8.3.4, eqns. (8.3.19) and (8.3.20), we obtain the following theorem.

Theorem 8.3.1

Let **A1–A5**, *eqns.* (8.2.18), (8.3.1) *for* $k \leq p+1$ *and eqn.* (8.3.7) *hold. There exists the error bound,*

$$\overline{\|u_h - u_{I,L}\|}_H$$

$$\leq \varepsilon = C \Bigg\{ h^{p+2} \|u\|_{p+3,S_1} (1 + L^\nu + hL^{2\nu} + h^{\sigma-1})$$

$$+ |R_L|_{1,S_2} + h^{p+2} |f|_{p+1,S_1} + (L^\nu + hL^{2\nu} + h^{-\sigma}) \overline{\|R_L\|}_{0,\Gamma_0}$$

$$+ h^\sigma \overline{\left\| \frac{\partial R_L}{\partial n} \right\|}_{0,\Gamma_0} + h^{p+2} L^{(p+1)\nu} \left| \frac{\partial u}{\partial n} \right|_{1,\Gamma_0} + h^{p+1+\sigma} \left| \frac{\partial u}{\partial n} \right|_{p+1,\Gamma_0} \Bigg\}.$$

Corollary 8.3.1

Let all conditions in Theorem 8.3.1 hold. Suppose $u \in H^{p+3}(S_1)$, $f \in H^{p+1}(S_1)$, $\sigma \geq 1$, and $L = O(|\ln h|)$ such that

$$|R_L|_{1,S_2} = O(h^{p+2}), \qquad \overline{\|R_L\|_{0,\Gamma_0}} = O(h^{p+2+\sigma}),$$

$$\overline{\left\|\frac{\partial R_L}{\partial n}\right\|_{0,\Gamma_0}} = O(h^{p+2-\sigma}),$$

there exists the superclose

$$\overline{\|u_h - u_{I,L}\|_H} = O(h^{p+2-\delta}),$$

where $0 < \delta \ll 1$.

In Theorem 8.3.1 and Corollary 8.3.1, the assumption $u \in H^{p+3}(S_1)$ is severe in application. In Huang [210], the local superconvergence of the combination is explored, by following the arguments in Xu and Zhou [474].

Remark 8.3.1

For Theorem 8.3.1, two inverse inequalities, the eqn. (8.3.1) for $k \leq p+1$ and the eqn. (8.3.7), hold for a polygon S_2. Here, we only give a brief argument of their validity. For polynomials υ^+ of order L, the eqn. (8.3.1) holds for $\nu = 2$. Let $\Gamma_0 = \cup_i \Gamma_0^i$, where Γ_0^i is a straight line segment, and let υ^+ on Γ_0^i be expanded into the Legendre polynomials. Based on their orthogonality, the eqn. (8.3.1) is proved for $\nu = 2$ in Ref. [280], p. 161. Next, suppose that S_2 is a polygon. The argument for eqn. (8.3.7) is given in Section 5.5 of Chapter 5 for polynomials, to prove $\nu = 2$. For a sectorial domain S_2, when the particular solutions υ^+ are chosen suitably, the eqns. (8.3.1) and (8.3.7) can be shown with $\nu = 1$. Therefore, we may assume that in general, the eqns. (8.3.1) and (8.3.7) hold with $1 \leq \nu \leq 2$.

8.4 Adini's elements

In this section, we discuss again the penalty plus hybrid method, but choose Adini's elements in S_1 instead of p-rectangles. We add two more assumptions.

(A6) Let $S_1 = \bigcup_{ij} \square_{ij}$, where all $\square_{ij}$ are *uniform* rectangles in 2×2 fashion. The rectangles $\square_{ij}$ are said to be *uniform* if $\square_{ij}$ are quasi-uniform and $h_i = h$ and $k_j = k$, where we assume $h \geq k$.

(A7) Choose the finite-dimensional space

$$v_p \in V_A \equiv \{v \in H^1(S), v\,|_{\square_{ij}} \in \widehat{P}_3, v_x, v_y \text{ are continuous at all vertices of } \square_{ij} \text{ and } v\,|_{\partial S_1 \cap \Gamma} = 0\},$$

where

$$\widehat{P}_3 = span\{1, x, y, x^2, y^2, xy, x^3, y^3, x^3y, xy^2, x^2y, xy^3\}.$$

The spaces V_h^* and V_h^0 are similarly defined as eqn. (8.2.9), where v_p is given in **A7**. The polynomial interpolant u_I^A of u in S_1 can be formulated by u, u_x, and u_y at four corners. Define the following interpolant of solution u,

$$u_{I,L}^A = \begin{cases} u_I^- = u_I^A, & \text{in } \overline{S}_1, \\ u_L^+ = f_L(a_1, \ldots, a_L), & \text{in } \overline{S}_2. \end{cases}$$

The penalty plus hybrid combined method is designed to seek $u_h^A \in V_h^*$ such that

$$\widehat{A}_h(u_h^A, v) = \hat{f}(v), \quad \forall v \in V_h^0,$$

where $\widehat{A}_h(u, v)$ is given in eqn. (8.2.12). The rule of integration is given by

$$\widehat{\int_{\Gamma_0}} uv = \int_{\Gamma_0} \hat{u}\hat{v}, \tag{8.4.1}$$

where $\hat{u}$ is the piecewise cubic Hermite polynomial interpolant of u.

Theorem 8.4.1

Let **A1–A7**, *eqns.* (8.2.18), (8.3.1) *for* $k \le 4$, *and eqn.* (8.3.7) *hold. Then there exists the bound,*

$$\overline{\|u_h^A - u_{I,L}\|_H} \le C \left\{ (h^{3.5} + h^4 L^{4\nu} + h^{3+\sigma}) \|u\|_{5,S_1} + h^4 |f|_{4,S_1} + |R_L|_{1,S_2} + (L^\nu + hL^{2\nu} + h^{-\sigma}) \overline{\|R_L\|_{0,\Gamma_0}} + h^\sigma \overline{\left\| \frac{\partial R_L}{\partial n} \right\|_{0,\Gamma_0}} \right\}. \tag{8.4.2}$$

Proof.

We obtain from Theorem 8.2.1,

$$\begin{aligned} &\overline{\|u_h^A - u_{I,L}^A\|_H} \\ &\le C \sup_{w \in V_h^0} \frac{1}{\|w\|_H} \times \left\{ \left| \iint_{S_1} \nabla(u - u_I^A) \cdot \nabla w \right| + \left| \iint_{S_2} \nabla(u - u_L) \cdot \nabla w \right| \right. \\ &\quad + \left| \left(\iint_{S_1} - \widehat{\iint_{S_1}} \right) fw \right| + \left| \left(\widehat{\int_{\Gamma_0}} - \int_{\Gamma_0} \right) \frac{\partial u}{\partial n}(w^+ - w^-) \right| \\ &\quad \left. + |\widehat{D}_h(u - u_{I,L}^A, w)| + |\widehat{E}_h(u - u_{I,L}^A, w)| \right\}. \end{aligned} \tag{8.4.3}$$

We cite from Refs. [213, 311],

$$\left|\iint_{S_1} \nabla(u-u_I^A)\cdot\nabla w\right| \le Ch^4\|u\|_{5,S_1}\left\{\|w\|_{1,S_1}+\left\|\frac{\partial w^-}{\partial n}\right\|_{0,\partial S_1}\right\}$$
$$\le Ch^{3.5}\|u\|_{5,S_1}\|w\|_{1,S_1} \le Ch^{3.5}\|u\|_{5,S_1}\overline{\|w\|}_H. \quad (8.4.4)$$

Also there exist the bounds,

$$\left|\iint_{S_2}\nabla(u-u_L)\cdot\nabla w\right| \le |R_L|_{1,S_2}\overline{\|w\|}_H, \quad (8.4.5)$$
$$\left|\left(\iint_{S_1}-\widehat{\iint_{S_1}}\right)fw\right| \le Ch^4|f|_{4,S_1}\overline{\|w\|}_H.$$

Next, from eqn. (8.3.1) for $k\le 4$ we have

$$\left|\left(\widehat{\int_{\Gamma_0}}-\int_{\Gamma_0}\right)\frac{\partial u}{\partial n}(w^+-w^-)\right|$$
$$=\left|\int_{\Gamma_0}\left[\frac{\partial u}{\partial n}(w^+-w^-)-\widehat{\frac{\partial u}{\partial n}}(\hat{w}^+-w^-)\right]\right|$$
$$=\left|\int_{\Gamma_0}\frac{\partial u}{\partial n}(w^+-\hat{w}^+)+\int_{\Gamma_0}\left(\frac{\partial u}{\partial n}-\widehat{\frac{\partial u}{\partial n}}\right)(\hat{w}^+-w^-)\right|$$
$$\le\left\|\frac{\partial u}{\partial n}\right\|_{0,\Gamma_0}\|w^+-\hat{w}^+\|_{0,\Gamma_0}+\left\|\frac{\partial u}{\partial n}-\widehat{\frac{\partial u}{\partial n}}\right\|_{0,\Gamma_0}\overline{\|w^+-w^-\|_{0,\Gamma_0}}$$
$$\le C\left\{h^4\left\|\frac{\partial u}{\partial n}\right\|_{0,\Gamma_0}|w^+|_{4,\Gamma_0}+h^3\left|\frac{\partial u}{\partial n}\right|_{3,\Gamma_0}\cdot h^\sigma\overline{\|w\|}_H\right\}$$
$$\le C\left\{h^4L^{4\nu}\left\|\frac{\partial u}{\partial n}\right\|_{0,\Gamma_0}+h^{3+\sigma}\left|\frac{\partial u}{\partial n}\right|_{3,\Gamma_0}\right\}\overline{\|w\|}_H$$
$$\le C\left\{h^4L^{4\nu}+h^{3+\sigma}\right\}\|u\|_{5,S_1}\overline{\|w\|}_H. \quad (8.4.6)$$

Therefore, the desired result, i.e., eqn. (8.4.2) is obtained from eqns. (8.4.3)–(8.4.6) and Lemma 8.3.3. ■

Corollary 8.4.1

Let the all conditions of Theorem 8.4.1 hold. Suppose $u\in H^5(S_1)$, $f\in H^4(S_1)$, $\sigma\ge 1$, and $L=O(|\ln h|)$ such that

$$|R_L|_{1,S_2}=O(h^{3.5}), \qquad \overline{\|R_L\|}_{0,\Gamma_0}=O(h^{3.5+\sigma}),$$
$$\overline{\left\|\frac{\partial R_L}{\partial n}\right\|}_{0,\Gamma_0}=O(h^{3.5-\sigma}).$$

Then there exists the superclose,

$$\overline{\|u_h^A - u_{I,L}^A\|_H} = O(h^{3.5}).$$

Construct the *a posteriori* interpolation

$$\Pi_p u_h^A = \begin{cases} \Pi_{2h}^5 u_h^A, & \text{in } \overline{S}_1, \\ u_h^A, & \text{in } \overline{S}_2, \end{cases} \tag{8.4.7}$$

where $\Pi_{2h}^5 u_h^A$ is the polynomial interpolant of order five based on the known u_h^A in $\Box_{2i+1,2j+1}^{2\times 2}$ in fig. 8.2; the explicit polynomials $\Pi_{2h}^5 u_h^A$ are given in Ref. [211]. We obtain the following corollary.

Corollary 8.4.2

Let all conditions in Corollary 8.4.1 hold. Then there exists the superconvergence:

$$\overline{\|u - \Pi_p u_h^A\|_H} = O(h^{3.5}).$$

Proof.

For the operation Π_{2h}^5, we have (see Refs. [213, 311])

$$\left\|\Pi_{2h}^5 v\right\|_{1,S_1} \leq C\,\|v\|_{1,S_1}.$$

Since $\Pi_{2h}^5 u_I$ is an interpolant from the nodal values of u_I, we have from the rule, i.e., eqn. (8.4.1)

$$\overline{\|u^+ - \Pi_{2h}^5 u_{I,L}^+ - (u^- - \Pi_{2h}^5 u_{I,L}^-)\|_{0,\Gamma_0}} = \overline{\|u - u_L\|_{0,\Gamma_0}} = \|\widehat{R}_L\|_{0,\Gamma_0}.$$

Hence, from Corollary 8.4.1 we obtain

$$\begin{aligned} \overline{\|u - \Pi_{2h}^5 u_h^A\|_H} &\leq \overline{\|u - \Pi_{2h}^5 u_{I,L}\|_H} + \overline{\|\Pi_{2h}^5 (u_{I,L} - u_h^A)\|_H} \\ &\leq \|u - \Pi_{2h}^5 u_I^A\|_{1,S_1} + \|u - u_L\|_{1,S_2} \\ &\quad + h^{-\sigma}\overline{\|u^+ - \Pi_{2h}^5 u_{I,L}^+ - (u^- - \Pi_{2h}^5 u_{I,L}^-)\|_{0,\Gamma_0}} + C\overline{\|u_{I,L} - u_h^A\|_H} \\ &\leq Ch^4 |u|_{5,S_1} + |R_L|_{1,S_2} + h^{-\sigma}\overline{\|R_L\|_{0,\Gamma_0}} \\ &\quad + C\overline{\|u_{I,L} - u_h^A\|_H} = O(h^{3.5}). \end{aligned}$$

■

Note that $O(h^{3.5})$ is the best rate of superclose and superconvergence of Adini's elements for Poisson's equation with the Dirichlet boundary condition, see Refs. [213, 311]. In the recent study [296], we may use the penalty TM coupled with FEM, to achieve the same convergence rates and the same growth of effective condition number in this chapter.

9 Eigenvalue problems

For the Laplace eigenvalue problems, this chapter presents new algorithms of the Trefftz method (TM), which solve the Helmholtz equation, and then use an iteration process to yield approximate eigenvalues and eigenfunctions. The new iterative method has superlinear convergence rates and gives a better performance in numerical testing, compared with the other popular methods of rootfinding. Piecewise particular solutions are used for a basic model of eigenvalue problems on the unit square with the Dirichlet condition. Moreover, error estimates are derived for the approximate eigenvalues and eigenfunctions. Numerical experiments are also conducted for the eigenvalue problems with singularities. Our new algorithms using piecewise particular solutions are well suited to seek very accurate solutions of eigenvalue problems, in particular those with multiple singularities, interfaces, and those on unbounded domains. Using piecewise particular solutions also has the advantage to solve complicated problems because uniform particular solutions may not always exist for the entire solution domain.

9.1 Introduction

There exist a number of numerical methods and their parallel implementation for solving eigenvalue problems, see Birkhoff and Lynch [45], Golub and van Loan [168], Kuttler and Sigilloto [261], Ortega [354], and Strang and Fix [426]. The use of the finite element method (FEM) has been surveyed by Babuska and Osborn [18], and the important treatises on eigenvalue problems include Courant and Hilbert [108], Wilkinson [472], Parlett [356] and Golub and van Loan [168]. The approaches using particular solutions are given in Bergman [31], Eisenstat [134], Fox, Henrici, and Moler [148], Mathon and Sermer [330], and Vekua [448]. We will follow Part I to employ the TM. In the TM, the solution domain is divided into several subdomains, different particular solutions on subdomains (i.e., piecewise particular solutions) are used, and an approximation of the solutions is then obtained by satisfying only the interior and exterior boundary conditions. An important advantage of the TM is that high accuracy of solutions can be achieved with a modest effort in computation.

In this chapter, we seek the eigenvalues λ_l and the non-zero eigenfunction ϕ_l satisfying

$$\begin{cases} -\Delta\phi_l = \lambda_l \phi_l & \text{in } \Omega, \\ \phi_l \mid_\Gamma = 0, \end{cases} \tag{9.1.1}$$

where $\Delta = \frac{\partial^2}{\partial x^2} + \frac{\partial^2}{\partial y^2}$, and Ω is a polygonal domain with the exterior boundary Γ. For simplicity, only the Dirichlet boundary condition is investigated here; other boundary conditions are treated in Section 9.5 (also see Refs. [280, 306, 315]). Denote the eigenvalues in an ascending order:

$$0 < \lambda_1 \le \lambda_2 \le \cdots \le \lambda_l \le \cdots, \qquad \lambda_1 = \lambda_{\min}. \tag{9.1.2}$$

The normalized eigenfunctions will satisfy the orthogonality property:

$$(\phi_i, \phi_j) = \iint_\Omega \phi_i \phi_j \, d\Omega = \delta_{ij} = \begin{cases} 1 & i = j, \\ 0 & i \ne j. \end{cases} \tag{9.1.3}$$

Let the solution domain Ω be divided into two subdomains, Ω^+ and Ω^-, by a piecewise straight line Γ_0. Then the eigenfunction ϕ_l must satisfy the continuity conditions across the interface Γ_0

$$\begin{cases} -\Delta\phi_l = \lambda_l \phi_l & \text{in } \Omega^+ \text{ and } \Omega^-, \\ \phi_l^+ = \phi_l^-, \quad \dfrac{\partial \phi_l^+}{\partial \nu} = \dfrac{\partial \phi_l^-}{\partial \nu} & \text{on } \Gamma_0, \\ \phi_l \mid_\Gamma = 0, \end{cases} \tag{9.1.4}$$

where ν is the unit normal to Γ_0, and ϕ_l^+ and ϕ_l^- are the values of ϕ_l on the two sides of Γ_0.

Let us first recall from Ortega [354], p. 40, that the matrix eigenvalue problem $\mathbf{A}\mathbf{x} = \lambda \mathbf{x}$ can be solved by the linear algebraic equations,

$$(\mathbf{A} - k^2 \mathbf{I})\mathbf{x} = \mathbf{b}, \tag{9.1.5}$$

where $\mathbf{I}$ is the identity matrix, $k(> 0)$ is chosen, and $\mathbf{b}$ is a non-zero vector. If eqn. (9.1.5) is ill-conditioned (called degeneracy in this chapter), then k^2 and $\mathbf{x}$ can be regarded approximately as an eigenvalue and the corresponding eigenvector of matrix $\mathbf{A}$, respectively. Note that the non-zero vector $\mathbf{b}$ can be chosen rather arbitrarily, and not necessarily to be small.

Analogously, for solving eqn. (9.1.4), we may seek the following Helmholtz solutions instead,

$$\begin{cases} -\Delta u = k^2 u & \text{in } \Omega^+ \text{ and } \Omega^-, \\ u^+ = u^-, \quad \dfrac{\partial u^+}{\partial \nu} = \dfrac{\partial u^-}{\partial \nu} & \text{on } \Gamma_0, \\ u \mid_\Gamma = g, \end{cases} \tag{9.1.6}$$

where $k > 0$, $g \in H^{\frac{1}{2}}(\Gamma)$ is a given non-zero function, and $H^{\frac{1}{2}}(\Gamma)$ is the Sobolev space on Γ. Consequently, when k is suitably chosen so as to lead to a degeneracy (or ill-conditioning) of eqn. (9.1.6), k^2 and u can be regarded approximately as an eigenvalue and its corresponding eigenfunction of eqn. (9.1.4), respectively.

Define the smallest relative distance between k^2 and λ_i by

$$\delta = \min_i \left| \frac{k^2 - \lambda_i}{k^2} \right| .$$

When k^2 approaches one of λ_i, the solution u of eqn. (9.1.6) will approach an eigenfunction [148]. Besides, when $k^2 = \lambda_l$ (i.e., $\delta = 0$) and $u = \phi_l$, the non-homogeneous term g must be zero. It seems to be a paradox due to the assumption $g \neq 0$ in eqn. (9.1.6). How can we clarify this paradox? In practical computation, we have either $\delta > 0$ or $u \neq \phi_l$ due to rounding errors in computer or truncation errors in numerical algorithms. The eqn. (9.1.6) could be very ill-conditioned, but never be exactly singular. A further clarification is deferred to Section 9.2.2. Therefore, the Helmholtz solutions can be solved by some numerical methods, e.g., by the TM and the collocation Trefftz method (CTM) in Part I. In fact, such a degeneracy of eqn. (9.1.6) can grant a very high accuracy of eigenvalues and eigenfunctions from the new algorithms developed below.

This chapter is organized as follows. In Section 9.2, we present new algorithms. In Sections 9.3 and 9.4, error bounds are derived for the eigenvalues and eigenfunctions by the TM. In Section 9.5, we test a simple eigenvalue problem to demonstrate the effectiveness of the algorithms proposed in this chapter. In Section 9.6, the eigenvalue model with singularity is studied, and its numerical experiments are reported in [303]. In the last section, summaries and discussions are given. The materials of this chapter are adapted from Refs. [284, 303].

9.2 New numerical algorithms for eigenvalue problems

To expose clearly our methods for eigenvalues and eigenfunctions, we first introduce the TM in Section 9.2.1 for eqn. (9.1.6), and then present some heuristic ideas in Section 9.2.2. The new algorithms are developed in Section 9.2.3. Since the degeneracy of eqn. (9.1.6) involves non-linear solutions k to the minimal eigenvalue of the associated matrix, a specific iteration method is designed, and its superlinear convergence rates are also proven in Ref. [303].

9.2.1 The Trefftz methods for eqn. (9.1.6)

First, the TM is applied to seek u of eqn. (9.1.6) under a fixed g and a given k. For simplicity, we split the solution domain into two subdomains Ω^+ and Ω^-. Some definitions of norms are needed to describe the TM. We define a space

$$H = \{v \in L_2(\Omega) \mid v \in H^1(\Omega^+), v \in H^1(\Omega^-) \text{ and } \Delta v + k^2 v = 0 \text{ in } \Omega^+ \text{ and } \Omega^-\},$$

and a functional

$$I(v) = \int_{\Gamma} (v-g)^2\, ds + \int_{\Gamma_0} (v^+ - v^-)^2\, ds + \sigma^2 \int_{\Gamma_0} (v_\nu^+ - v_\nu^-)^2\, ds,$$

where $H^1(\Omega^+)$ and $H^1(\Omega^-)$ are the Sobolev spaces, and σ is a positive weight. Define a bilinear form $[u, v]$ on $H \times H$ by

$$[u, v] = \int_{\Gamma} uv\, ds + \int_{\Gamma_0} (u^+ - u^-)(v^+ - v^-)\, ds + \sigma^2 \int_{\Gamma_0} (u_\nu^+ - u_\nu^-)(v_\nu^+ - v_\nu^-)\, ds,$$

and the induced norm

$$|v|_B = \sqrt{[v, v]} = \{|v|_{0,\Gamma}^2 + |v^+ - v^-|_{0,\Gamma_0}^2 + \sigma^2 |v_\nu^+ - v_\nu^-|_{0,\Gamma_0}^2\}^{\frac{1}{2}}. \tag{9.2.1}$$

The norms $\|v\|_H$ and $|v|_H$ over H are defined by

$$\|v\|_H = \{\|v\|_{1,\Omega^+}^2 + \|v\|_{1,\Omega^-}^2\}^{\frac{1}{2}}, \qquad |v|_H = \{|v|_{1,\Omega^+}^2 + |v|_{1,\Omega^-}^2\}^{\frac{1}{2}},$$

where $\|v\|_{1,\Omega^\pm}^2$ and $|v|_{1,\Omega^\pm}^2$ are the Sobolev norms.

Also define the finite-dimensional spaces $S_{m,n} \subseteq H$ such that

$$S_{m,n} = \left\{ v \mid v = v^+ = \sum_{i=1}^{m} c_i \Psi_i^+ \text{ in } \Omega^+, \text{ and } v = v^- = \sum_{i=1}^{n} d_i \Psi_i^- \text{ in } \Omega^- \right\}, \tag{9.2.2}$$

where $\{\Psi_i^\pm\}$ are the complete particular solutions of eqn. (9.1.6) in $\Omega^\pm$, and c_i and d_i are the coefficients to be sought. A TM approximation $u_{m,n} \in S_{m,n}$ to the problem, i.e., eqn. (9.1.6) can then be found by

$$I(u_{m,n}) = \min_{v \in S_{m,n}} I(v), \tag{9.2.3}$$

which can also be presented in a weak form

$$[u_{m,n}, v] = \int_{\Gamma} gv\, ds, \quad \forall v \in S_{m,n}. \tag{9.2.4}$$

Consequently, a system of linear algebraic equations

$$\mathbf{Ax} = \mathbf{A}(k)\mathbf{x} = \mathbf{b}, \tag{9.2.5}$$

can be obtained from eqn. (9.2.3) or eqn. (9.2.4), where $\mathbf{x}$ is the unknown vector consisting of all the expansion coefficients c_i and d_i in eqn. (9.2.2), and the normal matrix $\mathbf{A}(k)$ is non-negative definite and symmetric, given by

$$[u_{m,n}, u_{m,n}] = \frac{1}{2}\mathbf{x}^T \mathbf{A}(k)\mathbf{x}. \tag{9.2.6}$$

From eqn. (9.2.3), we may also solve $u_{m,n}$ by the least squares solution using the QR method or the singular value decomposition (see Golub and van Loan [168], and Atkinson [9], p. 636), with smaller condition number

$$\text{Cond} = \left[\frac{\lambda_{\max}(\mathbf{A}(k))}{\lambda_{\min}(\mathbf{A}(k))}\right]^{\frac{1}{2}}, \tag{9.2.7}$$

where $\lambda_{\max}(\mathbf{A}(k))$ and $\lambda_{\min}(\mathbf{A}(k))$ are the maximal and minimal eigenvalues of $\mathbf{A}(k)$, respectively.

9.2.2 Heuristic ideas of degeneracy of eqn. (9.1.6)

Below, we explain why the solutions of eqn. (9.1.6) in degeneracy will lead to solutions of eigenvalue problem, i.e., eqn. (9.1.4). Let μ_ℓ be the eigenvalues of $\mathbf{A}$ in eqn. (9.2.5),

$$\mathbf{A}(k)\overline{\mathbf{x}}_l = \mu_l \overline{\mathbf{x}}_l, \qquad \overline{\mathbf{x}}_l^T \overline{\mathbf{x}}_j = \delta_{l,j} = \begin{cases} 1, & l = j, \\ 0, & l \neq j. \end{cases}$$

Then the non-negative eigenvalues μ_i can be arranged in an ascending order

$$0 \le \mu_1 \le \mu_2 \le \mu_3 \le \cdots$$

The matrix $\mathbf{A}$ is said to be degenerate if the minimal eigenvalue

$$\mu_1 = \lambda_{\min}(\mathbf{A}(k)) \to 0. \tag{9.2.8}$$

Such a degeneracy implies that

$$\exists \ell \quad \text{such that } k^2 \approx \lambda_\ell \quad \text{and} \quad u \approx \phi_\ell. \tag{9.2.9}$$

Here, we only give a preliminary argument to eqn. (9.2.9). If the coefficients of the Helmholtz solution $\overline{u}_{m,n}$ are approximated to $\overline{\mathbf{x}}_1$, where $\overline{u}_{m,n}$ is a normalization of $u_{m,n}$ in eqn. (9.2.2), we have from eqns. (9.2.6) and (9.2.8)

$$\begin{aligned} |\overline{u}_{m,n}|_B^2 &= \frac{1}{2}\mathbf{x}^T \mathbf{A}(k)\mathbf{x} \approx \frac{1}{2}\overline{\mathbf{x}}_1^T \mathbf{A}(k)\overline{\mathbf{x}}_1 \\ &= \frac{1}{2}\lambda_{\min}(\mathbf{A}(k))\|\overline{\mathbf{x}}_1\|^2 = \frac{1}{2}\lambda_{\min}(\mathbf{A}(k)) \to 0, \end{aligned}$$

where $\|\cdot\|$ is the Euclidean norm. By noting the definition, i.e., eqn. (9.2.1) of the B-norm, the following equations are satisfied approximately

$$\overline{u}\,|_\Gamma = 0, \qquad \overline{u}^+ = \overline{u}^-, \qquad \text{and} \qquad \overline{u}_\nu^+ = \overline{u}_\nu^- \quad \text{on } \Gamma_0.$$

Also since $\overline{u}_{m,n}(\in S_{m,n})$ satisfies

$$\Delta \overline{u} + k^2 \overline{u} = 0 \quad \text{in } \Omega^+ \text{ and } \Omega^-,$$

k^2 and $\overline{u}_{m,n}$ are the approximations of certain eigenvalue λ_l and eigenfunction ϕ_l, respectively. Hence, the eqn. (9.2.9) holds. Based on the above ideas, k^2 will be chosen to decrease $\lambda_{\min}(\mathbf{A}(k))$ as much as possible.

Given a non-zero function g on the exterior boundary Γ, let us suppose

$$(g, \phi_l) \neq 0,$$

where $(g, v) = \iint_\Omega gv \, d\Omega$. The solutions $u_{m,n}$ can be obtained from eqn. (9.1.6) by the TM in Section 9.2.1. Since $\alpha\phi_l$ with any real $\alpha \neq 0$ also solves eqn. (9.1.4), a scaling condition is needed for the unique eigenfunction such as

$$\overline{u}_{m,n} = \frac{u_{m,n}}{\|\mathbf{x}_{m,n}\|},$$

and $\mathbf{x}_{m,n}$ is the vector in eqn. (9.2.6). We may, however, use a simpler scaling condition,

$$\overline{u}_{m,n} = \frac{u_{m,n}}{c_1}, \qquad \overline{\mathbf{x}}_{m,n} = \frac{\mathbf{x}_{m,n}}{c_1}, \quad c_1 \neq 0, \tag{9.2.10}$$

where c_1 is the leading coefficient of $u^+_{m,n}$ in Ω^+.

Based on Lemma 9.3.6 below, when $\lambda_{\min}(\mathbf{A}) \to 0$ the leading coefficient $c_1 \to \infty$ in (9.3.17). For the scaled solution, i.e., eqn. (9.2.10), the exterior boundary condition in eqn. (9.1.6) leads to

$$\overline{u}\,|_\Gamma = \frac{1}{c_1} u\,|_\Gamma = \frac{1}{c_1} g = O(g\sqrt{\lambda_{\min}(\mathbf{A})}) \to 0,$$

where function g is bounded, but not necessarily small. Therefore, the solutions of eqn. (9.1.6) approach the eigenfunctions in eqn. (9.1.4), i.e., eqn. (9.1.1). This gives a clarification to the paradox raised in the Section 9.1.

The same conclusions can be made from the error bounds in the next two sections,

$$\frac{|k^2 - \lambda_l|}{k^2} \leq C(K_{m,n} + \sigma^{-1}) \frac{|u_{m,n}|_B}{|u_{m,n}|_{0,\Omega}},$$

and

$$\frac{\|u_{m,n} - a_l \phi_l\|_H}{|u_{m,n}|_{0,\Omega}} \leq C\lambda_l^{\frac{3}{2}} (K_{m,n} + \sigma^{-1}) \frac{|u_{m,n}|_B}{|u_{m,n}|_{0,\Omega}},$$

where a_l is a constant, C is a bounded constant independent of k, λ_l, m and n, but $K_{m,n}$ may depend upon m and n, and the ratio

$$\rho = \frac{|u_{m,n}|_B}{|u_{m,n}|_{0,\Omega}} \leq C|\overline{u}_{m,n}|_B \leq C|g|_{0,\Gamma}\sqrt{\lambda_{\min}(\mathbf{A})}.$$

Therefore, we conclude that if $\lambda_{\min}(\mathbf{A}) \to 0$, then $k^2 \to \lambda_l$ *and* $u_{m,n} \to a_l\phi_l$ in the H-norms even when $g = O(1)$. So, we may simply let $g\,|_\Gamma = 1$ in the examples given below.

9.2.3 New iteration algorithms

Based on the above ideas that a degeneracy of eqn. (9.1.6) implies the infinite small values of $\lambda_{\min}(\mathbf{A}(k))$, we propose algorithms (A) and (B) given below.

Suppose $k^2 = \lambda_l$, i.e., $\delta = 0$. Then $u_{m,n} \to \phi_l$ as $m \to \infty$ and $n \to \infty$, since the eigenfunctions ϕ_l can be spanned by the complete particular solutions $\{\Psi_i^{\pm}\}$. Therefore, we may choose k_i such that

$$\lambda_{\min}(\mathbf{A}(k_i)) = \min_k \lambda_{\min}(\mathbf{A}(k)), \tag{9.2.11}$$

as $m = m_i \to \infty$ and $n = n_i \to \infty$. We now present the following algorithm to find the sequence $\{k_i\}$.

Algorithm (A) for seeking eigenvalues and eigenfunctions:

Step 1. Choose suitable term numbers, m and n, and good initial values of k^2 near to the target eigenvalue λ_l.

Step 2. Based on the local particular solutions $\{\Psi_i^{\pm}\}$, form the admissible functions $u_{m,n}$ from eqn. (9.2.2).

Step 3. For solving problem, i.e., eqn. (9.1.6), the solution $u_{m,n}$ is obtained from the TM by the least squares method (LSM) using the singular value decomposition [9, 168]. Then the scaled solution $\overline{u}_{m,n}$ is computed by eqn. (9.2.10), and the minimal eigenvalue is given by

$$f(k) = \lambda_{\min}(\mathbf{A}(k)),$$

where $\lambda_{\min}$ is the smallest eigenvalue of $\mathbf{A}(k)$.

Step 4. If $f(k)$ is satisfactorily small, the values k^2 can be regarded as a good approximation to λ_l; so $\overline{u}_{m,n}$ in eqn. (9.2.10) is close to ϕ_l, in view of the analysis in Section 9.2.2. Otherwise, we obtain a new value of k to minimize $f(k)$ in eqn. (9.2.11), based on algorithm (B) given in the next subsection, and then return back to Step 2. If $f(k)$ cannot be reduced sufficiently after many iterations through Steps 2–3, we should reasonably increase the term numbers, m and n, and then go to Step 1 for a new trial computation.

9.2.4 Solution of non-linear equations

From eqn. (9.2.11), a real value of k^* should be chosen such that

$$f(k^*) = \min_k f(k) = 0, \tag{9.2.12}$$

where

$$f(k) = \lambda_{\min}(\mathbf{A}(k)) \geq 0. \tag{9.2.13}$$

The condition ≥ 0 in eqn. (9.2.13) may not really happen in computation due to rounding errors. For instance, the very small negative values of $\lambda_{\min}(\mathbf{A}(\overline{k}))$ may even occur. Also when m and n are too small,

$$\min_k \lambda_{\min}(\mathbf{A}(k)) \geq \varepsilon > 0.$$

Therefore, minimization of $f(k)$ should be taken into account in algorithm (A) instead of eqn. (9.2.12).

Suppose that

$$f(k) \in C^3, \quad \text{when } k \in (0, \infty), \tag{9.2.14}$$

where C^n is the space of the functions having continuous nth-order derivatives. Consider $f(k^*) = \min_k f(k)$, thus having $f'(k^*) = 0$.

An interpolatory quadratic polynomial $P_2(k)$ to $f(k)$ can be formulated through three pairs of the values

$$(k_i, f(k_i)), \quad i = 0, 1, 2,$$

where k_i are distinct. Hence, a new value, k_3, is found by

$$P_2'(k_3) = 0, \tag{9.2.15}$$

in order to minimize $P_2(k)$. In fact, the quadratic function can be formed by the Newton divided differences (see Ref. [9]):

$$P_2(k) = f(k_2) + \overline{w}(k - k_2) + f[k_2, k_1, k_0](k - k_2)^2, \tag{9.2.16}$$

where

$$\overline{w} = f[k_2, k_1] + (k_2 - k_1) f[k_2, k_1, k_0], \tag{9.2.17}$$

and the divided differences are defined by

$$f[k_2, k_1] = \frac{(f(k_2) - f(k_1))}{(k_2 - k_1)},$$

$$f[k_2, k_1, k_0] = \frac{(f[k_2, k_1] - f[k_1, k_0])}{(k_2 - k_0)}.$$

Consequently, we obtain from eqns. (9.2.16) and (9.2.17)

$$k_3 = k_2 - \overline{w}/(2f[k_2, k_1, k_0]) = \frac{k_2 + k_1}{2} - \frac{1}{2}\frac{f[k_2, k_1]}{f[k_2, k_1, k_0]}. \tag{9.2.18}$$

Let us summarize the above approaches as the following algorithm with three steps.

Algorithm (B) for minimizing $f(k)$:

Step 1. Give three good distinct initial values

$$k_i \approx \sqrt{\lambda_i}, \quad i = 0, 1, 2, \tag{9.2.19}$$

and evaluate $f(k_i)$, i.e., the target eigenvalue by some numerical methods, for instance, by the QR method or the singular value decomposition [9, 168].

Step 2. Compute

$$k_{n+1} = \frac{k_n + k_{n-1}}{2} - \frac{1}{2} \frac{f[k_n, k_{n-1}]}{f[k_n, k_{n-1}, k_{n-2}]}, \quad n \geq 2. \tag{9.2.20}$$

Step 3. Stop if k_{n+1} is satisfactory, otherwise return to Step 1.

To integrate algorithm (B) with algorithm (A), we may embed eqn. (9.2.20) into Step 4 of algorithm (A), and should also supply three good guesses, i.e., eqn. (9.2.19) in Step 1. We now provide superlinear convergence rates of algorithm (B).

Definition 9.2.1

The sequence $\{k_i\}$ has superlinear convergence rates if the errors $\varepsilon_i = |k_i - k^*|$ satisfy

$$\varepsilon_{n+1} \leq \alpha_n \varepsilon_n + \beta_n \varepsilon_{n-1}, \quad n = 1, 2, \ldots \tag{9.2.21}$$

where the constants $\alpha_n \geq 0$ and $\beta_n \geq 0$ such that

$$\alpha_n \to 0 \quad \text{and} \quad \beta_n \to 0 \quad \text{as } n \to \infty. \tag{9.2.22}$$

We note that the definition of superlinear convergence given in Ref. [2] is a special case of eqn. (9.2.21) when $\beta_n = 0$. In fact, based on the recurrence solutions, we can easily show that when $n \to \infty$ the errors ε_n diminish faster than those of any linear convergence rates. We provide a proposition, whose proof is given in Ref. [303].

Proposition 9.2.1

Let eqn. (9.2.11), $f(k) \in C^4$, and $|f''(k^*)| \geq \delta > 0$ hold, and suppose that the sequence $\{k_i\}$ from algorithm (B) converges. Then $\{k_i\}$ has a superlinear convergence rate.

Let us compare algorithm (B) with other popular methods of non-linear equations. In Muller's method (see Ref. [9]), a closer root is sought from the same quadratic eqn. (9.2.16). Since $k > 0$, the real k_i are chosen, then the eqn. (9.2.18) leads to a modification, called the modified Muller's method. If the secant method is applied to eqn. (9.2.12), then

$$k_{n+1} = k_n - p\frac{f(k_n)}{f[k_n, k_{n-1}]}, \quad n \geq 1, \tag{9.2.23}$$

where p is multiplicity of the root k^*, and $p \geq 2$ always. The secant method, i.e., eqn. (9.2.23) also has a superlinear convergence rate (see Ref. [2]). When p is unknown and let $p = 1$, the sequence $\{k_n\}$ from eqn. (9.2.23) has only a linear convergence rate. However, we may then employ the Aitken extrapolation to speed up convergence rates. Algorithm (B) has been proven to be very effective by our many computational experiments.

9.3 Error bounds of eigenvalues

In the above algorithms, the magnitude and the accuracy of $\lambda_{\min}(\mathbf{A})$ are an important criterion to measure accuracy of numerical eigenvalues and eigenfunctions. This fact will be justified by the *a posteriori* error analysis below. The eigenvalue problem, i.e., eqn. (9.1.1) can be presented in a weak form: Seek $\lambda \in R$ and $0 \neq u \in H_0^1(\Omega)$ such that

$$(\nabla u, \nabla v) = \lambda(u, v), \quad \forall v \in H_0^1(\Omega), \tag{9.3.1}$$

where $\nabla = \left(\frac{\partial}{\partial x}, \frac{\partial}{\partial y}\right)^T$, $(\nabla u, \nabla v) = \frac{\partial u}{\partial x}\frac{\partial v}{\partial x} + \frac{\partial u}{\partial y}\frac{\partial v}{\partial y}$, and the Sobolev space $H_0^1(\Omega) = \{v \mid v \in H^1(\Omega), v\mid_\Gamma = 0\}$. Define another space H_0^* such that

$$H_0^* = \{v \mid v \in H^1(\Omega^+), v \in H^1(\Omega^-), v^+ = v^- \text{ in } \Gamma_0, v_\nu^+ = v_\nu^- \text{ in } \Gamma_0, v\mid_\Gamma = 0\}.$$

Problem, i.e., eqn. (9.1.4) can also be written in a weak form: Seek $\lambda \in R, 0 \neq u \in H_0^*$ such that

$$\langle \nabla u, \nabla v \rangle = \lambda(u, v), \quad \forall v \in H_0^1(\Omega), \tag{9.3.2}$$

where

$$\langle u, v \rangle = \iint_{\Omega^+} uv\, d\Omega + \iint_{\Omega^-} uv\, d\Omega.$$

By following Hall and Porsching [182], we can prove the following lemma.

Lemma 9.3.1

The weak forms, i.e., eqns. (9.3.1) and (9.3.2) are equivalent to each other, and

$$\langle \nabla \phi_i, \nabla \phi_j \rangle = (\nabla \phi_i, \nabla \phi_j) = \lambda \delta_{i,j}. \tag{9.3.3}$$

From Lemma 9.3.1, we conclude that any function $v(\in H_0^*)$ can be expressed by the eigenfunctions $\{\phi_i\}$, i.e.,

$$v = \sum_{i=1}^{\infty} \alpha_i \phi_i, \tag{9.3.4}$$

with the real expansion coefficients α_i. In fact, when $v \in H_0^*$, we have $v^+ = v^-$ and $v^+ v_\nu^+ = v^- v_\nu^-$, to give $|v|^2_{1,\Omega^+} + |v|^2_{1,\Omega^-} = |v|^2_{1,\Omega} + \int_{\Gamma_0} (v^+ v_\nu^+ - v^- v_\nu^-) = |v|^2_{1,\Omega}$. Hence, $v \in H_0^*$ implies $v \in H_0^1(\Omega)$, and v can be expanded by the complete set of eigenfunctions $\phi_i (\in H_0^1(\Omega))$.

Suppose that there exist jumps ϵ_1 and ϵ_2 of v and v_ν across the interface Γ_0, then the Helmholtz eqn. (9.1.6) is reduced to

$$\begin{cases} \Delta u + k^2 u = 0 & \text{in } \Omega^+ \text{ and } \Omega^-, \\ [u]_{\Gamma_0} = \epsilon_1, & [u_\nu]_{\Gamma_0} = \epsilon_2, \\ u \,|\, \Gamma = g, \end{cases} \tag{9.3.5}$$

where the notation $[u]_{\Gamma_0} = (u^+ - u^-) \,|\, \Gamma_0$. We use an auxiliary function w defined below:

$$\begin{cases} \Delta w = 0 & \text{in } \Omega^+ \text{ and } \Omega^-, \\ [w]_{\Gamma_0} = \epsilon_1, & [w_\nu]_{\Gamma_0} = \epsilon_2, \\ w \,|\, \Gamma = g. \end{cases} \tag{9.3.6}$$

Note that the function w is only for analysis but not for real computation. Now, we have the following lemma.

Lemma 9.3.2

Let k^2 be close to a target eigenvalue λ, and let u and w be the solutions of eqns. (9.3.5) and (9.3.6) satisfying

$$|w|_{0,\Omega} \leq \frac{1}{2} |u|_{0,\Omega}. \tag{9.3.7}$$

Then there exists an eigenvalue λ_l such that

$$\frac{|k^2 - \lambda_l|}{k^2} \leq 2 \frac{|w|_{0,\Omega}}{|u|_{0,\Omega}}. \tag{9.3.8}$$

When k^2 is close to a target eigenvalue, u is close to its eigenfunction. Hence, $\epsilon = \max\{|g|, |\epsilon_1|, |\epsilon_2|\}$ is small, and $|u|_{0,\Omega} = O(1)$ for some kinds of normalization. On the other hand, the maximal value of Laplace's solution w occurs only on the boundary, and then $|w|_{0,\Omega} \leq \epsilon\, Area(\Omega)$. Hence, the assumption, i.e., eqn. (9.3.7) can be made.

Proof.

Let $v = u - w$, then

$$\begin{cases} \Delta v + k^2 v = -k^2 w & \text{in } \Omega^+ \text{ and } \Omega^-, \\ [v]_{\Gamma_0} = 0, & [v_\nu]_{\Gamma_0} = 0, \\ v\,|_{\Gamma} = 0. \end{cases} \tag{9.3.9}$$

So $v \in H_0^*$, and the function v can be expressed by eqn. (9.3.4). We obtain from eqns. (9.1.4), (9.3.9), and (9.3.4) that

$$|w|_{0,\Omega}^2 = \frac{1}{k^4}|\Delta v + k^2 v|_{0,\Omega}^2 = \sum_{i=1}^{\infty}\left(\frac{k^2 - \lambda_i}{k^2}\right)^2 \alpha_i^2. \tag{9.3.10}$$

Also from eqns. (9.1.3), (9.3.4), and assumption (9.3.7)

$$\sum_{i=1}^{\infty}\alpha_i^2 = |v|_{0,\Omega}^2 = |u - w|_{0,\Omega}^2 \geq (|u|_{0,\Omega} - |w|_{0,\Omega})^2 \geq \frac{1}{4}|u|_{0,\Omega}^2. \tag{9.3.11}$$

Therefore, combining eqns. (9.3.10) and (9.3.11) yields

$$\min_i \left|\frac{k^2 - \lambda_i}{k^2}\right|^2 \leq \frac{\sum_{i=1}^{\infty}\left(\frac{k^2-\lambda_i}{k^2}\right)^2 \alpha_i^2}{\sum_{i=1}^{\infty}\alpha_i^2} \leq 4\frac{|w|_{0,\Omega}^2}{|u|_{0,\Omega}^2}.$$

The desired bound, i.e., eqn. (9.3.8) is obtained. ∎

The bounds, i.e., eqn. (9.3.8) can also be derived from Kuttler and Sigilloto [261] on the entire solution domain. We cite two lemmas from Chapter 1.

Lemma 9.3.3

Suppose that the auxiliary function of eqn. (9.3.6) satisfies the following inverse properties

$$|w_\nu|_{0,\Gamma} \leq K_w \|w\|_H, \qquad |w_\nu^+|_{0,\Gamma_0} \leq K_w \|w\|_H, \quad \forall w \in H,$$

where the constant K_w may depend on w. Then for any $\sigma > 0$ there exists a constant C independent of w such that

$$\|w\|_H \leq C(K_w + \sigma^{-1})|w|_B.$$

Lemma 9.3.4

Let u be the solution of eqn. (9.1.6). Then for $\sigma > 0$ there exists a unique function $u_{m,n} \in S_{m,n}$ by the TM such that

$$|u_{m,n}|_B \leq |g|_{0,\Gamma}, \qquad |u - u_{m,n}|_B \leq C \inf_{v \in S_{m,n}} |u - v|_B.$$

Now, let us prove a new theorem.

Theorem 9.3.1

Let u be the piecewise particular solution of the Helmholtz eqn. (9.3.5). Suppose that all conditions in Lemmas 9.3.2 and 9.3.3 hold. Then $\exists \lambda_l$ such that

$$\frac{|k^2 - \lambda_l|}{k^2} \leq C(K_w + \sigma^{-1}) \frac{|u|_B}{|u|_{0,\Omega}}, \tag{9.3.12}$$

where C in eqn. (9.3.12) and the following equations is a bounded constant independent of k, λ_l, and u. Moreover, let $u_{m,n} \in S_{m,n}$, then

$$\frac{|k^2 - \lambda_l|}{k^2} \leq C(K_w + \sigma^{-1}) \frac{|u_{m,n}|_B}{|u_{m,n}|_{0,\Omega}}. \tag{9.3.13}$$

Proof.

From Lemmas 9.3.2 and 9.3.3, we obtain

$$\frac{|k^2 - \lambda_l|}{k^2} \leq 2\frac{|w|_{0,\Omega}}{|u|_{0,\Omega}} \leq 2\frac{\|w\|_H}{|u|_{0,\Omega}} \leq C(K_w + \sigma^{-1}) \frac{|w|_B}{|u|_{0,\Omega}}. \tag{9.3.14}$$

Recall that the functions u and w have the same values on Γ and Γ_0; by comparing eqn. (9.3.5) with eqn. (9.3.6) the desired result, i.e., eqn. (9.3.12) is obtained, and so is eqn. (9.3.13) by letting $u = u_{m,n}$. ■

It is worth pointing out that the ratio in eqn. (9.3.13)

$$\rho = \frac{|u_{m,n}|_B}{|u_{m,n}|_{0,\Omega}} \tag{9.3.15}$$

plays an important role in error estimates for both eigenvalues and eigenfunctions. Note that $u_{m,n}$ satisfies the Helmholtz eqn. (9.1.6) approximately under a given g on Γ. From Lemma 9.3.4 we directly have the following lemma.

Lemma 9.3.5

Let $u_{m,n}(\in S_{m,n})$ be the solution to eqn. (9.1.6) from the TM. Suppose that there exists a constant $\rho_{10}(> 0)$ independent of m and n such that

$$|u_{m,n}|_{0,\Omega} \geq \rho_{10}|c_1|,$$

where ρ_{10} may depend on k. Then the ratio, i.e., eqn. (9.3.15) has the bounds

$$\rho \leq \frac{1}{\rho_{10}}|\overline{u}_{m,n}|_B, \qquad \rho \leq \frac{1}{\rho_{10}}\frac{1}{|c_1|}|g|_{0,\Gamma}, \tag{9.3.16}$$

where the scaled solution $\overline{u}_{m,n}$ is given by eqn. (9.2.10).

Let us consider the stiffness matrix $\mathbf{A}$ in eqn. (9.2.6). Denote the eigenvalues μ_i and eigenvectors $\overline{\mathbf{x}}_i$, then $A\overline{\mathbf{x}}_i = \mu_i \overline{\mathbf{x}}_i$, where $0 < \mu_1 \leq \mu_2 \leq \cdots \leq \mu_N$, $N = m+n$, and $\overline{\mathbf{x}}_i^T \overline{\mathbf{x}}_j = \delta_{ij}$.

We can also prove the following lemma by following Ref. [9], p. 604.

Lemma 9.3.6

Let $\overline{\mathbf{x}}_{m,n}$ be vector of the coefficients of the TM solution $\overline{u}_{m,n}$ for eqn. (9.1.6) using the LSM. Suppose that $\mu_1 = \lambda_{\min}(\mathbf{A}) \ll 1$, the next minimal eigenvalue $\mu_2 = \lambda_{\text{next}}(\mathbf{A}) = O(1)$, and $\mathbf{x}_1$ is the leading eigenvector of $\mathbf{A}(k)$ corresponding to μ_1 such that

$$(\overline{\mathbf{x}}_{m,n}, \mathbf{e}_1) = (\alpha \mathbf{x}_1, \mathbf{e}_1) = c_1 \neq 0,$$

where $\mathbf{e}_1$ is the N-dimensional unit vector, $\mathbf{e}_1 = (1,0,\ldots,0)^T$. Then there exist the bounds

$$c_1 = O\left(\frac{1}{\sqrt{\lambda_{\min}(\mathbf{A})}}\right), \tag{9.3.17}$$

and

$$\|\overline{\mathbf{x}}_{m,n} - \alpha \mathbf{x}_1\| = O\left(\sqrt{\frac{\lambda_{\min}(\mathbf{A})}{\lambda_{\text{next}}(\mathbf{A})}}\right),$$

with a suitable constant $\alpha \neq 0$.

Applying eqns. (9.3.13), (9.3.16), and (9.3.17) leads to the following corollary.

Corollary 9.3.1

Let all conditions in Theorem 9.3.1 and Lemmas 9.3.5 and 9.3.6 hold. Then $\exists \lambda_l$ such that

$$\left|\frac{k^2 - \lambda_l}{k^2}\right| \leq C\frac{(K_w + \sigma^{-1})}{\rho_{10}}|g|_{0,\Gamma}\sqrt{\lambda_{\min}(\mathbf{A})}. \tag{9.3.18}$$

Note that the function $g \mid_\Gamma$ in eqn. (9.1.6) may not be necessarily small. In fact, let $g = O(1)$, we can still conclude that if $\lambda_{\min}(\mathbf{A}) \to 0$ then $k^2 \to \lambda_l$. Also bounds of K_w can be derived by following Chapter 1 to give $K_w \leq C\sqrt{\max\{m,n\}}$ for a circular domain Ω^+.

9.4 Error bounds of eigenfunctions

In the algorithms in Section 9.2, the solution $\overline{u}_{m,n}$ in eqn. (9.2.10) from the TM can also be regarded as an approximation to the eigenfunctions ϕ_l. First, let us assume the distinct eigenvalues

$$0 < \lambda_1 < \lambda_2 < \cdots < \lambda_l < \cdots$$

Also the values of k^2 should be chosen to be close to a target eigenvalue λ_l. We provide the following lemma.

Lemma 9.4.1

Let u and w be the solutions of eqns. (9.3.5) and (9.3.6), respectively, and suppose that

$$|\lambda_i - \lambda_j| \geq \beta > 0, \quad i \neq j, \tag{9.4.1}$$

$$|w|_{0,\Omega} < \min\left\{\frac{1}{2}, \frac{\beta}{4k^2}\right\} |u|_{0,\Omega}. \tag{9.4.2}$$

Then there exists the bound,

$$|k^2 - \lambda_l| < \frac{1}{2}\beta. \tag{9.4.3}$$

Proof.

From eqns. (9.4.1), (9.4.2), and (9.3.14), we have

$$|k^2 - \lambda_l| = \min_i |k^2 - \lambda_i| = k^2 \min_i \frac{|k^2 - \lambda_i|}{k^2} \leq k^2 \frac{2|w|_{0,\Omega}}{|u|_{0,\Omega}} < \frac{\beta}{2},$$

where we have used eqn. (9.3.14) under assumption, i.e., eqn. (9.3.7). ■

Theorem 9.4.1

Let the conditions in Lemmas 9.3.3 and 9.4.1 hold. Then there exists a real constant $a_l \neq 0$ such that

$$|u - a_l\phi_l|_{0,\Omega} \leq C\frac{\lambda_l}{\beta}(K_w + \sigma^{-1})|u|_B.$$

Proof.

Let $v = u - w$, then v satisfies eqn. (9.3.9). The functions $v(\in H_0^*)$ can also be expressed by eqn. (9.3.4). Since the coefficients can be obtained explicitly from the orthogonality eqn. (9.3.3), we have

$$\alpha_i = -\frac{k^2}{k^2 - \lambda_i}(w, \phi_i). \tag{9.4.4}$$

Then the solution u of eqn. (9.3.5) is given by

$$u = w + v = w - \sum_{i=1}^{\infty} \frac{k^2}{k^2 - \lambda_i}(w, \phi_i)\phi_i$$

$$= w + a_l\phi_l - \sum_{i \neq l}^{\infty} \frac{k^2}{k^2 - \lambda_i}(w, \phi_i)\phi_i, \tag{9.4.5}$$

where $a_l = -\alpha_l = \frac{k^2}{k^2 - \lambda_l}(w, \phi_l)$. Since $\min_{i \neq l} |k^2 - \lambda_i| \geq \frac{\beta}{2}$, we obtain from eqn. (9.4.5) and the Parceval's inequality in Courant and Hilbert [108]

$$|u - a_l\phi_l|_{0,\Omega} \leq |w|_{0,\Omega} + \sqrt{\sum_{\substack{i=1 \\ i \neq l}}^{\infty} \left(\frac{k^2}{k^2 - \lambda_i}\right)^2 (w, \phi_i)^2}$$

$$\leq |w|_{0,\Omega} + \frac{2k^2}{\beta}\sqrt{\sum_{i=1}^{\infty} (w, \phi_i)^2} \leq \left(1 + \frac{2k^2}{\beta}\right)|w|_{0,\Omega}. \tag{9.4.6}$$

Also it follows from eqn. (9.4.3) that

$$k^2 \leq \lambda_l + |k^2 - \lambda_l| \leq \lambda_l + \frac{\beta}{2}. \tag{9.4.7}$$

Finally, by applying eqns. (9.4.6), (9.4.7), and Lemma 9.3.3,

$$|u - a_l\phi_l|_{0,\Omega} \leq C\frac{\lambda_l}{\beta}|w|_{0,\Omega} \leq C\frac{\lambda_l}{\beta}\|w\|_H$$

$$\leq C\frac{\lambda_l}{\beta}(K_w + \sigma^{-1})|w|_B = C\frac{\lambda_l}{\beta}(K_w + \sigma^{-1})|u|_B. \quad \blacksquare$$

Theorem 9.4.2

Let all conditions in Theorem 9.4.1 hold. Then there exists a real constant a_l such that

$$\|u - a_l\phi_l\|_H \leq C\frac{\lambda_l^{\frac{3}{2}}}{\beta}(K_w + \sigma^{-1})|u|_B. \tag{9.4.8}$$

Proof.

Let $v = u - a_l\phi_l$, we have from eqns. (9.3.5), (9.1.4), and (9.1.6)

$$|v|_H^2 \leq (-\Delta v, v) + C(K_w + \sigma^{-1})^2|v|_B^2$$

$$= (k^2u - \lambda_l a_l\phi_l, v) + C(K_w + \sigma^{-1})^2|v|_B^2$$

$$= (\lambda_l v + (k^2 - \lambda_l)u, v) + C(K_w + \sigma^{-1})^2|v|_B^2,$$

where we have used that $a_l\phi_l = u - v$. Hence, we obtain

$$|v|_H^2 \le \lambda_l |v|_{0,\Omega}^2 + |k^2 - \lambda_l||u|_{0,\Omega}|v|_{0,\Omega} + C(K_w + \sigma^{-1})^2 |v|_B^2. \tag{9.4.9}$$

Moreover, from Theorem 9.3.1,

$$|k^2 - \lambda_l||u|_{0,\Omega} \le k^2(K_w + \sigma^{-1})|u|_B. \tag{9.4.10}$$

Consequently, we can conclude from Theorem 9.4.1, eqns. (9.4.7), (9.4.9), and (9.4.10) that

$$\begin{aligned} \|v\|_H^2 &= |v|_{0,\Omega}^2 + |v|_H^2 \\ &\le (1+\lambda_l)|v|_{0,\Omega}^2 + k^2(K_w + \sigma^{-1})|u|_B|v|_{0,\Omega} + C(K_w + \sigma^{-1})^2|v|_B^2. \end{aligned} \tag{9.4.11}$$

From Theorem 9.4.1,

$$|v|_{0,\Omega} = |u - a_l\phi_l|_{0,\Omega} \le C\frac{\lambda_l}{\beta}(K_w + \sigma^{-1})|u|_B,$$

the eqn. (9.4.11) is reduced to

$$\|v\|_H^2 \le C\left\{(1+\lambda_l)\left(\frac{\lambda_l}{\beta}\right)^2(K_w + \sigma^{-1})^2 + k^2(K_w + \sigma^{-1})^2\frac{\lambda_l}{\beta}\right\}|u|_B^2.$$

The desired results, i.e., eqn. (9.4.8) are obtained immediately by noting $v = u - a_l\phi_l$. ■

Theorem 9.4.3

Let all the conditions in Theorem 9.4.1 hold, and let $u(= u_{m,n} \in S_{m,n})$ be the solution of eqn. (9.1.6) by the TM. Then there exists real $a_l \ne 0$ such that

$$\frac{|u_{m,n} - a_l\phi_l|_{0,\Omega}}{|u_{m,n}|_{0,\Omega}} \le C\frac{\lambda_l}{\beta}(K_w + \sigma^{-1})\frac{|u_{m,n}|_B}{|u_{m,n}|_{0,\Omega}}, \tag{9.4.12}$$

and

$$\frac{\|u_{m,n} - a_l\phi_l\|_H}{|u_{m,n}|_{0,\Omega}} \le C\frac{\lambda_l^{\frac{3}{2}}}{\beta}(K_w + \sigma^{-1})\frac{|u_{m,n}|_B}{|u_{m,n}|_{0,\Omega}}. \tag{9.4.13}$$

Compared with eqn. (9.3.13), the error bounds, i.e., eqns. (9.4.12) and (9.4.13) for eigenfunctions contain the same ratios ρ of eqn. (9.3.15). Therefore, other results as eqn. (9.3.18) can be similarly provided from Theorem 9.4.3 and Lemmas 9.3.5 and 9.3.6.

To close this section, we consider the eigenvalues with multiplicity $r \geq 1$:

$$\cdots < \lambda_{l-1} < \lambda_l = \lambda_{l+1} = \cdots = \lambda_{l+r-1} < \lambda_{l+r} < \cdots$$

By similar arguments as the above, we can conclude that when $u_{\min}$ is an approximation for an eigenvalue of λ_l, there exists a linear combination of the eigenfunctions, $\phi_l, \phi_{l+1}, \ldots \phi_{l+r-1}$, such that $\phi_l^* = \sum_{j=0}^{r} a_{l+j}\phi_{l+j}$ with real coefficients a_{l+j}. There also exist the error bounds

$$\frac{|u_{m,n} - \phi_l^*|_{0,\Omega}}{|u_{m,n}|_{0,\Omega}} \leq C\frac{\lambda_l}{\beta}(K_w + \sigma^{-1})\frac{|u_{m,n}|_B}{|u_{m,n}|_{0,\Omega}},$$

$$\frac{\|u_{m,n} - \phi_l^*\|_H}{|u_{m,n}|_{0,\Omega}} \leq C\frac{\lambda_l^{\frac{3}{2}}}{\beta}(K_w + \sigma^{-1})\frac{|u_{m,n}|_B}{|u_{m,n}|_{0,\Omega}}.$$

9.5 Computational models and numerical experiments

In order to show the effectiveness of the new algorithms proposed, we first introduce in Section 9.5.1 a basic sample of eigenvalue problems, and seek their particular solutions. We will describe the detailed algorithms in Section 9.5.2, and then investigate the behavior of the non-linear function $f(k)$ in Section 9.5.3. Numerical experiments are carried out in Section 9.5.4.

9.5.1 A basic sample of eigenvalue problems and particular solutions

Consider a sample of eigenvalue problems (see fig. 9.1) and their corresponding Helmholtz equation

$$\begin{cases} \Delta u + \lambda u = 0 & \text{in } \Omega^*, \\ u\,|_\Gamma = 0, \end{cases} \qquad \begin{cases} \Delta u + k^2 u = 0 & \text{in } \Omega^*, \\ u\,|_\Gamma = 1, \end{cases} \tag{9.5.1}$$

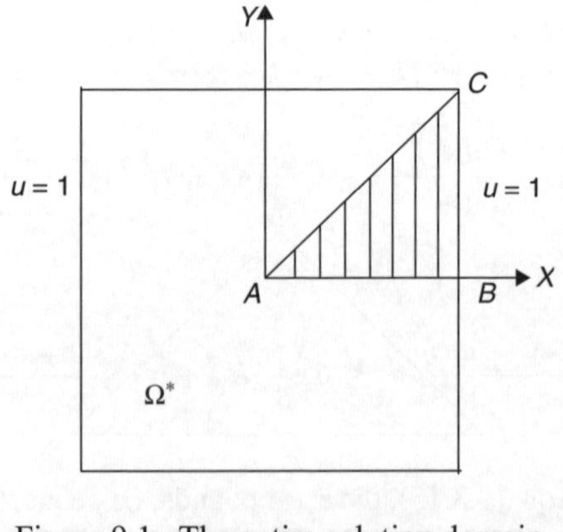

Figure 9.1: The entire solution domain.

where Ω^* is the square solution domain $\{(x,y)\,|\,-1 < x < 1, -1 < y < 1\}$. The minimal eigenvalue and its eigenfunction in eqn. (9.5.1) are our main task. For simplicity, we apply the symmetry and seek the solution only in Ω, one-eighth, of Ω^* (see fig. 9.2)

$$\begin{cases} \Delta u + k^2 u = 0 & \text{in } \Omega, \\ u_\nu \,|\, \overline{AC} = 0, & u_\nu \,|\, \overline{AB} = 0, \\ u \,|\, \overline{BC} = 1. \end{cases} \tag{9.5.2}$$

In eqn. (9.5.2), we only consider the symmetry on $\overline{AC}$, which covers the most important minimal eigenvalue of the eigenvalue problem in Ω^*. For other leading eigenvalues, different subdomains may be chosen with different interior boundary conditions.

Choose the admissible functions for the TM,

$$v_m = \sum_{i=0}^{m} \hat{c}_i J_i(kr) \cos i\theta, \tag{9.5.3}$$

where $\hat{c}_i$ are the coefficients to be sought, (r,θ) are the polar coordinates with the origin O, and $J_i(z)$ is the Bessel function defined by Refs. [2, 173]

$$J_\mu(r) = \sum_{i=0}^{\infty} \frac{(-1)^i}{\Gamma(i+1)\Gamma(i+\mu+1)} \left(\frac{r}{2}\right)^{2i+\mu}. \tag{9.5.4}$$

Based on the study in Section 1.5 of Chapter 1, the following partition of Ω is beneficial to numerical stability:

$$\Omega = \Omega_0 \cup \Omega_1 \cup \Omega_2 \cup \Omega_3,$$

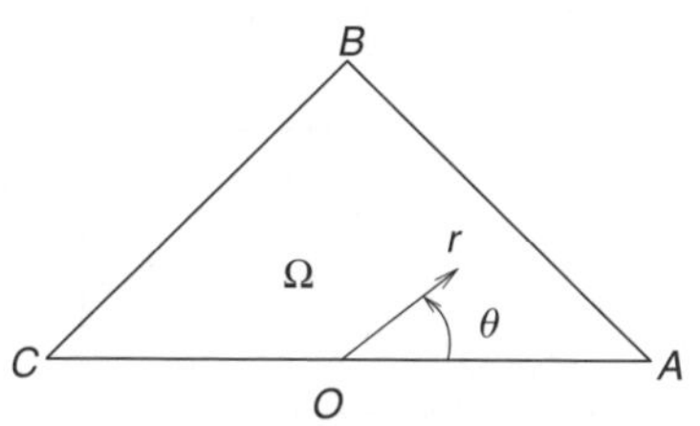

Figure 9.2: One-eighth of fig. 9.1 in partition I.

where the interface Γ_0 is composed of the piecewise straight lines as shown in figs. 9.3 and 9.4. The piecewise particular solutions can be found as follows:

$$\begin{aligned}
v_m^{(0)} &= \sum_{i=0}^{M} \hat{c}_i J_i(kr) \cos i\theta \quad \text{in } \Omega_0, \\
v_k^{(1)} &= 1 + \sum_{i=0}^{K} \hat{d}_i J_{2(2i+1)}(k\rho) \sin 2(2i+1)\phi \quad \text{in } \Omega_1, \\
v_n^{(2)} &= 1 + \sum_{i=0}^{N} \hat{b}_i J_{2i+1}(k\xi) \sin (2i+1)w \quad \text{in } \Omega_2, \\
v_l^{(3)} &= \sum_{i=0}^{L} \hat{a}_i J_{4i}(k\eta) \cos 4i\psi \quad \text{in } \Omega_3.
\end{aligned} \tag{9.5.5}$$

In eqn. (9.5.5) $\hat{a}_i, \hat{b}_i, \hat{c}_i$, and $\hat{d}_i$ are the unknown coefficients, and $(r,\theta), (\rho,\phi)$, (ξ, w), *and* (η, ψ) are the polar coordinates at the origins O, C, B, A, respectively.

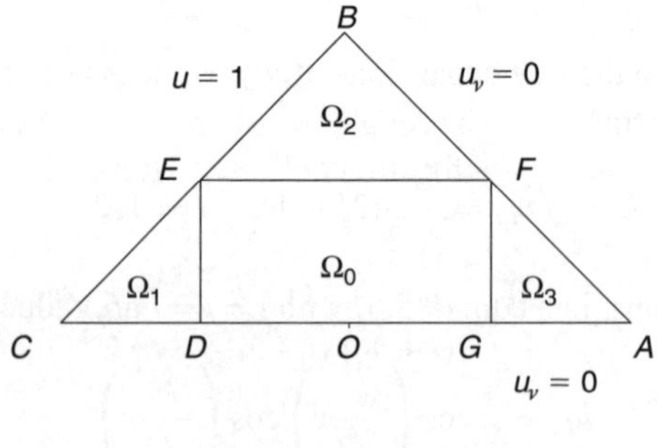

Figure 9.3: One-eighth of fig. 9.1 in partition II.

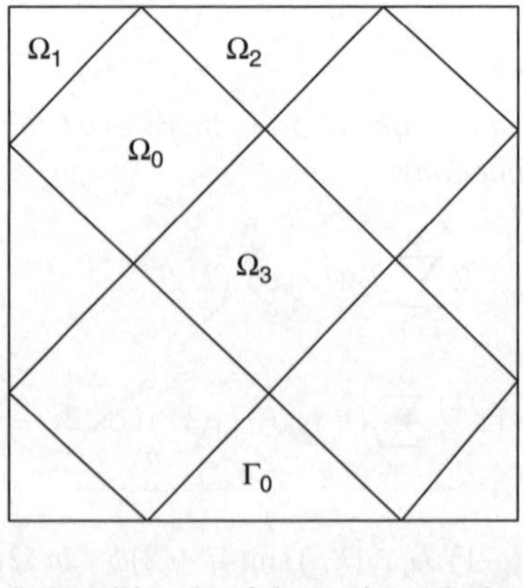

Figure 9.4: Partition II of the entire solution domain.

The division using the piecewise particular solutions in eqn. (9.5.5) is called partition II, and the division in fig. 9.1 using eqn. (9.5.3) in the entire solution domain is called partition I. Note that for the non-homogeneous boundary condition $u\,|_\Gamma = 1$, there exists a mild singularity $O(\rho^2 \ln \rho)$ at the corner C (i.e., the corners in fig. 9.2), and some singular solutions should be added for solving the Helmholtz equation exactly (see Ref. [304]). However, for the homogeneous boundary condition $u\,|_\Gamma = 0$, such a mild singularity does not exist. Since for the eigenvalue problem, only the homogeneous Dirichlet conditions are involved, we ignore the singular functions given in Ref. [304], which have no effects on the stiffness matrix **A**.

The eigenvalues and eigenfunctions of eqn. (9.5.1) are known as

$$\lambda_{i,j} = \frac{\pi^2}{4}[(2i-1)^2 + (2j-1)^2], \quad u_{i,j} = \cos\left(\frac{(2i-1)\pi}{2}x\right)\cos\left(\frac{(2j-1)\pi}{2}y\right). \tag{9.5.6}$$

The following leading eigenvalue and eigenfunction are of most interest in physical problems:

$$\lambda_{\min} = \lambda_{1,1} = \frac{\pi^2}{2}, \qquad u_{1,1} = \cos\left(\frac{\pi}{2}x\right)\cos\left(\frac{\pi}{2}y\right).$$

Below, let us provide the true expansions of $\hat{u}_{l,l}$ by means of the Bessel functions, i.e., eqn. (9.5.4). Denote

$$\hat{k} = \sqrt{\lambda_{l,l}} = \frac{\pi}{\sqrt{2}}(2l-1), \quad l = 1, 2, \ldots$$

and the eigenfunctions, i.e., eqn. (9.5.6) with $i = j = l$ are reduced to

$$\hat{u}_{l,l} = 2\cos\left(\frac{\hat{k}}{\sqrt{2}}x\right)\cos\left(\frac{\hat{k}}{\sqrt{2}}y\right). \tag{9.5.7}$$

We have the following lemma.

Lemma 9.5.1

For the eigenfunctions, i.e., eqn. (9.5.7), there exist the following expansions, spanned by the Bessel functions:

$$\hat{u}_{l,l} = J_0(kr) + 2\sum_{i=1}^{\infty} J_{2i}(\hat{k}r)\cos(2i)\theta$$
$$+ 2(-1)^{l+1}\sum_{i=0}^{\infty}(-1)^i J_{2i+1}(\hat{k}r)\cos(2i+1)\theta \quad \text{in } \Omega_0, \tag{9.5.8}$$

$$\hat{u}_{l,l} = 4\sum_{i=0}^{\infty}(-1)^i J_{4i+2}(\hat{k}\rho)\sin(4i+2)\phi \quad \text{in } \Omega_1, \tag{9.5.9}$$

$$\hat{u}_{l,l} = 2\sqrt{2}(-1)^{l+1}\sum_{i=0}^{\infty}(-1)^{\lfloor\frac{i+1}{2}\rfloor}J_{2i+1}(\hat{k}\xi)\sin(2i+1)w \quad in\ \Omega_2, \qquad (9.5.10)$$

$$\hat{u}_{l,l} = 2J_0(\hat{k}\eta) + 4\sum_{i=1}^{\infty}(-1)^i J_{4i}(\hat{k}\eta)\cos 4i\psi \quad in\ \Omega_3, \qquad (9.5.11)$$

where $\lfloor x \rfloor$ is the floor function of x.

Proof.

Firstly, to obtain eqn. (9.5.8), we have from eqn. (9.5.7)

$$\hat{u}_{l,l} = \cos\frac{\hat{k}}{\sqrt{2}}(x-y) + \cos\frac{\hat{k}}{\sqrt{2}}(x+y). \qquad (9.5.12)$$

By using the coordinate transformation

$$x = \frac{1}{2} + r\cos\left(\theta - \frac{3}{4}\pi\right), \qquad y = \frac{1}{2} + r\sin\left(\theta - \frac{3}{4}\pi\right),$$

we obtain after some manipulation

$$\hat{u}_{l,l} = \cos(\hat{k}r\sin\theta) + (-1)^{l+1}\sin(\hat{k}r\cos\theta).$$

Then, the expansion, i.e., eqn. (9.5.8) is obtained from the following formulas in Ref. [2], p. 361,

$$\cos(z\sin\theta) = J_0(z) + 2\sum_{i=1}^{\infty}J_{2i}(z)\cos 2i\theta,$$

$$\sin(z\cos\theta) = 2\sum_{i=0}^{\infty}(-1)^i J_{2i+1}(z)\cos(2i+1)\theta.$$

Secondly, other formulas in Ref. [1] such as

$$\cos(z\cos\theta) = J_0(z) + 2\sum_{i=1}^{\infty}(-1)^i J_{2i}(z)\cos 2i\theta,$$

$$\sin(z\sin\theta) = 2\sum_{i=0}^{\infty}J_{2i+1}(z)\sin(2i+1)\theta,$$

are also needed for proving eqns. (9.5.9)–(9.5.11). Based on the transformation

$$x = 1 - \xi\sin w, \qquad y = \xi\cos w,$$

we obtain from eqn. (9.5.12) similarly

$$
\begin{aligned}
\hat{u}_{l,l} &= (-1)^{l+1}\left[\sin\left(\hat{k}\xi\sin\left(w+\frac{\pi}{4}\right)\right) - \sin\left(\hat{k}\xi\cos\left(w+\frac{\pi}{4}\right)\right)\right] \\
&= (-1)^{l+1}2\sum_{i=0}^{\infty} J_{2i+1}(\hat{k}\xi)\left[\sin\left((2i+1)\left(w+\frac{\pi}{4}\right)\right)\right. \\
&\qquad\qquad\qquad \left. -(-1)^i\cos\left((2i+1)\left(w+\frac{\pi}{4}\right)\right)\right] \\
&= 2\sqrt{2}(-1)^{l+1}\sum_{i=0}^{\infty}(-1)^{\lfloor\frac{i+1}{2}\rfloor}J_{2i+1}(\hat{k}\xi)\sin(2i+1)w. \qquad (9.5.13)
\end{aligned}
$$

This gives eqn. (9.5.10).

Thirdly, the expansions, i.e., eqns. (9.5.9) and (9.5.11) can be obtained similarly by using the transformations $x=1-\rho\sin\phi, y=1-\rho\cos\phi$, and $x=\eta\cos\psi$, $y=\eta\sin\psi$, respectively. ■

Based on Lemma 9.5.1, we can find the true coefficients $\hat{c}_i, \hat{a}_i$, etc., in eqns. (9.5.3) and (9.5.6) of the eigenfunction $\hat{u}_{1,1}$:

$$
\begin{aligned}
&\{\hat{c}_i\} : 1, 2, 2, -2, 2, 2, 2, -2, \ldots \\
&\{\hat{a}_i\} : 2, -4, 4, -4, 4, -4, 4, -4, \ldots \\
&\{\hat{b}_i\} : 2\sqrt{2}, -2\sqrt{2}, -2\sqrt{2}, 2\sqrt{2}, 2\sqrt{2}, -2\sqrt{2}, -2\sqrt{2}, 2\sqrt{2}, \ldots \\
&\{\hat{d}_i\} : 4, -4, 4, -4, 4, -4, 4, -4, \ldots \qquad (9.5.14)
\end{aligned}
$$

Since the leading coefficient $\hat{c}_0 = 1$ in eqn. (9.5.14), the errors of numerical eigenfunctions can be easily discovered. It is worth pointing out that this basic sample with particular solutions given in this subsection may serve for testing other numerical methods.

9.5.2 Description of detailed algorithms for TM

In this subsection, we describe in detail the algorithms of the LSMs to solve eqn. (9.2.4) (i.e., eqn. (9.2.5)).

In computation, it is better to choose the scaled forms of eqn. (9.5.3)

$$v_m = \sum_{i=0}^{4M-1} c_i \frac{J_i(kr)}{J_i(kr_0)}\cos i\theta, \quad \text{if } J_i(kr_0) \neq 0, \qquad (9.5.15)$$

where $r_0 = \frac{1}{2}$ in computation. Hence, $\hat{c}_i = \frac{c_i}{J_i(kr_0)}$, where $\hat{c}_i$ are given in eqn. (9.5.3). Note that without such a scaling factor $\frac{1}{J_i(kr_0)}$, the convergence of algorithm (A) will deteriorate. The admissible function, i.e., eqn. (9.5.15) already satisfies the Helmholtz equation in Ω and the boundary condition $u_\nu|_{\overline{AC}}=0$. Hence, the coefficients c_i should be chosen to satisfy the rest of the boundary conditions in eqn. (9.5.2). Define a quadratic functional

$$I(c_i) = \widehat{\int_{\overline{BC}}} (v-1)^2 \, d\ell + \sigma^2 \widehat{\int_{\overline{AB}}} v_\nu^2 \, d\ell, \tag{9.5.16}$$

where $\sigma = \frac{1}{4M}$. In eqn. (9.5.16), $\widehat{\int_{\overline{BC}}}$ is an approximation of $\int_{\overline{BC}}$, and the central and the Gaussian rules may be chosen for integration quadrature. The CTM in partition I is designed for seeking the coefficients c_i such that

$$I(\widetilde{c_i}) = \min_{c_i} I(c_i).$$

The boundary errors are defined by

$$|\varepsilon|_B = |\varepsilon|_{\mathrm{I}} = (|\varepsilon|^2_{0,\overline{BC}} + \sigma^2 |\varepsilon_\nu|^2_{0,\overline{AB}})^{\frac{1}{2}},$$

where the error $\varepsilon = u - u_m$.

For partition II, the continuity conditions across Γ_0 should be supplied to eqn. (9.5.2), thus leading to

$$\begin{cases} \Delta u + k^2 u = 0 \quad \text{in } \Omega_0, \Omega_1, \Omega_2, \Omega_3, \\ u^+ = u^-, \qquad u_\nu^+ = u_\nu^- \quad \text{on } \Gamma_0, \\ u_\nu \,|_{\overline{AC}} = u_\nu \,|_{\overline{AB}} = 0, \\ u \,|_{\overline{BC}} = 1. \end{cases}$$

Similarly, the functions, i.e., eqn. (9.5.5) should be scaled as

$$\begin{aligned} v_m^{(0)} &= \sum_{i=0}^{4M-1} c_i \frac{J_i(kr)}{J_i(kr_0)} \cos i\theta \quad \text{in } \Omega_0, \\ v_k^{(1)} &= 1 + \sum_{i=0}^{K} d_i \frac{J_{2(2i+1)}(k\rho)}{J_{2(2i+1)}(k\rho_0)} \sin(2i+1)\phi \quad \text{in } \Omega_1, \\ v_n^{(2)} &= 1 + \sum_{i=0}^{N} b_i \frac{J_{2i+1}(k\xi)}{J_{2i+1}(k\xi_0)} \sin(2i+1)w \quad \text{in } \Omega_2, \\ v_l^{(3)} &= \sum_{i=0}^{L} a_i \frac{J_{4i}(k\eta)}{J_{4i}(k\eta_0)} \cos 4i\psi \quad \text{in } \Omega_3, \end{aligned} \tag{9.5.17}$$

where $r_0 = \rho_0 = \psi_0 = \eta_0 = \frac{1}{2}$ in computation, and all the denominators in eqn. (9.5.18) are assumed to be non-zero. The admissible functions, i.e., eqn. (9.5.18) satisfy the Helmholtz equations in Ω_i and the exterior boundary condition on $\partial\Omega$ already, i.e., $u_\nu|_{\overline{AC}} = u_\nu|_{\overline{AB}} = 0$, and $u|_{\overline{BC}} = 1$. Hence, the CTM in partition II is designed for seeking the coefficients c_i, d_i, b_i, a_i so as to minimize the functional

$$\mathrm{II}(\widetilde{a_i}, \widetilde{b_i}, \widetilde{c_i}, \widetilde{d_i}) = \min_{a_i, b_i, c_i, d_i} \mathrm{II}(a_i, b_i, c_i, d_i),$$

where $\mathrm{II}(\widetilde{a}_i, \widetilde{b}_i, \widetilde{c}_i, \widetilde{d}_i)$ involves only the interior boundary conditions,

$$\mathrm{II}(\widetilde{a}_i, \widetilde{b}_i, \widetilde{c}_i, \widetilde{d}_i) = \widehat{\int_{\Gamma_0}} (v^+ - v^-)^2 \, d\ell + \sigma^2 \widehat{\int_{\Gamma_0}} (v_\nu^+ - v_\nu^-)^2 \, d\ell,$$

with $\sigma \,|\, \overline{DE} = \frac{1}{\max(4M, 4K+2, 2N+1, 4L)}$. Also the error norm on the boundary is defined by

$$|\varepsilon|_{\mathrm{II}} = (|v^+ - v^-|^2_{0,\Gamma_0} + \sigma^2 |u_\nu^+ - u_\nu^-|^2_{0,\Gamma_0})^{\frac{1}{2}}.$$

9.5.3 Investigation of behavior for function $f(k)$ as $k^2 \to \lambda_1$

To support algorithm (B) we first study the function, $f(k) = \lambda_{\min}(\mathbf{A}(k))$, as $k^2 = \lambda_{\min}$, and $k^2 \to \lambda_{\min}$, respectively, where $\lambda_{\min} = \pi^2/2$.

Given $k^2 = \lambda_{\min}$, the CTM in partitions I and II are used to obtain the solutions, u_m and $u_{m,n,k,l}$. Studies on term distribution in partition II may be referred to Chapter 1. When different terms are chosen, the calculated results are provided in table 9.1, where $\lambda_{\min}(\mathbf{A})$, $\lambda_{\mathrm{next}}(\mathbf{A})$, and $\lambda_{\max}(\mathbf{A})$ denote the minimal, next minimal, and maximal eigenvalues of the associated matrix $\mathbf{A}$, respectively. The negative values of $\lambda_{\min}(\mathbf{A})$ may result from rounding errors. The condition numbers are computed from eqn. (9.2.7); the values $\sqrt{\frac{|\lambda_{\min}(\mathbf{A})|}{\lambda_{\mathrm{next}}(\mathbf{A})}}$ are also valuable to measure an approximate degree of eigenvectors (see Lemma 9.3.6). The errors $|\varepsilon^*|_{\mathrm{I}} = |\varepsilon|_{\mathrm{I}}/c_0$, and $|\varepsilon^*|_{\mathrm{II}} = |\varepsilon|_{\mathrm{II}}/c_0$, where c_0 is the leading coefficient of the eigenfunction of $\lambda_{\min}$. From table 9.1, we can see that $M = 1$ is inappropriate, but $M = 3 \sim 5$ will lead to a good approximation of $\lambda_{\min}$.

Table 9.1: The calculated results for partitions I and II with $k = \sqrt{\lambda_1}$.

M				$\|\epsilon\|_{\mathrm{I}}$	Cond	$\lambda_{\min}(\mathbf{A})$	$\lambda_{\mathrm{next}}(\mathbf{A})$	$\lambda_{\max}(\mathbf{A})$	$\sqrt{\frac{\lambda_{\min}(\mathbf{A})}{\lambda_{\mathrm{next}}(\mathbf{A})}}$
1				0.170	0.125(3)	0.207(−3)	1.02	3.23	0.0143
2				0.304(−3)	0.298(7)	0.290(−11)	0.314	25.8	0.304(−5)
3				0.209(−7)	0.190(9)	−0.746(−14)	0.146	26.9	0.226(−6)
4				0.103(−13)	0.293(11)	−0.372(−17)	0.0838	0.314(4)	0.666(−8)
5				0.122(−13)	0.1076(12)	−0.109(−18)	0.0541	0.403(5)	0.142(−8)
L	M	N	K	$\|\epsilon\|_{\mathrm{II}}$	Cond	$\lambda_{\min}(\mathbf{A})$	$\lambda_{\mathrm{next}}(\mathbf{A})$	$\lambda_{\max}(\mathbf{A})$	$\sqrt{\frac{\lambda_{\min}(\mathbf{A})}{\lambda_{\mathrm{next}}(\mathbf{A})}}$
1	1	2	1	0.337	706.	0.426(−5)	0.0667	0.212	0.799(−2)
2	2	3	2	0.827(−5)	0.216(8)	0.459(−14)	0.0148	2.14	0.557(−6)
3	3	5	3	0.166(−9)	0.212(9)	−0.480(−16)	0.245(−2)	2.15	0.140(−6)
4	4	7	4	0.121(−12)	0.197(9)	−0.555(−16)	0.387(−3)	2.16	0.379(−6)
5	5	9	5	0.789(−13)	0.140(9)	0.111(−15)	0.599(−4)	2.17	0.136(−5)

Table 9.2: The calculated results for partition I $(M=2)$ and partition II $(L=M=K=2, N=3)$ as $k(\delta^*) \to \sqrt{\lambda_1}$.

δ^*	Difference: $k(\delta^*) - \sqrt{\lambda_1}$	Partition I		Partition II	
		$\lvert\epsilon\rvert_{\text{I}}$	$\lambda_{\min}(\mathbf{A})$	$\lvert\epsilon\rvert_{\text{II}}$	$\lambda_{\min}(\mathbf{A})$
0.5	1.63	0.139(−2)	0.104	0.157	0.0150
0.1	0.407	0.184(−2)	0.548(−2)	0.772(−2)	0.0139
0.01	0.0440	0.111(−3)	0.497(−4)	0.333(−2)	0.121(−4)
0.1(−2)	0.444(−2)	0.931(−5)	0.493(−6)	0.304(−4)	0.119(−6)
0.1(−3)	0.444(−3)	0.154(−5)	0.492(−8)	0.309(−5)	0.119(−8)
0.1(−4)	0.444(−4)	0.206(−5)	0.521(−10)	0.405(−6)	0.119(−10)
0.1(−5)	0.444(−5)	0.534(−5)	0.338(−11)	0.182(−6)	0.124(−12)
0.1(−6)	0.444(−6)	0.432(−4)	0.290(−11)	0.374(−6)	0.586(−14)
0.1(−7)	0.444(−7)	0.190(−3)	0.289(−11)	0.518(−5)	0.457(−14)
0.1(−8)	0.444(−8)	0.302(−3)	0.289(−11)	0.112(−4)	0.463(−14)
0.1(−9)	0.444(−9)	0.287(−3)	0.289(−11)	0.849(−5)	0.471(−14)
0.0	0.0	0.304(−3)	0.290(−11)	0.827(−5)	0.459(−14)

To discover the behavior of $\lambda_{\min}(\mathbf{A})$ as $k^2 \to \lambda_1$, we choose

$$k(\delta^*) = [\lambda_{\min} + \delta^*(\lambda_{\text{next}} - \lambda_{\min})]^{\frac{1}{2}}, \quad \lambda_{\min} = \frac{\pi^2}{2}, \quad \lambda_{\text{next}} = 2.5\pi^2,$$

where $\delta^* = 0.5$ and $\delta^* = (0.1)^i, i = 1, 2, \ldots, 9$. The results are listed in table 9.2 for two partitions. When $\delta^* \to 0$, we can see

$$\lambda_{\min}(\mathbf{A}) \to 10^{-11} \text{ in partition I}, \qquad \lambda_{\min}(\mathbf{A}) \to 10^{-14} \text{ in partition II}.$$

From fig. 9.5, we can see that the $\lambda_{\min}$ can be obtained by minimizing $\lambda_{\min}(\mathbf{A}(k))$, whose accuracy may reach to $O(10^{-8})$ and $O(10^{-9})$ in partitions I and II, respectively. Obviously, the sensitivity of $\lambda_{\min}(\mathbf{A})$ on δ^* in partition II is higher than that in partition I. We then expect that partition II will have a better performance in seeking eigenvalues. Moreover, straight lines with the slope -2 in fig. 9.5 demonstrate the asymptotic behavior

$$\delta = O(\delta^*) = O(\sqrt{\lambda_{\min}(\mathbf{A})}),$$

which also agrees with eqns. (9.3.16) and (9.3.17).

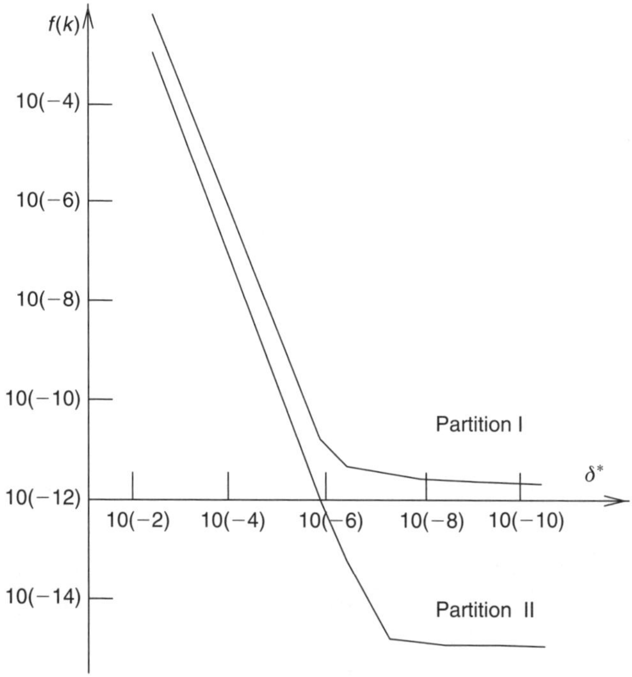

Figure 9.5: The curves of $f(k) = \lambda_{\min}(\mathbf{A}(k))$ versus δ^* from table 9.2.

9.5.4 Minimal eigenvalue and corresponding eigenfunction

Let us apply algorithms (A) and (B) to seek $\lambda_{\min}$ and u_{11}. The initial values of k are chosen as

$$k_0 = 2, \qquad k_1 = 2.001, \qquad k_2 = 2.002, \tag{9.5.18}$$

and the iterative results are given in tables 9.3, 9.4, 9.6, and 9.7. In table 9.4(b), N_1, N_2, and N_3 denote the numbers of subintervals of the central rule on $\overline{EF}, \overline{DE}$, and $\overline{FG}$ in fig. 9.3, respectively. All data in the tables in this chapter are carried out by Fortran programs in double precision. Excellent eigenvalues and eigenfunctions have been obtained. Since the relative errors

$$\delta = \left| \frac{k^2 - \lambda_l}{k^2} \right| \approx \frac{2}{k} \Delta \widetilde{\lambda_l} \quad \text{as } k \to \lambda_l,$$

we can see from table 9.3

$$\delta \approx 1.7 \times 10^{-8}, \quad \text{and} \quad 2.2 \times 10^{-9}$$

at the 10th iteration for partitions I and II, respectively. On comparing the coefficients in table 9.4 with the true values in eqn. (9.5.14), the following coefficient errors of the eigenfunctions can be observed

$$\Delta\hat{c}_0 = 0, \qquad \Delta\hat{c}_i \approx 2 \sim 7 \times 10^{-3}, \quad i = 1,2,3,4$$

for partition I, and

$$\Delta\hat{c}_0 = 0, \qquad \Delta\hat{c}_i \approx 2 \times 10^{-4}, \quad i = 1,2,3,4,$$
$$\Delta\hat{a}_0 \approx 1 \times 10^{-4}, \qquad \Delta\hat{b}_0 \approx 2 \times 10^{-4}, \qquad \Delta\hat{d}_0 \approx 3 \times 10^{-4}$$

for partition II. Evidently, partition II has a better performance, as expected.

Table 9.3: The iteration solution for partitions I and II to seek the minimal eigenvalue.

(a) For partition I ($M = 2$).

n	$\sqrt{\widetilde{\lambda}_1^{(n)}}$	$\Delta\sqrt{\widetilde{\lambda}_1^{(n)}}$	$\lambda_{\min}(\mathbf{A})$	$\|\epsilon\|_{\mathrm{I}}$	Cond	$\sqrt{\frac{\|\lambda_{\min}(\mathbf{A})\|}{\lambda_{\mathrm{next}}(\mathbf{A})}}$
1	2.0		0.1054(−2)			
2	2.001		0.1045(−2)			
3	2.002		0.1036(−2)			
4	2.2852315962	0.0638	0.1059(−3)	0.166(−3)	0.493(3)	0.0186
5	2.2286927857	0725(−2)	0.1317(−5)	0.166(−5)	0.442(4)	0.205(−2)
6	2.2161309892	−0.531(−2)	0.7004(−6)	0.141(−5)	0.607(4)	0.149(−2)
7	2.2214700664	0.286(−4)	0.2325(−10)	0.216(−5)	0.105(7)	0.860(−5)
8	2.2214285150	−0.130(−4)	0.7088(−11)	0.227(−5)	0.191(7)	0.475(−5)
9	2.2214414601	−0.898(−8)	0.2894(−11)	0.352(−5)	0.299(7)	0.303(−5)
10	2.2214414879	0.185(−7)	0.2893(−11)	0.242(−4)	0.299(7)	0.303(−5)

(b) For partition II ($L = M = K = 2, N = 3$).

n	$\sqrt{\widetilde{\lambda}_1^{(n)}}$	$\Delta\sqrt{\widetilde{\lambda}_1^{(n)}}$	$\lambda_{\min}(\mathbf{A})$	$\|\epsilon\|_{\mathrm{II}}$	Cond	$\sqrt{\frac{\|\lambda_{\min}(\mathbf{A})\|}{\lambda_{\mathrm{next}}(\mathbf{A})}}$
1	2.0		0.2490(−3)			
2	2.001		0.2469(−3)			
3	2.002		0.2449(−3)			
4	2.2994029817	0.0780	0.3900(−4)	0.644(−3)	237.	0.0513
5	2.2269757991	0.553(−2)	0.1856(−6)	0.380(−4)	0.339(4)	0.354(−2)
6	2.2141050490	−0.734(−2)	0.3229(−6)	0.486(−4)	0.257(4)	0.467(−2)
7	2.2213725668	−0.689(−4)	0.2865(−10)	0.412(−6)	0.273(6)	0.440(−4)
8	2.2214255292	−0.159(−4)	0.1538(−11)	0.144(−6)	0.118(7)	0.102(−4)
9	2.2214417167	−0.248(−6)	0.5031(−14)	0.641(−6)	0.206(8)	0.583(−6)
10	2.2214414716	0.252(−8)	0.4711(−14)	0.970(−6)	0.313(8)	0.564(−6)

Table 9.4: The calculated coefficients at the 10th iteration in seeking the minimal eigenvalue.

(a) Partition I with $M=2$, $N_1=36$, $k=2.2214414879$.

i	c_i	$\hat{c}_i = \frac{c_i/J_i(k/2)}{c_0/J_0(k/2)}$
0	−0.51189625231336(3)	1.0000000000000
1	−0.68056228480074(3)	2.0041823521466
2	−0.19959301558621(3)	2.0050241968523
3	0.37819229035445(2)	−1.9986819732447
4	−0.53197729344948(1)	1.9933725915138
5	−0.60452403655422	2.0182850091578
6	−0.53549502477247(−1)	1.9172637624130
7	0.35021805606486(−2)	−1.5717345445127

(b) For partition II with $L=K=M=2$, $N=3$, $N_1=36$, $N_2=N_3=16$, $k=2.2214414716$.

i	c_i	$\hat{c}_i = \frac{c_i/J_i(k/2)}{c_0/J_0(k/2)}$
0	0.12316821988238(5)	1.0000000000000
1	0.16339609495803(5)	1.9998357077860
2	0.47900157048407(4)	1.9998365842009
3	−0.91051947498290(3)	−1.9998782636634
4	0.12841347393427(3)	1.9998125149512
5	0.14411011276208(2)	1.9996163638668
6	0.13356894789306(1)	1.9875370595556
7	−0.97611894579621(−1)	−1.8206510555810
i	a_i	$\hat{a}_i = \frac{a_i/J_{4i}(k/2)}{c_0/J_0(k/2)}$
0	0.24632611565408(5)	1.9999161787782
1	−0.25681467518868(3)	−3.9994339046411
2	0.15246233907367(−1)	4.0787852268429
i	b_i	$\hat{b}_i = \frac{b_i/J_{2i+1}(k/2)}{c_0/J_0(k/2)}$
0	0.23108363990284(5)	2.8282763714859
1	−0.12875804596318(4)	−2.8280605134600
2	−0.20375848721408(2)	−2.8272742106772
3	0.16014021904186	2.9869255186080
i	d_i	$\hat{d}_i = \frac{d_i/J_{4i+2}(k/2)}{c_0/J_0(k/2)}$
0	0.95801465838770(4)	3.9997212537494
1	−0.27769257940907(1)	−4.1321302701358
2	−0.26956118786962(−1)	−0.20912262144594(4)

Table 9.5: Comparisons on different non-linear methods to seek the minimal eigenvalue λ_1.

Partitions	Methods	Iteration numbers	$\Delta\sqrt{\tilde{\lambda}_1}$	$\lambda_{\min}(\mathbf{A})$
Partition I ($M=2$)	Secant method with $p=2$	11	0.856(−5)	0.473(−11)
	Secant method with $p=1$ plus Aitken's method	26	0.375(−5)	0.324(−11)
	Brent's method	32	0.603(−7)	0.289(−11)
	Modified Muller's method	11	0.303(−6)	0.289(−11)
	Method(B)	9	0.897(−8)	0.289(−11)
Partition II ($M=K=2, N=3$)	Method(B)	10	0.251(−8)	0.471(−14)

To display the effectiveness of algorithm (B), let us compare it with other popular non-linear methods, such as Brent's method in Ref. [9], the Secant method, the Modified Muller's method, etc., described in Section 9.2.4. Numerical experiments under the same initial values in eqn. (9.5.18) are used and results are summarized in table 9.5. Algorithm (B) exhibits the best performance by giving the most accurate solutions with the fewest iterations. Take partition II as an example,

$$\Delta\sqrt{\tilde{\lambda}_{\min}} = \left|\sqrt{\tilde{\lambda}_{\min}} - \sqrt{\lambda_{\min}}\right| \approx 2.5 \times 10^{-9}, \qquad \lambda_{\min}(A) \approx 4.7 \times 10^{-15}$$

by 10 iterations including the three initial values. In contrast, the modified Muller's method and the Secant method can reach only the errors

$$\Delta\sqrt{\tilde{\lambda}_{\min}} \approx 3 \times 10^{-7} \qquad \text{and} \qquad 8 \times 10^{-6},$$

and they may, sometimes, have troubles in seeking other eigenvalues, based on trial computation. Furthermore, we have chosen other kinds of functions instead of eqn. (9.2.13), such as

$$f(k) = \sqrt{\lambda_{\min}(\mathbf{A})}, \qquad \text{or} \qquad \det|\mathbf{A}(k)| \quad \text{as in Ref. [148]}$$

to carry out algorithms (A) and (B). Unfortunately, computation shows that the sequences obtained are either divergent or very slowly convergent.

Table 9.6: The minimal eigenvalues calculated for partitions I and II.

M				$\sqrt{\widetilde{\lambda}_1}$	$\Delta\sqrt{\widetilde{\lambda}_1}$	$\lambda_{\min}(\mathbf{A})$	$\|\epsilon\|_{\mathrm{I}}$	$\sqrt{\frac{\|\lambda_{\min}(\mathbf{A})\|}{\lambda_{\mathrm{next}}(\mathbf{A})}}$
1				2.2170240851	−0.442(−2)	0.2053(−3)	1.74	0.0142
2				2.2214414878	0.187(−7)	0.289(−11)	0.242(−3)	0.303(−5)
3				2.2214414537	−0.154(−7)	0.933(−14)	0.375(−10)	0.253(−6)
4				2.2214417873	0.318(−6)	0.886(−13)	0.446(−10)	0.103(−5)
5				2.2214403561	−0.111(−5)	−0.289(−12)	0.590(−10)	0.231(−5)
L	M	N	K	$\sqrt{\widetilde{\lambda}_1}$	$\Delta\sqrt{\widetilde{\lambda}_1}$	$\lambda_{\min}(\mathbf{A})$	$\|\epsilon\|_{\mathrm{II}}$	$\sqrt{\frac{\|\lambda_{\min}(\mathbf{A})\|}{\lambda_{\mathrm{next}}(\mathbf{A})}}$
1	1	2	1	2.2218067548	0.365(−3)	0.426(−5)	0.627	0.799(−2)
2	2	3	2	2.2214414716	0.252(−8)	0.471(−14)	0.970(−5)	0.564(−6)
3	3	5	3	2.2214414665	−0.254(−8)	0.748(−17)	0.252(−11)	0.552(−7)
4	4	7	4	2.2214414676	−0.148(−8)	−0.111(−15)	0.653(−12)	0.535(−6)
5	5	9	5	2.2214414702	0.112(−8)	−0.111(−15)	0.219(−12)	0.136(−5)

From data of partition I in tables 9.6 and 9.7(a), we can discover that the following results at $M = 3$ are the best:

$$\begin{aligned} &\Delta\widetilde{\lambda}_{\min} \approx 1.5\times 10^{-8}, \quad \text{i.e., } \delta \approx 1.3\times 10^{-8},\\ &\lambda_{\min}(\mathbf{A}) \approx 9\times 10^{-15}, \quad |\varepsilon^*|_{\mathrm{I}} \approx 3.7\times 10^{-11},\\ &\Delta\hat{c}_0 = 0, \quad \Delta\hat{c}_1 \approx 3\times 10^{-8}, \quad \Delta\hat{c}_2 \approx 2\times 10^{-8}. \end{aligned}$$

From tables 9.6 and 9.7(b), however, at $M = 3 \sim 5$ in partition II the solutions are highly accurate. Take $L = M = K = 3$ and $L = 5$ as an example, we have

$$\begin{aligned} &\Delta\widetilde{\lambda}_{\min} \approx 2.5\times 10^{-9}, \quad \text{i.e., } \delta \approx 2\times 10^{-9},\\ &\lambda_{\min}(\mathbf{A}) \approx 7\times 10^{-18}, \quad |\varepsilon^*|_{\mathrm{II}} \approx 2.5\times 10^{-12},\\ &\Delta\hat{c}_0 = 0, \quad \Delta\hat{c}_1 \approx 5\times 10^{-9}, \quad \Delta\hat{c}_2 \approx 3\times 10^{-9},\\ &\Delta\hat{a}_0 \approx 8\times 10^{-9}, \quad \Delta\hat{b}_0 \approx 4\times 10^{-9}, \quad \Delta\hat{d}_0 \approx 5\times 10^{-9}. \end{aligned}$$

Consequently, using piecewise particular solutions in partition II is preferable. Note that the leading coefficients directly from the CTM may be extremely large due to degeneracy, i.e.,

$$\widetilde{c}_0 \approx 4.2\times 10^{7} \quad \text{and} \quad 2.5\times 10^{8} \quad \text{as } M = 3$$

for partitions I and II, respectively (also see Lemma 9.3.6).

Table 9.7: The leading coefficients calculated for λ_1.

(a) For partition I.

M	c_0	$\hat{c}_0$	$\hat{c}_1$	$\hat{c}_2$
1	−0.1370683964(1)	1.0	1.6633952166	1.830289310
2	−0.5118962523(3)	1.0	2.004182352	2.005024196
3	0.4190401959(8)	1.0	1.999999969	1.999999979
4	−0.2021863209(7)	1.0	2.000000642	2.000000439
5	0.5780419280(6)	1.0	1.999997753	1.999998465

(b) For partition II.

L	M	N	K	c_0	$\hat{c}_0$	$\hat{c}_1$	$\hat{c}_2$
1	1	2	1	−0.3109134477	1.0	8.295716631	8.278741850
2	2	3	2	0.3036872873(3)	1.0	1.993336522	1.999337114
3	3	5	3	0.2544534880(9)	1.0	1.999999995	1.999999997
4	4	7	4	0.4262991335(9)	1.0	1.999999997	1.999999998
5	5	9	5	−0.5606163341(9)	1.0	2.000000002	2.000000002
L	M	N	K	$\hat{a}_0$		$\hat{b}_0$	$\hat{d}_0$
1	1	2	1	5.331766814		8.512259240	0.1475997688(2)
2	2	3	2	1.996600014		2.822313087	3.988691347
3	3	5	3	1.999999959		2.828427121	3.999999995
4	4	7	4	1.999999998		2.828427123	3.999999997
5	5	9	5	2.000000002		2.828427126	4.000000002

9.6 Eigenvalues for the singularity problem

Let us consider a new eigenvalue model with singularity for the crack problem (see fig. 9.6)

$$\begin{aligned} -\Delta u &= \lambda u \quad \text{in} \quad \Omega, \\ u &= 0 \quad \text{on} \quad \overline{OD}, \\ u &= 0 \quad \text{on} \quad \overline{BC}, \\ u_\nu &= 0 \quad \text{on} \quad \overline{OA} \cup \overline{AB} \cup \overline{CD}, \end{aligned}$$

where $\Omega = (-1, 1) \times (0, 1)$. We may solve the Helmholtz problem

$$\begin{aligned} -\Delta u &= k^2 u \quad \text{in} \quad \Omega, \\ u &= 0 \quad \text{on} \quad \overline{OD}, \\ u &= 1 \quad \text{on} \quad \overline{BC}, \\ u_\nu &= 0 \quad \text{on} \quad \overline{OA} \cup \overline{AB} \cup \overline{CD}. \end{aligned}$$

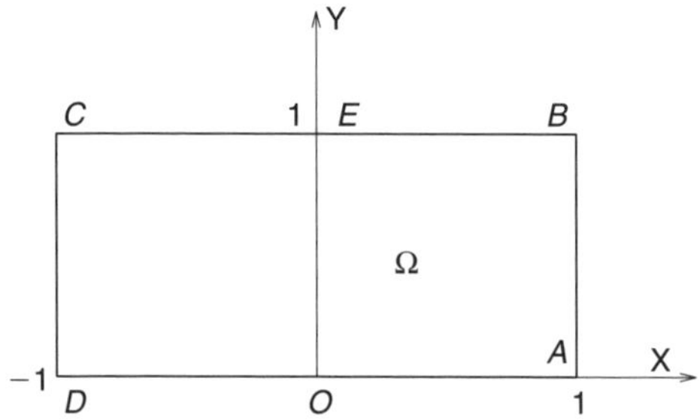

Figure 9.6: The singularity eigenvalue problem.

The particular solutions are given by

$$v^+ = \sum_{i=1}^{\infty} \hat{c}_i J_{i-\frac{1}{2}}(kr) \cos\left(i - \frac{1}{2}\right)\theta, \tag{9.6.1}$$

where $\hat{c}_i$ are expansion coefficients. In computation, we choose

$$v^+ = \sum_{i=1}^{L} c_i \frac{J_{i-\frac{1}{2}}(kr)}{J_{i-\frac{1}{2}}(kr_0)} \cos\left(i - \frac{1}{2}\right)\theta,$$

where r_0 is a parameter chosen as $r_0 = 1$ in computation. There exists the relation between $\hat{c}_i$ and c_i

$$\hat{c}_i = \frac{c_i}{J_{i-\frac{1}{2}}(kr_0)}.$$

In algorithm (B), a good initial guess of $\sqrt{\lambda_{\min}}$ (or $\sqrt{\lambda_{\text{next}}}$) is important to its convergence. Let us derive a bound of $\lambda_{\min}$. First consider an auxiliary eigenvalue problem (fig. 9.7(a))

$$-\Delta u = \lambda u \quad \text{in } \hat{S}, \tag{9.6.2}$$

$$u = 0 \quad \text{on } y = \pm 1, \tag{9.6.3}$$

where $\hat{S} = \{(x,y) \mid -\infty < x < \infty,\ -1 < y < 1\}$. The eigenfunctions of eqns. (9.6.2) and (9.6.3) are

$$u = \cos\left\{\frac{(2i-1)\pi}{2} y\right\}.$$

Hence, the minimal eigenvalue is found as

$$\hat{\lambda}_{\min} = \frac{\pi^2}{4}. \tag{9.6.4}$$

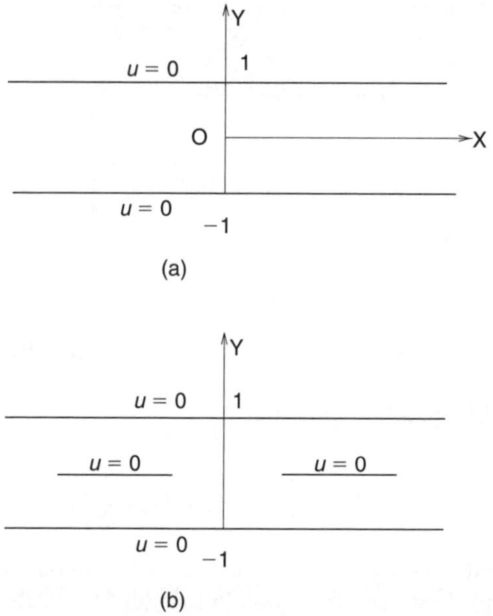

Figure 9.7: (a) The auxiliary eigenvalue problem, and (b) the expanded eigenvalue problem.

In fact, the minimal eigenvalue of the crack problem is also that for the domain in the fig. 9.7(b), which has the additional Dirichlet condition on the middle broken sections. Hence, based on Ref. [108], there exists the bound

$$\frac{\pi^2}{4} = \hat{\lambda}_{\min} \le \lambda_{\min}.$$

Next, consider another auxiliary eigenvalues on $S^+ = (-1, 1) \times (-\frac{1}{2}, \frac{1}{2})$ with the Dirichlet condition on the entire boundary ∂S^+. Since their eigenfunctions are

$$u = \cos\left\{\frac{(2i-1)\pi}{2}x\right\} \cos(2j-1)\pi y,$$

the minimal eigenvalue is given by

$$\lambda^+_{\min} = \frac{\pi^2}{4} + \pi^2 = \frac{5}{4}\pi^2.$$

For the crack eigenvalue problem in eqn. (9.6.1), there exists some Neumann condition on a part of ∂S^+. Hence, we have the bound from Ref. [108]

$$\lambda_{\min} \le \lambda^+_{\min} = \frac{5}{4}\pi^2. \tag{9.6.5}$$

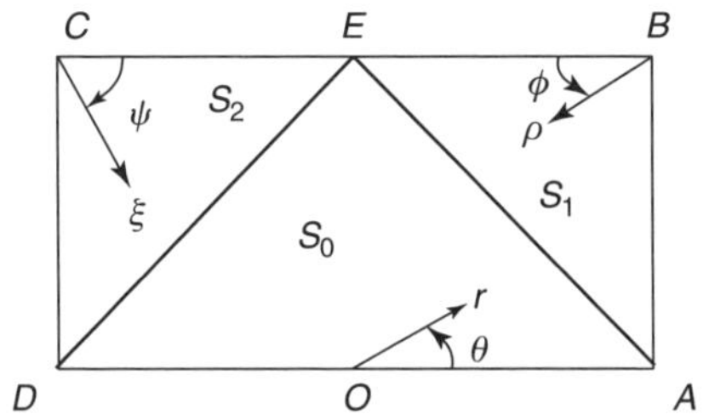

Figure 9.8: A partition for the singularity eigenvalue problem.

Combining eqns. (9.6.4) and (9.6.5) gives

$$\frac{\pi^2}{4} \le \lambda_{\min} \le \frac{5}{4}\pi^2, \qquad \frac{\pi}{2} \le \sqrt{\lambda_{\min}} \le \frac{\sqrt{5}}{2}\pi. \tag{9.6.6}$$

Based on the bound of eqn. (9.6.6), we may easily find good initial values of k for $\lambda_{\min}$. By increasing k, we can find good initial values of k for λ_{next} of the crack eigenvalue problem.

Let $\Omega = S_0 \cup S_1 \cup S_2$ in fig. 9.8. We choose the piecewise particular solutions,

$$v_L = \sum_{i=1}^{L} c_i \frac{J_{i-\frac{1}{2}}(kr)}{J_{i-\frac{1}{2}}(kr_0)} \cos\left(i - \frac{1}{2}\right)\theta \quad \text{in } S_0,$$

$$v_M = 1 + \sum_{i=1}^{M} a_i \frac{J_{2i-1}(k\rho)}{J_{2i-1}(k\rho_0)} \sin(2i-1)\phi \quad \text{in } S_1,$$

$$v_N = 1 + \sum_{i=1}^{N} b_i \frac{J_{2i-1}(k\xi)}{J_{2i-1}(k\xi_0)} \sin(2i-1)\psi \quad \text{in } S_2,$$

where c_i, a_i, and b_i are the unknown coefficients to be sought, and the parameters $r_0 = \rho_0 = \xi_0 = \frac{\sqrt{2}}{2}$. The polar coordinates $(r, \theta), (\rho, \phi)$, and (ξ, ψ) are shown in fig. 9.8. Numerical experiments of the singularity problem for the partitions in figs. 9.6 and 9.8 are provided in Ref. [303].

9.7 Summaries and discussions

To close this chapter, let us summarize the nature and novelties of the new methods for eigenvalue problems, and give a few concluding remarks.

1. The new algorithms for eigenvalue problems lie in solutions of the Helmholtz eqn. (9.1.6) by modifying k to lead to a degeneracy. The degeneracy is measured

by the infinitesimal values of the minimal eigenvalue of the associate matrix $\mathbf{A}(k)$ in eqn. (9.2.5), and the modification to k is realized by algorithm (B). Algorithms (A) and (B) are based on the fact that the eigenfunctions of eqn. (9.1.4) will dominate the solutions of eqn. (9.1.6) when a degeneracy occurs.

2. The degeneracy plays an important role in our algorithms. The leading coefficients such as c_1 are very large, and the scale solutions, i.e., eqn. (9.2.10) are recommended due to the simplicity and high accuracy of c_1. It is interesting to note that arbitrariness of the bounded function $g(\neq 0)$ in eqn. (9.1.6) does not influence much the final solutions of the algorithms. Hence, we simply choose $g|_\Gamma = 1$ in the computational models.
3. Finding the minimum of an interpolatory quadratic polynomial, a specific iteration of algorithm (B) is designed to find the eigenvalues and eigenfunctions effectively. Superlinear convergence rates are proven in Ref. [303], and comparison of numerical experiments is made in table 9.5. Algorithm (B) may be regarded as a variation of Muller's method for real roots. Since matrix $\mathbf{A}$ is symmetric, the eigenvalues of $\lambda_{\min}(\mathbf{A}(k))$ are all real. Of course, algorithm (B) is expected to be better than Muller's method. Table 9.5 already reveals the best performance of algorithm (B).
4. A basic sample of eigenvalue problems accompanied with piecewise particular solutions is given in Section 9.5.1; it can be adopted to test the effectiveness of numerical methods. Algorithms (A) and (B) are also applied to the eigenvalue problems with singularity. Since the eigenvalues and the expansion coefficients of eigenfunctions are very accurate, they can be treated as the true solution, to evaluate the true errors of solutions by other numerical methods, e.g., FEM, FDM, FVM, BEM, etc.
5. Only individual eigenvalues and eigenfunctions are sought by the methods given in this chapter. We may apply them to seek the leading eigenvalues in eqn. (9.1.2), which are of most interest in engineering problems.
6. However, the TM needs the explicit particular functions. This limitation confines these methods to linear eigenvalue problems, where local particular solutions can be found in textbooks or from some asymptotic analysis as done in Section 9.5.1 and in Ref. [304].
7. This chapter may be regarded as a further development of Fox, Henrici, and Moler [148] by using piecewise particular solutions. The methods in Ref. [148] use *uniform* particular solutions to seek the eigenvalues of eqn. (9.1.1). The algorithms in this chapter adopt *piecewise* particular solutions, thus leading to a wide range of application of complicated eigenvalue problems, for instance those with multiple singularities. Hence, we may partition the solution domain into finite subdomains; local particular solutions can be employed in the subdomains. This yields a framework of the CTM for eigenvalue problems.

10 The Helmholtz equation

The Trefftz method (TM) [441] is developed to solve the Helmholtz equation, $\Delta u + k^2 u = 0$, where k^2 is not exactly equal (but may be very close) to an eigenvalue of the Laplace operator $-\Delta$. Piecewise particular solutions are chosen and then matched together in order to satisfy the exterior and interior boundary conditions. Error analysis is presented to estimate error bounds in the entire solution domain. Let δ be the smallest relative distance between k^2 and the eigenvalues of $-\Delta$. We prove that the error asymptote of the solutions by the TM is $O(\frac{1}{\delta})$ asymptotically as $\delta \to 0$, which is called degeneracy in this chapter. Such an asymptote $O(\frac{1}{\delta})$ has been verified by the numerical computations in Li [283]. We also explain why the exponential convergence rates of solutions can be obtained easily by splitting the solution domain into smaller subdomains.

10.1 Introduction

Studies on the Helmholtz solutions are of interest in both theory and application. Several important literatures on the Helmholtz equation should be mentioned here: particular solutions of Courant and Hilbert [108], the finite element method (FEM) of Babuska and Osborn [18], the finite difference methods (FDM) of Birkhoff and Lynch [45], the least squares method (LSM) of Chang [82], the capacitance matrix method of Proskurwski and Widlund [372], the method converted to an integral equation of Lin [310], the coupling methods of the boundary integral and FEM of Johnson and Nedelec [233] and Jiang and Li [225], and the preconditioned conjugate gradient method of Bayliss, Goldstein, and Turkel [26]. Moreover, studies have been done for the Helmholtz equation on the unbounded domains, see Aziz, Dorr, and Kellogg [11], Goldstein [166, 167], and Harari and Hughes [185]. Recently, the Helmholtz eigenvalue problems in a multiply connected domain by the boundary element method (BEM) are studied in Chen, Lin, and Hong [86] and Chen et al. [87].

In this chapter, we will follow Part I to use the TM [441], Stojek [424] and Cheung, Jin, and Zienkiewicz [97] to solve the Helmholtz equation. In such methods the solution domain is divided into several subdomains, different particular solutions on subdomains (i.e., piecewise particular solutions) are used, and an approximation of the solution is then obtained by satisfying only the interior and exterior boundary conditions. The methods using the entire particular solutions for solving the elliptic and eigenvalue problems can be found in Bergman [31], Eisenstat [134], Fox, Henrici, and Moler [148], Mathon and Sermer [330], and Vekua [448]. The approaches in this chapter, however, use *piecewise* particular solutions for solving the Helmholtz equation. The TM is advantageous for solving elliptic problems with multiple singularities, multiple interfaces, or those on unbounded domains, which may cause some difficulty if using the standard, FEM and FDM (see Refs. [45, 103, 426]). Furthermore, an important advantage of the TM is that high accuracy of solutions with exponential convergence rates can be achieved with a modest effort in computation. Note that the approaches in this chapter of the TM are similar to those of the ultra weak variational formulation (UWVF) in Cessenat and Despres [80], since both methods employ local particular solutions of the Helmholtz equation as basis functions of the solution. Also the approaches of the TM satisfying the interior and exterior boundary conditions in this chapter are of the least squares method.

Let us consider the Helmholtz equation in two dimensions

$$\begin{cases} \Delta u + k^2 u = 0 & \text{in } \Omega, \\ u = f & \text{on } \Gamma, \end{cases} \tag{10.1.1}$$

where $k > 0$, Δ the Laplace operator: $\Delta = \frac{\partial^2}{\partial x^2} + \frac{\partial^2}{\partial y^2}$, and Ω a polygonal domain with the exterior boundary Γ. This equation is also called the *reduced wave equation* by Birkhoff and Lynch [45]. We assume in this chapter that k^2 is not *exactly* equal (but may be very close) to any eigenvalue of the following eigenvalue problem [18, 108, 472]:

$$\begin{cases} -\Delta \phi_l = \lambda_l \phi_l & \text{in } \Omega, \\ \phi_l \mid_\Gamma = 0 & \text{on } \Gamma. \end{cases} \tag{10.1.2}$$

When $k^2 \to \lambda_l$, the eigenvalue problem, i.e., eqn. (10.1.2) can be solved, see Chapter 9.

Let the eigenvalues $\{\lambda_l\}$ be arranged in an ascending order, i.e.,

$$0 < \lambda_1 \leq \lambda_2 \leq \cdots \leq \lambda_n \leq \lambda_{n+1} \leq \cdots.$$

The complete eigenfunctions $\{\phi_l\}$ are orthonormal,

$$(\phi_i, \phi_j) = \iint_\Omega \phi_i \phi_j \, d\Omega = \delta_{ij} = \begin{cases} 0 & \text{if } i \neq j, \\ 1 & \text{if } i = j. \end{cases} \tag{10.1.3}$$

Another Helmholtz equation is also given in Ref. [45]

$$\begin{cases} \Delta u + k^2 u = g & \text{in } \Omega, \\ u = f & \text{on } \Gamma. \end{cases} \tag{10.1.4}$$

Let the solution u^* be a particular solution to the equation $\Delta u^* + k^2 u^* = g$. The eqn. (10.1.1) is then obtained from eqn. (10.1.4) by a transformation $v = u - u^*$ so that we only discuss eqn. (10.1.1).

We first provide an expansion of solution to eqn. (10.1.1). Let w denote the solution of the *Laplace equation*

$$\begin{cases} \Delta w = 0 & \text{in } \Omega, \\ w = f & \text{on } \Gamma. \end{cases}$$

By the transformation $v = u - w$, the eqn. (10.1.1) is reduced to the following problem

$$\begin{cases} \Delta v + k^2 v = -k^2 w & \text{in } \Omega, \\ v = 0 & \text{on } \Gamma. \end{cases} \tag{10.1.5}$$

Consequently, the solution v of the above problem can be found by Span$\{\phi_i\}$,

$$v = \sum_{i=1}^{\infty} a_i \phi_i, \tag{10.1.6}$$

where a_i are expansion coefficients. Substituting (10.1.6) into (10.1.5) yields the identity:

$$\sum_{i=1}^{\infty} a_i (k^2 - \lambda_i)\phi_i = -k^2 w.$$

Since k^2 is not an eigenvalue λ_l by the assumption, the coefficients a_i can be defined uniquely from the orthogonality of eqn. (10.1.3)

$$a_i = -\frac{k^2 (w, \phi_i)}{k^2 - \lambda_i},$$

thus giving a unique solution of eqn. (10.1.1)

$$u = w - \sum_{i=1}^{\infty} \frac{k^2}{k^2 - \lambda_i} (w, \phi_i)\phi_i. \tag{10.1.7}$$

Denote by δ the smallest relative distance between k^2 and λ_i, where $k > 0$

$$\delta = \min_i \left| \frac{k^2 - \lambda_i}{k^2} \right| > 0. \tag{10.1.8}$$

Based on the eqn. (10.1.7) we can conclude the following:

1. When $\delta > 0$, the Helmholtz solution is unique.
2. When $\delta = 0$, the eigenvalue problem of eqn. (10.1.2) should be solved instead, see Chapter 9.
3. When $\delta \to 0$, the Helmholtz solution approaches to an eigenfunction of the eqn. (10.1.2). This is called the case of degeneracy in this chapter.

Our questions for the last degenerate case, however, are

1. When $0 < \delta \ll 1$, can we find an approximate solution to the Helmholtz equation?
2. If we can, how much error will be as $\delta \to 0$?

Since the maximum principle can no longer be applied to the Helmholtz solution, a new analysis to Chapter 1 will be provided in this chapter to justify the TM. To our knowledge, there seems to be no literature yet to provide error analysis on the degenerate solutions (i.e., as $\delta \to 0$) of the Helmholtz equation.

The rest of the chapter is organized as follows. In the next two sections, we shall describe the TM, and then derive new error bounds. In the last section, a few remarks are made. The materials in this chapter are adapted from Li [283].

10.2 The Trefftz method

Let the solution domain Ω be divided by a piecewise straight line Γ_0 into two subdomains Ω^+ and Ω^-. Consider the piecewise Helmholtz equation

$$\Delta u + k^2 u = 0, \quad \text{in } \Omega^+ \text{ and } \Omega^-, \tag{10.2.1}$$

with the interior and exterior boundary conditions:

$$u^+ = u^- \quad \text{on } \Gamma_0, \qquad u^+_\nu = u^-_\nu \quad \text{on } \Gamma_0 \qquad u = f \quad \text{on } \Gamma, \tag{10.2.2}$$

where u_ν is the unit normal derivative of u on Γ_0. Evidently, the solution u to eqn. (10.1.1) also satisfies eqns. (10.2.1) and (10.2.2). Define a space

$$H = \{v \in L_2(\Omega) \,|\, v \in H^1(\Omega^+), v \in H^1(\Omega^-) \text{ and } \Delta v + k^2 v = 0 \text{ in } \Omega^+ \text{ and } \Omega^-\},$$

and a functional

$$I(v) = \int_\Gamma (v - f)^2\, ds + \int_{\Gamma_0} (v^+ - v^-)^2\, ds + \sigma^2 \int_{\Gamma_0} (v^+_\nu - v^-_\nu)^2\, ds, \tag{10.2.3}$$

where σ is a positive weight. Define a finite-dimensional space $S_{m,n} \subseteq H$ such that

$$S_{m,n} = \left\{ v \,|\, v = v^+_m = \sum_{i=1}^m c_i \psi_i^+ \quad \text{on } \Omega^+, \quad \text{and } v = v^-_n = \sum_{i=1}^n d_i \psi_i^- \quad \text{on } \Omega^- \right\},$$

where $\{\psi_i^+\}$ and $\{\psi_i^-\}$ are complete sets of local particular solutions of eqn. (10.2.1) in Ω^+ and Ω^-, respectively, and c_i and d_i are the coefficients to be determined. A TM approximation $u_{m,n} \in S_{m,n}$ for the problem of eqns. (10.2.1) and (10.2.2) can then be found by

$$I(u_{m,n}) = \min_{v \in S_{m,n}} I(v). \tag{10.2.4}$$

Denote by **x** the vector consisting of all unknown coefficients c_i and d_i, then it follows from eqn. (10.2.3) that

$$I(u_{m,n}) = \frac{1}{2}\mathbf{x}^T\mathbf{A}\mathbf{x} - \mathbf{b}^T\mathbf{x} + \int_\Gamma f^2\, ds,$$

where

$$\frac{1}{2}\mathbf{x}^T\mathbf{A}\mathbf{x} = [u_{m,n}, u_{m,n}], \qquad \mathbf{b}^T\mathbf{x} = 2\int_\Gamma u_{m,n} f\, ds. \tag{10.2.5}$$

In eqn. (10.2.5), the bilinear form $[u, v]$ on $H \times H$ is defined by

$$[u, v] = \int_\Gamma uv\, ds + \int_{\Gamma_0} (u^+ - u^-)(v^+ - v^-)\, ds + \sigma^2 \int_{\Gamma_0} (u_\nu^+ - u_\nu^-)(v_\nu^+ - v_\nu^-)\, ds.$$

The stiffness matrix **A** in eqn. (10.2.5) is symmetric and positive definite. Therefore, a system of linear algebraic equations can be reduced from eqn. (10.2.4)

$$\mathbf{A}\mathbf{x} = \mathbf{b}. \tag{10.2.6}$$

In computation, $u_{m,n}$ is solved from the LSM in Chapter 2 by using the QR method or the singular value decomposition (SVD) instead of solving eqn. (10.2.6) directly, in order to achieve the smaller condition number given by (see Refs. [9, 168])

$$\text{Cond} = \left(\frac{\lambda_{\max}(\mathbf{A})}{\lambda_{\min}(\mathbf{A})}\right)^{\frac{1}{2}},$$

where $\lambda_{\max}(\mathbf{A})$ and $\lambda_{\min}(\mathbf{A})$ are the maximal and minimal eigenvalues of **A**, respectively.

10.3 Error analysis

The analysis in Chapter 1 cannot apply directly to eqn. (10.1.1); but some error bounds for the Laplace solutions may be used to derive error bounds for the Helmholtz solutions. New analysis should display the error behavior of the Helmholtz solutions as $\delta \to 0$ and $k \to \infty$.

For error analysis, we define the induced norm on the interior and exterior boundary as

$$|v|_{\mathrm{B}} = \sqrt{[v, v]} = \{|v|^2_{0,\Gamma} + |v^+ - v^-|^2_{0,\Gamma_0} + \sigma^2|v_\nu^+ - v_\nu^-|^2_{0,\Gamma_0}\}^{\frac{1}{2}},$$

and the norms $\|v\|_H$ and $|v|_H$ over H by

$$\|v\|_H = \{\|v\|^2_{1,\Omega^+} + \|v\|^2_{1,\Omega^-}\}^{\frac{1}{2}}, \qquad |v|_H = \{|v|^2_{1,\Omega^+} + |v|^2_{1,\Omega^-}\}^{\frac{1}{2}},$$

where $\|v\|_{1,\Omega^+}$ and $|v|_{1,\Omega^+}$ are the Sobolev norms in Refs. [103, 348] defined by

$$\|v\|_{m,\Omega} = \left\{ \sum_{|\alpha|\le m} \int_\Omega |D^\alpha v|^2 \, d\Omega \right\}^{\frac{1}{2}}, \qquad |v|_{m,\Omega} = \left\{ \sum_{|\alpha|=m} \int_\Omega |D^\alpha v|^2 \, d\Omega \right\}^{\frac{1}{2}}.$$

10.3.1 Preliminary lemmas

Let us first prove several lemmas.

Lemma 10.3.1

Let w satisfy the piecewise Laplace equation (see Dautray and Lions [120])

$$\Delta w = 0 \quad \text{in } \Omega^+ \text{ and } \Omega^-, \tag{10.3.1}$$

$$\begin{cases} w^+ - w^- = \epsilon_1, \qquad w_\nu^+ - w_\nu^- = \epsilon_2 & \text{on } \Gamma_0, \\ w = f & \text{on } \Gamma. \end{cases} \tag{10.3.2}$$

Suppose that the following inverse properties hold,

$$\begin{cases} |w_\nu|_{0,\Gamma} \le K_w \|w\|_H, \\ |w_\nu^+|_{0,\Gamma_0} \le K_w \|w\|_H, \end{cases} \tag{10.3.3}$$

where K_w is a constant which may be unbounded (see Refs. [280, 306]). Then there exists a bounded constant C independent of w such that

$$\|w\|_H \le C\{K_w(|w|_{0,\Gamma} + |w^+ - w^-|_{0,\Gamma_0}) + |w_\nu^+ - w_\nu^-|_{0,\Gamma_0}\}. \tag{10.3.4}$$

Proof.

By using the Green's theorem and eqn. (10.3.1), we obtain

$$\begin{aligned} |w|_H^2 &= \int_{\partial\Omega^+} w_\nu^+ w^+ \, ds + \int_{\partial\Omega^-} w_\nu^- w^- \, ds \\ &= \int_\Gamma w w_\nu \, ds + \int_{\Gamma_0} [(w_\nu^+ - w_\nu^-)w^- + w_\nu^+(w^+ - w^-)] \, ds. \end{aligned}$$

The following bounds can be found from the Schwarz inequality and the inverse properties, i.e., eqn. (10.3.3),

$$\left| \int_\Gamma w w_\nu \, ds \right| \le |w|_{0,\Gamma} |w_\nu|_{0,\Gamma} \le K_w |w|_{0,\Gamma} \|w\|_H, \tag{10.3.5}$$

$$\left| \int_{\Gamma_0} w_\nu^+ (w^+ - w^-) \, ds \right| \le K_w |w^+ - w^-|_{0,\Gamma_0} \|w\|_H. \tag{10.3.6}$$

From the Sobolev imbedding theorem (see Ref. [103]), we obtain

$$\left|\int_{\Gamma_0} (w_\nu^+ - w_\nu^-)w^- \, ds\right| \le |w_\nu^+ - w_\nu^-|_{0,\Gamma_0}|w^-|_{0,\Gamma_0} \le C|w_\nu^+ - w_\nu^-|_{0,\Gamma_0}\|w\|_H. \tag{10.3.7}$$

Therefore, we have from eqns. (10.3.5)–(10.3.7) and the Poincare–Friedrichs inequality [103],

$$\|w\|_H^2 \le C|w|_H^2 \le \{K_w(|w|_{0,\Gamma} + |w^+ - w^-|_{0,\Gamma_0}) + C|w_\nu^+ - w_\nu^-|_{0,\Gamma_0}\}\|w\|_H. \tag{10.3.8}$$

The desired result, i.e., eqn. (10.3.4) is obtained by dividing $\|w\|_H$ in both sides of inequality eqn. (10.3.8). ■

Lemma 10.3.2

Let λ_l and ϕ_l be the eigenvalue and its eigenfunction in eqn. (10.1.2). Then λ_l and ϕ_l are also the eigenvalue and its eigenfunction of the following piecewise eigenvalue problem,

$$\Delta u + \lambda u = 0 \quad \text{in } \Omega^+ \text{ and } \Omega^-, \tag{10.3.9}$$

$$u^+ = u^- \quad \text{on } \Gamma_0, \qquad u_\nu^+ = u_\nu^- \quad \text{on } \Gamma_0, \qquad u = 0 \quad \text{on } \Gamma. \tag{10.3.10}$$

Also the inner product

$$\langle -\Delta\phi_i, \phi_j\rangle = \lambda_i\delta_{i,j}, \tag{10.3.11}$$

where

$$\langle u, v\rangle = \iint_{\Omega^+} uv\, d\Omega + \iint_{\Omega^-} uv\, d\Omega \;=\; (u, v). \tag{10.3.12}$$

Proof.

The eqns. (10.3.9) and (10.3.10) follow from eqn. (10.1.2) directly. From eqns. (10.3.9) and (10.3.10), we have

$$\begin{aligned}\langle -\Delta\phi_i, \phi_j\rangle &= \iint_{\Omega^+} (-\Delta\phi_i)\phi_j\, d\Omega + \iint_{\Omega^-} (-\Delta\phi_i)\phi_j\, d\Omega \\ &= \lambda_i\left[\iint_{\Omega^+} \phi_i\phi_j\, d\Omega + \iint_{\Omega^-} \phi_i\phi_j\, d\Omega\right] = \lambda_i(\phi_i, \phi_j) = \lambda_i\delta_{i,j}.\end{aligned}$$

■

Lemma 10.3.3

Let w and u satisfy the piecewise Laplace eqns. (10.3.1) and (10.3.2), and the following piecewise Helmholtz equation, respectively:

$$\Delta u + k^2 u = 0 \quad \text{in } \Omega^+ \text{ and } \Omega^-, \tag{10.3.13}$$

$$u^+ - u^- = \epsilon_1 \quad \text{on } \Gamma_0, \qquad u_\nu^+ - u_\nu^- = \epsilon_2 \quad \text{on } \Gamma_0, \qquad u = f \quad \text{on } \Gamma. \tag{10.3.14}$$

Then there exists the bound,

$$|u|_{0,\Omega} \le \left(1 + \frac{1}{\delta}\right) |w|_{0,\Omega}, \tag{10.3.15}$$

where δ is given in eqn. (10.1.8).

Proof.

Let $v = u - w$. Then the solution v satisfies

$$\Delta v + k^2 v = -k^2 w \quad \text{in } \Omega^+ \text{ and } \Omega^-,$$

and the homogeneous boundary conditions, i.e., eqn. (10.3.10). We conclude that v can be expressed by Span$\{\phi_i\}$:

$$v = \sum_{i=1}^{\infty} a_i \phi_i \quad \text{in } \Omega.$$

In fact, based on Lemma 10.3.2 and an analogous argument in Section 10.1 we obtain (cf. eqn. (10.1.7)),

$$u = w - \sum_{i=1}^{\infty} \frac{k^2}{k^2 - \lambda_i} (w, \phi_i)\phi_i.$$

By using the Parceval's inequality in Ref. [108] and the definition of δ in eqn. (10.1.8), we have

$$\begin{aligned} |u|_{0,\Omega} &\le |w|_{0,\Omega} + \frac{1}{\delta} \left| \sum_{i=1}^{\infty} (w, \phi_i)\phi_i \right|_{0,\Omega} \\ &\le |w|_{0,\Omega} + \frac{1}{\delta} \left[\sum_{i=1}^{\infty} (w, \phi_i)^2 \right]^{\frac{1}{2}} \\ &\le |w|_{0,\Omega} + \frac{1}{\delta} |w|_{0,\Omega} = \left(1 + \frac{1}{\delta}\right) |w|_{0,\Omega}. \end{aligned} \tag{10.3.16}$$

■

Remark 10.3.1

The bounds, i.e., eqn. (10.3.15) have also been derived by Kuttler and Sigilloto [261] in the entire solution domain.

10.3.2 Error bounds

Theorem 10.3.1

Let eqn. (10.3.3) and $v \in S_{m,n}$ satisfy the following inverse properties:

$$|v_\nu|_{0,\Gamma} \leq K_{m,n}\|v\|_H, \qquad |v_\nu^+|_{0,\Gamma} \leq K_{m,n}\|v\|_H, \tag{10.3.17}$$

where the constant $K_{m,n}$ may be unbounded as $m,n \to \infty$. For any $\sigma > 0$, then there exists a bounded constant C independent of m,n,k,δ, and v such that

$$\|v\|_H \leq C\left(1+\frac{k}{\delta}\right)(K_{u,w}+\sigma^{-1})|v|_{\mathrm{B}}, \tag{10.3.18}$$

where $K_{u,w} = \max(K_w, K_{m,n})$, and the constant K_w is given in eqn. (10.3.3).

Proof.

By using the Green's theorem, eqns. (10.3.13) and (10.3.14), we obtain

$$\|v\|_H^2 = \int_{\partial\Omega^+} v_\nu^+ v^+ \, ds + \int_{\partial\Omega^-} v_\nu^- v^- \, ds + (1+k^2)|v|_{0,\Omega}^2.$$

By means of a similar argument in the eqns. (10.3.5)–(10.3.8), we have for $v \in S_{m,n}$

$$\begin{aligned}
&\left|\int_{\partial\Omega^+} v_\nu^+ v^+ \, ds + \int_{\partial\Omega^-} v_\nu^- v^- \, ds\right| \\
&\quad \leq \{K_{m,n}(|v|_{0,\Gamma} + |v^+ - v^-|_{0,\Gamma_0}) + C|v_\nu^+ - v_\nu^-|_{0,\Gamma_0}\}\|v\|_H \\
&\quad \leq C\left(K_{m,n}+\frac{1}{\sigma}\right)|v|_{\mathrm{B}}\|v\|_H.
\end{aligned}$$

Hence, we obtain

$$\|v\|_H^2 \leq C\left(K_{m,n}+\frac{1}{\sigma}\right)|v|_{\mathrm{B}}\|v\|_H + (1+k^2)|v|_{0,\Omega}^2. \tag{10.3.19}$$

Let $x = \|v\|_H$, then eqn. (10.3.19) is an inequality of order 2 with respect to x,

$$x^2 \leq px + q, \tag{10.3.20}$$

where $x,p,q>0$, and

$$p=C\left(K_{m,n}+\frac{1}{\sigma}\right)|v|_{\rm B},\qquad q=(1+k^2)|v|^2_{0,\Omega}.$$

Then, we have the following bounds by solving eqn. (10.3.20)

$$\|v\|_H=x\le p+\sqrt{q}\le C\left(K_{m,n}+\frac{1}{\sigma}\right)|v|_{\rm B}+(1+k)|v|_{0,\Omega}. \tag{10.3.21}$$

Next, let w be the piecewise Laplace solution in the eqns. (10.3.1) and (10.3.2) such that $w=v$ on Γ and Γ_0. Then, we obtain from Lemmas 10.3.3 and 10.3.1,

$$\begin{aligned}|v|_{0,\Omega}&\le\left(1+\frac{1}{\delta}\right)|w|_{0,\Omega}\le\left(1+\frac{1}{\delta}\right)\|w\|_H\\&\le C\left(1+\frac{1}{\delta}\right)\{K_w(|w|_{0,\Gamma}+|w^+-w^-|_{0,\Gamma_0})+|w^+_\nu-w^-_\nu|_{0,\Gamma_0}\}\\&=C\left(1+\frac{1}{\delta}\right)\{K_w(|v|_{0,\Gamma}+|v^+-v^-|_{0,\Gamma_0})+|v^+_\nu-v^-_\nu|_{0,\Gamma_0}\}\\&\le C\left(1+\frac{1}{\delta}\right)\left(K_w+\frac{1}{\delta}\right)|v|_{\rm B}.\end{aligned} \tag{10.3.22}$$

Combining eqns. (10.3.21) and (10.3.22) yields the desired results, i.e., eqn. (10.3.18). ■

Theorem 10.3.2

Let the inverse inequalities, i.e., eqns. (10.3.3) and (10.3.17) be given, and $u(\in H^1(\Omega))$ be the Helmholtz solution of eqn. (10.1.1). Then for any $\sigma>0$, there exists a unique function, $u_{m,n}\in S_{m,n}$ by the TM, eqn. (10.2.4), such that

$$[u_{m,n},v]=\int_\Gamma fv\,ds\quad\forall v\in S_{m,n}, \tag{10.3.23}$$

$$[u-u_{m,n},v]=0\qquad\forall v\in S_{m,n}, \tag{10.3.24}$$

$$|u-u_{m,n}|_{\rm B}\le C\inf_{v\in S_{m,n}}|u-v|_{\rm B}, \tag{10.3.25}$$

$$|u_{m,n}|_{\rm B}\le|f|_{0,\Gamma}, \tag{10.3.26}$$

$$\|u_{m,n}\|_H\le C\left(1+\frac{k}{\delta}\right)(K_{u,w}+\sigma^{-1})|f|_{0,\Gamma}. \tag{10.3.27}$$

Also, $u_{m,n}$ minimizes $I(v)$ over $v\in S_{m,n}$, that is, the eqn. (10.2.4) holds, if and only if the eqn. (10.3.16) holds.

Proof.

The eqn. (10.3.27) follows from eqn. (10.3.26) and Theorem 10.3.1. Other proofs are analogous to Chapter 1. ■

From eqns. (10.3.27) and (10.3.26), we can see that when f is bounded, the solutions $u_{m,n}$ are also bounded on the boundary Γ and Γ_0, but not necessarily bounded in Ω as $\delta \to 0$ or $k \to \infty$. This is a key difference between the solutions in this chapter and those in Chapter 1. Now, we have a new theorem.

Theorem 10.3.3

Let $u(\in H^1(\Omega))$ be the solution to eqn. (10.1.1) and $u_{m,n}$ be the approximation from the TM (i.e., the BAM), see eqn. (10.3.23) or eqn. (10.2.4). Suppose that the inverse properties, i.e., eqns. (10.3.3) and (10.3.17) hold for all functions in $S_{m,n}$. Then for any $\sigma > 0$, there exists a constant C independent of m, n, k, δ, and u such that

$$\|u - u_{m,n}\|_H \le \inf_{v \in S_{m,n}} \left\{ \|u - v\|_H + C\left(1 + \frac{k}{\delta}\right)(K_{u,w} + \sigma^{-1})|u - v|_{\mathrm{B}} \right\}.$$

Proof.

Let $\xi = v - u_{m,n}$ where $v \in S_{m,n}$. Applying the orthogonality property, i.e., eqn. (10.3.24) gives

$$|\xi|_{\mathrm{B}}^2 = [\xi, \xi] = [v - u, \xi] \le |u - v|_{\mathrm{B}} |\xi|_{\mathrm{B}}.$$

Hence, $|\xi|_{\mathrm{B}} \le |u - v|_{\mathrm{B}}$. We have from Theorem 10.3.1

$$\begin{aligned} \|u - u_{m,n}\|_H &\le \|u - v\|_H + \|\xi\|_H \\ &\le \|u - v\|_H + C\left(1 + \frac{k}{\delta}\right)(K_{u,w} + \sigma^{-1})|\xi|_{\mathrm{B}}. \\ &\le \|u - v\|_H + C\left(1 + \frac{k}{\delta}\right)(K_{u,w} + \sigma^{-1})|u - v|_{\mathrm{B}}. \end{aligned}$$ ■

By following the proof of Theorem 10.3.1, and letting $v = u - u_{m,n}$, we give an *a posteriori* error estimate.

Theorem 10.3.4

Let all the conditions in Theorem 10.3.3 hold. Suppose that the inverse property of eqn. (10.3.17) also holds for the differences $u - u_{m,n}$, then for any $\sigma > 0$ there exists a constant C independent of m, n, k, δ and u such that

$$\|u - u_{m,n}\|_H \le C\left(1 + \frac{k}{\delta}\right)(K_{u,w} + \sigma^{-1})|u - u_{m,n}|_{\mathrm{B}}. \qquad (10.3.28)$$

Moreover, suppose that

$$K_{u,w} = CN^{\frac{1}{2}}, \qquad N = \max\{m,n\}.$$

Then by choosing $\sigma^{-1} = O(N^{\frac{1}{2}})$ *there exists the bound of the ratios of the solution errors,*

$$\frac{\|u - u_{m,n}\|_H}{|u - u_{m,n}|_{\mathrm{B}}} = O\left(N^{\frac{1}{2}}\left(1 + \frac{k}{\delta}\right)\right). \tag{10.3.29}$$

The new estimates, i.e., eqns. (10.3.28) and (10.3.29) are useful in practical application because the values $|u - u_{m,n}|_{\mathrm{B}}$ can be obtained naturally from the TM algorithms in Section 10.2. In the above error bounds, the factor $1/\delta$ plays an important role. Hence, the questions arising in Section 10.1 for the degenerate Helmholtz solutions have been answered by Theorems 10.3.1–10.3.4, thus to fill in the analysis gap between the cases $\delta = 0$ and $\delta > 0$.

10.3.3 Exponential rates of convergence

Let us give the series solutions of eqns. (10.1.1) and (10.2.1). The Helmholtz solution of eqn. (10.1.1) can be expanded by the following particular solutions [438]:

$$u(r,\theta) = a_0 J_0(kr) + \sum_{i=1}^{\infty} J_i(kr)[a_i \cos i\theta + b_i \sin i\theta],$$

where (r,θ) are the polar coordinates, a_i and b_i are the expansion coefficients, and $J_i(z)$ is the Bessel functions defined by

$$J_\mu(r) = \sum_{i=0}^{\infty} \frac{(-1)^i}{\Gamma(i+1)\Gamma(i+\mu+1)} \left(\frac{r}{2}\right)^{2i+\mu}.$$

We may choose the following piecewise particular solutions:

$$u_{m,n} = \begin{cases} u_m^+ = a_0 J_0(kr) + \displaystyle\sum_{i=1}^{m} J_i(kr)[a_i \cos i\theta + b_i \sin i\theta] & \text{in } \Omega^+, \\ u_n^- = c_0 J_0(k\rho) + \displaystyle\sum_{i=1}^{n} J_i(k\rho)[c_i \cos i\phi + d_i \sin i\phi] & \text{in } \Omega^-, \end{cases} \tag{10.3.30}$$

where a_i, b_i, c_i, and d_i are the coefficients. We then have the following theorem.

Theorem 10.3.5

Suppose that all true expansion coefficients of the true solution are bounded:

$$|\overline{a}_i|, |\overline{b}_i|, |\overline{c}_i|, |\overline{d}_i| < C, \tag{10.3.31}$$

and the term numbers m and n in eqn. (10.3.30) satisfy

$$\left(\frac{ekr_{\max}^{+}}{2m}\right) \leq \alpha_{+} < 1, \qquad \left(\frac{ek\rho_{\max}^{-}}{2n}\right) \leq \alpha_{-} < 1, \tag{10.3.32}$$

where $r_{\max}^{+} = \max_{\Omega^+} r$ *and* $\rho_{\max}^{-} = \max_{\Omega^-} \rho$. *Then there exist the exponential convergence rates*

$$|u - u_{m,n}|_{\mathrm{B}} \leq C[1 + \sigma \max(m,n)] \left\{ \frac{\alpha_{+}^{m+1}}{(m+1)^{\frac{1}{2}}} + \frac{\alpha_{-}^{n+1}}{(n+1)^{\frac{1}{2}}} \right\}. \tag{10.3.33}$$

When the condition, i.e., eqn. (10.3.31) is relaxed by $|\overline{a}_i|, |\overline{b}_i|, |\overline{c}_i|, |\overline{d}_i| < C_i^{l+\frac{1}{2}}$ with integer $l \geq 0$, the convergence rates are derived in Ref. [283].

Proof.

Let $\overline{u}_{m,n}$ be the piecewise particular solution, i.e., eqn. (10.3.30), but with the true coefficients $\overline{a}_i$, $\overline{b}_i$, $\overline{c}_i$, $\overline{d}_i$. Then $u = \overline{u}_{m,n} + R_{m,n}$, where the remainder

$$R_{m,n} = \begin{cases} R_m^{+} = \sum_{i=m+1}^{\infty} J_i(kr)[\overline{a}_i \cos i\theta + \overline{b}_i \sin i\theta] & \text{in } \Omega^{+}, \\ R_n^{-} = \sum_{i=n+1}^{\infty} J_i(k\rho)[\overline{c}_i \cos i\phi + \overline{d}_i \sin i\phi] & \text{in } \Omega^{-}. \end{cases} \tag{10.3.34}$$

We have from eqn. (10.3.25)

$$\begin{aligned} |u - u_{m,n}|_{\mathrm{B}} &\leq C \inf_{v \in S_{m,n}} |u - v|_{\mathrm{B}} \leq C|u - \overline{u}_{m,n}|_{\mathrm{B}} \leq C|R_{m,n}|_{\mathrm{B}} \\ &\leq C \left\{ |R_m^{+}|_{0,\partial\Omega^+} + |R_n^{-}|_{0,\partial\Omega^-} + \sigma \left[\left|\frac{\partial R_m^{+}}{\partial \nu}\right|_{0,\Gamma_0} + \left|\frac{\partial R_n^{-}}{\partial \nu}\right|_{0,\Gamma_0} \right] \right\}. \end{aligned}$$

Since there exists the asymptotic formula in Ref. [2], p. 365

$$J_n(z) \approx \frac{1}{(2n\pi)^{\frac{1}{2}}} \left(\frac{ez}{2n}\right)^n, \quad \text{when } n \text{ is large,} \tag{10.3.35}$$

we can obtain from eqn. (10.3.34) and the assumptions, i.e., eqns. (10.3.31) and (10.3.32) that

$$|R_m^+|_{0,\partial\Omega^+} \leq [\text{Length}(\partial\Omega^+)]^{\frac{1}{2}} |R_m^+|_{\infty,\partial\Omega^+} \leq C \max_{r\in\partial\Omega^+} \sum_{i=m+1}^{\infty} |J_i(kr)|$$

$$\leq C \sum_{i=m+1}^{\infty} \frac{1}{(2\pi i)^{\frac{1}{2}}} \left(\frac{ekr_{\max}^+}{2i}\right)^i \leq C \frac{1}{(m+1)^{\frac{1}{2}}} \alpha_+^{m+1}.$$

Next, based on the derivative formulas along the direction ν in Ref. [2], p. 361

$$\frac{\partial R_m^+}{\partial \nu}(r,\theta) = \frac{\partial R_m^+}{\partial r} \cos(\nu, r) + \frac{1}{r} \frac{\partial R_m^+}{\partial \theta} \cos(\nu, \theta),$$

$$\frac{dJ_i(kr)}{dr} = \frac{i}{r} J_i(kr) - kJ_{i+1}(kr), \tag{10.3.36}$$

we can obtain from eqns. (10.3.35) and (10.3.36) similarly,

$$\left|\frac{\partial R_m^+}{\partial \nu}\right|_{0,\partial\Omega^+} \leq C \max_{r\in\partial\Omega^+} \left\{ \sum_{i=m+1}^{\infty} \frac{i}{r} |J_i(kr)| + k \sum_{i=m+2}^{\infty} |J_i(kr)| \right\} \leq C(m+1) \cdot \frac{1}{(m+1)^{\frac{1}{2}}} \cdot \alpha_+^{m+1}.$$

Bounds for the terms $|R_n^-|_{0,\partial\Omega^-}$ and $\left|\frac{\partial R_n^-}{\partial \nu}\right|_{0,\partial\Omega^-}$ can be found similarly, and the desired results, i.e., eqn. (10.3.33) are derived. ■

Corollary 10.3.1

Let all the conditions in Theorems 10.3.4 and 10.3.5 hold. Then

$$\|u - u_{m,n}\|_H \leq C \frac{N^2}{\delta} \left[\frac{\alpha_+^{m+1}}{(m+1)^{\frac{1}{2}}} + \frac{\alpha_-^{n+1}}{(n+1)^{\frac{1}{2}}} \right].$$

The condition, i.e., eqn. (10.3.32) is important for choosing the term numbers m, n for exponential convergence rates. Since

$$r_{\max}^+, \rho_{\max}^- \leq r_{\max} = \max_{\Omega} r,$$

using piecewise particular solutions may give better accuracy of numerical solutions, or reduce numbers m and n. Fewer terms of expansions are desirable for reducing condition numbers in TM. Hence, better approximation of solutions may be obtained if the solution domain Ω is spilt into suitably smaller subdomains Ω_i.

10.3.4 Estimates on bounds of constant $K_{m,n}$

Since the bound of constant $K_{u,w} = \max\{K_u, K_{m,n}\}$ is important to Theorems 10.3.1–10.3.4, we will focus on analyzing on the bound of $K_{m,n}$; the analysis on those of K_u in eqn. (10.3.3) for harmonic functions can be easily found (also see Chapter 1). In this subsection, we estimate bounds of K_m for u_m and $u - u_m$ only on a circular domain Ω^+, and proofs for sectorial domains are similar (see Refs. [280, 306]).

Lemma 10.3.4

Let $u \in H_{2,\Omega^+}$ be the Helmholtz solution to eqn. (10.2.1) in Ω^+ and satisfy

$$\|u\|_{\frac{3}{2},\Gamma_0} \leq C_0 \|u\|_{\frac{1}{2},\Gamma_0} \tag{10.3.37}$$

with a constant C_0. Then there exists a bounded constant C independent of k and u such that

$$\left\| \frac{\partial u}{\partial \nu} \right\|_{0,\Gamma_0} \leq C\sqrt{C_0}\|u\|_{1,\Omega^+}. \tag{10.3.38}$$

Proof.

Since $\Delta + k^2 I$ is an elliptic operator, based on the trace theorem and the interpolation theory in Babuska and Aziz [15] (p. 25 and p. 32), we have the bounds

$$\left\| \frac{\partial u}{\partial \nu} \right\|_{0,\Gamma_0} \leq C\|u\|_{\frac{1}{2},\Omega^+} \leq C\{\|u\|_{1,\Omega^+}\|u\|_{2,\Omega^+}\}^{\frac{1}{2}}. \tag{10.3.39}$$

Next based on eqn. (10.3.37) and the regularity theorem in Oden and Reddy [348], p. 176,

$$\|u\|_{2,\Omega^+} \leq C\{\|\Delta u + k^2 u\|_{0,\Omega^+} + \|u\|_{\frac{3}{2},\Gamma_0} + \|u\|_{0,\Omega^+}\} \leq C\{C_0\|u\|_{\frac{1}{2},\Gamma_0} + \|u\|_{0,\Omega^+}\}. \tag{10.3.40}$$

Also from the trace theorem

$$\|u\|_{\frac{1}{2},\Gamma_0} \leq C\|u\|_{1,\Omega^+},$$

the eqn. (10.3.40) leads to

$$\|u\|_{2,\Omega^+} \leq CC_0(\|u\|_{\frac{1}{2},\Gamma_0} + \|u\|_{0,\Omega^+}) \leq CC_0\|u\|_{1,\Omega^+}. \tag{10.3.41}$$

The desired results, i.e., eqn. (10.3.38) follow from eqns. (10.3.39) and (10.3.41). ■

Lemma 10.3.5

Let Ω^+ be a circular domain with the boundary $\Gamma_0 = \{(r,\theta) \mid r \text{ is given}, \theta \in (0,2\pi)\}$. Suppose that the admissible functions are chosen as

$$u_m = a_0 J_0(kr) + \sum_{i=1}^{m} J_i(kr)[a_i \cos i\theta + b_i \sin i\theta] \quad in\ \Omega^+,$$

where a_i and b_i are arbitrary constants. Then there exists a constant C independent of m and k such that

$$\|u_m\|_{\frac{3}{2},\Gamma_0} \le Cm\|u_m\|_{\frac{1}{2},\Gamma_0} \tag{10.3.42}$$

Proof.

Based on the interpolation theory again (see Babuska and Aziz [15], p. 25), for any integer $l \ge 2$ there exists the bound,

$$\|u\|_{\frac{3}{2},\Gamma_0} \le C\|u\|_{0,\Gamma_0}^{\frac{l-\frac{3}{2}}{l-\frac{1}{2}}} \|u\|_{l,\Gamma_0}^{\frac{1}{l-\frac{1}{2}}}, \tag{10.3.43}$$

where the norms

$$\|u\|^2_{l,\Gamma_0} = \sum_{j=0}^{l} \left\| \frac{\partial^j u}{\partial s^j} \right\|^2_{0,\Gamma_0}. \tag{10.3.44}$$

By applying the orthogonality of trigonometric functions, we obtain

$$\begin{aligned}
\left\| \frac{\partial^j u_m}{\partial s^j} \right\|^2_{0,\Gamma_0} = \left\| \frac{\partial^j u_m}{r^j \partial \theta^j} \right\|^2_{0,\Gamma_0} &= \pi r \sum_{i=1}^{m} \frac{i^{2j}}{r^{2j}} (a_i^2 + b_i^2) J_i^2(kr) \\
&\le \left(\frac{m}{r}\right)^{2j} \left\{ 2\pi r a_0^2 J_0^2(kr) + \pi r \sum_{i=1}^{m} (a_i^2 + b_i^2) J_i^2(kr) \right\} \\
&\le \left(\frac{m}{r}\right)^{2j} \|u_m\|^2_{0,\Gamma_0}.
\end{aligned} \tag{10.3.45}$$

Since $\frac{m}{r} < 1$ due to large m and $\|u_m\|_{0,\Gamma_0} \le C\|u_m\|_{\frac{1}{2},\Gamma_0}$, we obtain from eqns. (10.3.44) and (10.3.45)

$$\begin{aligned}
\|u_m\|^2_{l,\Gamma_0} &= \sum_{j=0}^{l} \left\| \frac{\partial^j u_m}{\partial s^j} \right\|^2_{0,\Gamma_0} = \left\{ \sum_{j=0}^{l} \left(\frac{m}{r}\right)^{2j} \right\} \|u_m\|^2_{0,\Gamma_0} = \left| \frac{1 - \left(\frac{m}{r}\right)^{2(l+1)}}{1 - \frac{m}{r}} \right| \|u_m\|^2_{0,\Gamma_0} \\
&\le C\left(\frac{m}{r}\right)^{2(l+1)} \|u_m\|^2_{0,\Gamma_0} \le C\left(\frac{m}{r}\right)^{2(l+1)} \|u_m\|^2_{\frac{1}{2},\Gamma_0}.
\end{aligned}$$

Hence, the eqn. (10.3.43) leads to

$$\|u_m\|_{\frac{3}{2},\Gamma_0} \leq C\left(\frac{m}{r}\right)^{\frac{l+1}{l-\frac{1}{2}}} \|u_m\|_{\frac{1}{2},\Gamma_0}. \tag{10.3.46}$$

The desired result, i.e., eqn. (10.3.42) follows from eqn. (10.3.46) as $l \to \infty$. ■

Let the true solution

$$u = \overline{u}_m + R_m \quad \text{in } \Omega^+,$$

where

$$\overline{u}_m = \overline{a}_0 J_0(kr) + \sum_{i=1}^{m} J_i(kr)[\overline{a}_i \cos i\theta + \overline{b}_i \sin i\theta] \quad \text{in } \Omega^+,$$

and the remainder term

$$R_m = R_m^+ = \sum_{i=m+1}^{\infty} J_i(kr)[\overline{a}_i \cos i\theta + \overline{b}_i \sin i\theta] \quad \text{in } \Omega^+,$$

with the true coefficients $\overline{a}_i$ and $\overline{b}_i$. From Lemmas 10.3.4 and 10.3.5, we have the following proposition.

Proposition 10.3.1

Let all the conditions in Lemmas 10.3.4 and 10.3.5 hold. There exists a constant C independent of k and m such that

$$\left\|\frac{\partial u_m}{\partial \nu}\right\|_{0,\Gamma_0} = \left\|\frac{\partial u_m}{\partial r}\right\|_{0,\Gamma_0} \leq C\sqrt{m}\|u_m\|_{1,\Omega^+}.$$

From Proposition 10.3.1 we have $K_{m,n} = C(\max\{m,n\})^{\frac{1}{2}}$. Similarly, we can show $K_w = C(\max\{m,n\})^{\frac{1}{2}}$, to lead to $K_{u,w} = C(\max\{m,n\})^{\frac{1}{2}}$. Note that these bounds are sharp, compared to those in Chapter 1.

10.4 Summaries and discussions

To close this chapter, let us summarize the novelties of this chapter and give a few concluding remarks.

1. We have developed the TM using *piecewise* particular solutions for solving the Helmholtz equation $\Delta u + k^2 u = f$, and derived a new error analysis, which

reveals the error bounds, particularly at the degeneracy case, i.e., $k^2 \to \lambda_l$. Theorems 10.3.3–10.3.5 with the asymptote

$$\|u - u_{m,n}\|_H = O\left(\frac{1}{\delta}\right) \quad \text{as } k^2 \to \lambda_l$$

indicate that the Helmholtz solutions quickly deteriorate as $O\left(\frac{1}{\delta}\right)$ when $\delta \to 0$. The analysis given in this chapter fills up the analysis gap between the Helmholtz solutions at $\delta > 0$ and the eigenfunction solutions at $\delta = 0$.

2. In this chapter we choose $\sigma^{-1} = N = \max\{m, n\}$ and $N = O(k)$. Theorem 10.3.4 leads to

$$\frac{\|u - u_{m,n}\|_H}{|u - u_{m,n}|_B} = O\left(\frac{k^{\frac{3}{2}}}{\delta}\right) \quad \text{as } k \to \infty. \tag{10.4.1}$$

The significance of error ratios in eqn. (10.4.1) to practical application is that the Sobolev norms over the entire domain may be evaluated from the errors on the boundary.

3. Let us recall the solutions of the following algebraic equations

$$(\mathbf{A} - k^2)\mathbf{x} = \mathbf{b}, \qquad (\mathbf{A} - k^2)\widetilde{\mathbf{x}} = \widetilde{\mathbf{b}},$$

where $\mathbf{A}$ is symmetric and positive definite, k real, $\mathbf{x}$ and $\mathbf{b}$ are unknown and known vectors, respectively, and $\widetilde{\mathbf{x}}$ and $\widetilde{\mathbf{b}}$ their perturbation vectors. Denote the matrix eigenvalue problem $\mathbf{A}\mathbf{x}_l = \lambda_l \mathbf{x}_l$ with positive eigenvalues λ_l and the normalized eigenvectors $\mathbf{x}_l$. By following the arguments in Section 10.1, we can obtain easily

$$\|\mathbf{x} - \widetilde{\mathbf{x}}\| \le \frac{\|\mathbf{b} - \widetilde{\mathbf{b}}\|}{\min_l |\lambda_l - k^2|} \le \frac{k^2}{\delta}\|\mathbf{b} - \widetilde{\mathbf{b}}\|,$$

where $\|\cdot\|$ is the Euclidean norm. The error asymptote $O\left(\frac{k^2}{\delta}\right)$ is similar to $O\left(\frac{k^{\frac{3}{2}}}{\delta}\right)$. However, the eqn. (10.4.1) includes not only the solution errors but also the derivative errors on the entire solution domain Ω.

4. The advantage of the TM is high accuracy of the Helmholtz solutions with exponential convergence rates, which are proven by the analysis in Section 10.3.3.
5. This chapter may be regarded as a development of Chapter 1, where the TM is used for the Laplace equation. On the other hand, when $k \to 0$, the piecewise Laplace eqn. (10.3.1) is reduced from eqn. (10.3.13). Since the factor $\frac{k}{\delta} = \max_i \left\{\frac{k^3}{|k^2 - \lambda_i|}\right\} \to 0$ as $k \to 0$, all the error bounds in Chapter 1 can also be deduced from Theorems 10.3.2–10.3.5.

11 Explicit harmonic solutions of Laplace's equation

In this chapter, the harmonic functions of Laplace's equations on sectors are derived explicitly for the Dirichlet and the Neumann boundary conditions. These harmonic functions are more explicit than those of Volkov [452], and easier to expose the mild singularity of the Laplace solutions at the domain corners. Moreover, the particular solutions of Poisson's equation on the polygon are provided. We also explore in detail the singularities of the solutions on sectors with the boundary angles $\Theta = \frac{\pi}{2}, \frac{3\pi}{2}, \pi$, and 2π, which often occur in many testing models. In addition to the popular singularity models, Motz's and its variants in Chapter 2, we design two new singularity models, one with discontinuous singularity, and the other with crack plus mild singularities. The collocation Trefftz method (CTM), the Schwarz alternating method (SAM), and their combinations may be chosen to seek their solutions with high accuracy, which may be used as the exact solution to test other numerical methods. The explicit harmonic solutions of the Laplace's equations and their singularities are crucial to algorithms and error analysis. In this chapter, the harmonic solutions are called for Laplace's solutions based on the technique of separation of variables, to distinguish from the fundamental functions, and the particular solutions are referred to Poisson's solutions in Section 11.4.1 as in Ref. [85].

11.1 Introduction

When a solution domain can be split into several subdomains, the local harmonic solutions in each subdomain may be found in this chapter. Several methods for Laplace's equations with highly accurate solutions may then be chosen: (1) the CTM, (2) the combinations of the CTM and the SAM, and (3) other methods such as the block method in Volkov [451, 452], Volvok and Kornoukhov [453], and Dosiyev [130]. To reach high accuracy of numerical solutions, care must be taken in the numerical methods for not only the angular singularities but also the mild singularities such as $O(r^k \ln r)$ $(k = 1, 2, \dots)$ if existing. The p-version of FEM is studied for mild singularities in Babuska and Guo [16]. This chapter is devoted to the CTM dealing with both the angular and the mild singularities.

This chapter is organized as follows. In the next section, the harmonic solutions are derived for Laplace's equations in sectors with the Dirichlet boundary conditions, and their explicit formulas are provided for special angles Θ. In Section 11.3, the harmonic solutions of those involving the Neumann boundary conditions are discussed. In Section 11.4, the particular solutions of Poisson's equation are provided, and those are developed for the cases that the Dirichlet and Neumann boundary conditions are not smooth. Especially, the singularities of the solutions on the boundary angles $\Theta = \frac{\pi}{2}, \frac{3\pi}{2}, \pi$, and 2π are analyzed. In Section 11.5, two new models with discontinuity, and with angular plus mild singularities are designed, and the collocation TM are used to provide the highly accurate solutions. The materials of this chapter are adapted from Li et al. [301].

11.2 Harmonic functions

Consider the Laplace equation with the Dirichlet boundary conditions, see fig. 11.1,

$$\Delta u = \frac{\partial^2 u}{\partial x^2} + \frac{\partial^2 u}{\partial y^2} = 0 \quad \text{in } S^*,$$

$$u = g \quad \text{on } \partial S^*,$$

where S^* is a polygon.

For each angle, we seek the harmonic solutions in a sectorial domain, see fig. 11.2, $S = \{(r, \theta) \,|\, 0 < r < R, 0 < \theta < \Theta\}$,

$$\Delta u = 0 \quad \text{in } S.$$

We suppose that the function g is highly smooth that it can be expressed by the power functions

$$u\,|_{\overline{OA}} = g\,|_{\overline{OA}} = \sum_{i=0}^{\infty} \alpha_i r^i, \quad 0 \le r \le R, \ \theta = 0, \tag{11.2.1}$$

$$u\,|_{\overline{OB}} = g\,|_{\overline{OB}} = \sum_{i=0}^{\infty} \beta_i r^i, \quad 0 \le r \le R, \ \theta = \Theta, \tag{11.2.2}$$

where β_i and α_i are known coefficients. In fact, when the function $g\,|_{\overline{OB}} = g_1(r)$ is highly smooth, it can be expanded by Taylor's series

$$g_1(r) = \sum_{i=0}^{\infty} \frac{g_1^{(i)}(0) r^i}{i!}.$$

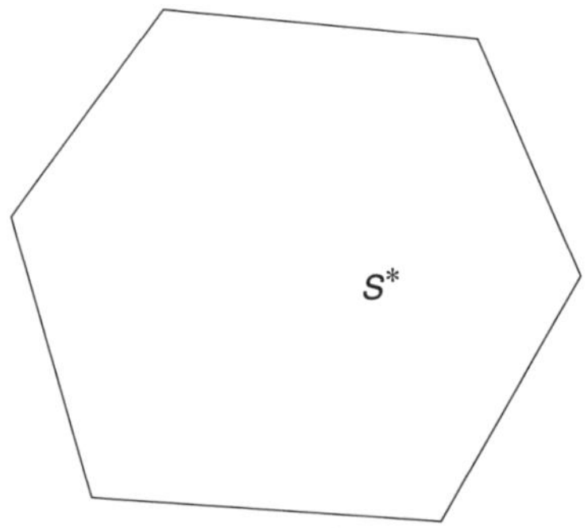

Figure 11.1: A polygonal domain.

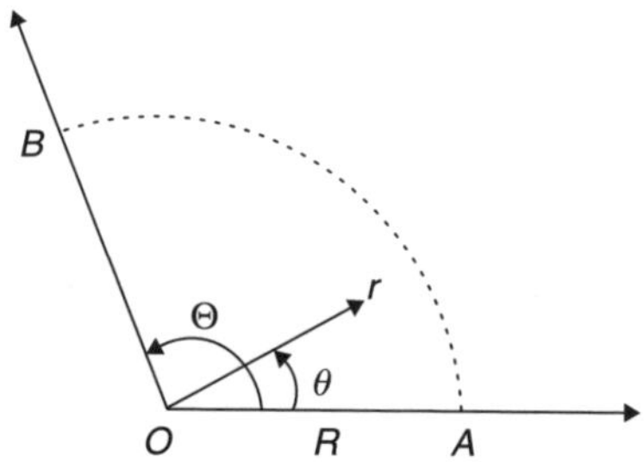

Figure 11.2: A sectorial domain.

Then, $\beta_i = g_1^{(i)}(0)/i!$. Similarly, for $g\,|_{\overline{OA}} = g_0(r)$, we have

$$g_0(r) = \sum_{i=0}^{\infty} \frac{g_0^{(i)}(0)r^i}{i!}.$$

Hence, for any smooth Dirichlet boundary condition $u = g$ on ∂S, we may simply consider the following case in S, see fig. 11.3. In this chapter, we assume that the corresponding series are also convergent. Otherwise, we may consider only

$$u\,|_{\overline{OA}} = g\,|_{\overline{OA}} = \sum_{i=0}^{M} \alpha_i r^i, \quad 0 \le r \le R, \;\; \theta = 0,$$

$$u\,|_{\overline{OB}} = g\,|_{\overline{OB}} = \sum_{i=0}^{N} \beta_i r^i, \quad 0 \le r \le R, \;\; \theta = \Theta,$$

where β_i and α_i are known coefficients. This is the simple case of eqns. (11.2.1) and (11.2.2) because we may let $\beta_i = 0$ as $i > N$ and $\alpha_i = 0$ as $i > M$.

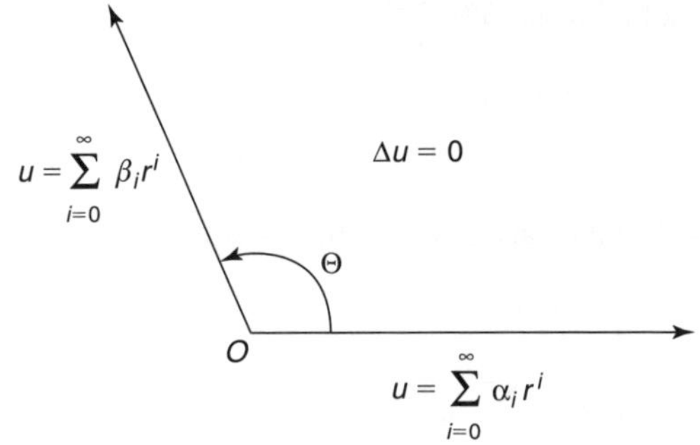

Figure 11.3: The Dirichlet boundary conditions.

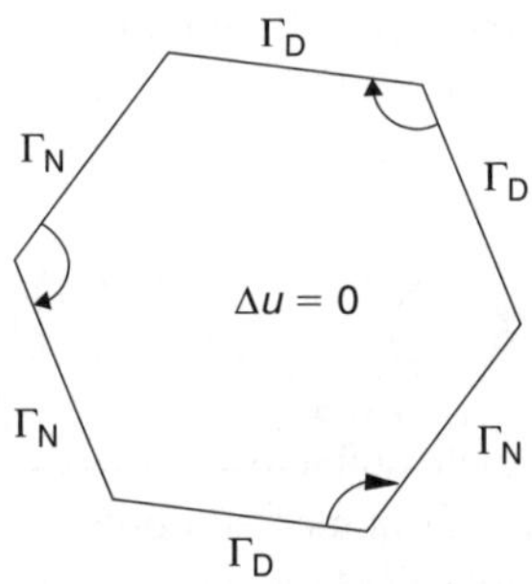

Figure 11.4: A polygon.

Let us consider the mixed Dirichlet–Neumann boundary conditions in fig. 11.4:

$$\Delta u = 0 \quad \text{in } S,$$

$$u = g_D \quad \text{on } \Gamma_D, \qquad \frac{\partial u}{\partial n} = g_N \quad \text{on } \Gamma_N,$$

where $\partial S = \Gamma_D \cup \Gamma_N$ and n is the outward normal of ∂S. There are four types of mixed boundary conditions on two adjacent edges of a corner: (1) the D–D type, (2) the N–D type, (3) the D–N type, and (4) the N–N type. The harmonic solutions of D–D type will be derived in this section, and those of the N–D, D–N, and N–N types in the next section.

11.2.1 General cases

The general solutions of Laplace's equation satisfying eqns. (11.2.1) and (11.2.2) can be split into $\overline{u}$ and u_g

$$u = \overline{u} + u_g,$$

where the general solutions u_g satisfy

$$\triangle u_g = 0 \quad \text{in } S, \qquad u_g = 0, \qquad \theta = 0, \qquad \text{and} \qquad \theta = \Theta, \tag{11.2.3}$$

where $0 < \Theta \leq 2\pi$, and the harmonic solution $\overline{u}$ satisfies

$$\triangle \overline{u} = 0 \quad \text{in } S, \tag{11.2.4}$$

$$\overline{u}\,|_{\theta=0} = \sum_{i=0}^{\infty} \alpha_i r^i, \qquad \overline{u}\,|_{\theta=\Theta} = \sum_{i=0}^{\infty} \beta_i r^i. \tag{11.2.5}$$

Note that in eqn. (11.2.3), the boundary condition on ℓ_k $(r = R, 0 \leq \theta \leq \Theta)$ has not been given yet. Hence, the general solutions u_g are not unique.

First for eqn. (11.2.3), the general solutions are $\phi_i = r^{\sigma_i} \sin \sigma_i \theta$, such that $\phi_i\,|_{\theta=0} = 0$ holds automatically, and $\phi_i\,|_{\theta=\Theta} = 0$ leads to $\sin \sigma_i \Theta = 0$. Hence, $\sigma_i \Theta = i\pi$, i.e., $\sigma_i = \frac{i\pi}{\Theta}$. We obtain the general solutions

$$u_g = \sum_{i=0}^{\infty} c_i r^{\frac{i\pi}{\Theta}} \sin\left(\frac{i\pi}{\Theta}\theta\right),$$

where c_i are the coefficients to be defined.

Second, we seek the harmonic solutions involving mild singularity $r^k \ln r,\ k = 1, 2, \ldots$. These are called the *mild singularity* in this chapter, to compare with the rather strong angular singularity $O(r^\gamma),\ 0 < \gamma < 1$. Let $z = x + iy = re^{i\theta}$. The real and imaginary parts of the complex functions, $z^p \ln z$, are harmonic, where p is real. We then have

$$\begin{aligned} z^p \ln z &= (r^p \cos p\theta + i r^p \sin p\theta) \times (\ln r + i\theta) \\ &= r^p\{\ln r \cos p\theta - \theta \sin p\theta\} + i r^p\{\ln r \sin p\theta + \theta \cos p\theta\}. \end{aligned}$$

Hence, the following functions are also harmonic

$$\begin{aligned} \varphi_p &= \varphi_p(r,\theta) = r^p\{\ln r \sin p\theta + \theta \cos p\theta\}, \\ \psi_p &= \psi_p(r,\theta) = r^p\{\ln r \cos p\theta - \theta \sin p\theta\}, \end{aligned} \tag{11.2.6}$$

where p is real. When p is positive integer, $p = k,\ k = 1, 2, \ldots$, we denote

$$\varphi_k = \varphi_k(r,\theta) = r^k\{\ln r \sin k\theta + \theta \cos k\theta\}, \tag{11.2.7}$$

$$\psi_k = \psi_k(r,\theta) = r^k\{\ln r \cos k\theta - \theta \sin k\theta\}. \tag{11.2.8}$$

Define the functions

$$\Phi_i = \Phi_i(r,\theta) = \begin{cases} \frac{r^i \sin i\theta}{\sin i\Theta}, & \text{if } i\Theta \neq k\pi,\ k = 1, 2, \ldots \\ \frac{(-1)^k}{\Theta}\varphi_i(r,\theta), & \text{if } i\Theta = k\pi \quad \text{for some } k, \end{cases} \tag{11.2.9}$$

where $i \geq 1$ and $\varphi_i(r,\theta)$ is given in eqn. (11.2.7). Hence,

$$\Phi_i \,|_{\theta=0} = 0, \qquad \Phi_i \,|_{\theta=\Theta} = r^i, \quad \forall r > 0, \; i = 1, 2, \ldots$$

Choose the harmonic solutions for eqns. (11.2.4) and (11.2.5) as the following form

$$\bar{u} = \sum_{i=0}^{\infty} A_i r^i \cos i\theta + B_0\theta + \sum_{i=1}^{\infty} B_i \Phi_i(r,\theta), \tag{11.2.10}$$

where A_i and B_i are the coefficients to be determined below. When $\theta = 0$ we have $A_i = \alpha_i$ from the boundary condition in eqn. (11.2.5), and when $\theta = \Theta$,

$$A_0 + B_0\Theta = \beta_0, \qquad A_i \cos i\Theta + B_i = \beta_i, \quad i = 1, 2, \ldots$$

Hence, we obtain

$$B_0 = \frac{\beta_0 - \alpha_0}{\Theta}, \qquad B_i = \beta_i - \alpha_i \cos i\Theta, \quad i = 1, 2, \ldots \tag{11.2.11}$$

Substituting eqn. (11.2.11) into eqn. (11.2.10) gives the harmonic solutions

$$\bar{u} = \frac{\beta_0 - \alpha_0}{\Theta}\theta + \sum_{i=0}^{\infty} \alpha_i r^i \cos i\theta + \sum_{i=1}^{\infty} (\beta_i - \alpha_i \cos i\Theta)\Phi_i(r,\theta). \tag{11.2.12}$$

In Volkov [452], the following forms of harmonic solutions for eqns. (11.2.4) and (11.2.5) are given

$$\bar{u} = \alpha_0 + \frac{\beta_0 - \alpha_0}{\Theta}\theta + \sum_{i=1}^{\infty} \alpha_i \overline{\Phi}_i(r,\theta) + \sum_{i=1}^{\infty} \beta_i \Phi_i(r,\theta), \tag{11.2.13}$$

where

$$\overline{\Phi}_i = \overline{\Phi}_i(r,\theta) = \Phi_i(r, \Theta - \theta), \tag{11.2.14}$$

to satisfy

$$\overline{\Phi}_i \,|_{\theta=\Theta} = 0, \qquad \overline{\Phi}_i \,|_{\theta=0} = r^i, \quad \forall r > 0, \; i = 1, 2, \ldots$$

First, let us show the equivalence between eqns. (11.2.12) and (11.2.13). When $\Theta \neq \frac{k\pi}{i}$ for $i, k \geq 1$, we have from eqn. (11.2.13),

$$\bar{u} = \alpha_0 + \frac{\beta_0 - \alpha_0}{\Theta}\theta + \sum_{i=1}^{\infty} \alpha_i r^i \frac{\sin i(\Theta - \theta)}{\sin i\Theta} + \sum_{i=1}^{\infty} \beta_i \frac{r^i \sin i\theta}{\sin i\Theta}.$$

Since $\sin i(\Theta-\theta) = \sin i\Theta \cos i\theta - \sin i\theta \cos i\Theta$, we obtain from the above equation

$$\overline{u} = \frac{\beta_0 - \alpha_0}{\Theta}\theta + \sum_{i=0}^{\infty} \alpha_i r^i \cos i\theta + \sum_{i=1}^{\infty} (\beta_i - \alpha_i \cos i\Theta) r^i \frac{\sin i\theta}{\sin i\Theta}.$$

This is the eqn. (11.2.12) for $\Theta \neq \frac{k\pi}{i}$.

Second, when $i\Theta = k\pi$ for some i and k, to confirm the equivalence between eqns. (11.2.12) and (11.2.13), it suffices to show

$$\overline{\Phi}_i = r^i \cos i\theta - \cos i\Theta\, \Phi_i. \tag{11.2.15}$$

In fact, we have from eqns. (11.2.14) and (11.2.9)

$$\begin{aligned}\overline{\Phi}_i(r,\theta) &= \Phi_i(r, \Theta-\theta)\\ &= r^i \frac{(-1)^k}{\Theta}[\ln r \sin i(\Theta-\theta) + (\Theta-\theta)\cos i(\Theta-\theta)].\end{aligned} \tag{11.2.16}$$

When $i\Theta = k\pi$, there exist the equalities

$$\sin i(\Theta-\theta) = \sin i\Theta \cos i\theta - \cos i\Theta \sin i\theta = (-1)^{k+1} \sin i\theta, \tag{11.2.17}$$

$$\cos i(\Theta-\theta) = \cos i\Theta \cos i\theta + \sin i\Theta \sin i\theta = (-1)^k \cos i\theta. \tag{11.2.18}$$

Substituting eqns. (11.2.17) and (11.2.18) into eqn. (11.2.16) gives the desired result, i.e., eqn. (11.2.15).

Third, Volkov in Ref. [452] also considers the case of $i\Theta \neq k\pi$ but $i\Theta \approx k\pi$ so that the ratio $\frac{\sin i\theta}{\sin i\Theta}$ becomes very large. The basis functions are also introduced in Ref. [452] for the case that $0 < |\sin i\Theta| < \frac{1}{2}$. This is interesting for theory but not for application. In practical engineering problems, usually we may assume $\Theta = \frac{K}{L}\pi, 0 < K \leq 2L$, and integers L and K are relatively prime. Then, we have

$$\min_i |\sin i\Theta| = \min_i \left|\sin \frac{iK}{L}\pi\right| = \left|\sin \frac{\pi}{L}\right| \approx \frac{\pi}{L}.$$

Hence, the ratio

$$\max_i \left|\frac{\sin i\theta}{\sin i\Theta}\right| \leq \frac{L}{\pi},$$

will not be very large. Consequently, we omit the case of $i\Theta \approx k\pi$ in this chapter.

Fourth, let us compare the formulas of the functions, i.e., eqns. (11.2.12), and (11.2.13) of Volkov [452]. The eqn. (11.2.13) has a symmetric form with respect to $\theta = \frac{\Theta}{2}$. In contrast, we may rewrite eqn. (11.2.12) as

$$\overline{u} = \alpha_0 + \frac{\beta_0 - \alpha_0}{\Theta}\theta + \sum_{i=1}^{\infty} \alpha_i [r^i \cos i\theta - \cos i\Theta \Phi_i(r,\theta)] + \sum_{i=1}^{\infty} \beta_i \Phi_i(r,\theta), \qquad (11.2.19)$$

which is not exactly symmetric, indeed. From the viewpoint of computation, both eqns. (11.2.12) and (11.2.13) are effective. However, the eqns. (11.2.12) and (11.2.19) are more explicit. In particular, the eqn. (11.2.12) displays explicitly the mild singularity. For instance, when $i\Theta = k\pi$ and $\beta_i \neq \alpha_i \cos i\Theta$, there does exist a mild singularity of $O(r^i \ln r)$. More exploration on the mild singularity is provided in Section 11.4.

Remark 11.2.1

It is assumed that the series in eqn. (11.2.19) are convergent. Otherwise, we may consider the finite terms in eqn. (11.2.5),

$$\overline{u}\,|_{\theta=0} = \sum_{i=0}^{M} \alpha_i r^i, \qquad \overline{u}\,|_{\theta=\Theta} = \sum_{i=0}^{N} \beta_i r^i.$$

Hence, the solutions with finite terms

$$\overline{u} = \alpha_0 + \frac{\beta_0 - \alpha_0}{\Theta}\theta + \sum_{i=1}^{M} \alpha_i [r^i \cos i\theta - \cos i\Theta \Phi_i(r,\theta)] + \sum_{i=1}^{N} \beta_i \Phi_i(r,\theta)$$

are obtained to replace eqn. (11.2.19). For all infinite series given below, it is always assumed that they are convergent. Otherwise, a suitable modification should be made correspondingly.

11.2.2 Formulas for special Θ

Based on eqns. (11.2.12) and (11.2.9), we list the harmonic solutions which are often used in application.

1. When $\Theta = \pi$,

$$\overline{u} = \frac{\beta_0 - \alpha_0}{\pi}\theta + \sum_{i=0}^{\infty} \alpha_i r^i \cos i\theta + \sum_{i=1}^{\infty} \frac{(-1)^i \beta_i - \alpha_i}{\pi} \varphi_i(r,\theta). \qquad (11.2.20)$$

2. When $\Theta = 2\pi$,

$$\overline{u} = \frac{\beta_0 - \alpha_0}{2\pi}\theta + \sum_{i=0}^{\infty} \alpha_i r^i \cos i\theta + \sum_{i=1}^{\infty} \frac{\beta_i - \alpha_i}{2\pi} \varphi_i(r,\theta). \qquad (11.2.21)$$

3. When $\Theta = \frac{\pi}{2}$,

$$\overline{u} = \frac{2(\beta_0 - \alpha_0)}{\pi}\theta + \sum_{i=0}^{\infty} \alpha_i r^i \cos i\theta + \sum_{j=0}^{\infty} (-1)^j \beta_{2j+1} r^{2j+1} \sin(2j+1)\theta$$

$$+ \sum_{j=1}^{\infty} \frac{2}{\pi}[(-1)^j \beta_{2j} - \alpha_{2j}]\varphi_{2j}(r,\theta). \tag{11.2.22}$$

4. When $\Theta = \frac{3\pi}{2}$,

$$\overline{u} = \frac{2(\beta_0 - \alpha_0)}{3\pi}\theta + \sum_{i=0}^{\infty} \alpha_i r^i \cos i\theta + \sum_{j=0}^{\infty} (-1)^{j+1} \beta_{2j+1} r^{2j+1} \sin(2j+1)\theta$$

$$+ \sum_{j=1}^{\infty} \frac{2}{3\pi}[(-1)^j \beta_{2j} - \alpha_{2j}]\varphi_{2j}(r,\theta). \tag{11.2.23}$$

Note that the formulas of the particular solutions for $\Theta = \pi$ and $\Theta = 2\pi$ are very similar, and so are those for $\Theta = \frac{\pi}{2}$ and $\Theta = \frac{3\pi}{2}$. Except the different angles Θ, the only difference is that the sign $(-1)^j$ may change in the series of β_j.

5. When $\Theta = \frac{\pi}{3}, \frac{2\pi}{3}, \frac{4\pi}{3}, \frac{5\pi}{3}$,

$$\begin{aligned}\overline{u} &= \frac{\beta_0 - \alpha_0}{\Theta}\theta + \sum_{i=0}^{\infty} \alpha_i r^i \cos i\theta \\ &+ \sum_{j=0}^{\infty} \frac{\beta_{3j+1} - \alpha_{3j+1}\cos(3j+1)\Theta}{\sin(3j+1)\Theta} r^{3j+1} \sin(3j+1)\theta \\ &+ \sum_{j=0}^{\infty} \frac{\beta_{3j+2} - \alpha_{3j+2}\cos(3j+2)\Theta}{\sin(3j+2)\Theta} r^{3j+2} \sin(3j+2)\theta \\ &+ \sum_{j=1}^{\infty} \frac{1}{\Theta}(\beta_{3j}\cos 3j\Theta - \alpha_{3j})\varphi_{3j}(r,\theta).\end{aligned}$$

6. When $\Theta = \frac{\pi}{4}, \frac{3\pi}{4}, \frac{5\pi}{4}, \frac{7\pi}{4}$,

$$\begin{aligned}\overline{u} &= \frac{\beta_0 - \alpha_0}{\Theta}\theta + \sum_{i=0}^{\infty} \alpha_i r^i \cos i\theta \\ &+ \sum_{j=0}^{\infty}\sum_{k=1}^{3} \frac{\beta_{4j+k} - \alpha_{4j+k}\cos(4j+k)\Theta}{\sin(4j+k)\Theta} r^{4j+k} \sin(4j+k)\theta \\ &+ \sum_{j=1}^{\infty} \frac{1}{\Theta}(\beta_{4j}\cos 4j\Theta - \alpha_{4j})\varphi_{4j}(r,\theta).\end{aligned}$$

7. When $\Theta = \frac{K}{L}\pi$, $0 < K < 2L$, and integers K and L are relatively prime,

$$\overline{u} = \frac{\beta_0 - \alpha_0}{\Theta}\theta + \sum_{i=0}^{\infty} \alpha_i r^i \cos i\theta$$
$$+ \sum_{j=0}^{\infty}\sum_{k=1}^{L-1} \frac{\beta_{Lj+k} - \alpha_{Lj+k}\cos(Lj+k)\Theta}{\sin(Lj+k)\Theta} r^{Lj+k} \sin(Lj+k)\theta$$
$$+ \sum_{j=1}^{\infty} \frac{1}{\Theta}(\beta_{Lj}\cos Lj\Theta - \alpha_{Lj})\varphi_{Lj}(r,\theta).$$

11.3 Harmonic solutions involving Neumann conditions

11.3.1 The case of N–D type

Consider

$$\Delta u = 0 \quad \text{in } S,$$
$$\left.\frac{\partial u}{\partial n}\right|_{\theta=0} = \sum_{i=0}^{\infty} \alpha_i r^i, \qquad u\,|_{\theta=\Theta} = \sum_{i=0}^{\infty} \beta_i r^i.$$

Let $u = \overline{u} + u_g$, where the general solutions are given by

$$u_g = \sum_{k=0}^{\infty} d_k r^{\sigma_k} \cos(\sigma_k\theta),$$

when $\sigma_k = \left(k + \frac{1}{2}\right)\pi/\Theta$. The harmonic solution $\overline{u}$ satisfies

$$\Delta\overline{u} = 0 \quad \text{in } S, \tag{11.3.1}$$
$$\left.\frac{\partial \overline{u}}{\partial n}\right|_{\theta=0} = \sum_{i=0}^{\infty} \alpha_i r^i, \qquad \overline{u}\,|_{\theta=\Theta} = \sum_{i=0}^{\infty} \beta_i r^i.$$

Define the functions

$$\Psi_i = \Psi_i(r,\theta) = \begin{cases} \frac{r^i \cos i\theta}{\cos i\Theta}, & \text{if } i\Theta \neq \left(k+\frac{1}{2}\right)\pi, \ k = 0,1,\ldots \\ \frac{(-1)^{k+1}}{\Theta}\psi_i(r,\theta), & \text{if } i\Theta = \left(k+\frac{1}{2}\right)\pi \quad \text{for some } k, \end{cases}$$

where $\psi_i(r,\theta)$ is given in eqn. (11.2.8). Hence,

$$\left.\frac{\partial \Psi_i}{\partial n}\right|_{\theta=0} = -\left.\frac{\partial \Psi_i}{r\,\partial\theta}\right|_{\theta=0} = 0, \qquad \Psi_i\,|_{\theta=\Theta} = r^i, \quad \forall r > 0, \ i = 1,2,\ldots$$

Choose the harmonic solutions of eqn. (11.3.1)

$$\overline{u} = \sum_{i=1}^{\infty} A_i r^i \sin i\theta + B_0 + \sum_{i=1}^{\infty} B_i \Psi_i(r,\theta), \tag{11.3.2}$$

where A_i and B_i are the coefficients. When $\theta = 0$, we have

$$\sum_{i=0}^{\infty} \alpha_i r^i = \left.\frac{\partial \overline{u}}{\partial n}\right|_{\theta=0} = -\sum_{i=1}^{\infty} A_i i r^{i-1}.$$

Then

$$A_i = -\frac{\alpha_{i-1}}{i}, \quad i = 1, 2, \ldots \tag{11.3.3}$$

Also when $\theta = \Theta$,

$$\sum_{i=1}^{\infty} \beta_i r^i = \overline{u}\,|_{\theta=\Theta} = \sum_{i=1}^{\infty} A_i r^i \sin i\Theta + B_0 + \sum_{i=1}^{\infty} B_i r^i.$$

This gives

$$\begin{aligned} B_0 &= \beta_0, \\ B_i &= \beta_i - A_i \sin i\Theta = \beta_i + \frac{\alpha_{i-1}}{i} \sin i\Theta, \quad i = 1, 2, \ldots \end{aligned} \tag{11.3.4}$$

Hence, from eqns. (11.3.2)–(11.3.4) we obtain the harmonic solutions

$$\overline{u} = -\sum_{i=1}^{\infty} \frac{\alpha_{i-1}}{i} r^i \sin i\theta + \beta_0 + \sum_{i=1}^{\infty} \left(\beta_i + \frac{\alpha_{i-1}}{i} \sin i\Theta\right) \Psi_i(r,\theta). \tag{11.3.5}$$

The eqn. (11.3.5) is explicit for computation, in particular for directly displaying the existence of mild singularity when $i\Theta = \left(k + \frac{1}{2}\right)\pi$ and $\beta_i + \frac{\alpha_{i-1}}{i} \sin i\Theta \neq 0$.

Below, we also list the useful formulas of the harmonic solutions for some special Θ from eqn. (11.3.5).

1. When $\Theta = \frac{\pi}{2}$,

$$\overline{u} = -\sum_{i=1}^{\infty} \frac{\alpha_{i-1}}{i} r^i \sin i\theta + \sum_{j=0}^{\infty} \frac{2}{\pi} \left[(-1)^{j+1} \beta_{2j+1} - \frac{\alpha_{2j}}{2j+1}\right] \psi_{2j+1}(r,\theta)$$
$$+ \sum_{j=0}^{\infty} (-1)^j \beta_{2j} r^{2j} \cos 2j\theta. \tag{11.3.6}$$

2. When $\Theta = \frac{3\pi}{2}$,

$$\bar{u} = -\sum_{i=1}^{\infty} \frac{\alpha_{i-1}}{i} r^i \sin i\theta + \sum_{j=0}^{\infty} \frac{2}{3\pi} \left[(-1)^j \beta_{2j+1} - \frac{\alpha_{2j}}{2j+1} \right] \psi_{2j+1}(r,\theta)$$
$$+ \sum_{j=0}^{\infty} (-1)^j \beta_{2j} r^{2j} \cos 2j\theta. \tag{11.3.7}$$

In Volkov [452], there are the different but equivalent formulas of eqn. (11.3.5). The discussions and comparisons between eqn. (11.3.5) and Volkov's are similar as the above.

11.3.2 The case of D–N type

Now, we consider the mixed D–N type

$$\Delta u = 0 \quad \text{in } S,$$
$$u \mid_{\theta=0} = \sum_{i=0}^{\infty} \alpha_i r^i, \qquad \left. \frac{\partial u}{\partial n} \right|_{\theta=\Theta} = \sum_{i=0}^{\infty} \beta_i r^i.$$

Let $u = \bar{u} + u_g$. We have the general solutions

$$u_g = \sum_{k=0}^{\infty} d_k r^{\sigma_k} \sin(\sigma_k \theta), \qquad \sigma_k = \frac{\left(k+\frac{1}{2}\right)}{\Theta} \pi.$$

The harmonic solution satisfies

$$\Delta \bar{u} = 0 \quad \text{in } S, \tag{11.3.8}$$
$$\bar{u} \mid_{\theta=0} = \sum_{i=0}^{\infty} \alpha_i r^i, \qquad \left. \frac{\partial \bar{u}}{\partial n} \right|_{\theta=\Theta} = \sum_{i=0}^{\infty} \beta_i r^i.$$

Define the functions

$$\widehat{\Phi}_i = \widehat{\Phi}_i(r,\theta) = \begin{cases} \dfrac{r^i \sin i\theta}{i \cos i\Theta}, & \text{if } i\Theta \neq \left(k+\frac{1}{2}\right)\pi, \ k = 1,2,\ldots \\[2ex] \dfrac{(-1)^{k+1}}{i\Theta} \varphi_i(r,\theta), & \text{if } i\Theta = \left(k+\frac{1}{2}\right)\pi \quad \text{for some } k. \end{cases}$$

Hence,

$$\widehat{\Phi}_i \mid_{\theta=0} = 0, \qquad \left. \frac{\partial \widehat{\Phi}_i}{\partial n} \right|_{\theta=\Theta} = \left. \frac{\partial \widehat{\Phi}_i}{r\, \partial \theta} \right|_{\theta=\Theta} = r^{i-1}, \quad \forall r > 0, \ i = 1,2,\ldots$$

Choose the harmonic solutions to eqn. (11.3.8) as

$$\overline{u}=\sum_{i=0}^{\infty}A_i r^i\cos i\theta+\sum_{i=1}^{\infty}B_i\widehat{\Phi}_i(r,\theta), \tag{11.3.9}$$

where A_i and B_i are the coefficients. When $\theta=0$, we have $A_i=\alpha_i$, and when $\theta=\Theta$,

$$\sum_{i=0}^{\infty}\beta_i r^i=\left.\frac{\partial\overline{u}}{\partial n}\right|_{\theta=\Theta}=-\sum_{i=1}^{\infty}A_i r^{i-1}\,i\,\sin i\Theta+\sum_{i=1}^{\infty}B_i r^{i-1}.$$

This gives

$$B_i=\beta_{i-1}+i\alpha_i\sin i\Theta.$$

Hence, we obtain from the harmonic solutions, i.e., eqn. (11.3.9)

$$\overline{u}=\sum_{i=0}^{\infty}\alpha_i r^i\cos i\theta+\sum_{i=1}^{\infty}(\beta_{i-1}+i\alpha_i\sin i\Theta)\widehat{\Phi}_i(r,\theta). \tag{11.3.10}$$

Below, we also list the useful formulas from eqn. (11.3.10) for some special Θ.

1. When $\Theta=\frac{\pi}{2}$,

$$\overline{u}=\sum_{i=0}^{\infty}\alpha_i r^i\cos i\theta+\sum_{j=0}^{\infty}\frac{2}{\pi}\left[(-1)^{j+1}\frac{\beta_{2j}}{2j+1}-\alpha_{2j+1}\right]\varphi_{2j+1}(r,\theta)$$
$$+\sum_{j=1}^{\infty}(-1)^j\frac{\beta_{2j-1}}{2j}r^{2j}\sin 2j\theta.$$

2. When $\Theta=\frac{3\pi}{2}$,

$$\overline{u}=\sum_{i=0}^{\infty}\alpha_i r^i\cos i\theta+\sum_{j=0}^{\infty}\frac{2}{3\pi}\left[(-1)^{j}\frac{\beta_{2j}}{2j+1}-\alpha_{2j+1}\right]\varphi_{2j+1}(r,\theta)$$
$$+\sum_{j=1}^{\infty}(-1)^j\frac{\beta_{2j-1}}{2j}r^{2j}\sin 2j\theta.$$

11.3.3 The case of N–N type

Consider

$$\Delta u=0\quad\text{in } S,$$
$$\left.\frac{\partial u}{\partial n}\right|_{\theta=0}=\sum_{i=0}^{\infty}\alpha_i r^i,\qquad\left.\frac{\partial u}{\partial n}\right|_{\theta=\Theta}=\sum_{i=0}^{\infty}\beta_i r^i.$$

Let $u=\overline{u}+u_g$, where $u_g=\sum_{i=0}^{\infty} r^{\sigma_i}\cos\sigma_i\theta$ with $\sigma_i=\frac{i\pi}{\Theta}$. The harmonic solution satisfies

$$\Delta\overline{u}=0 \quad \text{in } S, \tag{11.3.11}$$

$$\left.\frac{\partial\overline{u}}{\partial n}\right|_{\theta=0}=\sum_{i=0}^{\infty}\alpha_i r^i, \qquad \left.\frac{\partial\overline{u}}{\partial n}\right|_{\theta=\Theta}=\sum_{i=0}^{\infty}\beta_i r^i. \tag{11.3.12}$$

Define the functions

$$\widehat{\Psi}_i=\widehat{\Psi}_i(r,\theta)=\begin{cases}\dfrac{-r^i\cos i\theta}{i\sin i\Theta}, & \text{if } i\Theta\neq k\pi,\ k=1,2,\ldots\\[2ex] \dfrac{(-1)^{k+1}}{i\Theta}\psi_i(r,\theta), & \text{if } i\Theta=k\pi \quad \text{for some } k.\end{cases}$$

Hence,

$$\left.\frac{\partial\widehat{\Psi}_i}{\partial n}\right|_{\theta=0}=0, \qquad \left.\frac{\partial\widehat{\Psi}_i}{\partial n}\right|_{\theta=\Theta}=r^{i-1}, \quad \forall r>0,\ i=1,2,\ldots$$

Choose the harmonic solutions

$$\overline{u}=\sum_{i=1}^{\infty}A_i r^i\sin i\theta+\sum_{i=1}^{\infty}B_i\widehat{\Psi}_i(r,\theta),$$

with the coefficients A_i and B_i. When $\theta=0$, we have

$$\sum_{i=0}^{\infty}\alpha_i r^i=\left.\frac{\partial\overline{u}}{\partial n}\right|_{\theta=0}=-\left.\frac{\partial\overline{u}}{r\,\partial\theta}\right|_{\theta=0}=-\sum_{i=1}^{\infty}A_i i r^{i-1}.$$

Then $A_i=-\frac{\alpha_{i-1}}{i}$, $i=1,2,\ldots$. Next, when $\theta=\Theta$,

$$\sum_{i=0}^{\infty}\beta_i r^i=\left.\frac{\partial\overline{u}}{\partial n}\right|_{\theta=\Theta}=\left.\frac{\partial\overline{u}}{r\,\partial\theta}\right|_{\theta=\Theta}=\sum_{i=0}^{\infty}A_i r^{i-1}i\cos i\Theta+\sum_{i=0}^{\infty}B_i r^{i-1}.$$

This gives

$$B_i=\beta_{i-1}+\alpha_{i-1}\cos i\Theta, \quad i=1,2,\ldots$$

Hence, we obtain from the harmonic solutions of eqns. (11.3.11) and (11.3.12)

$$\overline{u}=-\sum_{i=1}^{\infty}\frac{\alpha_{i-1}}{i}r^i\sin i\theta+\sum_{i=1}^{\infty}(\beta_{i-1}+\alpha_{i-1}\cos i\Theta)\widehat{\Psi}_i(r,\theta). \tag{11.3.13}$$

From eqn. (11.3.13), we also list the useful formulas.

1. When $\Theta = \pi$,

$$\overline{u} = -\sum_{i=1}^{\infty} \frac{\alpha_{i-1}}{i} r^i \sin i\theta + \sum_{i=1}^{\infty} \frac{1}{i\pi}[(-1)^{i+1}\beta_{i-1} - \alpha_{i-1}]\psi_i(r,\theta). \tag{11.3.14}$$

2. When $\Theta = 2\pi$,

$$\overline{u} = -\sum_{i=1}^{\infty} \frac{\alpha_{i-1}}{i} r^i \sin i\theta - \sum_{i=1}^{\infty} \frac{1}{2i\pi}(\beta_{i-1} + \alpha_{i-1})\psi_i(r,\theta). \tag{11.3.15}$$

3. When $\Theta = \frac{\pi}{2}$,

$$\overline{u} = -\sum_{i=1}^{\infty} \frac{\alpha_{i-1}}{i} r^i \sin i\theta + \sum_{j=0}^{\infty} (-1)^{j+1} \frac{\beta_{2j}}{2j+1} r^{2j+1} \cos(2j+1)\theta$$
$$+ \sum_{j=1}^{\infty} \frac{1}{j\pi}[(-1)^{j+1}\beta_{2j-1} - \alpha_{2j-1}]\psi_{2j}(r,\theta). \tag{11.3.16}$$

4. When $\Theta = \frac{3\pi}{2}$,

$$\overline{u} = -\sum_{i=1}^{\infty} \frac{\alpha_{i-1}}{i} r^i \sin i\theta + \sum_{j=0}^{\infty} (-1)^{j} \frac{\beta_{2j}}{2j+1} r^{2j+1} \cos(2j+1)\theta$$
$$+ \sum_{j=1}^{\infty} \frac{1}{3j\pi}[(-1)^{j+1}\beta_{2j-1} - \alpha_{2j-1}]\psi_{2j}(r,\theta). \tag{11.3.17}$$

Interestingly, when $\Theta = \frac{i\pi}{2}$, $i = 1,2,3,4$, for the N–D, D–N, and N–N types, the worst singularity of $\overline{u}$ is $O(r \ln r)$. For the D–D type, the worst singularity of $\overline{u}$ is $O\left(\frac{\theta}{\Theta}\right)$.

11.4 Extensions and analysis on singularity

11.4.1 Particular solutions for Poisson's equations

In this section, we consider the simple case of Poisson's equation

$$-\Delta\overline{u} = f \quad \text{in } S, \tag{11.4.1}$$

$$\overline{u} = g_D \mid \Gamma_D, \qquad \left.\frac{\partial \overline{u}}{\partial n}\right|_{\Gamma_N} = g_N, \tag{11.4.2}$$

where $\partial S = \Gamma = \Gamma_D \cup \Gamma_N$, $f = ax^iy^j$, $i, j = 0, 1, \ldots$, and a is a constant. Suppose that $i \geq j$ without loss of generality. We give a set of particular solutions of eqn. (11.4.1) in the following.

Case I. For $0 \leq j \leq 1$,

$$\overline{u} = -a\frac{x^{i+2}y^j}{(i+1)(i+2)}.$$

Case II. For $2 \leq j \leq 3$,

$$\overline{u} = -a\left\{\frac{x^{i+2}y^j}{(i+1)(i+2)} - \frac{x^{i+4}y^{j-2}j(j-1)}{(i+4)(i+3)(i+2)(i+1)}\right\}.$$

Case III. For $2k \leq j \leq 2k+1$, $k=0,1,\ldots$

$$\overline{u} = -a\frac{x^{i+2}y^j}{(i+1)(i+2)} + \frac{a}{(i+1)(i+2)}\sum_{\ell=1}^{k}(-1)^{\ell}b_{i,j,\ell}x^{i+2+2\ell}y^{j-2\ell},$$

where the coefficients

$$b_{i,j,\ell} = \Pi_{m=1}^{\ell}\frac{(j-2m)(j-2m-1)}{(i+2+2m)(i+1+2m)}.$$

Besides, Cheng, Lafe, and Grilli [95] gave a different approach for deriving the particular solution for the same functions f. Their particular solutions are given as follows

$$\overline{u} = \begin{cases} \displaystyle\sum_{k=1}^{\mathrm{Int}\left[\frac{j+2}{2}\right]} a(-1)^k\frac{i!j!x^{i+2k}y^{j-2k+2}}{(i+2k)!(j-2k+2)!}, & \text{for } i \geq j, \\ \displaystyle\sum_{k=1}^{\mathrm{Int}\left[\frac{i+2}{2}\right]} a(-1)^k\frac{i!j!x^{i-2k+2}y^{j+2k}}{(i-2k+2)!(j+2k)!}, & \text{for } i < j, \end{cases}$$

where Int[s] means the integer part of s. By the above arguments, we can obtain the particular solutions $\overline{u}$ for

$$-\Delta\overline{u} = f, \qquad f = \sum_{i=0}^{M}\sum_{j=0}^{N}a_{ij}x^iy^j.$$

Denote $\overline{v} = u - \overline{u}$, we obtain from eqns. (11.4.1) and (11.4.2)

$$\begin{aligned} \Delta\overline{v} &= 0 && \text{in } S, \\ \overline{v} &= g_D - \overline{u} && \text{on } \Gamma_D, \\ \frac{\partial\overline{v}}{\partial n} &= g_N - \frac{\partial\overline{u}}{\partial n} && \text{on } \Gamma_N. \end{aligned}$$

Hence, it is reduced to find the solutions for the Laplace equation with the Dirichlet–Neumann conditions, which have been provided in Sections 11.2 and 11.3. More discussions for the particular solutions of Poisson's equation are given in Chen, Hon, and Schaback [85].

11.4.2 Extensions to not smooth functions of g_D and g_N

In this subsection, we consider functions g_D and g_N are not highly smooth. First, consider the D–D type

$$\Delta \overline{u} = 0 \quad \text{in } S, \qquad (11.4.3)$$
$$\overline{u}\,|_{\theta=0} = ar^q, \qquad \overline{u}\,|_{\theta=\Theta} = br^p,$$

where p and q are real. For the solutions $u \in H^1(S)$ of the Laplace equation, the boundary functions of the Dirichlet and Neumann conditions have $g_D \in H^{\frac{1}{2}}(\Gamma_D)$ and $g_N \in H^{-\frac{1}{2}}(\Gamma_N)$ (see Babuska and Aziz [15]). Then we assume $p, q > -\frac{1}{2}$ in eqn. (11.4.3). Hence, p and q are not confined to be non-negative integers (cf. Sections 11.2 and 11.3). When $p\Theta, q\Theta \neq \pm k\pi$, the harmonic solutions are given by

$$\overline{u} = br^p \frac{\sin p\theta}{\sin p\Theta} + ar^q \frac{\sin q(\Theta - \theta)}{\sin q\Theta}.$$

For simplicity, here we only give one term on the right-hand side in the Dirichlet condition, i.e., eqn. (11.4.3). For more terms, the harmonic solutions can be obtained easily by linear superposition as in Sections 11.2 and 11.3. Since the solutions $O(r^p \ln r)$ for $p \in \left(-\frac{1}{2}, 1\right)$ have strong singularity, we use the formulas in symmetry of the particular solutions as those in Volkov [452].

Suppose that $p\Theta = \pm m\pi$ and $q\Theta = \pm \ell\pi$, where m and ℓ are non-negative integers. The harmonic solutions are given by

$$\begin{aligned}\overline{u} &= \frac{b}{\Theta}\frac{\varphi_p(r,\theta)}{\cos p\Theta} + \frac{a}{\Theta}\frac{\varphi_q(r,\Theta-\theta)}{\cos q\Theta} \\ &= \frac{(-1)^m}{\Theta} b\varphi_p(r,\theta) + \frac{(-1)^\ell}{\Theta} a\varphi_q(r,\Theta-\theta).\end{aligned}$$

When $p\Theta \neq \pm m\pi$ and $q\Theta \neq \pm \ell\pi$, the harmonic solutions can be easily obtained. Moreover, the function $\varphi_q(r, \Theta - \theta)$ is defined in eqn. (11.2.6), and may be further simplified, see Sections 11.2 and 11.3.

Next, consider the N–D type

$$\Delta \overline{u} = 0 \quad \text{in } S,$$
$$\left.\frac{\partial \overline{u}}{\partial n}\right|_{\theta=0} = ar^q, \qquad \overline{u}\,|_{\theta=\Theta} = br^p,$$

where real $p > -\frac{1}{2}$ and $q > -\frac{3}{2}$. For $p\Theta, (q+1)\Theta \neq \pm\left(k+\frac{1}{2}\right)\pi$, the harmonic solutions are

$$\overline{u} = br^p \frac{\cos p\theta}{\cos p\Theta} + \frac{a}{q+1} r^{q+1} \left\{ \frac{\sin(q+1)(\Theta-\theta)}{\cos(q+1)\Theta} \right\}.$$

For $p\Theta = \pm\left(m+\frac{1}{2}\right)\pi$ and $(q+1)\Theta = \pm\left(\ell+\frac{1}{2}\right)\pi$, where m and ℓ are non-negative integers, the harmonic solutions

$$\begin{aligned}\overline{u} &= -\frac{b}{\Theta}\frac{\psi_p(r,\theta)}{\sin p\Theta} - \frac{a}{(q+1)\Theta}\frac{\varphi_{q+1}(r,\Theta-\theta)}{\sin(q+1)\Theta} \\ &= -\frac{\pm(-1)^m}{\Theta}b\psi_p(r,\theta) - \frac{\pm(-1)^\ell}{(q+1)\Theta}a\varphi_q(r,\Theta-\theta).\end{aligned}$$

The harmonic solutions of D–N type can be obtained from those of N–D type by $\varphi = \Theta - \theta$.

Finally, we consider the N–N type

$$\begin{aligned}&\Delta\overline{u} = 0 \quad \text{in } S, \\ &\left.\frac{\partial\overline{u}}{\partial n}\right|_{\theta=0} = ar^q, \qquad \left.\frac{\partial\overline{u}}{\partial n}\right|_{\theta=\Theta} = br^p,\end{aligned}$$

where real $p,q > -\frac{3}{2}$. When $(p+1)\Theta, (q+1)\Theta \neq \pm k\pi$, $k = 0,1,2,\ldots$, the harmonic solutions are

$$\overline{u} = -\frac{b}{p+1}r^{p+1}\frac{\cos(p+1)\theta}{\sin(p+1)\Theta} - \frac{a}{q+1}r^{q+1}\frac{\cos(q+1)(\Theta-\theta)}{\sin(q+1)\Theta}.$$

When $(p+1)\Theta = \pm m\pi$ and $(q+1)\Theta = \pm\ell\pi$, where $m,\ell = 0,1,2,\ldots$, the harmonic solutions are

$$\begin{aligned}\overline{u} &= -\frac{b}{(p+1)\Theta}\frac{\psi_{p+1}(r,\theta)}{\cos(p+1)\Theta} - \frac{a}{(q+1)\Theta}\frac{\psi_{q+1}(r,\Theta-\theta)}{\cos(q+1)\Theta} \\ &= -\frac{(-1)^m}{(p+1)\Theta}b\psi_{p+1}(r,\theta) - \frac{(-1)^\ell}{(q+1)\Theta}a\psi_{q+1}(r,\Theta-\theta).\end{aligned}$$

Of course, we may derive the harmonic solutions for $\Theta = \frac{\pi}{2}, \frac{3\pi}{2}, 2\pi$, and π by following Sections 11.2 and 11.3.

11.4.3 Regularity and singularity of the solutions of $\Theta = \frac{\pi}{2}, \frac{3\pi}{2}, \pi, 2\pi$

From the analysis in Sections 11.2 and 11.3, when g_D and g_N are highly smooth on ∂S, the solutions u inside S are also smooth for $\Theta = \frac{\pi}{2}, \frac{3\pi}{2}, \pi, 2\pi$. However, the solution u near the corners may have the mild singularities $O(r^k \ln r)$, $k = 1,2,\ldots$. Since the analysis of regularity and singularity on general solutions can be found in textbooks (cf. Li [280]), we focus on the analysis for the harmonic solution $\overline{u}$. In particular, we consider when $\Theta = \frac{i\pi}{2}$, $i = 1,2,3,4$, which exist in Motz's and its variants, the L-shaped domain problems, and the general cracked domains in fig. 11.5, see Refs. [280, 315].

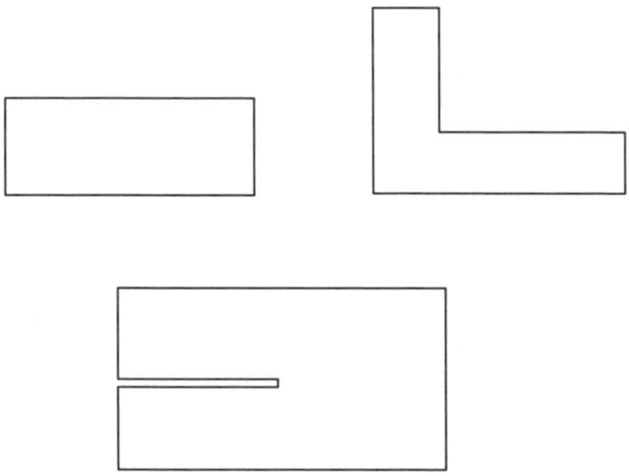

Figure 11.5: The popular domains in testing models with $\Theta = \frac{i\pi}{2}, i = 1, 2, 3, 4$.

11.4.3.1 For the case of $\Theta = \frac{\pi}{2}$

First, consider a simple case of the D–D type as $\Theta = \frac{\pi}{2}$

$$\begin{aligned} \Delta \overline{u} &= 0 && \text{in } S, \\ \overline{u} &= \alpha_0 + \alpha_1 r + \alpha_2 r^2 && \text{at } \theta = 0, \\ \overline{u} &= \beta_0 + \beta_1 r + \beta_2 r^2 && \text{at } \theta = \Theta, \end{aligned} \tag{11.4.4}$$

where β_i and α_i are constants. Only the quadratic polynomials in the Dirichlet conditions are chosen, because the resulted singularities are strongest among all kinds of mild singularities. We then obtain from eqn. (11.2.22)

$$\begin{aligned} \overline{u} = \alpha_0 &+ \frac{(\beta_0 - \alpha_0)\theta}{\Theta} + \beta_1 r \sin\theta - \frac{\beta_2 + \alpha_2}{\Theta}\varphi_2(r,\theta) \\ &+ \alpha_1 r \cos\theta + \alpha_2 r^2 \cos 2\theta. \end{aligned} \tag{11.4.5}$$

In eqn. (11.4.5), when $\beta_0 \neq \alpha_0$, the function $\theta \notin H^1(S)$ is called the discontinuity singularity. Moreover, when $\beta_2 + \alpha_2 \neq 0$, the function $r^2 \ln r \notin H^3(S)$ is called the mild singularity, compared with the angular singularity

$$u = O(r^{\frac{1}{2}}) \notin H^2(S).$$

Note that the coefficients of β_1 or α_1 do not effect singularities because $r \sin\theta = y$ and $r \cos\theta = x$.

Next, consider the N–D type

$$\begin{aligned} \Delta \overline{u} &= 0 && \text{in } S, \\ \frac{\partial \overline{u}}{\partial n} &= \alpha_0 + \alpha_1 r + \alpha_2 r^2 && \text{at } \theta = 0, \\ \overline{u} &= \beta_0 + \beta_1 r + \beta_2 r^2 && \text{at } \theta = \Theta. \end{aligned} \tag{11.4.6}$$

From eqn. (11.3.6), we obtain

$$\overline{u} = \beta_0 - \frac{(\beta_1+\alpha_0)}{\Theta}\psi_1(r,\theta) - \beta_2 r^2\cos 2\theta - \alpha_0 r\sin\theta$$
$$- \frac{\alpha_1}{2}r^2\sin 2\theta - \frac{\alpha_2}{3}\left\{r^3\sin 3\theta + \frac{1}{\Theta}\psi_3(r,\theta)\right\}. \tag{11.4.7}$$

From eqn. (11.4.7), when $\beta_1+\alpha_0 \neq 0$ which results from

$$\alpha_0 = \left.\frac{\partial\overline{u}}{\partial n}\right|_{\theta=0} = -\left.\frac{\partial\overline{u}}{\partial y}\right|_{\theta=0} \neq -\left.\frac{\partial\overline{u}}{\partial r}\right|_{\theta=\frac{\pi}{2}} = -\beta_1 \quad \text{at } r=0,$$

there exists a mild singularity $O(r\ln r)$, and when $\alpha_2 \neq 0$, there exists $O(r^3\ln r)$. Interestingly, β_0, β_2, and α_1 do not cause any singularity in the N–D type since $r^2\cos 2\theta = (x^2-y^2)$ and $r\sin 2\theta = 2xy$.

Similarly, we can draw the conclusion for the D–N type. Below, we only consider the N–N type

$$\Delta\overline{u} = 0 \qquad \text{in } S, \tag{11.4.8}$$
$$\frac{\partial\overline{u}}{\partial n} = \alpha_0+\alpha_1 r+\alpha_2 r^2 \quad \text{at } \theta=0,$$
$$\frac{\partial\overline{u}}{\partial n} = \beta_0+\beta_1 r+\beta_2 r^2 \quad \text{at } \theta=\Theta.$$

From eqn. (11.3.16), we obtain

$$\overline{u} = -\beta_0 r\cos\theta + \frac{(\beta_1-\alpha_1)}{2\Theta}\psi_2(r,\theta) + \frac{\beta_2}{3}r^3\cos 3\theta - \alpha_0 r\sin\theta$$
$$- \frac{\alpha_1}{2}r^2\sin 2\theta - \frac{\alpha_2}{3}r^3\sin 3\theta. \tag{11.4.9}$$

Only that $\beta_1-\alpha_1 \neq 0$ resulting from $\overline{u}_{xy} \neq \overline{u}_{yx}$ will cause the singularity $O(r^2\ln r)$.

We summarize the singularities at corner O for $\Theta = \frac{\pi}{2}$ in table 11.1. The discontinuity $\beta_0 \neq \alpha_0$ in the D–D type is the strongest. The next strongest singularity occurs in the N–D type of $\beta_1+\alpha_0 \neq 0$, and then the N–N type of $\beta_1 \neq \alpha_1$.

11.4.3.2 For the case of $\Theta = \pi$

Next, we consider the D–D type of eqn. (11.4.4) with $\Theta = \pi$. From eqn. (11.2.20), we obtain

$$\overline{u} = \alpha_0 + \frac{(\beta_0-\alpha_0)\theta}{\Theta} + \alpha_1 r\cos\theta + \alpha_2 r^2\cos\theta$$
$$- \frac{(\beta_1+\alpha_1)}{\Theta}\varphi_1(r,\theta) + \frac{(\beta_2-\alpha_2)}{\Theta}\varphi_2(r,\theta).$$

Table 11.1: The singularities of $\Theta = \frac{\pi}{2}$ for the harmonic solutions for the Dirichlet–Neumann conditions assigned by quadratic polynomials.

Types	Conditions	Solutions	Not in $H^p(S)$
D–D	$\alpha_0 \neq \beta_0$	$O\left(\frac{\theta}{\Theta}\right)$	$\notin H^1(S)$
	$\alpha_2 + \beta_2 \neq 0$	$O(r^2 \ln r)$	$\notin H^3(S)$
N–N	$\alpha_1 - \beta_1 \neq 0$	$O(r^2 \ln r)$	$\notin H^3(S)$
N–D	$\beta_1 + \alpha_0 \neq 0$	$O(r \ln r)$	$\notin H^2(S)$
	$\alpha_2 \neq 0$	$O(r^3 \ln r)$	$\notin H^4(S)$

When $\beta_0 \neq \alpha_0$, the solution of $O\left(\frac{\theta}{\Theta}\right)$ is discontinuous at origin O, and when $\beta_2 \neq \alpha_2$, the solutions of $O(r^2 \ln r)$ are obtained. Note that the case of $\beta_1 \neq -\alpha_1$ will also cause the singularity of $O(r \ln r)$. In fact, the case of $\beta_1 \neq -\alpha_1$ implies the existence of a piecewise boundary function on x-axis, because $r = x$ at $\theta = 0$ but $r = -x$ at $\theta = \pi$.

Consider the N–N type of eqn. (11.4.8) with $\Theta = \pi$. From eqn. (11.3.14) for $\Theta = \pi$

$$\begin{aligned}\bar{u} &= -\alpha_0 r \sin\theta - \frac{\alpha_1}{2} r^2 \sin 2\theta - \frac{\alpha_2}{3} r^3 \sin 3\theta \\ &+ \frac{(\beta_0 - \alpha_0)}{\Theta}\psi_1(r,\theta) - \frac{(\beta_1 + \alpha_1)}{2\Theta}\psi_2(r,\theta) + \frac{(\beta_2 - \alpha_2)}{3\Theta}\psi_3(r,\theta).\end{aligned}$$

When $\beta_0 \neq \alpha_0$, $\beta_1 \neq -\alpha_1$, and $\beta_2 \neq \alpha_2$, there exist the solutions with singularities, $u = O(r \ln r)$, $O(r^2 \ln r)$, and $O(r^3 \ln r)$, respectively.

Consider the N–D type of eqn. (11.4.6) with $\Theta = \pi$, which appears in Motz's and its variant problems in Chapter 2. The general solutions are $u_g = \sum_{i=0}^{L} d_i r^{i+\frac{1}{2}} \cos\left(i + \frac{1}{2}\right)\theta$, and the harmonic solutions are obtained from eqn. (11.3.5)

$$\bar{u} = \sum_{k=0}^{2} (-1)^k \beta_k r^k \cos k\theta - \sum_{k=1}^{3} \frac{\alpha_{k-1}}{k} r^k \sin k\theta.$$

Interestingly, the singularity results only from the general solutions of $O(r^{\frac{1}{2}})$, but not from $\beta_i \neq 0$ and $\alpha_i \neq 0$.

11.4.3.3 For the case of $\Theta = \frac{3\pi}{2}$

The boundary angle $\Theta = \frac{3\pi}{2}$ exists in a typical concave polygon, such as the L-shaped domains. First, we consider the D–D type of eqn. (11.4.4) with $\Theta = \frac{3\pi}{2}$.

The solution is $u = \overline{u} + u_g$, where $u_g = \sum_{i=1}^{L} d_i r^{\frac{2}{3}i} \sin\left(\frac{2i}{3}\theta\right)$, and the harmonic solutions from eqn. (11.2.23) are

$$\overline{u} = \alpha_0 + \frac{(\beta_0 - \alpha_0)\theta}{\Theta} - \beta_1 r \sin\theta - \frac{\beta_2 + \alpha_2}{\Theta}\varphi_2(r,\theta)$$
$$+\alpha_1 r \cos\theta + \alpha_2 r^2 \cos 2\theta.$$

On comparing the above function with that of $\Theta = \frac{\pi}{2}$ in eqn. (11.4.5), only the sign in front of the term $\beta_1 r \sin\theta$ is different. Note that when $d_1 \neq 0$, $u_g = O(r^{\frac{2}{3}}) \notin H^2(S)$ is the next strongest singularity to that of $O\left(\frac{\theta}{\Theta}\right)$.

Consider the N–N type of eqn. (11.4.8) with $\Theta = \frac{3\pi}{2}$. The general solutions are $u_g = \sum_{i=0}^{L} d_i r^{\frac{2}{3}i} \cos\left(\frac{2i}{3}\right)\theta$, and the harmonic solutions, i.e., eqn. (11.3.17) give

$$\overline{u} = \beta_0 r \cos\theta + \frac{(\beta_1 - \alpha_1)}{2\Theta}\psi_2(r,\theta) - \frac{\beta_2}{3} r^3 \cos 3\theta - \alpha_0 r \sin\theta$$
$$-\frac{\alpha_1}{2} r^2 \sin 2\theta - \frac{\alpha_2}{3} r^3 \sin 3\theta. \tag{11.4.10}$$

On comparing eqn. (11.4.10) with eqn. (11.4.9), only the sign in front of $\beta_0 r \cos\theta$ and $\frac{\beta_2}{3} r^3 \cos 3\theta$ is different.

Finally, consider the N–D type of eqn. (11.4.6) with $\Theta = \frac{3\pi}{2}$. The general solutions are $u_g = \sum_{i=0}^{L} d_i r^{\sigma_i} \cos(\sigma_i \theta)$, where $\sigma_i = \frac{2}{3}i + \frac{1}{3}$. The harmonic solutions of eqn. (11.3.7) give

$$\overline{u} = \beta_0 + \frac{(\beta_1 - \alpha_0)}{\Theta}\psi_1(r,\theta) - \beta_2 r^2 \cos 2\theta - \alpha_0 r \sin\theta$$
$$-\frac{\alpha_1}{2} r^2 \sin 2\theta - \frac{\alpha_2}{3}\left\{r^3 \sin 3\theta + \frac{1}{\Theta}\psi_3(r,\theta)\right\}.$$

The strongest singularity of u is $O(r^{\frac{1}{3}})$.

11.4.3.4 For the case of $\Theta = 2\pi$

The boundary angle $\Theta = 2\pi$ occurs for the domains with an inside crack without symmetry. First, we consider the D–D type of eqn. (11.4.4) with $\Theta = 2\pi$. The general solutions are given by $u_g = \sum_{i=0}^{L} d_i r^{\frac{i}{2}} \sin\left(\frac{i}{2}\right)\theta$, and the harmonic solutions from eqn. (11.2.21) are

$$\overline{u} = \alpha_0 + \frac{(\beta_0 - \alpha_0)\theta}{\Theta} + \sum_{k=1}^{2} \alpha_k r^k \cos k\theta + \sum_{k=1}^{2} \frac{(\beta_k - \alpha_k)}{\Theta}\varphi_k(r,\theta).$$

When $\beta_0 \neq \alpha_0$ the singularity $O\left(\frac{\theta}{\Theta}\right)$ is the strongest. The next strongest singularity results from $u_g = O(r^{\frac{1}{2}})$. Also when $\beta_i \neq \alpha_i$, $i = 1, 2$, the mild singularities $O(r^i \ln r)$ occur.

Consider the N–N type of eqn. (11.4.8) with $\Theta = 2\pi$. The general solutions $u_g = d_0 + \sum_{i=1}^{L} d_i r^{\frac{i}{2}} \cos\left(\frac{i}{2}\right)\theta$, where d_0 is an arbitrary constant. The particular solutions from eqn. (11.3.15) are

$$\bar{u} = -\sum_{k=1}^{3} \frac{\alpha_{k-1} + \beta_{k-1}}{k\Theta} \psi_k(r,\theta) - \sum_{k=1}^{3} \frac{\alpha_{k-1}}{k} r^k \sin k\theta.$$

Consider the N–D type of eqn. (11.4.6) with $\Theta = 2\pi$, $u_g = \sum_{i=0}^{L} d_i r^{\sigma_i} \cos \sigma_i \theta$, where $\sigma_i = \frac{i}{2} + \frac{1}{4}$. The harmonic solutions are from eqn. (11.3.5)

$$\bar{u} = \sum_{k=0}^{2} \beta_k r^k \cos k\theta - \sum_{k=1}^{3} \frac{\alpha_{k-1}}{k} r^k \sin k\theta.$$

Tables 11.2–11.4 list the overviews of the singularities of harmonic solutions for $\Theta = \frac{3\pi}{2}, \pi, 2\pi$.

Among all cases of $\Theta = \frac{\pi}{2}, \frac{3\pi}{2}, \pi, 2\pi$, the strongest singularity is still $O\left(\frac{\theta}{\Theta}\right)$, and the next singularity is $O(r^{\frac{1}{4}})$, resulting from the N–D type of $\Theta = 2\pi$. Based on the analysis for the harmonic solutions on the polygonal domains, we understand completely the regularity and singularity of Laplace's equations. Therefore, we may deliberately design the new models with different kinds of singularities. In this chapter, we consider the simplest rectangle in fig. 11.5 as the solution domain of testing models, and employ the harmonic solutions in Section 11.4. Of course, we may also design other testing models on the L-shaped and the inside cracked domains as shown in fig. 11.5, by means of Subsections 4.3.2–4.3.4.

Table 11.2: The singularities of $\Theta = \pi$ for the harmonic solutions for the Dirichlet–Neumann conditions assigned by quadratic polynomials.

Types	Conditions	Solutions	u_g
D–D	$\alpha_0 \neq \beta_0$	$O\left(\frac{\theta}{\Theta}\right)$	/
	$\alpha_1 + \beta_1 \neq 0$	$O(r \ln r)$	/
	$\alpha_2 \neq \beta_2$	$O(r^2 \ln r)$	/
N–N	$\alpha_0 \neq \beta_0$	$O(r \ln r)$	/
	$\alpha_1 + \beta_1 \neq 0$	$O(r^2 \ln r)$	/
	$\alpha_2 \neq \beta_2$	$O(r^3 \ln r)$	/
N–D	/	/	$O(r^{\frac{1}{2}})$

Table 11.3: The singularities of $\Theta = \frac{3\pi}{2}$ for the harmonic solutions for the Dirichlet–Neumann conditions assigned by quadratic polynomials.

Types	Conditions	Solutions	u_g
D–D	$\alpha_0 \neq \beta_0$	$O\left(\frac{\theta}{\Theta}\right)$	$O(r^{\frac{2}{3}})$
	$\alpha_2 + \beta_2 \neq 0$	$O(r^2 \ln r)$	$O(r^{\frac{2}{3}})$
N–N	$\alpha_1 - \beta_1 \neq 0$	$O(r^2 \ln r)$	$O(r^{\frac{2}{3}})$
N–D	$\beta_1 \neq \alpha_0$	$O(r \ln r)$	$O(r^{\frac{1}{3}})$
	$\alpha_2 \neq 0$	$O(r^3 \ln r)$	$O(r^{\frac{1}{3}})$

Table 11.4: The singularities of $\Theta = 2\pi$ for the harmonic solutions for the Dirichlet–Neumann conditions assigned by quadratic polynomials.

Types	Conditions	Solutions	u_g
D–D	$\alpha_0 \neq \beta_0$	$O\left(\frac{\theta}{\Theta}\right)$	$O(r^{\frac{1}{2}})$
	$\alpha_1 \neq \beta_1$	$O(r \ln r)$	$O(r^{\frac{1}{2}})$
	$\beta_2 \neq \beta_2$	$O(r^2 \ln r)$	$O(r^{\frac{1}{2}})$
N–N	$\alpha_0 + \beta_0 \neq 0$	$O(r \ln r)$	$O(r^{\frac{1}{2}})$
	$\alpha_1 + \beta_1 \neq 0$	$O(r^2 \ln r)$	$O(r^{\frac{1}{2}})$
	$\alpha_2 + \beta_1 \neq 0$	$O(r^3 \ln r)$	$O(r^{\frac{1}{2}})$
N–D	/	/	$O(r^{\frac{1}{4}})$

11.5 New models of singularities for Laplace's equation

11.5.1 Two models

The singularity models play an important role in studying numerical methods, because they can serve as benchmark problems to test the performance of different numerical methods. Two popular models, Motz's problem and its variant, Model A, have been explored in Chapter 2. In this chapter, we propose a new discontinuity model, called the discontinuity model, see fig. 11.6.

$$\Delta u = 0 \qquad \text{in } S, \tag{11.5.1}$$

$$u = 0 \qquad \text{on } \overline{OD},$$

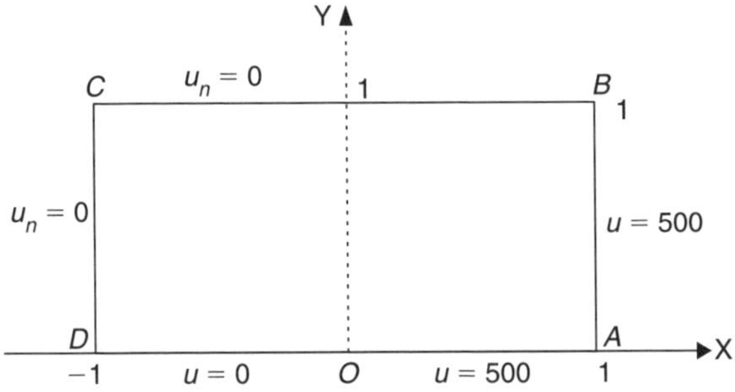

Figure 11.6: The discontinuity model.

$$u = 500 \quad \text{on } \overline{OA} \cup \overline{AB},$$
$$\frac{\partial u}{\partial n} = 0 \quad \text{on } \overline{BC} \cup \overline{CD}.$$

Note that the discontinuity model is different from Motz's problem only on the boundary condition on $\overline{OA}$, where the Dirichlet condition $u = 500$ is used to replace the Neumann condition $\frac{\partial u}{\partial n} = 0$. The solution u at the origin is discontinuous, having much stronger singularity than that of Motz's problem. We have the solution expansion,

$$v = \frac{500(\pi - \theta)}{\pi} + \sum_{i=1}^{L} c_i r^i \sin i\theta, \tag{11.5.2}$$

where c_i are the unknown coefficients to be sought. Since the function, i.e., eqn. (11.5.2) satisfies the boundary conditions on $\overline{DO}$ and $\overline{OA}$ already, c_i can be found to satisfy the rest of boundary conditions on ∂S as close as possible. We will solve it by the CTM described in Subsection 11.5.2 below, to solve the problem.

Consider the following model of the angular plus mild singularities of $\rho^k \ln \rho$, $k = 1, 2$, which are developed from Motz's problem, called Model B of variants of Motz's problem, or simply Model B, see fig. 11.7.

$$\Delta u = 0 \quad \text{in } S,$$
$$u = 0 \quad \text{on } \overline{OD},$$
$$u = 500 \quad \text{on } \overline{AB},$$
$$\frac{\partial u}{\partial n} = 0 \quad \text{on } \overline{OA} \cup \overline{CD},$$
$$u = 125(-x^2 + 2x + 3) \quad \text{on } \overline{BC}.$$

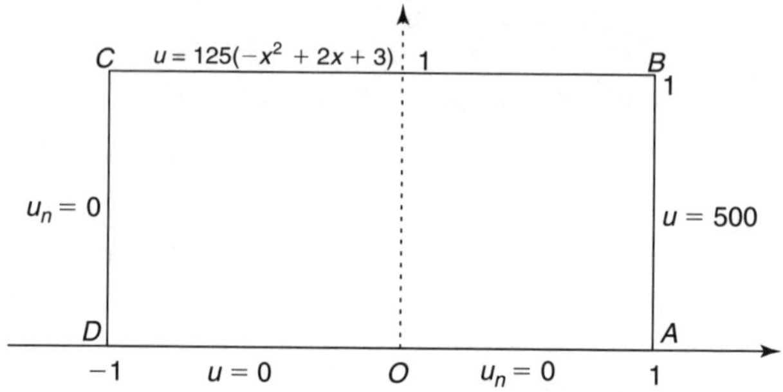

Figure 11.7: Model B of Motz's variant problems.

Since the boundary function on $\overline{BC}$ can be expressed by

$$u = 500(x+1) - 125(x+1)^2 = 500 - 125(x-1)^2 \quad \text{on } \overline{BC},$$

we obtain the solutions near the corners B and C,

$$v_1 = \overline{v}_1 + \sum_{i=1}^{M} a_i \rho^{2i} \sin 2i\phi \qquad \text{in } S_1,$$

$$v_2 = \overline{v}_2 + \sum_{i=0}^{M} b_i \xi^{2i+1} \sin(2i+1)\eta \quad \text{in } S_2,$$

where the coefficients a_i and b_i are unknowns, (ρ, ϕ) and (ξ, η) are the polar coordinates at corners B and C, respectively, and S_1 and S_2 are the subdomains (see fig. 11.8)

$$S_1 = \left\{(\rho, \phi) \,|\, 0 \le \rho \le \rho_1,\ 0 \le \phi \le \frac{\pi}{2}\right\},$$
$$S_2 = \left\{(\xi, \eta) \,|\, 0 \le \xi \le \xi_1,\ 0 \le \eta \le \frac{\pi}{2}\right\}.$$

The functions $\overline{v}_1$ and $\overline{v}_2$ can be found by following the harmonic solutions in Section 11.4

$$\overline{v}_1 = 500 - 125\rho^2 \cos 2\phi + \frac{125}{\Theta}\varphi_2(\rho, \phi),$$
$$\overline{v}_2 = -125\xi^2 \cos 2\eta + 500\xi \cos \eta - \frac{500}{\Theta}\varphi_1(\xi, \eta),$$

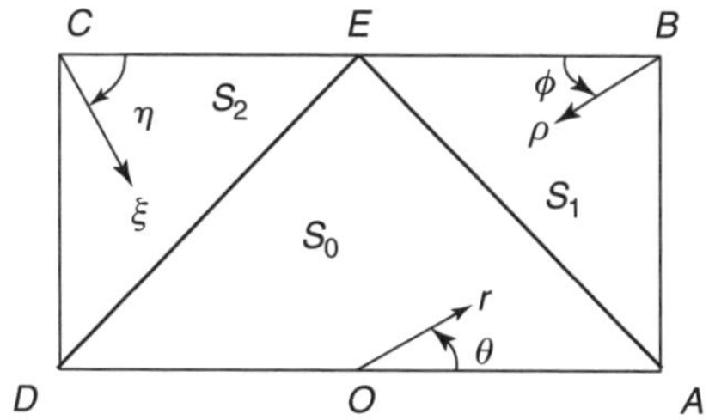

Figure 11.8: Partition of the rectangle.

where $\Theta = \frac{\pi}{2}$, and

$$\varphi_n(\rho, \phi) = \rho^n \{\ln \rho \sin n\phi + \phi \cos n\phi\}.$$

Let $S = S_0 \cup S_1 \cup S_2$ as shown in fig. 11.8. The piecewise admissible functions are given by

$$v = \begin{cases} v_0 = \sum_{i=0}^{L} d_i r^{i+\frac{1}{2}} \cos\left(i + \frac{1}{2}\right)\theta & \text{in } S_0, \\ v_1 = \overline{v}_1 + \sum_{i=1}^{M} a_i \rho^{2i} \sin 2i\phi & \text{in } S_1, \\ v_2 = \overline{v}_2 + \sum_{i=0}^{N} b_i \xi^{2i+1} \sin(2i+1)\eta & \text{in } S_2, \end{cases} \tag{11.5.3}$$

where

$$\overline{v}_1 = 500 - 125\rho^2 \cos 2\phi + \frac{125}{\Theta}\rho^2(\ln \rho \sin 2\phi + \phi \cos 2\phi),$$

$$\overline{v}_2 = -125\xi^2 \cos 2\eta + 500\xi \cos \eta - \frac{500}{\Theta}\xi(\ln \xi \sin \eta + \eta \cos \eta).$$

Note that the solutions at corners B and C have the mild singularities, $O(\rho^2 \ln \rho)$ and $O(\xi \ln \xi)$, respectively. In variants of Motz's problem, Model B involves the mild singularities, but Model A in Section 2.4 of Chapter 2 does not.

11.5.2 Trefftz methods

Based on the above analysis, we have found the local harmonic solutions near all the corners of S. If there exist the singularities, e.g., the discontinuity as $O\left(\frac{\theta}{\Theta}\right)$, the angular singularity as $O(r^p)$, $0 < p < 1$, and the mild singularity as

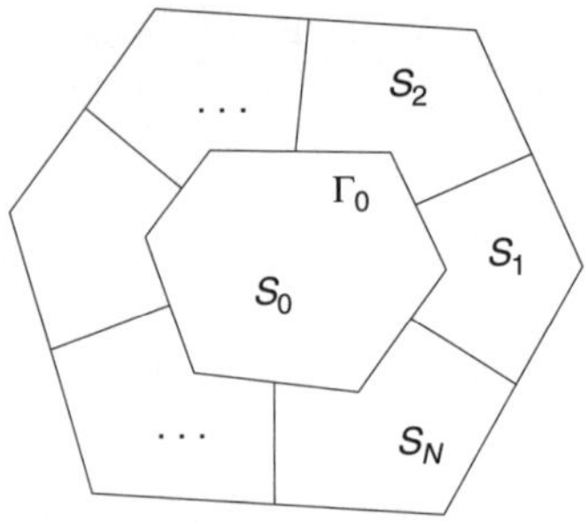

Figure 11.9: Partition for the collocation Trefftz method.

$O(r^k \ln r), k = 1, 2, \ldots$, we may split S by the interface Γ_0 into finite sub-polygons S_i, e.g., $S = \cup_{i=0}^{N} S_i$. In each S_i, there exists only one singularity point at one exterior corner, see fig. 11.9. We denote simply

$$v = v_i = \overline{v}_i + \sum_{k=0}^{N_i} c_k^{(i)} H_k^{(i)} \quad \text{in } S_i, \tag{11.5.4}$$

where $\overline{v}_i$ are the harmonic solutions, $H_k^{(i)}$ the known functions satisfying Laplace's equation and certain boundary conditions, and $c_k^{(i)}$ the unknown coefficients. In S_0 in fig. 11.9, the smooth solutions can be expressed by

$$u = \sum_{i=0}^{\infty} a_i r^i \cos i\theta + \sum_{i=1}^{\infty} b_i r^i \sin i\theta \quad \text{in } S_0,$$

where a_i and b_i are coefficients.

Suppose that the piecewise admissible functions, i.e., eqn. (11.5.4) satisfy eqn. (11.5.1) in S_i and the exterior Dirichlet–Neumann conditions. Then the coefficients $c_k = c_k^{(i)}$ may be sought by satisfying the interior continuity conditions

$$u^+ = u^-, \qquad \frac{\partial u^+}{\partial n} = \frac{\partial u^-}{\partial n} \quad \text{on } \Gamma_0.$$

Define the errors on Γ_0

$$\|v\|_B = \left\{ \int_{\Gamma_0} (v^+ - v^-)^2 \, d\ell + w^2 \int_{\Gamma_0} \left(\frac{\partial v^+}{\partial n} - \frac{\partial v^-}{\partial n} \right)^2 d\ell \right\}^{\frac{1}{2}},$$

where w is a suitable weight. For Model B, we choose $w = \max\{1/(L+1), 1/2M, 1/(2N+1)\}$ (see Chapters 1 and 3). Then, the coefficients $\widetilde{c}_k = \widetilde{c}_k^{(i)}$ are found by

$$I(\widetilde{c}_k) = \min_{c_k} I(c_k), \tag{11.5.5}$$

where

$$I(c_k) = \|v\|_B^2 = \int_{\Gamma_0} (v^+ - v^-)^2 \, d\ell + w^2 \int_{\Gamma_0} \left(\frac{\partial v^+}{\partial n} - \frac{\partial v^-}{\partial n} \right)^2 d\ell. \tag{11.5.6}$$

The eqn. (11.5.5) is called the TM. When the integrals in eqn. (11.5.6) involve numerical quadrature, we denote

$$\widehat{I}(c_k) = \widehat{\int_{\Gamma_0}} (v^+ - v^-)^2 \, d\ell + w^2 \widehat{\int_{\Gamma_0}} \left(\frac{\partial v^+}{\partial n} - \frac{\partial v^-}{\partial n} \right)^2 d\ell, \tag{11.5.7}$$

where $\widehat{\int_{\Gamma_0}}$ is evaluated by some rules. The CTM is to seek the coefficients $\hat{c}_k = \hat{c}_k^{(i)}$ by

$$\widehat{I}(\hat{c}_k) = \min_{c_k} \widehat{I}(c_k).$$

The detailed algorithms and error analysis are also provided in Chapter 2, and the exponential convergence rates can be achieved.

Numerical experiments for the discontinuity model and Model B by the CTM are provided in Ref. [301], to display the significance of the numerical techniques in this chapter.

11.6 Concluding remarks

To close this chapter, let us give a few remarks.

1. The explicit harmonic solutions of Poisson's and Laplace's equations on a polygon with the Dirichlet, the Neumann boundary conditions, and their mixed types are derived in detail. Although these solutions can be found in Volkov [451, 452], the solution formulas of harmonic functions given in this chapter are more explicit, and easier to expose the mild singularity at the domain corners than those in Ref. [452]. A number of useful harmonic solutions for the rectangular, the L-shaped, and the cracked domains are derived. Obviously, the analysis in this chapter is more comprehensive than that in Li [280] and Li and Lu [299]. Choosing the good basis functions, in particular those representing

singularities, is essential for the CTM. Hence, the analysis of harmonic solutions in this chapter will allow the CTM to solve the Poisson and the Laplace equations on a polygon efficiently.

2. The particular solutions can display a clear view of regularity and singularity. In this chapter, we provide the harmonic solutions on special angles, $\Theta = \frac{i\pi}{2}$, $\Theta = \frac{(2i-1)\pi}{4}$, $i = 1,2,3,4$. This solution behavior is important to the choices of numerical methods because different numerical methods need different regularities of the true solution. Take linear FEM and the finite difference method (FDM) as examples. If $u \in H^2(S)$, the optimal convergence rate can be obtained, where $H^k(S)$ is the Sobolev space (cf. Ciarlet [103]). If $u \in H^3(S)$, the superconvergence of FDM may be achieved, see Li, Yamamoto, and Fang [307]. Moreover, the existence of singularity may suggest whether the refinement of elements is needed or not, and where this refinement should take place. When multiple singularities occur, the division of S should also be considered, and different numerical treatments must be used. In summary, the harmonic solutions in this chapter are imperative in numerical methods not only for the CTM, but also for other methods, such as the combined method in Ref. [280] and the SAM.
3. Two new models are designed, to include the discontinuity and the angular plus the mild singularities of $O(r^k \ln r)$, $k = 1,2$. Their numerical experiments are reported in Ref. [301]. The high-accuracy solutions with exponential convergence rates are also provided by the CTM, which can be regarded as the "true" solution for testing other numerical methods. By using piecewise harmonic solutions in subdomains, not only can the condition numbers be reduced significantly, but the high accuracy of the solutions may be retained as well. This is a new discovery, compared with Ref. [280]. Moreover, the Gaussian rule with high order may raise the accuracy of the leading coefficients; this is also coincident with Chapter 2.
4. In Tang [433], a number of models for Laplace's equations with the Dirichlet–Neumann boundary conditions are computed on rectangle S. The uniform harmonic functions,

$$v = \sum_{i=0}^{L} d_i r^{i+\frac{1}{2}} \cos\left(i + \frac{1}{2}\right)\theta \quad \text{in } S, \tag{11.6.1}$$

are chosen. When there is only one angular singularity at the origin, the highly accurate solutions can be obtained by the CTM, and the exponential convergence rates are shown numerically. However, where there exists a mild singularity of $O(r^k \ln r), k = 1,2$ at some corners, the accuracy of the numerical solutions is declined significantly and only the polynomial convergences rates can be observed. Hence, we conclude that the divisions of S into three subdomains and the use of piecewise admissible functions as eqn. (11.5.3) are necessary to achieve the highly accurate solutions by the CTM, and to retain the exponential convergence rates. A recent study of error analysis reveals that the errors $\|\varepsilon\|_B = O\left(\frac{1}{L^{k+\frac{1}{2}-\delta}}\right)$, where $0 < \delta \ll 1$.

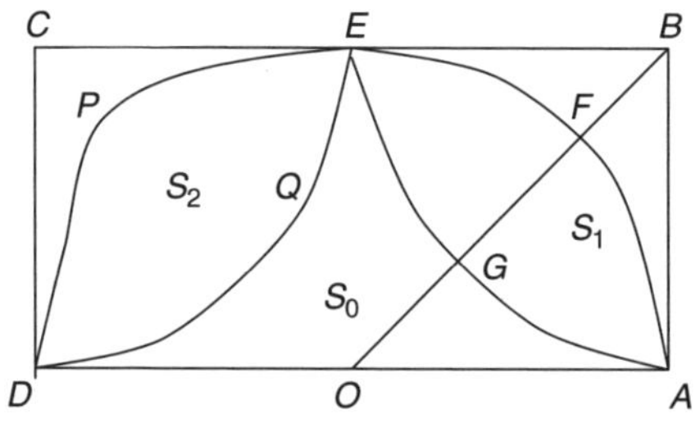

Figure 11.10: Overlapped subdomains of S.

5. Highly accurate CTM in Chapter 2 can be extended to the complicated problems by using the piecewise harmonic solutions as shown in Model B of Motz's variant problems, and by employing the SAM. For Model B, let S be divided into three overlapped subdomains S_0, S_1, and S_2 in fig. 11.10. We may carry out the CTM in each S_i including just one singularity, and use a few iterations to reach the solutions of Model B with the exponential convergence rates. In Li [277] and Li, Huang, and Chen [291], different interior boundary conditions in the SAM are investigated, to speed up the convergence rates.

Appendix Historic review of boundary methods

This appendix explores the rich heritage of boundary methods, which are not limited to the Trefftz method treated in this book. By "boundary methods" we refer to numerical methods for solving boundary value problems of partial differential equations (PDEs) that discretizes only the boundary of the solution domain. This is in contrast to the "domain methods" such as the finite element method (FEM) and the finite difference method (FDM) that require the discretization of the entire domain into discrete unknowns.

This reduction of dimension in discretization leads to a smaller size solution system, which not only reduces the computer storage requirement, but also results in more efficient computer solution. This effect is most pronounced when the domain is unbounded. In domain methods, unbounded domain must be truncated and approximated. Boundary methods, on the other hand, often automatically satisfy the condition at infinity and no additional effort is needed.

In the industrial setting, mesh generation is generally the most labor intensive part of numerical modeling, particularly in terms of the preparation of connectivity data for the FEM in Belytschko, Lu, and Gu [30]. Without the need of interior mesh, boundary methods are more cost-effective in mesh preparation. For problems involving moving boundaries, the frequent re-meshing is also much easier with boundary methods. Hence, boundary methods, whenever applicable, should be the numerical methods of the first choice.

In this appendix, we shall examine the mathematical foundation of boundary methods from the historic perspective, including the potential theory, boundary value problems, Green's functions, Green's identities, Fredholm integral equations, and Ritz and Trefftz methods. For the interest of the beginners of this field, biographical sketches of the pioneers of the field are provided. This appendix is adopted from the article by Cheng and Cheng [93].

A.1 Potential theory

The Laplace equation is one of the most widely used PDEs for modeling science and engineering problems. It typically comes from the physical consequence of combining a phenomenological gradient law (such as the Fourier law in heat conduction and the Darcy law in groundwater flow) with a conservation law (such as the heat energy conservation and the mass conservation of an incompressible substance). For example, Fourier law was presented by *Jean Baptiste Joseph Fourier* (1768–1830) in 1822, see Fourier [147]. It states that the heat flux in a thermal conducting medium is proportional to the spatial gradient of temperature distribution:

$$\mathbf{q} = -k\,\nabla T, \tag{A.1.1}$$

where $\mathbf{q}$ is the heat flux vector, k the thermal conductivity, and T the temperature. The steady-state heat energy conservation requires that at any point in space

$$\nabla \cdot \mathbf{q} = 0. \tag{A.1.2}$$

Combining eqns. (A.1.1) and (A.1.2) and assuming that k is a constant, we obtain the Laplace equation

$$\nabla^2 T = 0. \tag{A.1.3}$$

For groundwater flow, similar procedure leads to

$$\nabla^2 h = 0, \tag{A.1.4}$$

where h is the piezometric head. It is of interest to mention that the notation ∇ used in the above equation came from *William Rowan Hamilton*(1805–1865). The symbol ∇, known as "nabla", is a Hebrew stringed instrument that has a triangular shape in Gellert et al. [159].

The above theories are based on physical quantities. The second way that the Laplace equation arises is through the mathematical concept of finding a "potential" that has no direct physical meaning. In fluid mechanics, the velocity of an incompressible fluid flow satisfies the divergence equation

$$\nabla \cdot \mathbf{v} = 0, \tag{A.1.5}$$

which is again based on the mass conservation principle. For an inviscid fluid flow that is irrotational, it satisfies the equation of curl

$$\nabla \times \mathbf{v} = 0. \tag{A.1.6}$$

It can be shown mathematically that the identity, i.e., eqn. (A.1.6) guarantees the existence of a scalar potential ϕ such that

$$\mathbf{v} = \nabla\phi, \tag{A.1.7}$$

Combining eqns. (A.1.5) and (A.1.7) we again obtain the Laplace equation. We notice that ϕ, called the velocity potential, is a mathematical creation; it is not associated with any measurable physical quantity. In fact, the phrase "potential function" was coined by *George Green* (1793–1841) in his 1828 study [175] of electrostatics and magnetics: electric and magnetic potentials were used as convenient tools for manipulating the solution of electric and magnetic forces.

The original derivation of Laplace equation, however, was based on the study of gravitational attraction, based on the third law of motion of *Isaac Newton* (1643–1727):

$$\mathbf{F} = -\frac{G\, m_1\, m_2\, \mathbf{r}}{|\mathbf{r}|^3}, \tag{A.1.8}$$

where $\mathbf{F}$ is the force field, G the gravitational constant, m_1 and m_2 are two concentrated masses, and $\mathbf{r}$ is the distance vector between the two masses. *Joseph-Louis Lagrange* (1736–1813) in 1773 was the first to recognize the existence of a potential function that satisfied the above equation [264]

$$\phi = \frac{1}{r}, \tag{A.1.9}$$

whose spatial gradient gave the gravity force field

$$\mathbf{F} = G\, m_1\, m_2\, \nabla\phi. \tag{A.1.10}$$

Subsequently, *Pierre-Simon Laplace* (1749–1827) in his study of celestial mechanics demonstrated that the gravity potential satisfies the Laplace equation. The equation was first presented in polar coordinates in 1782, and then in the Cartesian form in 1787 [265] as:

$$\frac{\partial^2\phi}{\partial x^2} + \frac{\partial^2\phi}{\partial y^2} + \frac{\partial^2\phi}{\partial z^2} = 0. \tag{A.1.11}$$

The Laplace equation, however, had been used earlier in the context of hydrodynamics by *Leonhard Euler* (1707–1783) in 1755 [140], and by Lagrange in 1760 [263]. But Laplace was credited for making it a standard part of mathematical physics [51, 242]. We note that the gravity potential eqn. (A.1.9) satisfying eqn. (A.1.11) represents a concentrated mass. It is a "fundamental solution" of the Laplace equation.

Simeon-Denis Poisson (1781–1840) derived in 1813 [367] the equation of force potential for points interior to a body with density ρ as

$$\nabla^2\phi = -4\pi\rho. \tag{A.1.12}$$

This is known as the Poisson equation.

A.1.1 Euler

Leonhard Euler (1707–1783) was the son of a Lutheran pastor who lived near Basel, Switzerland. While studying theology at the University of Basel, Euler was attracted to mathematics by the leading mathematician at the time, *Johann Bernoulli* (1667–1748), and his two mathematician sons, *Nicolaus* (1695–1726) and *Daniel* (1700–1782). With no opportunity in finding a position in Switzerland due to his young age, Euler followed Nicolaus and Daniel to Russia. Later, at the age of 26, he succeeded Daniel as the chief mathematician of the Academy of St. Petersburg. Euler surprised the Russian mathematicians by computing in three days some astronomical tables whose construction was expected to take several months.

In 1741, Euler accepted the invitation of Frederick the Great to direct the mathematical division of the Berlin Academy, where he stayed for 25 years. The relation with the King, however, deteriorated toward the end of his stay; hence, Euler returned to St.Petersburg in 1766. Euler soon became totally blind after returning to Russia. By dictation, he published nearly half of his papers in the last 17 years of his life. In his words, "Now I will have less distraction." Without doubt, Euler was the most prolific and versatile scientific writer of all times. During his lifetime, he published more than 700 books and papers, and it took St. Petersburg's Academy next 47 years to publish the manuscripts he left behind [68]. The modern effort of publishing Euler's collected works, the *Opera Omnia* [141], begun in 1911. However, after 73 volumes and 25,000 pages, the work is unfinished to the present day.

Euler contributed to many branches of mathematics, mechanics, and physics, including algebra, trigonometry, analytical geometry, calculus, complex variables, number theory, combinatorics, hydrodynamics, and elasticity. He was the one who set mathematics into the modern notations. We owe Euler the notations of "e" for the base of natural logs, "π" for pi, "i" for $\sqrt{-1}$, "Σ" for summation, and the concept of functions.

Carl Friedrich Gauss (1777–1855) has been called the greatest mathematician in modern mathematics for his setting up the rigorous foundation for mathematics. Euler, on the other hand, was more intuitive and has been criticized by pure mathematicians as being lacking modern-day rigor. However, by the abundance of deep and indelible marks that Euler made in science and engineering, he can certainly earn the title of the greatest *applied* mathematician ever lived [131].

A.1.2 Lagrange

Joseph-Louis Lagrange (1736–1813), Italian by birth, German by adoption, and French by choice, was next to Euler, the foremost mathematician of the 18th century. At age 18 he was appointed professor of geometry at the Royal Artillery School in Turin. Euler was impressed by his work and arranged a prestigious position for him in Prussia. Despite the inferior condition in Turin, Lagrange only wanted to be able to devote his time to mathematics; hence, he declined the offer. However, in 1766, when Euler left Berlin for St. Petersburg, Frederick the Great arranged for Lagrange to fill the vacated post. Accompanying the invitation was a modest message saying, "It is necessary that the greatest geometer of Europe should live near the greatest of Kings." To D'Alembert, who recommended Lagrange, the king wrote, "To your care and recommendation am I indebted for having replaced a half-blind mathematician with a mathematician with both eyes, which will especially please the anatomical members of my academy."

After the death of Frederick, the situation in Prussia became unpleasant for Lagrange. He left Berlin in 1787 to become a member of the Académie des Sciences in Paris, where he remained for the rest of his career. Lagrange's contributions were mostly in the theoretical branch of mathematics. In 1788, he published the monumental work *Mécanique Analytique* that unified the knowledge of mechanics up to that time. He banished the geometric idea and introduced differential equations. In the preface, he proudly announced: "One will not find figures in this work. The methods that I expound require neither constructions, nor geometrical or mechanical arguments, but only algebraic operations, subject to a regular and uniform course" [68, 347].

A.1.3 Laplace

Pierre-Simon Laplace (1749–1827), born in Normandy, France, came from relatively humble origins. But with the help of *Jean le Rond D'Alembert* (1717–1783), he was appointed professor of mathematics at the Paris Ecole Militaire when he was only 20 years old. Some years later, as examiner of the scholars of the royal artillery corps, Laplace happened to examine a 16-year old sub-lieutenant named Napoleon Bonaparte. Fortunately for both their careers, the examinee passed. When Napoleon came to power, Laplace was rewarded: he was appointed the minister of interior for a short period of time, and later the president of the senate. Upon presenting the monumental work to Napoleon, the emperor teasingly chided Laplace for an apparent oversight: "They told me that you have written this huge book on the system of the universe without once mentioning its Creator." Whereupon Laplace drew himself up and bluntly replied, "I have no need for that hypothesis" [68].

Among Laplace's greatest achievement was the five-volume *Traité du Mécanique Céleste* that incorporated all the important discoveries of planetary system of the previous century, deduced from Newton's law of gravitation. He was eulogized by his disciple Poisson as "the Newton of France" [180]. Other important contributions of Laplace in mathematics and physics included probability, Laplace transform, celestial mechanics, the velocity of sound, and capillary action. He was considered more than anyone else to have set the foundation of the probability theory [163].

A.1.4 Fourier

Jean Baptiste Joseph Fourier (1768–1830), born in Auxerre, France, was the ninth of the 12 children of his father's second marriage. One of his letters showed that he really wanted to make a major impact in mathematics: "*Yesterday was my 21st birthday; at that age Newton and Pascal had already acquired many claims to immortality.*" In 1790, Fourier became a teacher at the Benedictine College, where he had studied earlier. Soon after, he was entangled in the French Revolution and joined the local revolutionary committee. He was arrested in 1794, and almost went to the guillotine. Only the political changes resulted in his being released. In 1794, Fourier was admitted to the newly established Ecole Normale in Paris, where he was taught by Lagrange, Laplace, and *Gaspard Monge* (1746–1818). In 1797, he succeeded Lagrange in being appointed to the chair of analysis and mechanics.

In 1787, Fourier joined Napoleon's army in its invasion of Egypt as a scientific advisor. It was there that he recorded many observations that later led to his work in heat diffusion. Fourier returned to Paris in 1801. Soon Napoleon appointed him as the Prefect of Isére, headquartered at Grenoble. Among his achievements in this administrative position included the draining of swamps of Bourgoin and the construction of a new highway between Grenoble and Turin. Some of his most important scientific contributions came during this period (1802–1814). In 1807, he completed his memoir *On the Propagation of Heat in Solid Bodies* in which he not only expounded his idea about heat diffusion, but also outlined his new method of mathematical analysis, which we now call Fourier analysis. This memoir however was never published, because one of its examiner, Lagrange, objected to his use of trigonometric series to express initial temperature. Fourier was elected to the Académie des Sciences in 1817. In 1822, he published *The Analytical Theory of Heat* [147], 10 years after its winning the Institut de France competition of the Grand Prize in Mathematics in 1812. The judges however criticized that he had not proven the completeness of the trigonometric (Fourier) series. The proof will come many years later by others [174].

A.1.5 Poisson

Simeon-Denis Poisson (1781–1840) was born in Pithiviers, France. In 1796, Poisson was sent to Fontainebleau to enroll in the Ecole Centrale there. He soon showed great talents for learning, especially mathematics. His teachers at the Ecole Centrale were extremely impressed and encouraged him to sit in the entrance examinations for the Ecole Polytechnique in Paris, the premiere institution at the time. Although he had far less formal education than most of the young men taking the examinations, he achieved the top place. His teachers Laplace and Lagrange quickly saw his mathematical talents. They were to become friends for life with their extremely able young student and they gave him strong support in a variety of ways. In his final year of study, he wrote a paper on the theory of equations and Bezout's theorem, and this was of such quality that he was allowed to graduate in 1800 without taking the final examination. He proceeded immediately to the position equivalent to the present-day assistant professor in the Ecole Polytechnique at the age of 19, mainly on the strong recommendation of Laplace. It was quite unusual for anyone to gain their first appointment in Paris, as most of the top mathematicians had to serve in the provinces before returning to Paris. Poisson was named associate professor in 1802, a position he held until 1806 when he was appointed to the professorship at the Ecole Polytechnique, which Fourier vacated when he was sent by Napoleon to

Grenoble. In 1813, in his effort to answer the challenge question for the election to the Académie des Sciences, he developed the Poisson's eqn. (A.1.12) to solve the electrical field caused by distributed electrical charges in a body.

Poisson made great contributions in both mathematics and physics. His name is attached to a wide variety of ideas, for example, Poisson's integral, Poisson equation, Poisson brackets in differential equations, Poisson's ratio in elasticity, and Poisson's constant in electricity [347].

A.1.6 Hamilton

William Rowan Hamilton (1805–1865) was an extremely precocious child. At the age of five, he read Greek, Hebrew, and Latin; at 10, he was acquainted with half a dozen oriental languages. He entered Trinity College, Dublin at the age of 18. His performance was so outstanding that he was appointed Professor of Astronomy and the Royal Astronomer of Ireland when he was still an undergraduate at Trinity. Hamilton was knighted at the age of 30 for the scientific work he had already achieved.

Among Hamilton's most important contributions is the establishment of an analogy between the optical theory of systems of rays and the dynamics of moving bodies. With the further development by *Carl Gustav Jacobi* (1804–1851), this theory is generally known as the Hamilton–Jacobi principle. By this construction, for example, it was possible to determine the 10 planetary orbits around the Sun, a feat that normally required the solution of 30 ordinary differential equations (ODEs), by merely two equations involving Hamilton's characteristic functions. However, this method was more elegant than practical; hence for almost a century, Hamilton's great method was more praised than used [349].

This situation however changed when *Irwin Schrödinger* (1887–1961) introduced the revolutionary wave-function model for quantum mechanics in 1926. Schrödinger had expressed Hamilton's significance quite unequivocally: "*The modern development of physics is constantly enhancing Hamilton's name. His famous analogy between optics and mechanics virtually anticipated wave mechanics, which did not have much to add to his ideas and only had to take them more seriously . . . If you wish to apply modern theory to any particular problem, you must start with putting the problem in Hamiltonian form*" [51].

A.2 Existence and uniqueness

The potential problems we solve are normally posed as boundary value problems. For example, given a closed region Ω with the boundary Γ and the boundary condition

$$\phi = f(\mathbf{x}), \quad \mathbf{x} \in \Gamma, \tag{A.2.1}$$

where $f(\mathbf{x})$ is a continuous function, we are asked to find a harmonic function (meaning a function satisfying the Laplace equation) $\phi(\mathbf{x})$ that fulfills the boundary condition, i.e., eqn. (A.2.1). This is known as the Dirichlet problem, named after *Johann Peter Gustav Lejeune Dirichlet* (1805–1859). The corresponding problem of finding a harmonic function with the normal derivative boundary condition,

$$\frac{\partial \phi}{\partial n} = g(\mathbf{x}), \quad \mathbf{x} \in \Gamma, \tag{A.2.2}$$

where n is the outward normal, is called the Neumann problem, after *Carl Gottfried Neumann* (1832–1925).

The question of whether a solution of a Dirichlet or a Neumann problem exists, and when it exists, whether it is unique or not, is of great importance in mathematics. Obviously, if we cannot guarantee the existence of a solution, the effort of finding it can be in vain. If a solution exists, but may not be unique, then we cannot tie the solution to the associated physical problem, because we believe the physical state is unique.

The question of uniqueness is easier to answer: if a solution exists, it is unique for the Dirichlet problem, and it is unique to within an arbitrary constant for the Neumann problem. (See, for example, the proof in the classical monograph *Foundations of Potential Theory* [242] by *Oliver Dimon Kellogg* (1878–1932).) The existence theorem, however, is more difficult to prove.

To a physicist, the existence question seems to be moot. We may argue that if the mathematical problem correctly describes a physical problem, then a mathematical solution must exist, because the physical state exists. For example, Green in his 1828 seminal work [175], in which he developed Green's identities and Green's functions, presented a similar argument. He reasoned that if for a given closed region Ω, there *exists* a harmonic function U (an assumption that will be justified later) that satisfies the boundary condition

$$U = -\frac{1}{r} \quad \text{on } \Gamma, \tag{A.2.3}$$

then one can define the function

$$G = \frac{1}{r} + U. \tag{A.2.4}$$

It is clear that G satisfy the Laplace equation everywhere except at the pole, where it is singular. Furthermore, G takes the null value at the boundary Γ. G is known as the Green's function. Green went on to prove that for a harmonic function ϕ, whose boundary value is given by a continuous function $\phi(\mathbf{x})$, $\mathbf{x} \in \Gamma$, its solution is represented by the boundary integral equation [242, 434]

$$\phi(\mathbf{x}) = -\frac{1}{4\pi} \iint_{\Gamma} \phi \frac{\partial G}{\partial n}\, dS, \quad \mathbf{x} \in \Omega, \tag{A.2.5}$$

where dS denotes surface integral. Since eqn. (A.2.5) gives the solution of the Dirichlet boundary value problem, hence the solution exists!

The above proof hinges on the existence of U, which is taken for granted at this point. How can we be sure that U exists for an arbitrary closed region Ω? Green argued that U is nothing but the electrical potential created by the charge on a grounded sheet conductor, whose shape takes the form of Γ, induced by a single charge located inside Ω. This physical state obviously exists; hence U must exist! It seems that the Dirichlet problem is proven. But is it?

In fact, mathematicians can construct counter examples for which a solution does not exist. An example was presented by *Henri Léon Lebesgue* (1875–1941) [268], which can be described as follows. Consider a deformable body whose surface is pushed inward by a sharp spine. If the tip of the deformed surface is sharp enough, for example, given by the revolution of the curve $y = \exp(-1/x)$, then the tip is an exceptional point and the Dirichlet problem is not always solvable. (See Kellogg [242] for more discussion.) Furthermore, if the deformed surface closes onto itself to become a single line protruding into the body, then a Dirichlet condition cannot be arbitrarily prescribed on this degenerated boundary, as it is equivalent to prescribing a value inside the domain!

Generally speaking, the existence and uniqueness theorem for potential problems has been proven for interior and exterior boundary value problems of the Dirichlet, Neumann, Robin, and mixed type, if the bounding surface and the boundary condition satisfy certain smoothness condition [223, 242]. (For interior Neumann problem, the uniqueness is only up to an arbitrarily additive constant.) For the existing proofs, the bounding surface Γ needs to be a "Liapunov surface," which is a surface of the $C^{1,\alpha}$ continuity class, where $0 \leq \alpha < 1$. Put it simply, the smoothness of the surface is such that on every point there exists a tangent plane and a normal, but not necessarily a curvature. Corners and edges, on which a tangent plane does not exist, are not allowed in this class. This puts great restrictions on the type of problems that one can solve. On the other hand, in numerical solutions such as the FEM and the boundary element method (BEM), the solution is often sought in the weak sense by minimizing an energy norm in some sense, such as the well-known Galerkin scheme. In this case, the existence theorem has been proven for surface Γ in the $C^{0,1}$ class [106], known as the Lipschitz surface, which is a more general class than the Liapunov surface, such that edges and corners are allowed.

A.2.1 Dirichlet

Johann Peter Gustav Lejeune Dirichlet (1805–1859) was born in Düren (present-day Germany), French Empire. He attended the Jesuit College in Cologne at the age of 14. There he had the good fortune to be taught by *Georg Simon Ohm* (1789–1854). At the age of 16, Dirichlet entered the Collège de France in Paris, where he had some of the leading mathematicians as teachers. In 1825, he published his first paper proving a case in Fermat's Last Theorem, which gained him instant fame. Encouraged by *Alexander von Humboldt* (1769–1859), who made recommendations on his behalf, Dirichlet returned to Germany the same year. From 1827 Dirichlet taught at the University of Breslau. Again with von Humboldt's help, he moved to Berlin in 1828 where he was appointed at the Military College. Soon after this he was appointed a professor at the University of Berlin where he remained from 1828 to 1855. Dirichlet was appointed to the Berlin Academy of Sciences in 1831. An improved salary from the university put him in a position to marry, and he married Rebecca Mendelssohn, one of the composer Felix Mendelssohn's two sisters. Dirichlet had a lifelong friendship with Jacobi, who taught at Königsberg, and the two exerted considerable influence on each other in their researches in number theory. Dirichlet had a high teaching load and in 1853 he complained in a letter to his pupil *Leopold Kronecker* (1823–1891) that he had 13 lectures a week to give, in addition to many other duties. It was therefore something of a relief when, on Gauss's death in 1855, he was offered his chair at Göttingen. Sadly he was not to enjoy the new life for long. He died in 1859 from a heart attack.

Dirichlet made great contributions to the number theory, and the analytic number theory may be said to begin with him. In mechanics, he investigated Laplace's problem on the stability of the solar system, which led him to the Dirichlet problem concerning harmonic functions with given boundary conditions. Dirichlet is also well known for his papers on conditions for the convergence of trigonometric series. Because of this work Dirichlet is considered the founder of the theory of Fourier series [347].

A.2.2 Neumann

Carl Gottfried Neumann (1832–1925) was the son of *Franz Neumann* (1798–1895), a famous physicist who made contributions in thermodynamics. His mother was a sister-in-law of *Friedrich Wilhelm Bessel* (1784–1846). Neumann was born in Königsberg where his father was the professor of physics at the university. Neumann entered the University of Königsberg and received his doctorate in 1855. He worked on his habilitation at the University of Halle in 1858. He taught several universities, including Halle, Basel, and Tübingen. Finally, he moved to a chair at the University of Leipzig in 1868, and would stay there until his retirement in 1911.

He worked on a wide range of topics in applied mathematics such as mathematical physics, potential theory, and electrodynamics. He also made important pure mathematical contributions. He studied the order of connectivity of Riemann surfaces. During the 1860s Neumann wrote papers on the Dirichlet principle and the "logarithmic potential" was a term he coined [347].

A.2.3 Kellogg

Oliver Dimon Kellogg (1878–1932) was born at Linnwood, Pennsylvania. His interest in mathematics was aroused as an undergraduate at Princeton University, where he received his B.A. in 1899. He was awarded a fellowship for graduate studies and obtained a Master degree in 1900 at Princeton. The same fellowship allowed him to spend the next year at the University of Berlin. He then moved to Göttingen to pursue his doctorate. He attended lectures by *David Hilbert* (1862–1943). At that time, *Erik Ivar Fredholm* (1866–1927) had just made progress in proving the existence of Dirichlet problem using integral equations. Hilbert was excited about the development and suggested Kellogg to undertake research on the Dirichlet problem for boundary containing corners, where Fredholm's solution did not apply. Kellogg however failed to answer the question satisfactorily in his thesis and several subsequent papers. With the realization of his errors, he never referred to these papers in his later work. Kellogg was hard to blame because similar errors were later made by both Hilbert and *Jules Henri Poincaré* (1854–1912), and to this date the proof of Dirichlet problem for boundary containing corners has not been accomplished.

Kellogg received his Ph.D. in 1903 and returned to the United States to take up a post of instructor in mathematics at Princeton. Two years later he joined the University of Missouri as an assistant professor. He spent the next 14 fruitful years at Missouri until he was called by Harvard University in 1919. Kellogg continued to work at Harvard until his death [44, 133]. His book "*Foundations of Potential Theory*" [242], first published in 1929, remains among the most authoritative work to this date.

A.3 Reduction in dimensions and Green's formula

A key to the success of BEM is the reduction of spatial dimension in its integral equation representation, leading to a more efficient numerical discretization. One of the most celebrated technique of this type is the divergence theorem, which transforms a volume integral into a surface integral,

$$\iiint_\Omega \nabla \cdot \mathbf{A}\, dV = \iint_\Gamma \mathbf{A} \cdot \mathbf{n}\, dS, \tag{A.3.1}$$

where $\mathbf{A}$ is a vector, $\mathbf{n}$ the unit outward normal of Γ, and dV stands for volume integral. Early development of this type was found in the work of Lagrange [263] and Laplace. Equation (A.3.1), also called Gauss' theorem, is generally attributed to Gauss [156]. However, Gauss in 1813 only presented a few special cases in the form [241]

$$\iint_\Gamma n_x\, dS = 0, \tag{A.3.2}$$

where n_x is the x-component of outward normal, and

$$\iint_\Gamma \mathbf{A} \cdot \mathbf{n}\, dS = 0, \tag{A.3.3}$$

where the components of $\mathbf{A}$ are given by $A_x = A_x(y,z)$, $A_y = A_y(x,z)$, and $A_z = A_z(x,y)$. The general theorem should be credited to *Mikhail Vasilevich Ostrogradski* (1801–1862), who in 1826 presented the following result to the Paris Académie des Sciences [241]

$$\iiint_\Omega \mathbf{a} \cdot \nabla\phi\, dV = \iint_\Gamma \phi\, \mathbf{a} \cdot \mathbf{n}\, dS, \tag{A.3.4}$$

where $\mathbf{a}$ is a constant vector.

Another useful formula is the Stokes' theorem, presented by *George Gabriel Stokes* (1819–1903), which transforms a surface integral into a contour integral [425]

$$\iint_S (\nabla \times \mathbf{A}) \cdot \mathbf{n}\, dS = \int_C \mathbf{A} \cdot d\mathbf{s}, \tag{A.3.5}$$

where S is an open, two-sided curve surface, C the closed contour bounding S, and $d\mathbf{s}$ denotes line integral.

The most important work related to the boundary integral equation solving potential problems came from George Green, whose groundbreaking work remained obscure during his lifetime. Green in 1828 [175] presented the three Green's identities. The first identity is

$$\iiint_\Omega (\phi\nabla^2\psi + \nabla\phi \cdot \nabla\psi)\, dV = \iint_\Gamma \phi \frac{\partial \psi}{\partial n}\, dS. \tag{A.3.6}$$

The above equation easily leads to the second identity

$$\iiint_\Omega (\phi\nabla^2\psi - \psi\nabla^2\phi)\, dV = \iint_\Gamma \left(\phi \frac{\partial \psi}{\partial n} - \psi \frac{\partial \phi}{\partial n}\right) dS. \tag{A.3.7}$$

Using the fundamental solution of Laplace equation $1/r$ in eqn. (A.3.7), the third identity is obtained

$$\phi = \frac{1}{4\pi} \iint_\Gamma \left[\frac{1}{r}\frac{\partial \phi}{\partial n} - \phi \frac{\partial (1/r)}{\partial n}\right] dS, \tag{A.3.8}$$

which is exactly the formulation of the present-day BEM.

A.3.1 Gauss

Carl Friedrich Gauss (1777–1855) was born an infant prodigy into a poor and unlettered family. According to a well-authenticated story, he corrected an error in his father's payroll calculations as a child of three. However, his early career was not very successful and he had to continue to rely on the support from his benefactor Duke Ferdinand of Braunschweig. At the age of 22, he published the most celebrated work as his doctoral thesis, the *Fundamental Theorem of Algebra*. In 1807, Gauss was finally able to secure a position as the director of the newly founded observatory at the Göttingen University, a job he held for the rest of his lifetime.

Gauss devoted more of his time in theoretical astronomy than in mathematics. This is considered a great loss for mathematics – just imagine how much more mathematics could have been accomplished. He devised a procedure for calculating the orbits planetoids that included the use of least square that he developed. Using his superior method, Gauss redid in an hour's time the calculation on which Euler had spent three days, and which sometimes was said to have led to Euler's loss of sight. Gauss remarked unkindly, "*I should also have gone blind if I had calculated in that fashion for three days.*"

Gauss not only adorned every branch of pure mathematics and was called the Prince of Mathematicians, but also pursued work in several related scientific fields, notably physics, mechanics, and astronomy. Together with *Wilhelm Eduard Weber* (1804–1891), he explored electromagnetism. They were the first to have successfully transmitted telegraph [67, 68].

A.3.2 Green

George Green (1793–1841) was virtually unknown as a mathematician during his lifetime. His most important piece of work was discovered posthumously. As the son of a semi-literate, but well-to-do Nottingham baker and miller, Green was sent to a private academy at the age of eight, and left school at nine. This was the only formal education that he received until adulthood. For the next 20 years after leaving primary school, no one knew how, and from whom Green could have acquainted himself to the advanced mathematics of his day in a backwater place like Nottingham. Even the whole country of England in those days was scientifically depressed as compared to the continental Europe. Hence, it was a mystery how Green could have produced as his first publication such a masterpiece.

The next time there existed a record about Green was in 1823. At the age of 30, he joined the Nottingham Subscription Library as a subscriber. In the library he had access to books and journals. Also he had the opportunity to meet with people from the higher society. The next five years were not easy for Green; he had to work full time in the mill, he had two daughters born (he had seven children with Jane Smith, but never married her), and his mother died in 1825. Despite these difficulties in life and his flimsy mathematical background, in 1828 he self-published one of the most important mathematical works of all times – *An Essay on the Application of Mathematical Analysis to the Theories of Electricity and Magnetism* [175]. The essay had 51 subscribers, each paid 7 shillings 6 pence, a sum equal to a poor man's weekly wage, for a work which they could hardly understood a word. One subscriber, Sir Edward Bromhead, however, was impressed by Green's prowess in mathematics. He encouraged and recommended Green to attend Cambridge University.

Several years later, Green finally enrolled at Cambridge University at the age of 40. From 1833 to 1836, Green wrote three more papers, two on electricity published by the Cambridge Philosophical Society and one on hydrodynamics published by the Royal Society of Edinburgh. After graduating in 1837, he stayed at Cambridge for a few years to work on his own mathematics and to wait for an appointment. In 1838 to 1839, he had two papers in hydrodynamics, two papers on reflection and refraction of light, and two papers on sound [176]. In 1839, he was elected to a Parse Fellowship in Cambridge, a junior position. Due to poor health, he had to return to Nottingham in 1840. He died in 1841 at the age of 47. At the time of his death, his work was virtually unknown.

A few weeks before Green's death, *William Thomson* (*Lord Kelvin*) (1824–1907) was admitted to Cambridge. While studying the subject of electricity as a part of preparation for his Senior Wrangler exam, he first noticed the existence of Green's paper in a footnote of a paper by Robert Murphy. He started to look for a copy, but no one knew about it. After his graduation in 1845, and before his departure to France to enrich his education, he mentioned it to his teacher *William Hopkins* (1793–1866). It happened that Hopkins had three copies. Thomson was immediately excited about what he had read in the paper. He brought the article to Paris and showed it to *Jacques Charles François Sturm* (1803–1855) and *Joseph Liouville* (1809–1882). Later Thomson republished Green's essay, rescuing it from sinking into permanent obscurity [70].

Green's 1828 essay profoundly influenced Thomson and *James Clerk Maxwell* (1831–1879) in their study of electrodynamics and magnetism. The methodology has also been applied to many classical fields of physics such as acoustics, elasticity, and hydrodynamics. During the bicentennial celebration of Green's birth in 1993, physicists *Julian Seymour Schwinger* (1918–1994) and *Freeman J. Dyson* (1923–) delivered speeches on the role of Green's functions in the development of 20th century quantum electrodynamics [70].

A.3.3 Ostrogradski

Mikhail Vasilevich Ostrogradski (1801–1862) was born in Pashennaya, Ukraine. He entered the University of Kharkov in 1816 and studied physics and mathematics. In 1822, he left Russia to study in Paris. Between 1822 and 1827 he attended lectures by Laplace, Fourier, *Adrien-Marie Legendre* (1752–1833), Poisson, and *Augustin-Louis Cauchy* (1789–1857). He made rapid progress in Paris and soon began to publish papers in the Paris Academy. His papers at this time showed the influence of the mathematicians in Paris and he wrote on physics and the integral calculus. These papers were later incorporated in a major work on hydrodynamics, which he published in Paris in 1832.

Ostrogradski went to St. Petersburg in 1828. He presented three important papers on the theory of heat, double integrals, and potential theory to the Academy of Sciences. Largely on the strength of these papers, he was elected an academician in the applied mathematics section. In 1840, he wrote on ballistics introducing the topic to Russia. He should also be considered as the founder of the Russian school of theoretical mechanics [347].

A.3.4 Stokes

George Gabriel Stokes (1819–1903) was born in Skreen, County Sligo, Ireland. In 1837, he entered Pembroke College of Cambridge University. He was coached by William Hopkins, who had among his students Thomson, Maxwell, and *Peter Guthrie Tait* (1831–1901), and had the reputation as the "senior wrangler maker." In 1841, Stokes graduated as Senior Wrangler (the top First Class degree) and was also the first Smith's prizeman. Pembroke College immediately gave him a fellowship.

Inspired by the recent work of Green, Stokes started to undertake research in hydrodynamics and published papers on the motion of incompressible fluids in 1842. After completing the research Stokes discovered that *Jean Marie Constant Duhamel* (1797–1872) had already obtained similar results for the study of heat in solids. Stokes continued his investigations, looking into the internal friction in fluids in motion. After he had deduced the correct equations of motion, Stokes discovered that again he was not the first to obtain the equations, since *Claude Louis Marie Henri Navier* (1785–1836), Poisson, and *Adhémar Jean Claude Barré de Saint-Venant* (1797–1886) had already considered the problem. Stokes decided that his results were sufficiently different and published the work in 1845. Today the fundamental equation of hydrodynamics is called the Navier–Stokes equations. The viscous flow in slow motion is called Stokes flow. The mathematical theorem that carries his name, Stokes theorem, first appeared in print in 1854 as an examination question for the Smith's Prize. It is not known whether any of the students answered the question at that time.

In 1849, Stokes was appointed the Lucasian Professor of Mathematics at Cambridge, the chair Newton once held. In 1851 Stokes was elected to the Royal Society, awarded the Rumford Medal in 1852, and appointed Secretary of the Society in 1854. He held the secretary position until 1885, and was President of the Society from 1885 to 1890. Stokes received the Copley Medal from the Royal Society in 1893, and served as the Master of Pembroke College in 1902–1903 [347].

A.4 Integral equations

Inspired by the use of influence functions as a method for solving beam deflection problems subject to distributed load, Fredholm investigated integral equations [159]. He proved in 1903 [151] the existence and uniqueness of solution of the linear integral equation known as the Fredholm integral equation of the second kind

$$\mu(x) - \lambda \int_a^b K(x,\xi)\,\mu(\xi)\,d\xi = f(x), \quad a \le x \le b, \tag{A.4.1}$$

where λ is a constant, $f(x)$ and $K(x,\xi)$ are given continuous functions, and $\mu(x)$ is the solution sought for.

By virtue of the Fredholm theorem, we can solve a Dirichlet problem by the following formula [423]

$$\phi(\mathbf{x}) = \mp 2\pi\mu(\mathbf{x}) + \iint_\Gamma K(\mathbf{x},\xi)\,\mu(\xi)\,dS(\xi), \quad \mathbf{x} \in \Gamma. \tag{A.4.2}$$

In the above equation, the upper sign corresponds to the interior problem, the lower sign the exterior problem, μ is the distribution density, Γ a closed Liapunov surface, $\phi(\mathbf{x})$ the Dirichlet boundary condition, and the kernel K is given by

$$K(\mathbf{x},\xi) = \frac{\partial}{\partial n(\xi)}\left[\frac{1}{r(\mathbf{x},\xi)}\right]. \tag{A.4.3}$$

The kernel is known as a dipole, or a "double-layer potential." The Fredholm theorem guarantees the existence and uniqueness of μ. Once the distribution density μ is solved from eqn. (A.4.2) by some technique, the full solution of the boundary value problem is given by

$$\phi(\mathbf{x}) = \iint_{\Gamma} \frac{\partial\,[1/r(\mathbf{x},\xi)]}{\partial n(\xi)}\,\mu(\xi)\,dS(\xi), \quad \mathbf{x} \in \Omega, \tag{A.4.4}$$

which is a continuous distribution of the double-layer potential on the boundary.

For the Neumann problem, we can utilize the following boundary equation:

$$\frac{\partial\phi(\mathbf{x})}{\partial n(\mathbf{x})} = \pm 2\pi\sigma(\mathbf{x}) + \iint_{\Gamma} K(\xi,\mathbf{x})\,\sigma(\xi)\,dS(\xi), \quad \mathbf{x} \in \Gamma. \tag{A.4.5}$$

Here again the upper and lower sign, respectively, correspond to the interior and exterior problems, σ is the distribution density, Γ the bounding Liapunov surface, $\partial\phi/\partial n$ the Neumann boundary condition, and the kernel is given by

$$K(\xi,\mathbf{x}) = \frac{\partial}{\partial n(\mathbf{x})}\left[\frac{1}{r(\mathbf{x},\xi)}\right]. \tag{A.4.6}$$

After solving for σ, the potential for the whole domain is given by

$$\phi(\mathbf{x}) = \iint_{\Gamma} \frac{1}{r(\mathbf{x},\xi)}\,\sigma(\xi)\,dS(\xi), \quad \mathbf{x} \in \Omega. \tag{A.4.7}$$

This is equivalent to a distribution of the source, or the "single-layer potential." Fredholm suggested a discretization procedure to solve the above equations. Without a fast computer, however, the idea was impractical; hence, further development of utilizing these equations was limited to analytical work.

For mixed boundary value problems, a pair of integral equations is needed. For the single-layer method applied to interior problems, the following pair

$$\phi(\mathbf{x}) = \iint_{\Gamma} \frac{1}{r(\mathbf{x},\xi)}\,\sigma(\xi)\,dS(\xi), \qquad \mathbf{x} \in \Gamma_{\phi}, \tag{A.4.8}$$

$$\frac{\partial\phi(\mathbf{x})}{\partial n(\mathbf{x})} = \iint_{\mathrm{CPV}} \frac{\partial[1/r(\mathbf{x},\xi)]}{\partial n(\mathbf{x})}\,\sigma(\xi)\,dS(\xi), \quad \mathbf{x} \in \Gamma_{q}, \tag{A.4.9}$$

can be selectively imposed at the Dirichlet part Γ_{ϕ} and the Neumann part Γ_{q} of the boundary. We note that eqn. (A.4.8) is a Fredholm integral equation of the first kind and is weakly singular. For solving Dirichlet problems, numerical modelers often prefer this formulation [223] to the theoretically more stable second-kind integral eqn. (A.4.2). We also notice that eqn. (A.4.9) contains a strong singularity as $\xi \to \mathbf{x}$. It cannot be integrated in the ordinary sense and needs to be interpreted in the "Cauchy principal values" sense, as denoted by CPV under the integral sign.

On a smooth part of the boundary, the result of the limit is just eqn. (A.4.5). This idea of interpreting the non-integrable singularity was introduced by Cauchy in 1814 [79].

A double-layer method can also be formulated to solve mixed boundary value problems, using the following pair of equations:

$$\phi(\mathbf{x}) = \iint_{\text{CPV}} \frac{\partial[1/r(\mathbf{x},\xi)]}{\partial n(\xi)} \mu(\xi)\, dS(\xi), \qquad \mathbf{x} \in \Gamma, \quad \text{(A.4.10)}$$

$$\frac{\partial \phi(\mathbf{x})}{\partial n(\mathbf{x})} = \iint_{\text{HFP}} \frac{\partial}{\partial n(\mathbf{x})} \left[\frac{\partial[1/r(\mathbf{x},\xi)]}{\partial n(\xi)} \right] \mu(\xi)\, dS(\xi), \quad \mathbf{x} \in \Gamma. \quad \text{(A.4.11)}$$

The integral in eqn. (A.4.11) contains a "hypersingularity" and is marked with HFP under the integral sign, standing for "Hadamard finite part." This concept was introduced by *Jacques Salomon Hadamard* (1865–1963) in 1908 [179].

In boundary element terminology, the single- and double-layer methods are referred to as the "indirect methods," because the distribution density, not the potential itself, is solved. The numerical method based on Green's third identity eqn. (A.3.8) is called the "direct method."

Similar formula exists in the complex variable domain. Cauchy in 1825 [78] presented one of the most important theorems in complex variable – the Cauchy integral theorem, from which came the Cauchy integral formula expressed as

$$f(z) = \frac{1}{2\pi i} \int_C \frac{f(\zeta)}{\zeta - z}\, d\zeta, \quad \text{(A.4.12)}$$

where z and ζ are complex variables, f is an analytic function, and C a smooth, closed contour in the complex plane. When z is located on the contour, $z \in C$, eqn. (A.4.12) can be exploited for the numerical solution of boundary value problems, a procedure known as the complex variable BEM.

A.4.1 Cauchy

Augustin-Louis Cauchy (1789–1857) was born in Paris during the difficult time of French Revolution. Cauchy's father was active in the education of young Augustin-Louis. Laplace and Lagrange were frequent visitors at the Cauchy family home, and Lagrange particularly took interest in young Cauchy's mathematical ability. In 1805, Cauchy took the entrance examination for the Ecole Polytechnique and was placed second. In 1807, he entered Ecole des Ponts et Chaussées to study engineering, specializing in highways and bridges, and finished school in two years. At the age of 20, he was appointed as a junior engineer to work on the construction of Port Napoléon in Cherbourg. He worked there for three years and performed excellently. In 1812, he became ill and decided to returned to Paris to seek a teaching position.

Although Cauchy continued to publish important pieces of mathematical work, his initial attempts in seeking academic appointment were unsuccessful. He lost in competition for academic position to Legendre, to *Louis Poinsot* (1777–1859), and to *André Marie Ampère* (1775–1836). But his mathematical output remained strong and in 1814 he published the memoir on definite integrals that later became the basis of his theory of complex functions. In 1815, Cauchy lost out to *Jacques Philippe Marie Binet* (1786–1856) for a mechanics chair at the Ecole Polytechnique, but then he was finally appointed assistant professor of analysis there. In 1816, he won the Grand Prix of the Académie des Sciences for a work on waves, and was later admitted to the Académie. In 1817, he was able to substitute for *Jean-Baptiste Biot* (1774–1862), chair of mathematical physics at the Collège de France, and later for Poisson. It was not until 1821 that he was able to obtain a full position replacing Ampère.

Cauchy was staunchly Catholic and was politically a royalist. By 1830 the political events in Paris forced him to leave Paris and he visited Switzerland. He soon lost all his positions in Paris. In 1831, Cauchy went to Turin and later accepted an offer to become a chair of theoretical physics. In 1833, Cauchy went from Turin to Prague, and returned to Paris in 1838. He regained his position at the Académie but not his teaching positions because he had refused to take an oath of allegiance to the new regime. Due to his political and religious views, he continued to have difficulty in winning important appointment.

Cauchy was probably next to Euler the most published author in mathematics, having produced five textbooks and over 800 articles. Cauchy and his contemporary Gauss were credited for introducing rigor into modern mathematics. It was said that when Cauchy read to the Académie des Sciences in Paris his first paper on the convergence of series, Laplace hurried home to verify that he had not made mistake of using any divergence series in his *Mécanique Céleste*. The formulation of elementary calculus in modern textbooks is essentially what Cauchy expounded in his three great treatises: *Cours d'Analyse de l'École Royale Polytechnique* (1821), *Résumé des Leçons sur le Calcul Infinitésimal* (1823), and *Leçons sur le Calcul Diffé rentiel* (1829). Cauchy was also credited for setting the mathematical foundation for complex variable and elasticity. The basic equation of elasticity is called the Navier–Cauchy equation [29, 172].

A.4.2 Hadamard

Jacques Salomon Hadamard (1865–1963) began his schooling at the Lycée Charlemagne in Paris, where his father taught. In his first few years at school, he was not good at mathematics; he wrote in 1936: "*... in arithmetic, until the fifth grade, I was last or nearly last.*" It was a good mathematics teacher who turned him around. In 1884, Hadamard was placed first in the entrance examination for École Normale Supérieure, where he obtained his doctorate in 1892. His thesis on functions of a complex variable was one of the first to examine the general theory of analytic functions; in particular it contained the first general work on singularities. In the same year, Hadamard received the Grand Prix des Sciences Mathématique for his paper "*Determination of the number of primes less than a given number.*" The topic proposed for the prize, concerning filling gaps in work of *Bernhard*

Riemann (1826–1866) on zeta functions, had been put forward by *Charles Hermite* (1822–1901) with his friend *Thomas Jan Stieltjes* (1856–1894) in mind. However, Stieltjes discovered a gap in his proof and never submitted an entry. The next four years Hadamard was first a lecturer at Bordeaux, and then promoted to professor of astronomy and rational mechanics in 1896. During this time he published his famous determinant inequality. Matrices satisfying this relation are today called Hadamard matrices, which are important in the theory of integral equations, coding theory, and other areas.

In 1897, he resigned his chair in Bordeaux and moved to Paris to take up posts in Sorbonne and Collège de France. His research turned more toward mathematical physics from the time he took up the posts in Paris, yet he always argued strongly that he was a mathematician, not a physicist. His famous 1898 work on geodesics on surfaces of negative curvature laid the foundations of symbolic dynamics. Among the other topics he considered were elasticity, geometrical optics, hydrodynamics, and boundary value problems. He introduced the concept of a well-posed initial value and boundary value problem. Hadamard continued to receive prizes for his research and he was further honored in 1906 with the election as the President of the French Mathematical Society. In 1909, he was appointed to the chair of mechanics at the Collège de France. In the following year, he published *Leçons sur le calcul des variations*, which helped lay the foundations of functional analysis (the word "functional" was introduced by him). Then in 1912 he was appointed as professor of analysis at the École Polytechnique. Near the end of 1912, Hadamard was elected to the Academy of Sciences to succeed Poincaré. After the start of World War II, when France fell to Germany in 1940, Hadamard, being a Jew, escaped to the United States where he was appointed to a visiting position at Columbia University. He left America in 1944 and spent a year in England before returning to Paris after the end of the war. He was lauded as one of the last universal mathematicians whose contributions broadly span the fields of mathematics. He died at the age of 98 year old [332, 347].

A.4.3 Fredholm

Erik Ivar Fredholm (1866–1927) was born in Stockholm, Sweden. After his baccalaureate, Fredholm enrolled in 1886 at the University of Uppsala, which was the only doctorate granting university in Sweden at that time. Through an arrangement he studied under *Magnus Gösta Mittag-Leffler* (1846–1927) at the newly founded University of Stockholm, and acquired his Ph.D. from the University of Uppsala in 1893. Fredholm's first publication "*On a special class of functions*" came in 1890. It impressed Mittag-Leffler so much that he sent a copy of the paper to Poincaré. In 1898, he received the degree of Doctor of Science from the same university.

Fredholm is best remembered for his work on integral equations and spectral theory. Although *Vito Volterra* (1860–1940) before him had studied the integral equation theory, it was Fredholm who provided a more thorough treatment. This work was accomplished during the months of 1899 which Fredholm spent in Paris studying the Dirichlet problem with Poincaré, *Charles Emile Picard* (1856–1941), and Hadamard. In 1900, a preliminary report was published and the work was completed in 1903 [151]. Fredholm's contributions quickly became well known. Hilbert immediately saw the importance and extended Fredholm's work to include a complete eigenvalue theory for the Fredholm integral equation. This work led directly to the theory of Hilbert spaces.

After receiving his Doctor of Science degree, Fredholm was appointed as a lecturer in mathematical physics at the University of Stockholm. He spent his whole career at the University of Stockholm being appointed to a chair in mechanics and mathematical physics in 1906. In 1909–1910, he was Pro-Dean and then Dean in Stockholm University.

Fredholm wrote papers with great care and attention, so he produced work of high quality that quickly gained him a high reputation throughout Europe. However, his papers required so much effort that he wrote only a few. In fact, his *Complete Works* in mathematics comprises of only 160 pages. After 1910, he wrote little beyond revisiting his earlier work [347].

A.5 Extended Green's formula

Green's formula, i.e., eqn. (A.3.8), originally designed to solve electrostatic problems, was such a success that the idea was followed to solve many other physical problems [434], For example, *Hermann Ludwig Ferdinand von Helmholtz* (1821–1894) in his study of acoustic problems presented the following equation in 1860 [187], known as the Helmholtz equation:

$$\nabla^2\phi + k^2\phi = 0, \tag{A.5.1}$$

where k is a constant known as the wave number. He also derived the fundamental solution of eqn. (A.5.1) as

$$\phi = \frac{\cos kr}{r}. \tag{A.5.2}$$

In the same paper, he established the equivalent Green's formula

$$\phi = \frac{1}{4\pi}\iint_\Gamma \left[\frac{\cos kr}{r}\frac{\partial\phi}{\partial n} - \phi\frac{\partial}{\partial n}\left(\frac{\cos kr}{r}\right)\right] dS, \tag{A.5.3}$$

which can be compared to eqn. (A.3.8).

For elasticity, an important step toward deriving Green's formula was made by *Enrico Betti* (1823–1892) in 1872, when he introduced the reciprocity theorem, one of the most celebrated relation in mechanics [36]. The theory can be stated as follows: given two independent elastic states in a static equilibrium, $(\mathbf{u}, \mathbf{t}, \mathbf{F})$ and $(\mathbf{u}', \mathbf{t}', \mathbf{F}')$, where $\mathbf{u}$ and $\mathbf{u}'$ the displacement vectors, $\mathbf{t}$ and $\mathbf{t}'$ the tractions on a closed surface Γ, and $\mathbf{F}$ and $\mathbf{F}'$ the body forces in the enclosed region Ω, they satisfy the following reciprocal relation

$$\iint_{\Gamma} (\mathbf{t}' \cdot \mathbf{u} - \mathbf{t} \cdot \mathbf{u}')\, dS = \iiint_{\Omega} (\mathbf{F} \cdot \mathbf{u}' - \mathbf{F}' \cdot \mathbf{u})\, dV. \tag{A.5.4}$$

The above theorem, known as the Betti–Maxwell reciprocity theorem, was a generalization of the reciprocal principle derived earlier by Maxwell [331] using trusses. *John William Strutt* (*Lord Rayleigh*) (1842–1919) further generalized the above theorem to elastodynamics in the frequency domain, and also extended the forces and displacements concept to generalized forces and generalized displacements [378, 379].

In the same sequence of papers [36, 37], Betti presented the fundamental solution known as the center of dilatation [314]

$$\mathbf{u}^* = \frac{1 - 2\nu}{8\pi G(1 - \nu)} \nabla \left(\frac{1}{r}\right), \tag{A.5.5}$$

where G is the shear modulus, and ν is the Poisson ratio. The use of eqn. (A.5.5) in eqn. (A.5.4) produced the integral representation for dilatation

$$e = \nabla \cdot \mathbf{u} = \iint_{\Gamma} (\mathbf{t} \cdot \mathbf{u}^* - \mathbf{t}^* \cdot \mathbf{u})\, dS + \iiint_{\Omega} \mathbf{F} \cdot \mathbf{u}^*\, dV, \tag{A.5.6}$$

where $\mathbf{t}^*$ is the boundary traction vector of the fundamental solution, i.e., eqn. (A.5.5).

The more useful formula that gives the integral equation representation of displacements, rather than dilatation, requires the fundamental solution of a point force in infinite space, which was provided by Kelvin in 1848 [246],

$$u^*_{ij} = \frac{1}{16\pi G(1 - \nu)} \frac{1}{r} \left[\frac{x_i x_j}{r^2} + (3 - 4\nu)\delta_{ij}\right], \tag{A.5.7}$$

where δ_{ij} is the Kronecker delta. In the above equation, we have switched to the tensor notation, and the second index in u^*_{ij} indicates the direction of the applied point force. Utilizing eqn. (A.5.7), *Carlo Somigliana* (1860–1955) in 1885 [418] developed the following integral representation for displacements:

$$u_j = \iint_{\Gamma} (t_i\, u^*_{ij} - t^*_{ij}\, u_i)\, dS + \iiint_{\Omega} F_i\, u^*_{ij}\, dV. \tag{A.5.8}$$

Equation (A.5.8), called the Somigliana identity, is the elasticity counterpart of Green's formula, i.e., eqn. (A.3.8).

Volterra in 1907 [455] presented the dislocation solution of elasticity, as well as other singular solutions such as the force double and the disclination, generally known as the nuclei of strain [314]. Further dislocation solutions were given by

Somigliana in 1914 [419] and 1915 [420]. For a point dislocation in unbounded three-dimensional space, the resultant displacement field is

$$u^*_{ijk} = \frac{1}{4\pi(1-\nu)}\frac{1}{r^2}\left[(1-2\nu)(\delta_{kj}x_i - \delta_{ij}x_k - \delta_{ik}x_j) - \frac{2}{r^2}x_i x_j x_k\right]. \tag{A.5.9}$$

This singular solution can be distributed over the boundary Γ to give the Volterra integral equation of the first kind [454]

$$u_k = \iint_\Gamma u^*_{kji}\, n_j\, \mu_i\, dS + \iiint_\Omega u^*_{ki}\, F_i\, dV, \tag{A.5.10}$$

where μ_i is the component of the distribution density vector μ, also known as the displacement discontinuity. Equation (A.5.10) is equivalent to eqn. (A.4.10) of the potential problem, and can be called the double-layer method. The counterpart of the single-layer method, i.e., eqn. (A.4.8) is given by the Somigliana integral equation

$$u_j = \iint_\Gamma u^*_{ji}\, \sigma_i\, dS + \iiint_\Omega u^*_{ji}\, F_i\, dV, \tag{A.5.11}$$

where σ_i is the component of the distribution density vector σ, known as the stress discontinuity.

Similar to Cauchy integral eqn. (A.4.12) for potential problems, the complex variable potentials and integral equation representation for elasticity, which was formulated by *Gury Vasilievich Kolosov* (1867–1936) in 1909 [253], exist. These were further developed by *Nikolai Ivanovich Muskhelishvili* (1891–1976) [343, 344].

We can derive the above extended Green's formulae in a unified fashion. Consider the generalized Green's theorem [340]

$$\int_\Omega (\mathbf{v}\,\mathcal{L}\{\mathbf{u}\} - \mathbf{u}\,\mathcal{L}^*\{\mathbf{v}\})\, d\mathbf{x} = \int_\Gamma (\mathbf{v}\,\mathcal{B}\{\mathbf{u}\} - \mathbf{u}\,\mathcal{B}^*\{\mathbf{v}\})\, d\mathbf{x}. \tag{A.5.12}$$

In the above, $\mathbf{u}$ and $\mathbf{v}$ are two independent vector functions, $\mathcal{L}$ a linear partial differential operator, $\mathcal{L}^*$ its adjoint operator, $\mathcal{B}$ the generalized boundary normal derivative, and $\mathcal{B}^*$ its adjoint operator. The right-hand side of eqn. (A.5.12) is the consequence of integration by parts of the left-hand side operators. Equation (A.5.12) may be compared with the Green's second identify, i.e., eqn. (A.3.7). If we assume that $\mathbf{u}$ is the solution $\mathbf{G}$ of the homogeneous equations

$$\mathcal{L}\{\mathbf{u}\} = 0 \quad \text{in } \Omega, \tag{A.5.13}$$

subject to certain boundary conditions, and $\mathbf{v}$ is replaced by the fundamental solution $\mathbf{G}$ of the adjoint operator satisfying

$$\mathcal{L}^*\{\mathbf{G}\} = \delta, \tag{A.5.14}$$

eqn. (A.5.12) becomes the boundary integral equation

$$\mathbf{u} = \int_{\Gamma} (\mathbf{u}\,\mathcal{B}^*\{\mathbf{G}\} - \mathbf{G}\,\mathcal{B}\{\mathbf{u}\})\,d\mathbf{x}. \tag{A.5.15}$$

As an example, we consider that the general second-order linear PDE is two-dimensional

$$\mathcal{L}\{u\} = A\frac{\partial^2 u}{\partial x^2} + 2B\frac{\partial^2 u}{\partial x \partial y} + C\frac{\partial^2 u}{\partial y^2} + D\frac{\partial u}{\partial x} + E\frac{\partial u}{\partial y} + Fu, \tag{A.5.16}$$

where the coefficients A, B, ..., and F are functions of x and y. The generalized Green's second identity in the form of eqn. (A.5.12) exists with the definition of the operators [177]

$$\mathcal{L}^*\{v\} = \frac{\partial^2 Av}{\partial x^2} + 2\frac{\partial^2 Bv}{\partial x \partial y} + \frac{\partial^2 Cv}{\partial y^2} - \frac{\partial Dv}{\partial x} - \frac{\partial Ev}{\partial y} + Fv, \tag{A.5.17}$$

$$\mathcal{B}\{u\} = \left(A\frac{\partial u}{\partial x} + 2B\frac{\partial u}{\partial y}\right) n_x + \left(C\frac{\partial u}{\partial y} + Eu\right) n_y, \tag{A.5.18}$$

$$\mathcal{B}^*\{v\} = \left(\frac{\partial Av}{\partial x} - Dv\right) n_x + \left(2\frac{\partial Bv}{\partial x} + \frac{\partial Cv}{\partial y}\right) n_y. \tag{A.5.19}$$

If we require u and v to satisfy eqns. (A.5.13) and (A.5.14), respectively, we then obtain the boundary integral equation formulation, i.e., eqn. (A.5.15).

A.5.1 Helmholtz

Hermann Ludwig Ferdinand von Helmholtz (1821–1894) was born in Potsdam, Germany. He attended Potsdam Gymnasium where his father was a teacher. His interests at school were mainly in physics. However, due to the financial situation of his family, he accepted a government grant to study medicine at the Royal Friedrich-Wilhelm Institute of Medicine and Surgery in Berlin. His research career began in 1841 when he worked on the connection between nerve fibers and nerve cells in his dissertation. He rejected the dominant physiology theory at that time based on vital forces and strongly argued for founding physiology on the principles of physics and chemistry. He graduated from the Medical Institute in 1843 and had to serve as a military doctor for 10 years. He spent all his spare time doing research. In 1847, he published the important paper "*Uber die Erhaltung der Kraft*" that established the law of conservation of energy. In the following year, Helmholtz was released from his obligation as an army doctor and became an assistant professor and director of the Physiological Institute at Königsberg. In 1855, he was appointed to the chair of anatomy and physiology in Bonn. Although at this time Helmholtz had gained a world reputation, complaints were made to the Ministry of Education from traditionalist that his lectures on anatomy were incompetent. Helmholtz reacted strongly to these criticisms and moved to Heidelberg in 1857 to set up a new Physiology

Institute. Some of his most important work was carried out during this time. In 1858, Helmholtz published his important paper on the motion of a perfect fluid by decomposing it into translation, rotation, and deformation. The study of vortex tube bore important consequences in the later study of turbulence in hydrodynamics, and knot theory in topology. Helmholtz also studied mathematical physics and acoustics, producing in 1863 "*On the Sensation of Tone as a Physiological Basis for the Theory of Music*" [188]. From around 1866 Helmholtz began to move away from physiology and towards physics. When the chair of physics in Berlin became vacant in 1871, he was able to negotiate a new Physics Institute under his control. In 1883, he was ennobled by William I. In 1888, he was appointed as the first President of the Physikalisch-Technische Reichsanstalt at Berlin, a post that he held for the rest of his life [69, 347, 471].

A.5.2 Betti

Enrico Betti (1823–1892) studied mathematics and physics at the University of Pisa. He graduated in 1846 and was appointed as an assistant at the university. He worked at the university at a time when political and military events in Italy were intensifying as the country came nearer to unification. In 1849, Betti returned to his hometown of Pistoia where he became a teacher of mathematics at a secondary school. In 1854, he moved to Florence where again he taught in a secondary school. He was appointed as professor of higher algebra at the University of Pisa in 1857. In the following year, Betti visited the mathematical centers of Europe, including Göttingen, Berlin, and Paris, making many important mathematical contacts. In particular, in Göttingen Betti met and became friendly with Riemann. Back in Pisa he moved in 1859 to the chair of analysis and higher geometry. In 1859, there was a war with Austria and by 1861 the Kingdom of Italy was formally created. Betti served in the government of the new country when he became a member of Parliament in 1862.

In 1863, Riemann left his post as professor in mathematics at Göttingen and moved to Pisa, hoping that warmer weather would cure his tuberculosis. Influenced by his friend Riemann, Betti started to work on potential theory and elasticity. His famous theory of reciprocity in elasticity was published in 1872.

Over quite a number of years Betti mixed political service with service for his university. He served a term as Rector of the University of Pisa and in 1846 became the Director of its teacher's college, the Scuola Normale Superiore, a position he held until his death. Under his leadership the Scuola Normale Superiore in Pisa became the leading Italian center for mathematical research and mathematical education. He served as an Undersecretary of State for education for a few months, and a Senator in the Italian Parliament in 1884 [347].

A.5.3 Kelvin

William Thomson (*Lord Kelvin*) (1824–1907) was well prepared by his father, James Thomson, professor of mathematics at the University of Glasgow, for his career.

He attended Glasgow University at the age of 10, and later entered Cambridge University at 17. It was expected that he would win the senior Wrangler position at graduation, but to his and his father's disappointment, he finished the second Wrangler in 1845. The fierce competition of the "tripos," an honors examination instituted at Cambridge in 1824, attracted many best young minds to Cambridge in those days. Among Thomson's contemporaries were Stokes, a senior Wrangler in 1841, and Maxwell, a second Wrangler in 1854. In 1846, the chair of natural philosophy in Glasgow became vacant. Thomson's father ran a successful campaign to get his son elected to the chair at the age of 22.

Thomson was foremost among the small group of British scientists who helped to lay the foundations of modern physics. His contributions to science included a major role in the development of the second law of thermodynamics, the absolute temperature scale (measured in "kelvins"), and the dynamical theory of heat, besides the mathematical analysis of electricity and magnetism, including the basic ideas for the electromagnetic theory of light, the geophysical determination of the age of the Earth, and fundamental work in hydrodynamics. His theoretical work on submarine telegraphy and his inventions of mirror-galvanometer for use on submarine cables aided Britain in laying the transatlantic cable, thus gaining the lead in world communication. His participation in the telegraph cable project earned him the knighthood in 1866, and a large personal fortune [405, 437].

A.5.4 Rayleigh

John William Strutt (*Lord Rayleigh*) (1842–1919) was the eldest son of the second Baron Rayleigh. After studying in a private school without showing extraordinary signs of scientific capability, he entered Trinity College, Cambridge, in 1861. As an undergraduate, he was coached by *Edward John Routh* (1831–1907), who had the reputation of being an outstanding teacher in mathematics and mechanics. Rayleigh (a title he did not inherit until he was 30 years old) was greatly influenced by Routh, as well as by Stokes. He graduated in 1865 with top honors garnering not only the Senior Wrangler title, but also the first Smith's prizeman. Rayleigh was faced with a difficult decision: knowing that he would succeed to the title of the third Baron Rayleigh, taking up a scientific career was not really acceptable to the members of his family. By this time, however, Rayleigh was determined to devote his life to science so that his social obligations would not stand in his way.

In 1866, Rayleigh was elected to a fellowship of Trinity College. Around that time he read Helmholtz's book *On the Sensations of Tone* [188], and became interested in acoustics. Rayleigh was married in 1871, and had to give up his fellowship at Trinity. In 1872, Rayleigh had an attack of rheumatic fever and was advised to travel to Egypt for his health. He took his wife and several relatives sailed down the Nile during the last months of 1872, returning to England in the spring of 1873. It was during that trip that he started his work on the famous two volume treatise *The Theory of Sound* [379], eventually published in 1877. Rayleigh's father died in 1873. He became the third Baron Rayleigh and had to devote part of his time supervising the estate. In 1879, Maxwell died, and Rayleigh was elected to the vacated post of Cavendish Professor of experimental physics at Cambridge. At the end of 1884, Rayleigh resigned his Cambridge professorship and settled in his estate. There in his "book-room" and self-funded laboratory, he continued his intensive scientific work to the end of his life. Rayleigh made many important scientific contributions including the first correct light scattering theory that explained why the sky is blue, the theory of soliton, the surface wave known as Rayleigh wave, the hydrodynamic similarity theory, and the Rayleigh–Ritz method in elasticity. In 1904, Rayleigh won a Nobel Prize for his 1895 discovery of argon gas. He also served many public functions including being the President of the London Mathematical Society (1876–1878), President of the Royal Society of London (1905–1908), and Chancellor of Cambridge University (1908–until his death) [380, 439].

A.5.5 Volterra

Vito Volterra (1860–1940) was born in Ancona, Italy, a city on the Adriatic Sea. When Volterra was two years old, his father died and he was raised by his uncle. Volterra was able to begin his studies at the Faculty of Natural Sciences of the University of Florence in 1878. In the following year, he won a competition to become a student at the Scuola Normale Superiore di Pisa, and it was at the University of Pisa where he completed his studies in mathematics and physics, graduating in 1882 with a doctorate in physics. Among his teachers were Betti, who held the chair of rational mechanics. Betti was so impressed by his student that upon graduation he appointed Volterra his assistant. In 1883, Volterra was given a professorship in rational mechanics at Pisa; following Betti's death in 1892 he was also in charge of mathematical physics. From 1893 until 1900, he held the chair of rational mechanics at the University of Torino. In 1900, he moved to the University of Rome, succeeding *Eugenio Beltrami* (1835–1900) as professor of mathematical physics. Volterra's work encompassed integral equations, the theory of functions of a line (called functionals since *Jacques Salomon Hadamard* (1865–1963)), the theory of elasticity, integro-differential equations, the description of hereditary phenomena in physics, and mathematical biology. Beginning in 1912, Volterra regularly lectured at the Sorbonne in Paris which became like a second home to him.

In 1922, when the Fascists seized power in Italy, Volterra – a Senator of the Kingdom of Italy since 1905 – was one of the few who spoke out against fascism, especially the proposed changes to the educational system. At that time (1923–1926) he was President of the Accademia Nazionale dei Lincei, and he was regarded as the most eminent man of science in Italy. As a direct result of his unwavering stand, especially his signing of the "Intellectual's Declaration" against fascism in 1926 and, five years later, his refusal to swear the oath of allegiance to the fascist government imposed on all university professors, Volterra was dismissed from his chair at the University of Rome in 1931. In the following year, he was deprived of all his memberships in the scientific academies and cultural institutes in Italy. From that time on he lectured and lived mostly abroad, in Paris, in Spain, and in other European countries. Volterra died in isolation on October 11, 1940 [64].

A.5.6 Somigliana

Carlo Somigliana (1860–1955) began his university study at Pavia, where he was a student of Beltrami. Later he transferred to Pisa and had Betti among his teachers, and Volterra among his contemporaries. He graduated from Scuola Normale Superiore di Pisa in 1881. In 1887, Somigliana began teaching as an assistant at the University of Pavia. In 1892, as the result of a competition, he was appointed as university professor of mathematical physics. Somigliana was called to Turin in 1903 to become the chair of mathematical physics. He held the post until his retirement in 1935, and moved to live in Milan. During the World War II, his apartment was destroyed. After the war, he retreated to his family villa in Casanova Lanza and stayed active in research until near his death in 1955.

Somigliana was a classical physicist–mathematician faithful to the school of Betti and Beltrami. He made important contributions in elasticity. The Somigliana integral equation for elasticity is the equivalent of Green's formula for harmonic functions. He is also known for the Somigliana dislocations. His other contributions included seismic wave propagation and gravimetry [443].

A.5.7 Kolosov

Gury Vasilievich Kolosov (1867–1936) was educated at the University of St. Petersburg. After working at Yurev University from 1902 to 1913, he returned to St. Petersburg where he spent the rest of his career. He studied the mechanics of solid bodies and the theory of elasticity, particularly the complex variable theory. In 1907, Kolosov derived the solution for stresses around an elliptical hole. It showed that the concentration of stress became far greater as the radius of curvature at an end of the hole became small compared with the overall length of the hole. Engineers have to understand Kolosov's results so that stresses can be kept to safe levels [347].

A.6 Pre-electronic computer era

Numerical efforts solving boundary value problems predate the emergence of digital computers. One important effort is the Ritz method proposed by *Walter Ritz* (1878–1909) in 1908 [384]. When applied to subdomains, the Ritz method is considered as the forerunner of the FEM [489]. Ritz's idea involves the use of variational method and trial functions to find approximate solutions of boundary value problems. For example, for the following functional

$$\Pi = \iiint_{\Omega} \frac{1}{2} (\nabla\phi)^2 \, dV - \iint_{\Gamma} \frac{\partial\phi}{\partial n}(\phi - f) \, dS, \tag{A.6.1}$$

finding its stationary value by variational method leads to

$$\delta\Pi = -\iiint_{\Omega} \delta\phi \, \nabla^2\phi \, dV - \iint_{\Gamma} \delta\left(\frac{\partial\phi}{\partial n}\right)(\phi - f) \, dS = 0. \tag{A.6.2}$$

Since the variation is arbitrary, the above equation is equivalent to the statement of Dirichlet problem

$$\nabla^2 \phi = 0 \quad \text{in } \Omega,$$

and

$$\phi = f(\mathbf{x}) \quad \text{on } \Gamma.$$

Ritz proposed to approximate ϕ using a set of trial functions ψ_i by the finite series

$$\phi \approx \sum_{i=1}^{n} \alpha_i \, \psi_i, \tag{A.6.3}$$

where α_i are constant coefficients to be determined. Equation (A.6.3) is substituted into the functional, i.e., eqn. (A.6.1) and the variation is taken with respect to the n unknown coefficients α_i. The domain and boundary integration were performed, often in the subdomains, to produce numerical values. This leads to a linear system that can be solved for α_i. The approximate solution is then evaluated as eqn. (A.6.3). The above procedure involves the integration over the solution domain; hence, it is considered as a domain method, not a boundary method.

Based on the same idea, *Erich Trefftz* (1888–1937) in his 1926 article "*A counterpart to Ritz method*" [441] devised the boundary method, known as the Trefftz method. Utilizing Green's first identity eqn. (A.3.6), we can write eqn. (A.6.1) in an alternate form

$$\Pi = -\iiint_{\Omega} \frac{1}{2} \phi \nabla^2 \phi \, dV - \iint_{\Gamma} \frac{\partial \phi}{\partial n} \left(\frac{1}{2} \phi - f \right) dS. \tag{A.6.4}$$

In making the approximation, i.e., eqn. (A.6.3), Trefftz proposed to use trial functions ψ_i that satisfy the governing differential equation

$$\nabla^2 \psi_i = 0, \tag{A.6.5}$$

but not necessarily the boundary condition. For Laplace equation, these could be the harmonic polynomials

$$\psi_i = \left\{ 1, x, y, z, x^2 - y^2, y^2 - z^2, z^2 - x^2, xy, yz, \ldots \right\}. \tag{A.6.6}$$

With the substitution of eqn. (A.6.3) into eqn. (A.6.4), the domain integral vanishes, and the functional is approximated as

$$\Pi \approx -\iint_{\Gamma} \sum_{i=1}^{n} \alpha_i \frac{\partial \psi_i}{\partial n} \left(\frac{1}{2} \sum_{i=1}^{n} \alpha_i \, \psi_i - f \right) dS. \tag{A.6.7}$$

Taking variation of eqn. (A.6.7) with respect to the undetermined coefficients α_j, and setting each part associated with the variations $\delta\alpha_j$ to zero, we obtain the linear system

$$\sum_{i=1}^{n} a_{ij}\,\alpha_i = b_j; \quad j = 1, \ldots, n, \tag{A.6.8}$$

where

$$a_{ij} = \frac{1}{2}\iint_\Gamma \frac{\partial \psi_i \psi_j}{\partial n}\, dS,$$
$$b_j = \iint_\Gamma f\,\frac{\partial \psi_j}{\partial n}\, dS. \tag{A.6.9}$$

Equation (A.6.8) can be solved for α_i.

The above procedure requires the integration of functions over the solution boundary. Even though this is only a boundary integral, it can still be a tedious job. In the present-day TM, a simpler procedure is often taken. Rather than minimizing the functional over the whole boundary, the boundary condition is enforced point-wise on a finite set of boundary points $\mathbf{x}_j$ as

$$\phi(\mathbf{x}_j) \approx \sum_{i=1}^{n} \alpha_i\,\psi_i(\mathbf{x}_j) = f(\mathbf{x}_j), \quad j = 1, \ldots, n \quad \text{and} \quad \mathbf{x}_j \in \Gamma. \tag{A.6.10}$$

This is a point collocation method and there is no integration involved. Equation (A.6.10) can also be derived in a weighted residual formulation using Dirac delta function as the test function. Such numerical approaches are called the collocation Trefftz method (CTM) in this book, and the exploration of TM and CTM is provided in Introduction and Chapters 1–11.

In the last decade, the study of TM becomes more active. In 1995, there was a special journal issue on TM in Advances in Engineering Software, edited by Kamiya and Kita [235, 250]. In 1996, the first International Workshop on Trefftz Method was held in Cracow, Poland, organized by Zieliński, with proceedings published in special issue [483], in which Zienkiewicz in an article [487] reported the history and development of TM. In 1999, the second International Workshop was held in Sintra, Portugal, with proceedings edited by Freitas and Almeida (see Ref. [152]); and in 2002, the third Workshop was held at the University of Exeter, U.K. Excellent examples of applying TM to various engineering problems were reported in these workshops and proceedings. In this book, we focus on error analysis of TM applied to Laplace's, biharmonic, and Helmholtz equations to provide theoretical support to the excellent performance of TM as observed in the literature. We hope that these results may stimulate the further development of TM.

Following the same spirit of the TM, one can use the fundamental solution as the trial function. Since fundamental solution satisfies the governing equation as

$$\mathcal{L}\{G(\mathbf{x}, \mathbf{x}')\} = \delta(\mathbf{x}, \mathbf{x}'), \tag{A.6.11}$$

where $\mathcal{L}$ is a linear partial differential operator, G the fundamental solution of that operator, and δ the Dirac delta function. It is obvious that the approximate solution

$$\phi(\mathbf{x}) \approx \sum_{i=1}^{n} \alpha_i \, G(\mathbf{x}, \mathbf{x}_i), \quad \mathbf{x} \in \Omega, \quad \mathbf{x}_i \notin \Omega, \tag{A.6.12}$$

satisfies the governing equation as long as the source points $\mathbf{x}_i$ are placed outside of the domain. To ensure that the boundary condition is satisfied, again the point collocation is applied

$$\phi(\mathbf{x}_j) \approx \sum_{i=1}^{n} \alpha_i \, G(\mathbf{x}_j, \mathbf{x}_i) = f(\mathbf{x}_j), \quad j = 1, \ldots, n \quad \text{and} \quad \mathbf{x}_j \in \Gamma. \tag{A.6.13}$$

This is called the method of fundamental solutions.

For exterior domain problems, this technique is well known in fluid mechanics. *William John Macquorn Rankine* (1820–1872) in 1864 [376] showed that the superposition of sources and sinks along an axis, together with a rectilinear flow, creates the flow field of a uniform flow around closed bodies known as the Rankine bodies. Various combinations were experimented to create different bodies. However, there was no control over the shape of the body. *Theodore von Kármán* (1881–1963) in 1927 [456] proposed a collocation procedure to create the desirable body shapes. He distributed $n+1$ sources and sinks of unknown strengths along the axis inside an axisymmetric body, together with a rectilinear flow

$$\phi(\mathbf{x}) \approx Ux + \sum_{i=1}^{n+1} \frac{\sigma_i}{4\pi r(\mathbf{x}, \mathbf{x}_i)}, \tag{A.6.14}$$

where U is the uniform flow velocity, $\mathbf{x}_i$ are located on x-axis, and σ_i are source/sink strengths. The strengths can be determined by forcing the normal flux to vanish at n specified points on the meridional trace of the axisymmetric body. An auxiliary condition

$$\sum_{i=1}^{n+1} \sigma_i = 0, \tag{A.6.15}$$

is needed to ensure the closure of the body. In fact, other singularities, such as doublets (dipoles) and vortices, can be distributed inside a body to create flow around arbitrarily shaped two- and three-dimensional bodies [444].

In 1930, von Kármán [457] further proposed the distribution of singularity along a line inside a two-dimensional streamlined body to generate the potential

$$\phi(\mathbf{x}) = -\int_L \ln r(\mathbf{x}, \xi)\, \sigma(\xi)\, ds(\xi), \quad \mathbf{x} \in \Omega_e, \tag{A.6.16}$$

where ϕ is the perturbed potential from the uniform flow field, σ the distribution density, L a line inside the body, and Ω_e the external domain (fig. A.1a). For

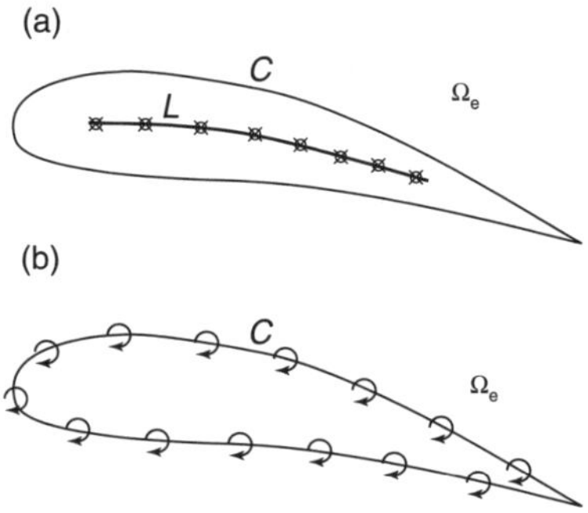

Figure A.1: Two methods of distributing sigularities: (a) sources, sinks, and doublets on a line L inside the airfoil, (b) vortices on the surface $\mathcal{C}$ of the airfoil.

vanishing potential at infinity, the following auxiliary condition is needed

$$\int_L \sigma(\xi)\, ds(\xi) = 0. \tag{A.6.17}$$

To find the distribution density, Neumann boundary condition is enforced on a set of discrete points $\mathbf{x}_i$, $i = 1, \ldots, n$, on the surface of the body

$$\frac{\partial \phi(\mathbf{x}_i)}{\partial n(\mathbf{x}_i)} = -\int_L \frac{\partial \ln r(\mathbf{x}_i, \xi)}{\partial n(\mathbf{x}_i)}\, \sigma(\xi)\, ds(\xi), \quad \mathbf{x}_i \in C, \tag{A.6.18}$$

where C is the boundary contour (fig. A.1a).

Prager [370] in 1928 proposed a different idea: vortices are distributed on the surface of a streamlined body (fig. A.1b) to generate the desirable potential. When this is written in terms of stream function ψ, the integral equation becomes

$$\psi(\mathbf{x}) = -\int_C \ln r(\mathbf{x}, \xi)\, \sigma(\xi)\, ds(\xi), \quad \mathbf{x} \in \Omega_e. \tag{A.6.19}$$

In this case, Dirichlet condition is enforced on the surface of the body.

Lotz in 1932 [313] proposed the discretization of Fredholm integral equation of the second kind on the surface of an axisymmetric body for solving external flow problems. The method was further developed by Vandrey in 1951 and 1960 [446, 447]. Other early efforts in solving potential flows around obstacles, prior to the invention of electronic computers, can be found in a review [199].

In 1937, Muskhelishvili derived the complex variable equations for elasticity and suggested to solve them numerically [342]. The actual numerical implementation was accomplished by Gorgidze and Rukhadze [169] in a procedure that resembled the present-day BEM: it divided the contour into elements, approximated the function within the elements, and formed a linear algebraic system consisting the unknown coefficients.

The above review demonstrates that finding approximate solutions of boundary value problems using boundary or boundary-like discretization is not a new idea. These early attempts of Trefftz, von Kármán, and Muskhelishvili existed prior to the electronic computers. However, despite these heroic attempts, without the aid of modern computing tools these calculations had to be performed by human or mechanical computers. The drudgery of computation was a hindrance for their further development; hence, these methods remained dormant for a while and had to wait for a later date to flourish.

A.6.1 Ritz

Walter Ritz (1878–1909) was born in Sion in the southern Swiss canton of Valais. As a specially gifted student, the young Ritz excelled academically at the Lycèe communal of Sion. In 1897, he entered the Polytechnic school of Zurich where he began studies in engineering. He soon found that he could not live with the approximations and compromises involved with engineering, so he switched to the more mathematically exacting studies in physics, where *Albert Einstein* (1879–1955) was one of his classmates. In 1901, he transferred to Göttingen, where his forming aspirations were strongly influenced by *Woldemar Voigt* (1850–1919) and Hilbert. Ritz's dissertation on spectroscopic theory led to what is known as the Ritz combination principle. In the next few years, he continued his work on radiation, magnetism, electrodynamics, and variational method. But in 1904 his health failed and he returned to Zurich. During the following three years, Ritz unsuccessfully tried to regain his health and was outside the scientific centers. In 1908, he relocated to Göttingen where he qualified as a Privat Dozent. There he produced his opus magnum *Recherches critiques sur l'Électrodynamique Générale*. In 1908–1909, Ritz and Einstein held a war in *Physikalische Zeitschrift* over the proper way to mathematically represent black-body radiation and over the theoretical origin of the second law of thermodynamics; it was judged in his favor. Six weeks after the publication of this series, Ritz died at age 31, leaving behind a short but brilliant career in physics [154].

A.6.2 Von Kármán

Theodore von Kármán (1881–1963) was born in Budapest, Hungary. He was trained as a mechanical engineer in Budapest and graduated in 1902. He did further graduate studies at Göttingen and earned his Ph.D. in 1908 under *Ludwig Prandtl* (1875–1953). In 1911, he made an analysis of the alternating double row of vortices behind a bluff in a fluid stream, known as Kármán's vortex street. In 1912, at the age of 31, he became Professor and Director of Aeronautical Institute at Aachen, where he built the world's first wind tunnel. In World War I, he was called into military service for the Austro–Hungarian Empire and became head of research in the air force, where he led the effort to build the first helicopter. After the war, he was instrumental in calling an international congress on aerodynamics and hydrodynamics at Innsbruck, Austria, in 1922. This meeting became the forerunner of the International Union of Theoretical and Applied Mechanics (IUTAM) with von Kármán as its honorary president. He first visited the United States in 1926. In 1930, he headed the Guggenheim Aeronautical Lab at the Caltech. In 1944, he was cofounder of the present NASA Jet Propulsion Laboratory and undertook America's first governmental long-range missile and space-exploration research program. His personal scientific work included contributions to fluid mechanics, turbulence theory, supersonic flight, mathematics in engineering, and aircraft structures. He is widely recognized as the father of modern aerospace science [458].

A.6.3 Trefftz

Erich Trefftz (1888–1937) was born on February 21, 1888 in Leipzig, Germany. In 1890, the family moved to Aachen. In 1906, he began his studies in mechanical engineering at the Technical University of Aachen, but soon changed to mathematics. In 1908, Trefftz transferred to Göttingen, at that time the Mecca of mathematics and physics. Here after Gauss, Dirichlet, and Riemann, now Hilbert, *Felix Christian Klein* (1849–1925), *Carle David Tolmé Runge* (1856–1927), and Prandtl contributed to continuous progress of first-class mathematical work. Trefftz's most important teachers were Runge (his uncle), Hilbert, and also Prandtl, the genius mechanician in modern fluid- and aero-dynamics. Trefftz spent one year at the Columbia University, New York, and then left Göttingen for Strassburg to study under the guidance of the famous Austrian applied mathematician *Richard von Mises* (1883–1953), who founded the GAMM (Gesellschaft für Angewandte Mathematik und Mechanik) in 1922 together with Prandtl, and *Hans Reissner* (1874–1967). The three of them served as the president (Prandtl), vice president (Reisnner), and secretary (von Mises) for many years. Mises was also the first editor of ZAMM (Zeitschrift für Angewandte Mathematik und Mechanik).

Trefftz's academic career began with his doctoral thesis in Strassburg in 1913, where he solved a mathematical problem of hydrodynamics. He was a soldier in the World War I, but already in 1919 he got his habilitation and became a full professor of mathematics in Aachen. In the year 1922, he got a call as a full professor with a chair in the faculty of mechanical engineering at the Technical University of Dresden. There he became responsible for teaching and research in strength of materials, theory of elasticity, hydrodynamics, aerodynamics, and aeronautics. In 1927, he moved from the engineering to the mathematical and natural science faculty, being appointed there as a chair in Technical (Applied) Mechanics.

Trefftz had a lifelong friendship with von Mises. In 1933, the political climate changed in Germany. The passage of the Civil Service Law provided means of removing Jewish teachers from the universities. Now von Mises and Reissner, being Jewish, offered to Prandtl to resign from GAMM. Informing Trefftz about their colleagues' intention, Prandtl offered to resign himself and suggested Trefftz to become the president. Trefftz replied that Prandtl should continue to be president, but he also wrote: "If we must exclude Jewish members, I would consider dissolution the worthiest action." Eventually Prantdl decided to preserve GAMM and Trefftz was asked to serve as vice president. Von Mises left Germany in 1933, and emigrated to the U.S. in 1939; and Reissner emigrated to the U.S. in 1938. Trefftz felt and showed outgoing solidarity and friendship to von Mises, and he clearly was in expressed distance to the Hitler regime until his early death. Feeling the responsibility for science, he took over the presidentship of GAMM and became the editor of ZAMM in 1933 [421].

Trefftz died in 1937, not quite 49 years old, in Dresden, of a malicious disease. An expressive bust (see page xiii of this book for a photograph) in the Willers building of the Technical University of Dresden is a reminiscence to a great applied mathematician of the 20th century.

A.6.4 Muskhelishvili

Nikolai Ivanovich Muskhelishvili (1891–1976) was a student at the University of St. Petersburg. He was naturally influenced by the glorious tradition of the St. Petersburg mathematical school, which began with Euler and continued by the prominent mathematicians such as Ostrogradsky, *Pafnuty Lvovich Chebyshev* (1821–1894), and *Aleksandr Mikhailovich Lyapunov* (1857–1918). As an undergraduate student, Muskhelishvili was greatly impressed by the lectures of Kolosov on the complex variable theory of elasticity. Muskhelishvili took this topic as his graduation thesis and performed brilliantly that Kolosov decided to publish these results as a coauthor with his student in 1915. In 1922, Muskhelishvili became a professor at the Tbilisi State University, where he remained until his death. In 1935, he published the masterpiece *Some Basic Problems of the Mathematical Theory of Elasticity*, which won him the Stalin Prize of the first degree. He held many positions such as Chair, Director, President, at Tbilisi and the Georgian Branch of USSR Academy of Sciences, and received many honors [490, 491].

A.7 Electronic computer era

Although electronic computers were invented in the 1940s, they did not become widely available to common researchers until the early 1960s. It is not surprising that the development of FEM started around that time [104]. A number of independent efforts of creating boundary methods also emerged in the early 1960s. Some of the more significant ones are reviewed below.

Friedman and Shaw [153] in 1962 solved the scalar wave equation

$$\nabla^2\phi - \frac{1}{c^2}\frac{\partial^2\phi}{\partial t^2} = 0, \tag{A.7.1}$$

where ϕ is the velocity potential and c is the wave speed, for the scattered wave field resulting from a shock wave impinging on a cylindrical obstacle. The use of the fundamental solution

$$G = -\frac{1}{r}\,\delta\left[\frac{r}{c} - (t - t_0)\right], \tag{A.7.2}$$

where δ is the Dirac delta function, in Green's second identity, i.e., eqn. (A.3.7) produces the boundary integral equation

$$\phi_s(\mathbf{x}, t) = \frac{1}{4\pi}\int_0^{t^+}\iint_\Gamma\left[G\frac{\partial\phi_s}{\partial n} - \phi_s\frac{\partial G}{\partial n}\right] dS\, dt_0, \tag{A.7.3}$$

where ϕ_s is the scattered wave field. Equation (A.7.3) was further differentiated with respect to time to create the equation for acoustic pressure. For a two-dimensional problem, the equation was discretized in space (boundary contour) and in time that resulted into a double summation. Variables were assumed to be constant over space and time subintervals, so that the integration could be performed exactly. Finite difference explicit time-stepping scheme was used and the resultant algebraic system required only successive, not simultaneous solution. The computation was performed by hand. The scattering due to a box-shaped rigid obstacle was solved. The work was extended in 1967 by Shaw [406] to handle different boundary conditions on the obstacle surface, and by Mitzner [339] using the retarded potential integral representation.

Banaugh and Goldsmith [21] in 1963 tackled the two-dimensional wave equation in the frequency domain, governed by the Helmholtz eqn. (A.5.1). The two-dimensional boundary integral equation counterpart to the three-dimensional version, i.e., eqn. (A.5.3) is

$$\phi = \frac{i}{4}\int_C\left[H_0^{(1)}(kr)\frac{\partial\phi}{\partial n} - \phi\,\frac{\partial H_0^{(1)}(kr)}{\partial n}\right] ds, \tag{A.7.4}$$

where $H_0^{(1)}$ is the Hankel function of the first kind of order zero, k the wave number, and $i = \sqrt{-1}$. Equation (A.7.4) is solved in the complex variable domain. Similar to Friedman and Shaw [153], the integration over a subinterval was made easy by assuming constant variation of the potential on the subinterval. The problem of a steady-state wave scattered from the surface of a circular cylinder was solved as a demonstration. An IBM 7090 mainframe computer was used for the numerical solution. It is of interest to observe that the discretization was restricted to 36 points, corresponding to 72 unknowns (real and imaginary parts), due to the memory restriction of the computer. A larger linear system would have required the read/write operation on the tape storage and special linear system solution algorithm.

In the same year (1963), Chen and Schweikert [89] solved the three-dimensional sound radiation problem in the frequency domain using the Fredholm integral equation of the second kind

$$\frac{\partial \phi(\mathbf{x})}{\partial n(\mathbf{x})} = -2\pi\sigma(\mathbf{x}) + \int\!\!\int_{\Gamma} \sigma(\xi) \frac{\partial}{\partial n(\mathbf{x})} \left(\frac{e^{ikr}}{r} \right) dS(\xi), \quad \mathbf{x} \in \Gamma. \tag{A.7.5}$$

Problems of vibrating spherical and cylindrical shells in infinite fluid domain were solved. The surface of the body was divided into triangular elements. An IBM 704 mainframe was used, which allowed up to 1,000 degrees of freedom to be modeled.

Maurice Aaron Jaswon (1922–) and Ponter [221] in 1963 employed Green's third identity, i.e., eqn. (A.3.8), but in two-dimensional form,

$$\phi(\mathbf{x}) = \frac{1}{\pi} \int_C \left[\phi(\xi) \frac{\partial \ln r(\mathbf{x}, \xi)}{\partial n(\xi)} - \ln r(\mathbf{x}, \xi) \frac{\partial \phi(\xi)}{\partial n(\xi)} \right] ds(\xi), \quad \mathbf{x} \in C, \tag{A.7.6}$$

for the numerical solution of prismatic bars subjected to torsion. The boundary conditions were Dirichlet type. Ponter [368] in 1966 extended it to multiple domain problems.

Jaswon [223] and Symm [431] in 1963 used the single-layer method, namely, the Fredholm equation of the first kind as shown in eqn. (A.4.8), but in two dimensions,

$$\phi(\mathbf{x}) = -\int_C \ln r(\mathbf{x}, \xi)\, \sigma(\xi)\, ds(\xi), \quad \mathbf{x} \in C, \tag{A.7.7}$$

for the solution of Dirichlet problems. When the logarithmic capacity (i.e., the transfinite diameter) $C_\Gamma \neq 1$, the unique solution of eqn. (A.7.7) exists, where C_Γ is the diameter if Γ is a circle, see Yan [475]. For Neumann problems, the Fredholm integral equation of the second kind

$$\frac{\partial \phi(\mathbf{x})}{\partial n(\mathbf{x})} = \pi\sigma(\mathbf{x}) - \int_C \frac{\partial \ln r(\mathbf{x}, \xi)}{\partial n(\mathbf{x})} \sigma(\xi)\, ds(\xi), \quad \mathbf{x} \in C, \tag{A.7.8}$$

was used. In the same paper [431], a mixed boundary value problem was solved using Green's formula, i.e., eqn. (A.7.6), rather than the Fredholm integral equations.

Hess and Smith [200] in 1964 utilized the single-layer method, i.e., eqn. (A.4.5) to solve problems of external potential flow about arbitrary three-dimensional bodies

$$\frac{\partial \phi(\mathbf{x})}{\partial n(\mathbf{x})} = -2\pi\sigma(\mathbf{x}) + \int\!\!\int_{\Gamma} \frac{\partial(1/r(\mathbf{x}, \xi))}{\partial n(\mathbf{x})} \sigma(\xi)\, dS(\xi), \quad \mathbf{x} \in \Gamma. \tag{A.7.9}$$

The formulation is the same as that of Lotz [313] and Vandrey [446, 447]. The surface of the body is discretized into quadrilateral elements and the source density is assumed to be constant on the element. This technique, called the surface source method, has developed into a powerful numerical tool for the aircraft industry [199].

Massonet [328] in 1965 discussed a number of ideas of using boundary integral equations solving elasticity problems. However, only in two cases numerical solutions were carried out. In the first case, Fredholm integral equation of the second kind was used to solve torsion problems

$$\phi(\mathbf{x}) = -\pi\mu(\mathbf{x}) - \int_C \frac{\partial \ln r(\mathbf{x}, \xi)}{\partial n(\xi)} \mu(\xi)\, ds(\xi), \quad \mathbf{x} \in C. \tag{A.7.10}$$

In the second case, plane elasticity problems were solved using the distribution of the radial stress field resulting from a half-plane point force on the boundary. The following Fredholm equation of the second kind was used:

$$\mathbf{t}(\mathbf{x}) = \mu(\mathbf{x}) - \frac{2}{\pi} \int_C \mu(\xi) \frac{\cos\varphi \cos\alpha}{r} \mathbf{e}_r\, ds(\xi), \quad \mathbf{x} \in C, \tag{A.7.11}$$

where $\mathbf{t}$ is the boundary traction vector, μ the intensity of the fictitious stress, and μ its magnitude, $\mathbf{e}_r$ the unit vector in the r direction, φ the angle between the two vectors μ and $\mathbf{e}_r$, and α the angle between $\mathbf{e}_r$ and the boundary normal. Solutions were found using the iterative procedure of successively approximating the function μ. Due to the half-plane kernel function used, this technique applies only to simply-connected domains.

During the first decade of the 20th century, the introduction of the Fredholm integral equation theorems puts the potential theory on a solid foundation. For elasticity problems, however, similar-level rigorousness was not accomplished for another 40 years. Started in the 1940s, a Georgian school of elasticians led by Muskhelishvili [490, 491] and followed by *Ilia Nestorovich Vekua* (1907–1977) [448], *Nikolai Petrovich Vekua* (1913–1993) [449], and *Viktor Dmitrievich Kupradze* (1903–1985) [258, 260], all associated with the Tbilisi State University, together with *Solomon Grigorevich Mikhlin* (1908–1991) [336] of St. Petersburg, made important progresses in the theory of vector potentials (elasticity) through the study of singular integral equations. The initial development, however, was limited to one-dimensional singular integral equations, which solve only two-dimensional problems. The development of multi-dimensional integral equations started in the 1960s [158].

Kupradze in 1964 [259] and 1965 [258] discussed a method for finding approximate solutions of potential and elasticity (static and dynamic) problems. He called the approach "method of functional equations." Numerical examples were given in two dimensions. For potential problems with the Dirichlet boundary condition

$$\phi = f(\mathbf{x}), \quad \mathbf{x} \in C,$$

where C is the boundary contour, the solution is represented by the pair of integral equations

$$\phi(\mathbf{x}) = \frac{1}{2\pi}\int_C f(\xi)\frac{\partial \ln r(\mathbf{x},\xi)}{\partial n(\xi)}\,ds(\xi) + \frac{1}{\pi}\int_C \sigma(\xi)\,\ln r(\mathbf{x},\xi)\,ds(\xi), \quad \mathbf{x}\in\Omega, \tag{A.7.12}$$

$$0 = \frac{1}{2\pi}\int_C f(\xi)\frac{\partial \ln r(\mathbf{x},\xi)}{\partial n(\xi)}\,ds(\xi) + \frac{1}{\pi}\int_C \sigma(\xi)\,\ln r(\mathbf{x},\xi)\,ds(\xi), \quad \mathbf{x}\in C'. \tag{A.7.13}$$

In the above, C' is an arbitrary auxiliary boundary that encloses C, and σ the distribution density, which needs to be solved from eqn. (A.7.13). We notice that the above equations involve the distribution of both the single-layer and the double-layer potential. Another observation is that in eqn. (A.7.13) the center of singularity $\mathbf{x}$ is located on C', which is *outside* of the solution domain Ω. Since C and C' are distinct contours, eqn. (A.7.13) is not an integral equation in the classical sense, in which the singularities are located on the boundary. This is why the term "functional equation" was used instead. In the numerical implementation, C' was chosen as a circle, upon which n nodes were selected to place the singularity. Since the singularities are not located on the boundary C, the integrals in eqn. (A.7.12) are regular and can be numerically evaluated using a simple quadrature rule. Gaussian quadrature with n nodes was used for the integration. The resultant linear system was solved for the n discrete σ values located at the quadrature nodes. Equation (A.7.12) was then used to find solution at any point in the domain.

For elasticity problem, the same technique was employed. For static, two-dimensional problems with prescribed boundary displacement,

$$u_i = f_i(\mathbf{x}), \quad \mathbf{x}\in C,$$

the following pair of vector integral equations solve the boundary value problem:

$$u_j(\mathbf{x}) = \frac{1}{\pi}\int_C \sigma_i(\xi)\,u^f_{ij}(\mathbf{x},\xi)\,ds(\xi) - \frac{1}{2\pi}\int_C f_i(\xi)\,u^d_{ij}(\mathbf{x},\xi)\,ds(\xi), \quad \mathbf{x}\in\Omega, \tag{A.7.14}$$

$$0 = \frac{1}{\pi}\int_C \sigma_i(\xi)\,u^f_{ij}(\mathbf{x},\xi)\,ds(\xi) - \frac{1}{2\pi}\int_C f_i(\xi)\,u^d_{ij}(\mathbf{x},\xi)\,ds(\xi), \quad \mathbf{x}\in C', \tag{A.7.15}$$

where σ_i is the distribution density (vector), and the kernel

$$u^f_{ij} = \frac{1}{4G(1-\nu)}\left[(3-4\nu)\delta_{ij}\ln r - \frac{x_i x_j}{r^2}\right], \tag{A.7.16}$$

is the fundamental solution due to a point force (single-layer potential) in the x_j direction, and

$$u_{ij}^{d} = \frac{1}{2(1-\nu)}\frac{1}{r}\left[(1-2\nu)\left(\frac{n_i x_j}{r} - \frac{n_j x_i}{r} + \delta_{ij}\frac{\partial r}{\partial n}\right) + 2\frac{x_i x_j}{r^2}\frac{\partial r}{\partial n}\right], \qquad \text{(A.7.17)}$$

is the fundamental solution due to a dislocation (double-layer potential) oriented in the x_j direction. Kupradze's method was closely followed in Russia under the name "potential method," particularly in the solution of shells [157, 445] and plates [257, 450].

Kupradze's technique of distributing fundamental solutions on an exterior, auxiliary boundary has also been considered as the origin of the "method of fundamental solutions" [48]. However, in a narrower definition, the method of fundamental solutions [143] often bypasses the integral equation formulation. It considers the distribution of admissible solutions of discrete and unknown density on an external auxiliary boundary, for example, in the form of eqn. (A.6.12). The boundary conditions are satisfied by collocating at a set of boundary nodes. Hence, the method of fundamental solutions can be viewed as a special case of the method of functional equations and in fact was independently developed. Oliveira in 1968 [351] proposed the use of fundamental solutions of point forces linearly distributed over linear segments to solve plan elasticity problems. For potential problems, its origin can be traced to Mathon and Johnston in 1977 [329].

Another type of problem that has traditionally used boundary methods involves discontinuities such as fractures, dislocations resulting from imperfections of crystalline structures, interface between dissimilar materials, and others. In these cases, certain physical quantities, such as displacements or stresses, suffer a jump. These discontinuities can be simulated by the distribution of singular solutions such as the Volterra [455] and Somigliana dislocations [419, 420] over the physical surface, which often results in integral equations [43, 139]. For example, integral equation of this type

$$A(x)\psi(x) + \frac{1}{\pi}\int_a^b \frac{B(x')\psi(x')}{x'-x}\,dx' + \int_a^b K(x,x')\psi(x')\,dx' = f(x), \quad a < x < b, \qquad \text{(A.7.18)}$$

and other types were numerically investigated by Erdogan and Gupta in 1972 [137, 138] using Chebyshev and Jacobi polynomials for the approximation. These type of one-dimensional singular integral equations have also been solved using piecewise, low-degree polynomials [162].

The review presented so far has focused on the solution of physical and engineering problems, and omitted the development of numerical solution of integral equations in the applied mathematics community. We seek the integral equation representations that allow the numerical schemes with reduction in mesh dimensions. The formulations often borrow the physical idea of distributing concentrated loads; hence, the integral equations are typically singular. Due to the multiple spatial and time dimensions present in physical problems, the integral equations are also multi-dimensional.

For the mathematical community, the effort of finding approximate solutions of integral equations existed since the presentation of fundamental existence theorems of Fredholm in the 1900s. Early efforts involved the finding of successive approximations of linear, one-dimensional, and non-singular integral equations. Different kinds of integral equations that may or may not have physical origin were investigated. One of the first monographs on numerical solution of integral equations was by Bückner in 1952 [66]. Another early monograph was by Mikhlin and Smolitsky [337] in 1967. The field flourished in the 1970s with the publication of several monographs – Kagiwada and Kalaba [234] in 1974, Atkinson [8] in 1976, Ivanov [219] in 1976, and Baker [19] in 1977. As mentioned above, mostly one-dimensional integral equations were investigated. Some integral equations have physical origin such as flow around hydrofoil, population competition, and quantum scattering [125], while others do not. The methods used include projection method, polynomial collocation, Galerkin method, least squares, and quadrature method, among others [165]. It is of interest to observe that the developments in the two communities, the applied mathematics and the engineering, seem to run parallel to each other, almost devoid of any cross citations, although it is clear that many of the techniques have much in common and cross-fertilization is needed. It is our hope that this book could stimulate a merge of research on the TM and boundary methods in both applied mathematics and engineering.

As seen from the review above, the "origin" of boundary numerical methods, as well as other numerical methods, can be traced to this period, during which many ideas sprouted. However, even though methods such as those by Jaswon and Kupradze started to receive attention, these efforts did not immediately coalesce into a single "movement" and grow rapidly. In the following sections, we shall review those significant events that led to the development of the boundary integral equation method and the BEM, and the ensuing movement that established these methods as one of the leading numerical methods.

A.7.1 Kupradze

Viktor Dmitrievich Kupradze (1903–1985) was born in the village of Kela, Russian Georgia. He was enrolled in the Tbilisi State University in 1922 and was awarded the diploma in mathematics in 1927. He stayed on as a lecturer in mathematical analysis and mechanics until 1930. In that year he entered the Steklov Mathematical Institute in Leningrad for postgraduate study and obtained his doctor of mathematics degree in 1935. In 1933, Muskhelishvili founded a research institute of mathematics, physics, and mechanics in Tbilisi. In 1935, Muskhelishvili and his closest associates Kupradze and I. Vekua transformed the institute to become affiliated with the Georgian Academy of Sciences, with Kupradze serving as its first director from 1935 to 1941. The institute was later known as A. Razmadze. From 1937 until his death, Kupradze served as the Head of Differential and Integral Equations Department at Tbilisi. Kupradze's research interest covers theory of PDEs and integral equations, and mathematical theory of elasticity and thermoelasticity. He received many honors, including political ones. He was elected as an

Academician of the Georgian Academy in 1946. From 1954 to 1958 he served as the Rector of Tbilisi University, and from 1954 to 1963 the Chairman of Supreme Soviet of the Georgian SSR.

A.7.2 Jaswon

Maurice Aaron Jaswon (1922–) was born in Dublin, Ireland. He was enrolled in the Trinity College, Dublin, and obtained his B.Sc. degree in 1944. He entered the University of Birmingham, U.K. and was awarded his Ph.D. degree in 1949. In the same year, he started his academic career as a Lecturer in Mathematics at the Imperial College, London. His early research was focused on the mathematical theory of crystallography and dislocation, which cumulated into a book published in 1965 [220], with an updated version in 1983 [222]. In 1957, Jaswon was promoted to the Reader position and stayed at the Imperial College until 1967. It was during this period that he started his seminal work on numerical solution of integral equations with his students George Thomas Symm [432] and Alan R.S. Ponter. In 1963–1964, Jaswon visited Brown University. In 1965–1966, he was a visiting professor at the University of Kentucky. His presence there was what initially made Rizzo aware of an opening position at Kentucky [387]. Upon Rizzo's arrival in 1966, they had a few months of overlapping before Jaswon's returning to England. In 1967, Jaswon left the Imperial College to take up a position as Professor and Head of Mathematics at the City University of London, where he stayed for the next 20 years until his retirement in 1987. He remains active as an Emeritus Professor at City University. Jaswon was considered by some as the founder of the boundary integral equation method based on his 1963 work [221] implementing Green's formula.

A.8 Boundary integral equation and boundary element methods

A turning point marking the rapid growth of numerical solutions of boundary integral equations, known as the *boundary integral equation method*, happened in 1967, when *Frank Joseph Rizzo* (1938–) published the article "*An integral equation approach to boundary value problems of classical elastostatics*" [386]. In this paper, a numerical procedure was applied for solving the Somigliana identity, i.e., eqn. (A.5.8) for elastostatics problems. The work was an extension of Rizzo's doctoral dissertation [385] at the University of Illinois, Urbana–Champaign, which described the numerical algorithm, yet without actual implementation. The work was heavily influenced by the earlier work of Jaswon [223] on potential problems.

In collaboration with David J. Shippy, Rizzo continued the work on boundary integral equation method by extending its application to elasticity problems with inclusions [388], plane anisotropic bodies [389], and by utilizing Laplace transform and the numerical Laplace inversion, to transient heat conduction [390], and quasi-static viscoelasticity problems [391].

Thomas Allen Cruse (1941–), suggested by Rizzo, completed a doctoral dissertation on boundary integral solution of elastodynamics [110], which was published in 1968 in two papers [111, 117]. In 1970 and 1971, Cruse published boundary integral solutions of three-dimensional fracture problems [112, 119]; these were among the first numerical solutions of three-dimensional fracture problems [115]. In 1971, Cruse in his work on elastoplastic flow [430] referred the methods that distributed single- and double-layer potential at fictitious densities, such as those based on the Fredholm integrals and Kupradze's method, as the "indirect potential methods", and the methods that utilized Green's formality, such as Green's third identity and the Somigliana integral, as the "direct potential methods."

In 1975, Cruse and Rizzo organized the first dedicated boundary integral equation method meeting under the auspices of the Applied Mechanics Division of the American Society of Mechanical Engineers (ASME) in Troy, New York. The proceedings of the meeting [118] reflected the rapid growth of the boundary integral equation method to cover a broad range of applications that included water waves [407], transient phenomena in solids (heat conduction, viscoelasticity, and wave propagation) [412], fracture mechanics [113], elastoplastic problems [334], and rock mechanics [3]. The next international meeting on boundary integral equation method was held in 1977 as the First International Symposium on Innovative Numerical Analysis in Applied Engineering Sciences, at Versailles, France, organized by Cruse and Lachat [116]. In the same year, Jaswon and Symm published the first book on numerical solution of boundary integral equations [223].

In the late 1960s, Hugh Tottenham and his students at the University of Southampton in U.K. started the investigation of integral equation method using Kupradze's [258] approach (indirect method). Doctoral dissertations based on indirect methods produced around this time included that by P.K. Banerjee [22] in 1970, and by J.O. Watson [462] and G.R. Tomlin [440] in 1973.

Up to 1977 the numerical method for solving integral equations had been called the "boundary integral equation method," following Cruse's naming. However, with the growing popularity of the FEM, it became clear that many of the finite element ideas can be applied to the numerical technique solving boundary integral equations. This is particularly demonstrated in the work of Lachat and Watson [262]. Furthermore, parallel to the theoretical development of FEM, it was shown that the weighted residual technique can be used to derive the boundary integral equations [58, 60]. The term "boundary element method," mirroring "finite element method," finally emerged in 1977.

Carlos Alberto Brebbia (1938–) presented the BEM using the weighted residuals formulation [57, 58, 60]. The development of solving boundary value problems using functions defined on local domains with low degree of continuity was strongly influenced by the development of extended variational principles and weighted residuals in the mid 1960s. Key players included Eric Reissner [381] and Kyuichiro Washizu [461], who pioneered the use of mixed variational statements that allowed the flexibility in choosing localized functions. To deal with non-conservative and time-dependent problems, the strategy shifted from the variational approach to the method of weighted residuals combined with the concept of weak forms.

Brebbia [58] showed that one could generate a spectrum of methods ranging from finite elements to boundary elements.

Consider a function ϕ satisfying the linear partial differential operator $\mathcal{L}$ in the following fashion

$$\mathcal{L}\{\phi\} = b(\mathbf{x}), \quad \mathbf{x} \in \Omega, \tag{A.8.1}$$

and subject to the essential and natural boundary conditions

$$\begin{aligned} S\{\phi\} &= f(\mathbf{x}), \quad \mathbf{x} \in \Gamma_1, \\ N\{\phi\} &= g(\mathbf{x}), \quad \mathbf{x} \in \Gamma_2, \end{aligned} \tag{A.8.2}$$

where S and N are the corresponding differential operators. Our goal is to find the approximate solution that minimizes the error with respect to a weighing function w in the following fashion:

$$\langle \mathcal{L}\{\phi\} - b, w\rangle_\Omega = \langle N\{\phi\} - g, S^*\{w\}\rangle_{\Gamma_2} - \langle S\{\phi\} - f, N^*\{w\}\rangle_{\Gamma_1}, \tag{A.8.3}$$

where S^* and N^* are the adjoint operators of S and N, and the angle brackets denote the inner product,

$$\langle \alpha, \beta\rangle_\gamma = \int_\gamma \alpha(\mathbf{x})\beta(\mathbf{x})\, d\mathbf{x}. \tag{A.8.4}$$

Equation (A.8.3) can be considered as the theoretical basis for a number of numerical methods [58]. For example, finite difference can be interpreted as a method using Dirac delta function as the weighing function and enforcing the boundary conditions exactly. The well-known Galerkin formulation in FEM uses of the basis function for w the same as that used for the approximation of ϕ.

For the boundary element formulation, we perform integration by parts on eqn. (A.8.3) for as many times as needed to obtain

$$\begin{aligned} \langle \phi, \mathcal{L}^*\{w\}\rangle_\Omega &= \langle S\{\phi\}, N^*\{w\}\rangle_{\Gamma_2} - \langle N\{\phi\}, S^*\{w\}\rangle_{\Gamma_1} \\ &\quad + \langle f, N^*\{w\}\rangle_{\Gamma_1} - \langle g, S^*\{w\}\rangle_{\Gamma_2} + \langle b, w\rangle_\Omega, \end{aligned} \tag{A.8.5}$$

where $\mathcal{L}^*$ is the adjoint operators of $\mathcal{L}$. The idea for the boundary method is to replace w by the fundamental solution G^*, which satisfies

$$\mathcal{L}^*\{G^*\} = \delta, \tag{A.8.6}$$

such that eqn. (A.8.5) reduces to

$$\begin{aligned} \phi &= \langle S\{\phi\}, N^*\{G^*\}\rangle_{\Gamma_2} - \langle N\{\phi\}, S^*\{G^*\}\rangle_{\Gamma_1} \\ &\quad + \langle f, N^*\{G^*\}\rangle_{\Gamma_1} - \langle g, S^*\{G^*\}\rangle_{\Gamma_2} + \langle b, G^*\rangle_\Omega. \end{aligned} \tag{A.8.7}$$

This is the weighted residual formulation for BEM. For the case of Laplace equation, which is self-adjoint, with the boundary conditions

$$\phi = f(\mathbf{x}), \quad \mathbf{x} \in \Gamma_1,$$
$$\frac{\partial \phi}{\partial n} = g(\mathbf{x}), \quad \mathbf{x} \in \Gamma_2, \tag{A.8.8}$$

eqn. (A.8.7) becomes

$$\phi = -\frac{1}{4\pi}\iint_{\Gamma_1} \frac{1}{r}\frac{\partial \phi}{\partial n}\, dS + \frac{1}{4\pi}\iint_{\Gamma_1} f\,\frac{\partial(1/r)}{\partial n}\, dS - \frac{1}{4\pi}\iint_{\Gamma_2} \frac{1}{r} g\, dS + \frac{1}{4\pi}\iint_{\Gamma_2} \phi\,\frac{\partial(1/r)}{\partial n}\, dS. \tag{A.8.9}$$

which is just eqn. (A.3.8) with boundary conditions substituted in.

In 1978, Brebbia published the first textbook on BEM "*The Boundary Element Method for Engineers*" [58]. In the same year, Brebbia organized the first conference dedicated to the BEM: the First International Conference on BEMs, at the University of Southampton [59]. This conference series has become an annual event.

A.8.1 Rizzo

Frank Joseph Rizzo (1938–) was born in Chicago, Illinois. After graduating from St. Rita High School in 1955, he attended the University of Illinois at Chicago. Two years later, he transferred to the Urbana campus and received his B.S. degree in 1960, M.S. degree in 1961, and Ph.D. in 1964. While pursuing the graduate degrees, he was employed as a half-time teaching staff in the Department of Theoretical and Applied Mechanics. In 1964, he began his career as an assistant professor at the University of Washington. Two years later, he left for the University of Kentucky, where he stayed for the next 20 years. In 1987, Rizzo moved to Iowa State University and served as the Head of the Department of Engineering Sciences and Mechanics, which later became a part of the Aerospace Engineering and Engineering Mechanics Department. In late 1989, he returned to his alma mater, the University of Illinois at Urbana–Champaign, to become the Head of the Department of Theoretical and Applied Mechanics. Near the end of 1991, he returned to the Iowa State University and remained there until his retirement in 2000. Rizzo's 1967 article "*An integral equation approach to boundary value problems of classical elastostatics*," which was cited more than 300 times as of 2003 based on the *Web of Science* search [492], is generally considered as the turning point that sets off the modern-day development of boundary element (boundary integral equation) method.

A.8.2 Cruse

Thomas Allen Cruse (1941–) was born in Anderson, Indiana. After graduation from Riverside Polytechnic High School, Cruse entered Stanford University, where he obtained a B.S. degree in Mechanical Engineering in 1963, and an M.S. in Engineering Mechanics in 1964. After a year working with the Boeing Company, he enrolled in 1965 at the University of Washington to pursue a Ph.D. degree, which he was awarded in 1967. In the same year, Cruse join Carnegie Mellon University as an assistant professor. In 1973, Cruse resigned from Carnegie Mellon and joined Pratt & Whitney Aircraft Group, where he spent the next 10 years. In 1983, he moved to the Southwest Research Institute at San Antonio, Texas, where he stayed until 1990. In that year Cruse returned to the academia by joining the Vanderbilt University as the holder of the H. Fort Flower Professor of Mechanical Engineering. He retired in 1999 as the Associate Dean for Research and Graduate Affairs of the College of Engineering at Vanderbilt University.

A.8.3 Brebbia

Carlos Alberto Brebbia (1938–) was born in Rosario, Argentina. He received a B.S. degree in civil engineering from the University of Litoral, Rosario, in 1962. He did early research on the application of Volterra equations to creep buckling and other problems. His mentor there was José Nestor Distefano, latterly of the University of California, Berkeley. Brebbia went to the University of Southampton, U.K. to carry out his Ph.D. study under Hugh Tottenham. During the whole 1966 and first six months in 1967, he visited MIT and conducted research under Eric Reissner and Jerry Connor. He attributed his success with FEM as well as BEM to these great teachers. Brebbia was granted his Ph.D. at Southampton in 1967. After a year's research at the U.K. Electricity Board Laboratories, in 1970 Brebbia started working as a Lecturer at Southampton. In 1975, he accepted a position as Associate Professor at Princeton University, where he stayed for over a year. He then returned to Southampton where he eventually became a Reader. In 1979, Brebbia was again in the U.S. holding a full professor position at the University of California, Irvine. In 1981, he moved back to the U.K. and founded the Wessex Institute of Technology as an international focus for BEM research. He has been serving as its director since.

References

[1] Abou-Dina, M.S., Implementation of Trefftz method for the solution of some elliptic boundary value problems, *Appl. Math. Comput.*, **127**, pp. 125–147, 2002.

[2] Abramowitz, M. & Stegun, I.A., *Handbook of Mathematical Functions with Formulas, Graphs and Mathematical Tables*, Devor Publications, Inc., New York, 1970.

[3] Altiero, N.J. & Sikarskie, D.L., An integral equation method applied to penetration problems in rock mechanics, *Boundary-Integral Equation Method: Computational Applications in Applied Mechanics*, AMD**11**, eds. T.A. Cruse & F.J. Rizzo, Springer-Verlag, New York, pp. 119–141, 1975.

[4] Arad, A., Yakhot, A. & Ben-Dor, G., A highly accurate numerical solution of a biharmonic equation, *Numer. Meth. PDE*, **13**, pp. 375–397, 1997.

[5] Arnold, D., An interior penalty finite element method with discontinuous elements, *SIAM J. Numer. Anal.*, **19**, pp. 742–760, 1982.

[6] Arnold, D.N. & Wendland, W.L., On the asymptotic convergence of collocation method, *Math. Comp.*, **41(164)**, pp. 349–381, 1983.

[7] Ascher, U.M., Mattheij, R.M. & Russell, R.D., *Numerical Solution of Boundary Value Problems for Ordinary Differential Equations*, Prentice Hall, Englewood Cliffs, NJ, 1988.

[8] Atkinson, K.E., *A Survey of Numerical Methods for the Solution of Fredholm Integral Equations of the Second Kind*, SIAM, Philadelphia, 1976.

[9] Atkinson, K.E., *An Introduction to Numerical Analysis*, John Wiley & Sons, New York, 1989.

[10] Atkinson, K.E., *The Numerical Solution of Integral Equations of the Second Kind*, Cambridge University Press, Cambridge, 1997.

[11] Aziz, A.K., Dorr, M.R. & Kellogg, R.B., An new approximation method for the Helmholtz equation method in an exterior domain, *SIAM J. Numer. Anal.*, **19**, pp. 899–908, 1982.

[12] Aziz, A.K., Kellogg, R.B. & Stephen, A.B., Least squares methods for elliptic systems, *Math. Comp.*, **44**, pp. 53–70, 1985.

[13] Aziz, A.K. & Liu, J.L., A weighted least-squares method for the backward–forward heat-equation, *SIAM J. Numer. Anal.*, **28**, pp. 156–167, 1991.

[14] Babuška, I., The finite element method with Lagrangian multipliers, *Numer. Math.*, **20**, pp. 179–192. 1973.

[15] Babuška, I. & Aziz, A.K., Survey lectures on the mathematical foundations of the finite element method, *The Mathematical Foundations of the Finite Element with Applications to Partial Differential Equations*, ed. A.K. Aziz, Academic Press, Inc., pp. 3–358, 1971.

[16] Babuška, I. & Guo, B., Direct and inverse approximation theorems for the p-version of the finite element method in the framework of weighted Besov spaces, Part I: Approximability of functions on the weighted Besov space, Technical Report, Department of Mathematics, University of Manitoba, 2001.

[17] Babuška, I. & Guo, B., Direct and inverse approximation theorems for the p-version of the finite element method is the framework of weighted Besov spaces, Part III: Inverse approximation theorems, Technical Report, Department of Mathematics, University of Manitoba, 2002.

[18] Babuška, I. & Osborn, J., Eigenvalue problems, *Handbook of Numerical Analysis*, **II**., *Finite Element Methods* (*Part I*), eds. P.G. Ciarlet & J.L. Lions, pp. 641–787, North-Holland, Amsterdam, 1991.

[19] Baker, C.T.H., *The Numerical Treatment of Integral Equations*, Oxford University Press, 1977.

[20] Baker, G.A., Finite element methods for elliptic equations using nonconforming elements, *Math. Comp.*, **31**, pp. 45–59, 1977.

[21] Banaugh, R.P. & Goldsmith, W., Diffraction of steady acoustic waves by surfaces of arbitrary shape, *J. Acoust. Soc. Am.*, **35**, pp. 1590–1601, 1963.

[22] Banerjee, P.K., *A contribution to the study of axially loaded pile foundations,* Doctoral dissertation, University of Southampton, 1970.

[23] Banoczi, J.M., Chiu, N.C., Cho, G.E. & Ipsen, I.C.P., The lack of influence of the right-hand side on the accuracy of linear system solution, *SIAM J. Sci. Stat. Comput.*, **20**, pp. 203–227, 1998.

[24] Barbosa, H.J.C. & Hughes, T.J.R., Boundary Lagrange multipliers in finite element methods: Error analysis in natural norms, *Numer. Math.*, **62**, pp. 1–15, 1992.

[25] Barrett, J.W. & Elliott, C.M., Finite element approximation of the Dirichlet problem using the boundary penalty method, *Numer. Math.*, **49**, pp. 343–366, 1986.

[26] Bayliss, A., Goldstein, C.I. & Turkel, E., Preconditioned conjugate gradient methods for the Helmholtz equation, *Elliptic Problems Solvers II*, eds. G. Birkhoff & A. Schoenstadt, Academic Press, New York, pp. 734–754, 1984.

[27] Bayliss, A., Gunzburger, M. & Turkel, E., Boundary conditions for the numerical solution of elliptic equations in exterior regions, *SIAM J. Appl. Math.*, **42(2)**, pp. 430–451, 1982.

[28] Begehr, H. & Gilbert, R.P., Transformations, transmutation, and kernel functions, *Longman Scientific & Technical*, **1**, pp. 108–127, 1992.

[29] Belhoste, B., *Augustin-Louis Cauchy. A Biography*, Springer-Verlag, New York, 1991.

[30] Belytschko, T., Lu, Y.Y. & Gu, L., Element-free Galerkin methods, *Int. J. Numer. Meth. Eng.*, **37**, pp. 229–256, 1994.

[31] Bergman, S., *Integral Operations in the Theory of Partial Differential Equations*, Springer-Verlag, New York, 1969.

[32] Bergman, S. & John, G.H., Application of the method of the kernel function for solving boundary value problems, *Numer. Math.*, **3**, pp. 209–225, 1961.

[33] Bergman, S. & John, G.H., Numerical solution of boundary value problems by the method of integral operator, *Numer. Math.*, **7**, pp. 42–65, 1965.

[34] Bernardi, C., Debit, N. & Maday, Y., Coupling finite element and spectral methods: First result, *Math. Comp.*, **54**, pp. 21–39, 1990.

[35] Bernardi, C. & Maday, Y., Spectral methods, *Handbook of Numerical Analysis*, **V**, *Techniques of Scienctific Computing* (*Part 2*), eds. P.G. Ciarlet & J.L. Lions, Elsevier Science, pp. 209–485, 1997.

[36] Betti, E., Teoria della elasticità, *Nuovo Cimento*, **2**, pp. 6–10, 1872.

[37] Betti, E., Sopra l'equazioni di equilibrio dei corpi solidi elastici, *Annali delle Università Toscane*, **10**, pp. 143–158, 1874.

[38] Bialecki, B., Convergence analysis of orthogonal spline collocation for elliptic boundary value problems, *SIAM J. Numer. Anal.*, **35**, pp. 617–631, 1998.

[39] Bialecki, B. & Cai, X.C., H^1 – norm error bounds for piecewise Hermite bicubic orthogonal spline collocation schemes for elliptic boundary value problems, *SIAM J. Numer. Anal.*, **31**, pp. 1128–1146, 1994.

[40] Bialecki, B. & Dryja, M., Multilevel additive and multiplicative methods for orthogonal spline collocation problems, *Numer. Math.*, **77**, pp. 35–58, 1997.

[41] Bialecki, B., Fairweather, G. & Kargeorghis, A., Matrix decomposition algorithms for modified spline collocation for Helmholtz problems, *SIAM J. Sci. Comput.*, **24**, pp. 1733–1753, 2003.

[42] Bialecki, B. & Fernandes, R.I., An orthogonal spline collocation alternating direction implicit Crank-Nicolson method for linear parabolic problems on rectangles, *SIAM J. Numer. Anal.*, **36**, pp. 1414–1434, 1999.

[43] Bilby, B.A., Continuous distributions of dislocations, *Progress in Solid Mechanics*, eds. I.N. Sneddon & R. Hill, North-Holland, Amsterdam, pp. 329–398, 1960.

[44] Birkhoff, G.D., Oliver Dimon Kellogg – in memoriam, *Bull. Amer. Math. Soc.*, **39**, pp. 171–177, 1933.

[45] Birkhoff, G. & Lynch, R.E., *Numerical Solution of Elliptic Equations*, SIAM, Philadelphia, 1984.

[46] Bjorck, A., Least squares methods, *Handbook of Numerical Analysis*, **1**, eds. P.G. Ciarlet, & J.L. Lions, North-Holland, Amsterdam, pp. 465–652, 1990.

[47] Bochev, P.B. & Gunzburger, M.D., Analysis of least squares finite element methods for the Stokes equations, *Math. Comp.*, **63(208)**, pp. 479–506, 1994.

[48] Bogomolny, A., Fundamental solutions method for elliptic boundary value problems, *SIAM J. Numer. Anal.*, **22**, pp. 644–669, 1985.

[49] Bohmer, K. & Locker, J., Asymptotic expansions for the discretization error of least squares solutions of linear boundary value problems, *Math. Comp.*, **51(183)**, pp. 75–91, 1988.

[50] Boyd, J.P., Chebyshev and Fourier spectral methods, *Lectures Notes in Engineering*, **49**, Springer, Berlin, 1989.

[51] Boyer, C.B., *A History of Mathematics*, 2nd ed., Wiley, Hoboken, 1991.

[52] Bramble, J.H., The Lagrange multiplier methods for Dirichlet's problem, *Math. Comp.*, **37(155)**, pp. 1–12, 1981.

[53] Bramble, J.H. & Nitsche, J.A., A generalized Ritz-least-squares method for Dirichlet problems, *SIAM J. Numer. Anal.*, **10**, pp. 81–93, 1973.

[54] Bramble, J.H. & Schatz, A.H., Rayleigh–Ritz–Galerkin methods for Dirichlet's problem using subspaces without boundary conditions, *Comm. Pure Appl. Math.*, **23**, pp. 653–675, 1970.

[55] Bramble, J.H. & Schatz, A.H., Least squares methods for $2m$th order elliptic boundary-value problems, *Math. Comp.*, **25(113)**, pp. 1–32, 1971.

[56] Bramble, J.H. & Schatz, A.H., Higher order local accuracy by averaging in the finite element method, *Math. Comp.*, **31**, pp. 94–111, 1977.

[57] Brebbia, C.A., Introduction to boundary element methods, *Recent Advances in Boundary Element Methods, Proc. of the 1st Int. Conf. Boundary Element Methods,* ed. C.A. Brebbia, University of Southampton, Pentech, pp. 1–42, 1978.

[58] Brebbia, C.A., *The Boundary Element Method for Engineering*, Pentch Press, London, 1978.

[59] Brebbia, C.A. (ed)., *Recent Advances in Boundary Element Methods, Proc. of the 1st Int. Conf. Boundary Element Methods,* University of Southampton, Pentech, 1978.

[60] Brebbia, C.A. & Dominguez, J., Boundary element methods for potential problems, *Appl. Math. Model.*, **1**, pp. 372–378, 1977.

[61] Brebbia, C.A. & Dominguez, J., *The Boundary Elements, In Introductory Course*, McGraw-Hill Book Company, New York, 1989.

[62] Brezzi, F. & Fortin, M., *Mixed and Hybrid Finite Element Methods*, Springer-Verlag, New York, 1991.

[63] Browder, F.E., Approximation by solutions of partial differential equations, *Amer. J. Math.*, **84**, pp. 134–160, 1962.

[64] Brunner, H., 1896–1996: One hundred years of Volterra integral equations of the first kind, *Appl. Numer. Math.*, **24**, pp. 83–93, 1997.

[65] Brunner, H., Pedas, A. & Vainikko, G., Piecewise polynomial collocation methods for linear Volterra integro-differential equations with weakly singular kernels, *SIAM J. Numer. Anal.*, **39**, pp. 957–982, 2001.

[66] Bückner, H.F., *Die praktische Behandlung von Integralgleichungen*, Springer-Verlag, Berlin, 1952.

[67] Bühler, W.K., *Gauss, A Biographical Study*, Springer-Verlag, Berlin, 1981.

[68] Burton, D.M., *The History of Mathematics – An Introduction*, Allyn & Bacon, Boston, 1985.

[69] Cahan, D. (ed)., *Hermann von Helmholtz and the Foundation of Nineteenth Century Science*, University of California Press, Berkeley, CA, 1993.

[70] Cannell, D.M., *George Green, Mathematician and Physicist 1793–1841, The Background to His Life and Work*, 2nd ed., SIAM, Philadelphia, 2001.

[71] Canuto, C., Hariharan, S.I. & Lustman, L., Spectral methods for exterior elliptic problems, *Numer. Math.*, **46**, pp. 505–520, 1985.

[72] Canuto, C., Hussaini, M.Y., Quarteroni, A. & Zhang, T.A., *Spectral Methods in Fluid Dynamics*, Springer-Verlag, New York, 1987.

[73] Canuto, C. & Quarteroni, A., Approximation results for orthogonal polynomials in Sobolev spaces, *Math. Comp.*, **52**, pp. 67–86, 1982.

[74] Cao, Y., Herdman, T. & Xu, Y., A hybrid collocation method for Volterra integral equations with weakly singular kernels, *SIAM J. Numer. Anal.*, **41**, pp. 364–381, 2003.

[75] Carey, G.F. & Oden, J.T., *Finite Elements, A Second Course, II*, Prentice Hall, Inc., Englewood Cliffs, NJ, 1983.

[76] Carey, G.F. & Shen, Y., Convergence studies of least-squares finite elements for first order systems, *Comm. Appl. Numer. Meth.*, **5**, pp. 427–434, 1989.

[77] Carstensen, C. & Praetorius, D., A posteriori error control in adaptive qualocation boundary element analysis for a logarithmic kernel integral equation of the first kind, *SIAM J. Sci. Comput.*, **25**, pp. 259–283, 2003.

[78] Cauchy, A.-L., *Leçons sur les applications de calcul infinitésimal à la géométries*, 2 vols., Paris, 1826 and 1828.

[79] Cauchy, A.-L., Mémoire sur les intégrales définies, *Mémoires des divers savants*, ser. 2, **1**, pp. 601–799 (written in 1814), 1827.

[80] Cessenat, O. & Despres, B., Application of an ultra weak variational formulation of elliptic PDEs to the two-dimensional Helmholtz problem, *SIAM J. Numer. Anal.*, **35**, pp. 255–299, 1998.

[81] Chan, T.F. & Foulser, D., Effectively well-conditioned linear systems, *SIAM J. Sci. Comput.*, **9**, pp. 963–969, 1988.

[82] Chang, C.L., A least-squares finite element method for the Helmholtz equation, *Comput. Method. Appl. Mech. Eng.*, **83**, pp. 1–7, 1990.

[83] Chang, J.R., Liu, R.F., Kuo, S.R. & Yeih, W.C., Application of symmetric indirect Trefftz method to free vibration problems in 2D, *Inter. J. Numer. Meth. Eng.*, **56(8)**, pp. 1175–1192, 2003.

[84] Chen, C.H., *Further study on Motz's problem*, Master thesis, Department of Applied Mathematics, National Sun Yat-sen University, Kaohsiung, 1998.

[85] Chen, C.S., Hon, Y.C. & Schaback R.A., *Scientfic Computing with Radius Basis Functions*, Book manuscript, Göttingen, 2005.

[86] Chen, J.T., Lin, L.W. & Hong, H.K., Spurious and true eigensolutions of Helmoholtz BIEs and BEMs for a multiply-connected problem, *Proc. R. Soc. Lond., Ser. A*, **459**, pp. 1891–1925, 2003.

[87] Chen, J.T., Lin, J.T., Kuo, S.R. & Chyuan, S.W., Boundary element analysis for the Helmholtz eigenvalue problems with a multiply connected domain, *Proc. R. Soc. Lond., Ser. A*, **457**, pp. 2521–2546, 2001.

[88] Chen, Z., Micchelli, C.A. & Xu, Y., Fast collocation methods for second kind integral equations, *SIAM J. Numer. Anal.*, **40**, pp. 344–375, 2002.

[89] Chen, L.H. & Schweikert, D.G., Sound radiation from an arbitrary body, *J. Acoust. Soc. Am.*, **35**, pp. 1626–1632, 1963.

[90] Chen, J.T., Wu, C.S., Chen, I.L. & Chen, K.H., On the equivalence of the Trefftz method and method of fundamental solutions for Laplace and biharmonic equations, *Comp. Math. Appl.*, **53**, pp. 851–879, 2007.

[91] Chen, G. & Zhou, J., *Boundary Element Methods*, Academic Press, New York, 1992.

[92] Cheney, E.W., *Introduction to Approximation Theory*, McGraw-Hill, New York, 1966.

[93] Cheng, A.H.-D. & Cheng, D.T., Heritage and early development of boundary element, *Eng. Anal. Bound. Elem.*, **29**, pp. 268–302, 2005.

[94] Cheng, A.H.-D., Golberg, M.A., Kansa, E.J. & Zammito G., Exponential convergence and H-c multiquadrics collocation method for partial differential equations, *Numer. Meth. PDE*, **19(5)**, pp. 571–594, 2003.

[95] Cheng, A.H.-D., Lafe, O. & Grilli, S., Dual-reciprocity BEM based on global interpolation functions, *Eng. Anal. Bound. Elem.*, **13**, pp. 303–311, 1994.

[96] Cheung, Y.K., Jin, W.G. & Zienkiewicz, O.C., Direct solution procedure for solution of Harmonic problems using complete, non-singular, Trefftz functions, *Commun. Appl. Numer. Meth.*, **5**, pp. 159–169, 1989.

[97] Cheung, Y.K., Jin, W.G. & Zienkiewicz, O.C., Solution of Helmholtz equation by Trefftz method, *Inter. J. Numer. Meth. Eng.*, **32**, pp. 63–78, 1991.

[98] Chien, W.Z., *Variational and Finite Element Methods* (in Chinese), East Asia Publishing, Taipei, 1989.

[99] Cho. H.A., Golberg, M.A., Muleshkov, A.S. & Li X., Trefftz methods for time dependent partial differential equations, *CMC*, **1**, pp. 1–37, 2004.

[100] Christiansen, S., Condition number of matrices derived from two classes of integral equations, *Math. Meth. Appl. Sci.*, **3**, pp. 364–392, 1981.

[101] Christiansen, S. & Hansen, P.C., The effective condition number applied to error analysis of certain boundary collocation methods, *J. Comput. Appl. Math.*, **54(1)**, pp. 15–36, 1994.

[102] Christiansen, S. & Saranen, J., The conditioning of some numerical methods for first kind boundary integral equations, *J. Comput. Appl. Math.*, **67(1)**, pp. 43–58, 1996.

[103] Ciarlet, P.G., Basic error estimates for elliptic problems, *Finite Element Methods* (*Part I*), eds. P.G. Ciarlet & J.L. Lions, North-Holland, Amsterdam, pp. 17–352, 1991.

[104] Clough, R.W., The finite element method in plane stress analysis, *Proc. 2nd ASCE Conf. Electronic Computation*, 1960.

[105] Comodi, M.I. & Mathon, R., A boundary approximation method for fourth order problems, *Math. Mod. Meth. Appl. Sci.,* **4**, pp. 432–445, 1991.

[106] Costabel, M., Boundary integral operators on Lipschitz domains: Elementary results, *SIAM J. Math. Anal.*, **19**, pp. 613–626, 1988.

[107] Costabel, M., Ervin V.J. & Stephan E.P., On the convergence of collocation methods for Symm's integral equation on open curves, *Math. Comp.*, **51**, pp. 167–179, 1988.

[108] Courant, R. & Hilbert, D., *Methods of Mathematical Physics*, **I**, Wiley Interscience Publishers, New York, 1953.

[109] Courant, R. & Hilbert, D., *Methods of Mathematical Physics,* **II**, *Partial Differential Equations*, Wiley Interscience Publishers, New York, 1962.

[110] Cruse, T.A., *The transient problem in classical elastodynamics solved by integral equations,* Doctoral dissertation, University of Washington, p. 117, 1967.

[111] Cruse, T.A., A direct formulation and numerical solution of the general transient elastodynamic problem–**II**, *J. Math. Anal. Appl.*, **22**, pp. 341–355, 1968.

[112] Cruse, T.A., Lateral constraint in a cracked, three-dimensional elastic body, *Int. J. Fracture Mech.*, **6**, pp. 326–328, 1970.

[113] Cruse, T.A., Boundary-integral equation fracture mechanics analysis, *Boundary-Integral Equation Method: Computational Applications in Applied Mechanics*, AMD **11**, eds. T.A. Cruse & F.J. Rizzo, New York, pp. 31–46, 1975.

[114] Cruse, T.A., *Boundary Element Analysis in Computational Fracture Mechanics*, Kluwer Academic Publishers, Dordrecht, 1988.

[115] Cruse, T.A., BIE fracture mechanics: 25 years of developments, *Comp. Mech.*, **18**, pp. 1–11, 1996.

[116] Cruse, T.A. & Lachat, J.-C., *Proceedings of the International Symposium on Innovative Numerical Analysis in Applied Engineering Sciences*, Versailles, France, 1977.

[117] Cruse, T.A. & Rizzo, F.J., A direct formulation and numerical solution of the general transient elastodynamic problem–I, *J. Math. Anal. Appl.*, **22**, pp. 244–259, 1968.

[118] Cruse, T.A. & Rizzo, F.J., *Boundary-Integral Equation Method: Computational Applications in Applied Mechanics,* AMD **11**, Applied Mechanics Conference, ASME, Rensselaer Polytechnic Institute, June 23–25, 1975.

[119] Cruse, T.A. & VanBuren, W., Three-dimensional elastic stress analysis of a fracture specimen with an edge crack, *Int. J. Fracture Mech.*, **7**, pp. 1–15, 1971.

[120] Dautray, R. & Lions, J.L., *Mathematical Analysis and Numerical Methods for Science and Technology,* **II**. *Functional and Variational Methods*, Springer-Verlag, Berlin Heidelberg, New York, 1988.

[121] de Freitas, J.A.T. & Leitao, V.M.A., A boundary integral Trefftz formulation with symmetric collocation, *Comp. Mech.*, **25(6)**, pp. 515–523, 2003.

[122] Delves, L.M., Global and regional methods, *A Survey of Numerical Methods for Partial Differential Equations*, eds. L. Gladwell & R. Wait, Clarendon Press, Oxford, pp. 106–127, 1979.

[123] Delves, L.M. & Freeman, T.L., *Analysis of Global Expansion Methods: Weakly Asymptotically Diagonal Systems*, Academic Press, London, New York, 1981.

[124] Delves, L.M. & Hall, C.A., An implicit matching principle for global element calculations, *J. Inst. Math. Appl.*, **23(2)**, pp. 223–234, 1979.

[125] Delves, L.M. & Mohamed, J.L., *Computational Methods for Integral Equations*, Cambridge University Press, Cambridge, 1985.

[126] Demmel, J.W., *Applied Numerical Linear Algebra*, SIAM, Philadelphia, p. 132, 1997.

[127] Diaz, J.C., A collocation-Galerkin method for the two point boundary value problem using continuous piecewise polynomial spaces, *SIAM J. Numer. Anal.*, **14**, pp. 844–858, 1977.

[128] Domingues, J.S., Portela, A. & de Castro, P.M.S.T., Trefftz boundary element method applied to fracrure mechanics, *Eng. Fract. Mech.*, **64**, pp. 67–85, 1999.

[129] Dong, C.Y., Lo, S.H., Cheung, Y.K. & Lee, K.Y., Anisotropic thin plate bending problem by Trefftz boundary collocation method, *Eng. Anal. Bound. Elem.*, **28**, pp. 1017–1024, 2004.

[130] Dosiyev, A.A., The high accurate block-grid method for solving Laplace's boundary problem with singularities, *SIAM J. Numer. Anal.*, **42**, pp. 153–178, 2004.

[131] Dunham, W., *Euler, The Master of Us All*, The Mathematical Association of America, Washington, DC, 1999.

[132] Dunn, R.J. & Wheeler, M.F., Some collocation-Galerkin methods for two-point boundary value problems, *SIAM J. Numer. Anal.*, **13**, pp. 720–733, 1977.

[133] Ehrlich, P., Kellogg, Bliss, Hedrick: Mizzou math pioneers, *Critical Points*, **4**, Dept. Mathematics, Univ. Missouri, 1999, http://www.math.missouri.edu/~news/issue4/pioneers.html.

[134] Eisenstat, S.C., On the note of convergence of the Bergman-Vekua methods for the numerical solution of elliptic boundary-value problems, *SIAM J. Numer. Anal.*, **11**, pp. 654–680, 1974.

[135] Elliotis, M., Georgiou, G. & Xenophontos, C., The singular function boundary integral method for a two-dimensional fracture problem, accepted by *Eng. Anal. Bound. Elem.*, **31**(**3**), pp. 209–215, 2007.

[136] Elschner, J. & Graham, I., An optimal order collocation method for first kind boundary integral equations on polygons, *Numer. Math.*, **70**, pp. 1–31, 1995.

[137] Erdogan, F. & Gupta, G.D., Numerical solution of singular integral-equations, *Quart. Appl. Math.*, **29**, pp. 525–534, 1972.

[138] Erdogan, F., Gupta, G.D. & Cook, T.S., Numerical solution of singular integral equations, *Mechanics of Fracture 1, Methods of Analysis and Solutions of Crack Problems*, ed. G.C. Sih, Noordhoff, Amsterdam, pp. 368–425, 1973.

[139] Eshelby, J.D., The continuum theory of lattice defects, *Solid State Phys.*, **3**, pp. 79–114, 1956.

[140] Euler, L., Principes géneraux du mouvement des fluides, *Mém. Acad. Sci. Berlin,* **11**, pp. 274–315, 1755.

[141] Euler, L., *Opera Omnia*, Series prima: Opera mathematica (29 volumes), Series secunda: Opera mechanica et astronomica (31 volumes), Series tertia: Opera physica, Miscellanea (12 volumes), Series quarta: A. Commercium epistolicum (10 volumes), edited by the Euler Committee of the Swiss Academy of Science, Birkhäuser, Basel, 2003.

[142] Fairweather, G., *Finite Element Galerkin Methods for Partial Differential Equations*, Marcel Dekker, Inc., New York, 1978.

[143] Fairweather, G. & Karageorghis, A., The method of fundamental solutions for elliptic boundary value problems, *Adv. Comput. Math.*, **9**, pp. 69–95, 1998.

[144] Feng, K. & Yu, D.H., A theorem for the natural integral operator of harmonic equation (in Chinese), *Math. Num. Sinica*, **16**, pp. 221–226, 1994.

[145] Fix, G.M., Hybrid finite element methods, *SIAM Review*, **18(3)**, pp. 460–485, 1976.

[146] Fix, G.J., Gulati, S. & Wakoff, G.I., On the use of singular functions with finite element approximations, *J. Comp. Phys.*, **13**, pp. 209–238, 1973.

[147] Fourier, J.B.J., *Théorie Analytique de la Chaleur* (*The Analytical Theory of Heat*), 1822.

[148] Fox, L., Henrici, P. & Moler, C., Approximations and bounds for eigenvalues of elliptic operators, *SIAM J. Numer. Anal.*, **4**, pp. 89–102, 1967.

[149] Franke, R., Scattered data interpolation tests of some methods, *Math. Comp.*, **38**, pp. 181–200, 1982.

[150] Franke, R. & Schaback, R., Solving partial differential equations by collocation using radial functions, *Appl. Math. Comp.*, **93**, pp. 73–82, 1998.

[151] Fredholm, I., Sur une classe d'équations fonctionelles, *Acta Math.*, **27**, pp. 365–390, 1903.

[152] Freitas, J.A.T. & Cismasiu, C., Developments with hybrid: Trefftz stress and displacement elements, *Computer Assisted Mechanics and Engineering Sciences* (*CAMES*), **8**, pp. 289–311, 2001.

[153] Friedman, M.B. & Shaw, R., Diffraction of pulse by cylindrical obstacles of arbitrary cross section, *J. Appl. Mech., Trans. ASME*, **29**, pp. 40–46, 1962.

[154] Fritzius, R.S., Abbreviated biographical sketch of Walter Ritz, Appendix, *Lorentz and Poincaré Invariance – 100 Years of Relativity*, eds. J.-P. Hsu & Y.-Z. Zhang, World Scientific, Singapore, 2003.

[155] Gatica, G.N., Harbrecht, H. & Schneider, R., Least squares methods for the coupling of FEM and BEM, *SIAM J. Numer. Anal.*, **41**, pp. 1974–1995, 2003.

[156] Gauss, C.F., Theoria attractionis corporum sphaeroidicorum ellipticorum homogeneorum methodo novo tractata, *Commentationes societatis regiae scientiarum Gottingensis recentiores*, **II**, pp. 2–5, 1813.

[157] Gavelya, S.P., Periodical problems for shallow shells of arbitrary curvature with apertures, *Rep. Ukr. Acad. Sci.*, **8**, pp. 703–708, 1969.

[158] Gegelia, T. & Jentsch, L., Potential methods in continuum mechanics, *Georgian Mathematical Journal*, **1(6)**, pp. 599–640, 1994.

[159] Gellert, W., Gottwald, S., Hellwich, M., Kästner, H. & Küstner, H. (eds.), *The VNR Concise Encyclopedia of Mathematics*, 2nd ed., Van Nostrand Reinhold, 1989.

[160] Georgiou, G.C., Boudouvis, A. & Poullikkas, A., Comparison of two methods for the computation of singular solution in elliptic problem, *J. Comput. Appl. Math.*, **79**, pp. 277–289, 1997.

[161] Georgiou, G.C., Olson, L.G. & Smyrlis, Y.S., A singular function boundary integral method for the Laplace equation, *Commun. Numer. Meth. Eng.*, **12**, pp. 127–134, 1996.

[162] Gerasoulis, A., The use of piecewise quadratic polynomials for the solution of singular integral equations of the Cauchy type, *Comp. Math. Appl.*, **8**, pp. 15–22, 1982.

[163] Gillispie, C.C., *Pierre-Simon Laplace, 1749–1827, A Life in Exact Science*, Princeton University Press, Princeton, NJ, 1997.

[164] Golberg, M., Recent developments in the numerical evaluation of partial solutions in the boundary element methods, *Appl. Math. Comp.*, **75**, pp. 91–101, 1996.

[165] Golberg, M.A., A survey of numerical methods for integral equations, Chap. 1, *Solution Methods for Integral Equations*, ed. M.A. Golberg, Plenum, New York, pp. 1–58, 1978.
[166] Goldstein, C.I., The infinite element method with nonuniform mesh sizes applied to the exterior Helmholtz problem, *Numer. Math.*, **38**, pp. 61–82, 1982.
[167] Goldstein, C.I., A finite element method for solving Helmholtz type equations in waveguides and other unbounded domains, *Math. Comp.*, **39**, pp. 309–324, 1982.
[168] Golub, G.H. & van Loan, C.F., *Matrix Computations*, 2nd ed., The Johns Hopkins University Press, Baltimore and London, 1989.
[169] Gorgidze, A.Y. & Rukhadze, A.K., On the numerical solution of the integral equations of the plane theory of elasticity, (in Russian), *Soobshchenia Gruz. Filiala A.N.S.S.S.R.*, **1**, pp. 255–258, 1940.
[170] Gottlieb, D. & Orszag, S.A., *Numerical Analysis of Spectral Methods: Theory and Applications*, SIAM, Philadelphia, 1977.
[171] Gourgeon, H. & Herrera, I., Boundary methods C-complete systems for the biharmonic equations, in *Boundary Element Methods* (Irvine, Calif., 1981), CML Publisher, Springer, Berlin, pp. 431–441, 1981.
[172] Grabiner, J.V., *The Origin of Cauchy's Rigorous Calculus*, MIT Press, Cambridge, MA, 1981.
[173] Gradshteyn, S. & Ryzhik, I.M., *Table of Integrals, Series, and Products, Corrected and Enlarged Edition*, Academic Press, New York, 1980.
[174] Grattan-Guinness, I., *Joseph Fourier, 1768–1830*, MIT Press, Boston, 1972.
[175] Green, G., *An Essay on the Application of Mathematical Analysis to the Theories of Electricity and Magnetism*, Printed for the Author by T. Wheelhouse, Nottingham, p. 72, 1828. Also in, *Mathematical Papers of George Green*, Chelsea Publishing Company, New York, pp. 1–115, 1970.
[176] Green, G., *Mathematical Papers of George Green*, ed. N.M. Ferrers, 1871, Chelsea Publishing Company, New York, 1970.
[177] Greenberg, M.D., *Application of Green's Functions in Science and Engineering*, Prentice-Hall, Englewood Cliffs, NJ, 1971.
[178] Grisvard, P., *Elliptic Problems in Non-smooth Domains*, Pitman Advanced Publishing Program, Boston, 1985.
[179] Hadamard, J., Theorie des équations aux dérivées partielles linéaires hyperboliques et du problème de Cauchy, *Acta Math.*, **31**, pp. 333–380, 1908.
[180] Hahn, R., *Laplace as a Newtonian Scientist*, a paper delivered at a Seminar on the Newtonian Influence, Clark Library, UCLA, 1967.
[181] Haidvogel, D.B., The accurate solution of Poisson's equation by expansion in Chebyshev polynomials, *J. Comp. Phys.*, **30**, pp. 167–180, 1979.
[182] Hall, C.A. & Porsching, T.A., *Numerical Analysis of Partial Differential Equations*, Prentice-Hall, Inc., Englewood Cliffs, New Jersey, 1990.
[183] Han, H.D., The numerical solutions of interface problems by infinite element method, *Numer. Math.*, **39**, pp. 39–50, 1982.
[184] Han, H.D. & Wu, X.N., Approximation of infinite boundary condition and its application to finite element methods, *J. Comp. Appl. Math.*, **3(2)**, pp. 179–192, 1985.
[185] Harari, I. & Hughes, T.J.R., Finite element methods for the Helmholtz equation in an exterior domain: Model problems, *Comput. Method. Appl. Mech. Eng.*, **87**, pp. 59–96, 1991.
[186] Hardy, R.L., Multiquadric equations of topography and other irregular surfaces, *J. Geo. Res.*, **76**, pp. 1905–1915, 1971.

[187] Helmholtz, H., Theorie der Luftschwingungen in Röhren mit offenen Enden, *Journal für die reine und angewandte Mathematik*, **57**, pp. 1–72, 1860.

[188] Helmholtz, H., *Die Lehre von den Tonempfindungen*, (On the Sensations of Tone), 1863.

[189] Herrera, I., Boundary methods: A criterion for completeness, *Proc. Nat. Acad. Sci., USA*, **77(8)**, pp. 4395–4398. 1980.

[190] Herrera, I., Boundary methods: Development of complete systems of solutions, *Finite Element in Flow Analysis*, eds. T. Kawai, University of Tokyo Press, Tokyo, pp. 897–906. 1982.

[191] Herrera, I., *Boundary Methods: An Algebraic Theory*, Pitman, Boston, 1984.

[192] Herrera, I., Unified formulation of numerical methods. Part I. Green's formulas for operators in discontinuous fields, *Numer. Meth. PDE*, **1**, pp. 25–44, 1985.

[193] Herrera, I., Trefftz-Herrera domain decomposition, *Adv. Eng. Softw.*, **24**, pp. 43–56, 1995.

[194] Herrera, I., Trefftz method: A general theory, *Numer. Meth. PDE*, **16(6)**, pp. 561–580, 2000.

[195] Herrera, I. & Diaz, M., Indirect methods of collocation: Trefftz-Herrera collocation, *Numer. Meth. PDE*, **15**, pp. 709–738, 1999.

[196] Herrera, I. & Diaz, M., General theory of domain decomposition: Indirect methods, *Numer. Meth. PDE*, **18**, pp. 296–322, 2002.

[197] Herrera, I. & Gourgeon H., Boundary methods: T-complete system for Stokes problems, *Comput. Method. Appl. Mech. Eng.*, **30**, pp. 225–241, 1982.

[198] Herrera, I. & Solano, J., A non-overlapping TH-domain decomposition, *Adv. Eng. Softw.*, **28(4)**, pp. 223–229, 1997.

[199] Hess, J.L., Review of integral-equation techniques for solving potential-flow problems with emphasis on the surface source method, *Comput. Method. Appl. Mech. Eng.*, **5**, pp. 145–196, 1975.

[200] Hess, J.L. & Smith, A.M.O., Calculations of nonlifting potential flow about arbitrary three-dimensional bodies, *J. Ship Res.*, **8**, pp. 22–44, 1964.

[201] Hon, Y.C. & Schaback, R., On Unsymmetric Collocation by Radial Basis Functions, Preprint Göttingen, 1999.

[202] Hsiao, G.C., *Boundary element methods for a class of exterior singular perturbation problems, in BAIL* **IV**, pp. 89–97, Boole, Dun Laoghaire, 1986.

[203] Hsiao, G.C., Boundary element methods for exterior problems in elasticity and fluid mechanics, *The Mathematics of Finite Elements and Applications VI*, ed. J.R. Whiteman, Academic Press, London, pp. 323–341, 1988.

[204] Hsu, C.H., *Biharmonic boundary value problem with singularity*, Master thesis, Department of Applied Mathematics, National Sun Yat-sen University, Kaohsiung, 2002.

[205] Hu, H.Y., *The Trefftz and collocation methods for elliptic equations*, Doctoral disseration, Department of Applied Mathematics, National Sun Yat-sen University, Kaohsiung, June, 2003.

[206] Hu, H.Y., Chen, J.S. & Hu, W., Weighted radial basis collocation method for boundary value problems, *Int. J. Numer. Meth. Eng.* **69**, pp. 2736–2757, 2007.

[207] Hu, H.Y. & Li, Z.C., Combinations of collocation and finite-element methods for Poisson's equations, *Comp. Math. Appl.*, **51**, pp. 1831–1853, 2006.

[208] Hu, H.Y. & Li, Z.C., Collocation methods for Poisson's equation, *Comput. Method. Appl. Mech. Eng.*, **195**, pp. 4139–4160, 2006.

[209] Hu, H.Y., Li, Z.C. & Cheng, A.H.-D., Radial basis collocation method for elliptic boundary value problem, *Comp. Math. Appl.*, **50**, pp. 289–320, 2005.

[210] Huang, H.T., *Global superconvergence of finite element methods for elliptic equations*, Doctoral dissertation, Department of Applied Mathematics, National Sun Yat-sen University, Kaohsiung, June, 2002.

[211] Huang, H.T. & Li, Z.C., Global superconvergence of Adini's elements coupled with the Trefftz method for singular problems, *Eng. Anal. Bound. Elem.*, **27(3)**, pp. 227–240, 2003.

[212] Huang, H.T. & Li, Z.C., Effective condition number and superconvergence of the Trefftz method coupled with high order FEM for singularity problems, *Eng. Anal. Bound. Elem.*, **30**, pp. 270–283, 2006.

[213] Huang, H.T., Li, Z.C. & Yan, N., New error estimates for Adini's elements, *Appl. Numer. Math.*, **50**, pp. 49–74, 2004.

[214] Huang, H.T., Li, Z.C. & Zhou, A., New error estimates of Lagrange bi-quadratic elements, *Appl. Numer. Math.*, **56**, pp. 712–744, 2006.

[215] Huang, J., *The mechanical quadrature methods with the high accuracy and splitting extrapolations for solving boundary integral equations for the first kind of scientific and engineering problems on nonsmooth domains*, Doctoral dissertation, Mathematical College, Sichuan University, ChengDu, China, 2004.

[216] Huang, J. & Lü, T., The mechanical quadrature methods and their extrapolation for solving BIE of Steklov eigenvalue problems, *J. Comput. Appl. Math.*, **22(5)**, pp. 719–726, 2004.

[217] Huang, S.C. & Shaw, R.P., The Trefftz method as an integral equation, *Adv. Eng. Softw.*, **24**, pp. 57–63, 1995.

[218] Huffel, S.V. & Vandewalle, J., *The Total Least Squares Problem: Computational Aspects and Analysis*, SIAM, Philadelphia, 1991.

[219] Ivanov, V.V., *The Theory of Approximate Methods and Their Application to the Numerical Solution of Singular Integral Equations*, Noordhoff, Amsterdam, (Russian edition 1968), 1976.

[220] Jaswon, M.A., *An Introduction to Mathematical Crystallography*, American Elsevier, New York, 1965.

[221] Jaswon, M.A. & Ponter, A.R., An integral equation solution of the torsion problem, *Proc. R. Soc. Lond., Ser. A*, **273**, pp. 237–246, 1963.

[222] Jaswon, M.A. & Rose, M.A., *Crystal Symmetry: Theory of Colour Crystallography*, Halsted Press, New York, 1983.

[223] Jaswon, M.A. & Symm, G.T., *Integral Equation Methods in Potential Theory and Elastostatics*, Academic Press, Amsterdam, 1977.

[224] Jiang, B.N., On the least squares method, *Comput. Method. Appl. Mech. Eng.*, **152(1–2)**, pp. 239–257, 1998.

[225] Jiang, S. & Li, K.T., The coupling of finite element method and boundary method for two-dimensional Helmholtz equation in an exterior domain, *J. Comput. Appl. Math.*, **5**, pp. 21–37, 1987.

[226] Jin, W.G. & Cheung, Y.K., Trefftz direct method, *Adv. Eng. Softw.*, **24(1–3)**, pp. 65–69, 1995.

[227] Jin, W.G., Cheung, Y.K. & Zienkiewicz, O.C., Trefftz method for Kirchoff plate bending problems, *Int. J. Numer. Meth. Eng.*, **36(5)**, pp. 765–781, 1993.

[228] Jirousek, J. & Guex, L., The hybrid-Trefftz finite element model and its application to plate bending, *Int. J. Numer. Meth. Eng.*, **23**, pp. 651–693, 1986.

[229] Jirousek, J. & Leon, N., A powerful finite element for plate blending, *Comput. Meth. Appl. Mech. Eng.*, **12**, pp. 77–96, 1977.

[230] Jirousek, J. & Wróblewski, A., T-elements: A finite element approach with advantages of boundary solution methods, *Adv. Eng. Softw.*, **24**, pp. 71–88, 1995.

[231] Jirousek, J. & Wróblewski, A., T-elements: State of the art and further trends, *Archin. Comput. Meth. Eng.*, **3**, pp. 323–424, 1995.

[232] Jirousek, J. & Zielinski, A.P., Survey of Trefftz-type element formulations, *Computers & Structures*, **64**, pp. 225–242, 1997.

[233] Johnson, C. & Nedelec, C., On the coupling of boundary integral and finite element methods, *Math. Comp.*, **35**, pp. 1063–1079, 1980.

[234] Kagiwada, H. & Kalaba, R.E., *Integral Equations via Imbedding Methods*, Addison-Wesley, Boston, 1974.

[235] Kamiya, N. & Kita, E., Trefftz method 70 years, *Adv. Eng. Softw.*, **24 (1–3)**, p. 1, 1995.

[236] Kansa, E.J., Multiqudrics: A scattered data approximation scheme with applications to computational fluid-dynamics – I Surface approximations and partial derivatives, *Comp. Math. Appl.*, **19(8–9)**, pp. 127–145, 1992.

[237] Kansa, E.J., Multiqudrics: A scattered data approximation scheme with applications to computational fluid-dynamics – II Solutions to parabolic, hyperbolic and elliptic partial differential equations, *Comp. Math. Appl.*, **19(8–9)**, pp. 147–161, 1992.

[238] Kantorovich, L.V. & Krylov, V.I., *Approximate Methods of Higher Analysis*, Wiley Interscience Publication, New York, 1958.

[239] Karageorghis, A., Modified methods of fundamental solutions for harmonic and biharmonic problems with boundary singularities, *Numer. Meth. PDE*, **8**, pp. 1–18, 1992.

[240] Katsurda, M. & Okamoto, H., The collocation points of the fundamental solution method for the potential problem, *Comp. Math. Appl.*, **31**, pp. 123–137, 1996.

[241] Katz, V.J., The history of Stokes' theorem, *Math. Mag.*, **52**, pp. 146–156, 1979.

[242] Kellogg, O.D., *Foundations of Potential Theory*, Dover, New York, 1953.

[243] Kellogg, R.B., Singularities in interface problems, *Numer. Solution of PDE, II (SYNSPADE 1970)*, Academic Press, New York, pp. 351–400, 1971.

[244] Kellogg, R.B., Higher order singularities for interface problems, *The Mathematical Foundations of the Finite Element Method with Applications to Partial Differential Equations*, ed. A.K. Aziz, Academic Press, New York, pp. 589–602, 1972.

[245] Kellogg, R.B., On the Poisson equations with intersecting interfaces, *Appl. Anal.*, **4**, pp. 101–129, 1975.

[246] Kelvin, W.T., Note on the integrations of the equations of equilibrium of an elastic solid, *Cambridge Dublin Math. J.*, **3**, 1848.

[247] Kim, S.D., Lee, H.C. & Shin, B.C., Pseudospectral least-squares methods for the second-order elliptic boundary value problem, *SIAM J. Numer. Anal.*, **41**, pp. 1370–1380, 2003.

[248] Kita, E., Ikeda, Y. & Kamiya, N., Indirect Trefftz method for boundary value problem of Poisson equation, *Eng. Anal. Bound. Elem.*, **27(8)**, pp. 825–833, 2003.

[249] Kita, E. & Kamiya, N., Trefftz method, An overview, *Adv. Eng. Softw.*, **24 (1–3)**, pp. 3–12, 1995.

[250] Kita, E., Kamiya, N. & Iio, T., Application of a direct Trefftz method with domain decomposition to 2D potential problems, *Eng. Anal. Bound. Elem.*, **23(7)**, pp. 539–548, 1999.

[251] Kita, E., Kamiya, N. & Ikeda, Y., Boundary-type sensitivity analysis scheme based on indirect Trefftz formulation, *Adv. Eng. Softw.*, **24**, pp. 89–96, 1995.

[252] Kolodziej, J., Review of application of boundary collocation methods in mechanics of continuous media, *SM Archives*, **12(4)**, pp. 187–231, 1987.

[253] Kolosov, G.V., *On an Application of Complex Function Theory to a Plane Problem of the Mathematical Theory of Elasticity*, (in Russian), Yuriev, 1909.

[254] Kreiss, H.O. & Oliger, J., Stability of Fourier method, *SIAM J. Numer. Anal.*, **16**, pp. 421–433, 1979.

[255] Křížek, M. & Neittaanmäki, P., Superconvergence phenomenon in the finite element method arising from averaging gradients, *Numer. Math.*, **45**, pp. 105–116, 1984.

[256] Křížek, M. & Neittaanmäki, P., On a global-superconvergence of the gradient of linear triangular elements, *J. Comput. Appl. Math.*, **18**, pp. 221–233, 1987.

[257] Kulakov, V.M. & Tolkachev, V.M., Bending of plates of an arbitrary shape, *Rep. Rus. Acad. Sci.*, **230**, 1976.

[258] Kupradze, V.D., *Potential Methods in the Theory of Elasticity*, Israeli Program for Scientific Translation, 1965. (Earlier edition: Kupradze, V.D., *Dynamical Problems in Elasticity*, **III** in *Progress in Solid Mechanics*, eds. I.N. Sneddon and R. Hills, North-Holland, Amsterdam, 1963.)

[259] Kupradze, V.D. & Aleksidze, M.A., The method of functional equations for the approximate solution of some boundary value problems, (in Russian), *Zh. vichisl. mat. i. mat. fiz.*, **4**, pp. 683–715, 1964.

[260] Kupradze, V.D., Gegelia, T.G., Basheleishvili, M.O. & Burchuladze, T.V., *Three-Dimensional Problems of the Mathematical Theory of Elasticity and Thermoelasticity*, (Russian edition 1976), North-Holland, Amsterdam, 1979.

[261] Kuttler, J.R. & Sigilloto, V.G., Eigenvalues of the Laplacian in two dimensions, *SIAM Review*, **26**, pp. 163–193, 1984.

[262] Lachat, J.C. & Watson, J.O., Effective numerical treatment of boundary integral equations: A formulation for three-dimensional elastostatics, *Int. J. Numer. Meth. Eng.*, **10**, pp. 991–1005, 1976.

[263] Lagrange, J.-L., *Miscellanea Taurinasia*, **2**, pp. 273, 1760–1761.

[264] Lagrange, J.-L., *Mémoires de l'Académie Royale des Sciences de Paris, Savants étrangèrs*, **VII**, 1773.

[265] Laplace, P.-S., *Histoire de l'Académie des Sciences de Paris*, p. 135, 1782/85, and p. 252, 1787/89.

[266] Laubin, P. & Baiwir, M., Spline collocation for a boundary integral equation on polygons with cuts, *SIAM J. Numer. Anal.*, **35**, pp. 1452–1472, 1998.

[267] Layton, A.T., A semi-Lagrangian collocation method for the shallow water equations on the sphere, *SIAM J. Sci. Comput.*, **24**, pp. 1433–1449, 2003.

[268] Lebesgue, H., Sur des cas d'impossibilité du problème de Dirichlet, *Comptes Rendus de la Société Mathématique de France*, **41**, p. 17, 1913.

[269] Lee, P., A Lagrange multiplier method for the interface equations from electromagnetic applications, *SIAM J. Numer. Anal.*, **30(2)**, pp. 478–506, 1993.

[270] Lefeber, D., *Solving Problems with Singularities Using Boundary Elements*, Computational Mechanics Publications, Southampton, 1989.

[271] Leitao, V.M.A., Application of multi-region Trefftz-collocation to fracture mechanics, *Eng. Anal. Bound. Elem.*, **22(3)**, pp. 251–256, 1998.

[272] Li, B., Fairweather, G. & Bialecki, B., Discrete-time orthogonal spline collocation methods for vibration problems, *SIAM J. Numer. Anal.*, **39**, pp. 2045–2065, 2002.

[273] Li, Z.C., A nonconforming combined method for solving Laplace's boundary value problems with singularities, *Numer. Math.*, **49**, pp. 475–497, 1986.

[274] Li, Z.C., *Numerical methods for elliptic boundary value problems with singularities*, Doctoral dissertation, Department of Mathematatics, University of Toronto, May 1986, the 1987 Annual Canadian Applied Mathematics(CAM) Doctoral Dissertation Award, 1987.

[275] Li, Z.C., A note on Kellogg's eigenfuctions of periodic Sturm-Liouville system, *Appl. Math. Lett.*, **1**, pp. 123–126, 1988.

[276] Li, Z.C., *Numerical Methods for Elliptic Equations with Singularities, Boundary Methods and Nonconforming Combinations*, World Scientific, Singapore, 1990.

[277] Li, Z.C., The Schwarz alternating method for the singularity problems, *SIAM J. Sci. Comput.*, **15(5)**, pp. 1064–1082, 1994.

[278] Li, Z.C., Parallel penalty combinations for singularity problems, *Eng. Anal. Bound. Elem.*, **18**, pp. 119–130, 1996.

[279] Li, Z.C., Penalty combinations of Ritz–Galerkin and finite difference method for singularity problems, *J. Comput. Appl. Math.*, **81**, pp. 1–17, 1997.

[280] Li, Z.C., *Combined Methods for Elliptic Equations with Singularities, Interfaces and Infinities*, Kluwer Academic Publishers, Boston, London, 1998.

[281] Li, Z.C., Lagrange multipliers and other coupling techniques for combined methods and other coupling techniques for combined methods for elliptic equations, *Inter. J. Information*, **1(2)**, pp. 5–21, 1998.

[282] Li, Z.C., Global superconvergence of simplified hybrid combinations for elliptic equations with singularities, **I**. Basic theory, *Computing*, **65**, pp. 27–44, 2000.

[283] Li, Z.C., The Trefftz method for the Helmholtz equation with degeneracy, accepted by *Appl. Numer. Math.*, 2006.

[284] Li, Z.C., Error analysis for boundary approximation methods solving eigenvalue problems, *J. Comput. Appl. Math.*, **200**, pp. 231–254, 2007.

[285] Li, Z.C. & Bui, T.D., Generalized hybrid-combined methods for singularity problems of homogeneous equations, *Int. J. Numer. Meth. Eng.*, **26**, pp. 785–803, 1988.

[286] Li, Z.C. & Bui, T.D., Six combinations of the Ritz–Galerkin and finite element methods for elliptic boundary value problems, *Numer. Meth. PDE*, **4**, pp. 197–218, 1988.

[287] Li, Z.C. & Bui, T.D., The simplified hybrid-combined methods for Laplace equation with singularities, *J. Comput. Appl. Math.*, **29**, pp. 171–193, 1990.

[288] Li, Z.C. & Bui, T.D., Coupling techniques in boundary-combined methods, *Eng. Anal. Bound. Elem.*, **10**, pp. 75–85, 1992.

[289] Li, Z.C., Chan, Y.L., Georgiou, G.C. & Xenophontos, C., Special boundary approximation methods for Laplace equation problems with boundary singularities: Applications to the Motz problem, *Comp. Math. Appl.*, **51**, pp. 115–142, 2006.

[290] Li, Z.C., Chien, C.S., & Huang, H.T., Effective condition number for finite difference method, *J. Comput. Appl. Math.*, **198**, pp. 208–235, 2007.

[291] Li, Z.C., Huang, C.S. & Chen, R.C., Interior boundary conditions in the Schwarz alternating method for the Trefftz method, *Eng. Anal. Bound. Elem.*, **29**, pp. 477–493, 2005.

[292] Li, Z.C. & Huang, H.T., Global superconvergence of simplified hybrid combinations of the Ritz–Galerkin and FEMs of elliptic equations with singularities, II. Lagrange elements and Adini's elements, *Appl. Numer. Math.*, **43**, pp. 253–273, 2002.

[293] Li, Z.C. & Huang, H.T., Hybrid and other Trefftz methods using piecewise particular solutions, Technical report, Department of Applied Mathematics, National Sun Yat-sen University, Kaohsiung, Taiwan, 2005.

[294] Li, Z.C., Huang, H.T., & Huang, J., Superconvergence and stability for boundary penalty techniques of finite difference methods, accepted by *Numer. Meth. PDE*, 2007.

[295] Li, Z.C., Huang, H.T., & Huang, J., Effective condition number of Hermite finite element methods for biharmonic equations, accepted by *Appl. Numer. Math.*, 2007.

[296] Li, Z.C., Huang, H.T., & Huang, J., Stability analysis and superconvergence for the penalty Trefftz method coupled with FEM for singularity problems, *Eng. Anal. Bound. Elem.*, **31**, pp. 631–645, 2007.

[297] Li, Z.C., Huang, H.T., Huang, J. & Ling, L., Stability analysis for the penalty plus hybrid and the direct Trefftz methods for singularity problems, *Eng. Anal. Bound. Elem.*, **31**, pp. 163–175, 2007.

[298] Li, Z.C. & Liang, G.P., On the simplified hybrid-combined method, *Math. Comp.*, **41**, pp. 13–25, 1983.

[299] Li, Z.C. & Lu, T.T., Singularities and treatments of elliptic boundary value problems, *Math. Comput. Model.*, **31**, pp. 79–145, 2000.

[300] Li, Z.C., Lu, T.T. & Hu, H.Y., Collocation Trefftz methods for biharmonic equations with crack singularities, *Eng. Anal. Bound. Elem.*, **28(1)**, pp. 79–96, 2004.

[301] Li, Z.C., Lu, T.T., Hu, H.Y. & Cheng, A.H.-D., Particular solutions of Laplace's equations on polygons and new models involving mild singularities, *Eng. Anal. Bound. Elem*, **29**, pp. 59–75, 2005.

[302] Li, Z.C., Lu, T.T., Huang, H.T. & Cheng, A.H.-D., Trefftz, collocation and other boundary methods: A comparison, *Numer. Meth. PDE*, **23**, 93–144, 2007.

[303] Li, Z.C., Lu, T.T., Tsai, H.S. & Cheng, A.H.-D., The Trefftz method for solving eigenvalue problems, *Eng. Anal. Bound. Elem.*, **30**, pp. 292–308, 2006.

[304] Li, Z.C. & Mathon, R., Boundary methods for elliptic problems on unbounded domains, *J. Comp. Phys.*, **89**, pp. 414–431, 1990.

[305] Li, Z.C. & Mathon, R., Error and stability analysis of boundary methods for elliptic problems with interfaces, *Math. Comp.*, **54**, pp. 41–61, 1990.

[306] Li, Z.C., Mathon, R. & Sermer, P., Boundary methods for solving elliptic problem with singularities and interfaces, *SIAM J. Numer. Anal.*, **24**, pp. 487–498, 1987.

[307] Li, Z.C., Yamamoto, T. & Fang, Q., Superconvergence of solution derivatives for the Shortly-Weller difference approximation of Poisson's equation, Part I, smoothness problems, *J. Comput. Appl. Math.*, **151**, pp. 307–333, 2003.

[308] Liang, G.P. & He, J.H., The non-conforming domain decomposition method for elliptic problems with Lagrangian multipliers (in Chinese), *Math. Numer. Sinica,* **14(2)**, pp. 207–215, 1992.

[309] Liang, G.P. & Liang, P., Non-conforming domain decomposition with hybrid method, *J. Comp. Math.*, **8(4)**, pp. 363–370, 1990.

[310] Lin, T.C., The numerical solution of the Helmholtz's equation for the exterior Dirichlet problems in three dimensions, *SIAM J. Numer. Anal.*, **22**, pp. 670–686, 1985.

[311] Lin, Q. & Yan, N., *The Construction and Analysis of High Efficient FEMs* (in Chinese), Hobei University Publishing, Baoding, 1996.

[312] Liu, J.L., Exact a posteriori error analysis of the least squares finite element method, *Appl. Math. Comp.*, **116(3)**, pp. 297–305, 2000.

[313] Lotz, I., *Calculation of potential flow past airship body in a yaw,* NACA TM 675, 1932.

[314] Love, A.E.H., *A Treatise on the Mathematical Theory of Elasticity*, 4th ed., Dover, New York, 1944.

[315] Lu, T.T., Hu, H.Y. & Li, Z.C., Highly accurate solutions of Motz's and the cracked beam problems, *Eng. Anal. Bound. Elem.*, **28**, pp. 1387–1403, 2004.

[316] Lu, T.T. & Li, Z.C., The cracked-beam problem by boundary approximation method, *Appl. Math. Lett.*, **28**, pp. 11–16, 2005.

[317] Lu, T.T., Wei, Y., Li, Z.C. & Huang, H.T., Effective condition number for weighted least squares problems, Technical report, Department of Mathematics, National Sun Yat-sen University, Kaohsiung, Taiwan, 2006.

[318] Lucas, T.R. & Oh, H.S., The method of auxiliary mapping for the finite element solutions of elliptic problems containing singularities, *J. Comp. Phys.*, **1108**, pp. 327–352, 1993.

[319] Lucas, T.R. & Reddien, G.W., Some collocation methods for nonlinear boundary value problems, *SIAM J. Numer. Anal.*, **9**, pp. 341–356, 1972.

[320] Ma, H. & Sun, W., A Legendre-Petrov-Galerkin and Chebyshev collocation method and the third-order differential equations, *SIAM J. Numer. Anal.*, **38**, pp. 1425–1438, 2000.

[321] MacKinnon, R.J. & Carey, G.F., Nodal superconvergence and solution enhancement for a class of finite-element and finite difference methods, *SIAM J. Sci. Stat. Comput.*, **11**, pp. 343–353, 1990.

[322] Madych, W.R., Miscellaneous error bounds for multiquadric and related interpolatory, *Comp. Math. Appl.*, **24(12)**, pp. 121–138, 1992.

[323] Madych, W.R. & Nelson, S.A., Multivariate interpolation and conditionally positive definite functions, I, *Math. Comp.*, **54**, pp. 211–230, 1990.

[324] Mai-Duy, N. & Tran-Cong, T., Numerical solution of differential equations using multiquadric radial basis function networks, *Neural Networks*, **14**, pp. 185–199, 2001.

[325] Mai-Duy, N. & Tran-Cong, T., Mesh-free radial basis function network methods with domain decomposition for approximation of functions and numerical solution of Poisson's equations, *Eng. Anal. Bound. Elem.*, **26**, pp. 133–156, 2002.

[326] Mandel, J. & Tezaur, R., Convergence of a substructuring method with Lagrange multipliers, *Numer. Math.*, **73**, pp. 473–487, 1996.

[327] Marti, J.T., *Introduction to Sobolev Spaces and Finite Element Solutions of Elliptic Boundary Value Problems*, Academic Press, London, 1986.

[328] Massonet, C.E., Numerical use of integral procedures (Chapter 10), *Stress Analysis*, eds. O.C. Zienkiewicz & G.S. Hollister, Wiley, New York, pp. 198–235, 1965.

[329] Mathon, R. & Johnston, R.L., The approximate solutions of elliptic boundary-value problems by fundamental solutions, *SIAM J. Numer. Anal.*, **14**, pp. 638–650, 1977.

[330] Mathon, R. & Sermer, P., Numerical solution of the Helmholtz equation, *Congressus Numerantium*, **34**, pp. 313–330, 1982.

[331] Maxwell, J.C., On the calculation of the equilibrium and stiffness of frames, *Phil. Mag.*, **27**, pp. 294–299, 1864.

[332] Maz'ya, V. & Shaposhnikova, T., *Jacques Hadamard, A Universal Mathematician*, American Mathematical Society, Providence, Rhode Island, 1998.

[333] Melenk, J.M. & Bubuska, I., Approximation with harmonic and generalized harmonic polynomials in the partition and unity method, *Computer Assisted Mechanics and Engineering Sciences* (*CAMES*), **4**, pp. 607–632, 1997.

[334] Mendelson, A. & Albers, L.U., Application of boundary integral equation to elasto-plastic problems, *Boundary-Integral Equation Method: Computational Applications in Applied Mechanics*, AMD **11**, eds. T.A. Cruse & F.J. Rizzo, New York, pp. 47–84, 1975.

[335] Mercier, B., *Numerical Analysis of Spectral Method*, Springer-Verlag, Berlin, 1989.

[336] Mikhlin, S.G., *Integral Equations and Their Applications to Certain Problems in Mechanics, Mathematical Physics and Technology*, (1st Russian edition 1949), 2nd rev. ed., Pergamon, Oxford, 1964.

[337] Mikhlin, S.G. & Smolitsky, J.L., *Approximation Methods for the Solution of Differential and Integral Equations*, Elsevier, Amsterdam, 1967.

[338] Miller, J.J.H., O'Roordan, E. & Shiskin, G.I., *Fitted Numerical Methods for Singular Perturbation Problems*, World Scientific, Singapore, 1996.

[339] Mitzner, K.M., Numerical solution for transient scattering from a hard surface of arbitrary shape – Retarded potential technique, *J. Acoust. Soc. Am.*, **42**, pp. 391–397, 1967.

[340] Morse, P.M. & Feshbach, H., *Methods of Theoretical Physics*, McGraw-Hill, New York, 1953.

[341] Motz, M., The treatment of singularities of partial differential equations by relaxation methods, *Q. Appl. Math.*, **4**, pp. 371–377, 1947.

[342] Muskhelishvili, N.I., On the numerical solution of the plane problems of the theory of elasticity, (in Georgian), *Trudy Tbilisskogo Matematicheskogo Instituta*, **1**, pp. 83–87, 1937.

[343] Muskhelishvili, N.I., *Singular Integral Equations*, (Russian edition 1946), Noordhoff, Amsterdam, 1953.

[344] Muskhelishvili, N.I., *Some Basic Problems of the Mathematical Theory of Elasticity*, (First Russian edition 1933), Noordhoff, Amsterdam, 1959.

[345] Nakao, M.T., Superconvergence of the gradient of Galerkin approximations for elliptic problems, *J. Comput. Appl. Math.*, **20(1)**, pp. 341–348, 1987.

[346] Nitsche, J.A., Uber ein variationsprinzip zur losung von Dirichlet probleme bei verwendung von teilraumen, die keinen randbedingungen unterworfen sind, *Adhes. Math. Sem. Univ.*, **36**, pp. 9–15, Hamburg, 1971.

[347] O'Connor, J.J. & Robertson, E.F., *MacTutor History of Mathematics Archive*, http://www-groups.dcs.st-and.ac.uk/history/, 2003.

[348] Oden, J.T. & Reddy, J.N., *An Introduction to the Mathematical Theory of Finite Elements*, John Wiley & Sons, New York, 1976.

[349] O'Donnell, S., *William Rowan Hamilton: Portrait of a Prodigy*, Boole Press, Dublin, 1983.

[350] Ohnimus, S., Rüter, M. & Stein, E., General aspects of Trefftz method and relations to error estimation of finite element approximations, *Computer Assisted Mechanics and Engineering Sciences* (*CAMES*), **8**, pp. 425–437, 2001.

[351] Oliveira, E.R.A., Plane stress analysis by a general integral method, *J. Eng. Mech. Div., ASCE*, **94**, pp. 79–101, 1968.

[352] Oliveira, F.A., Collocation and residual correction, *Numer. Math.*, **36**, pp. 27–31, 1980.

[353] Olson, L.G., Georgiou, G.C. & Schultz, W.W., An efficient finite element method for treating singularities in Laplace's equations, *J. Comp. Phys.*, **96**, pp. 319–410, 1991.

[354] Ortega, M., *Numerical Analysis, A Second Course*, SIAM, Philadelphia, 1990.

[355] Papamichael, N. & Whiteman, J.R., A numerical conformal transformation method for harmonic mixed boundary value problems in polygonal domain, *J. Appl. Math. Phys.* (*ZAMP*), **124**, pp. 304–316, 1973.

[356] Parlett, B.N., *The Symmetric Eigenvalue Problems*, Prentice Hall, Inc., New York, 1980.

[357] Parter, S.V., Preconditioning Legendre spectral collocation methods for elliptic problems I: Finite difference operators, *SIAM J. Numer. Anal.*, **39**, pp. 330–347, 2001.

[358] Parter, S.V., Preconditioning Legendre spectral collocation methods for elliptic problems II: Finite element operators, *SIAM J. Numer. Anal.*, **39**, pp. 348–362, 2001.

[359] Pathria, D. & Karniadakis, G.E., Spectral element methods for elliptic problems in nonsmooth domains, *J. Comp. Phys.*, **122**, pp. 83–95, 1995.

[360] Pehlivanov, A.I., Lazarov, R.D., Carey, C.F. & Chow, S.S., Superconvergence analysis of approximate boundary-flux calculations, *Numer. Math.*, **63**, pp. 483–501, 1992.

[361] Percell, P. & Wheeler, M.F., A local residual finite element procedure for elliptic equations, *SIAM J. Numer. Anal.*, **15**, pp. 705–714, 1978.

[362] Percell, P. & Wheeler, M.F., A C^1 finite element collocation method for elliptic equations, *SIAM J. Numer. Anal.*, **17**, pp. 605–622, 1980.

[363] Piltner, R., Recent developments in the Trefftz method for finite element and boundary element applications, *Adv. Eng. Softw.*, **24**, pp. 107–115, 1995.

[364] Piltner, R. & Taylor, R.L., The solution of plate blending problems with the aid of a boundary element algorithm based on singular complex function, in *Boundary Elements XII, Proc. 12nd World Conf. on BEM, Sapporo*, Japan, pp. 437–445, 1990.

[365] Pitkaranta, J., Boundary subspaces for the finite element method with Lagrange multipliers, *Numer. Math.*, **33**, pp. 273–289, 1979.

[366] Pitkaranta, J., The finite element method with Lagrange multipliers for domains with corners, *Math. Comp.*, **37(155)**, pp. 13–30, 1981.

[367] Poisson, P.-S., *Bulletin de in soclété philomatique*, **3**, p. 388, 1813.

[368] Ponter, A.R.S., An integral equation solution of the inhomogeneous torsion problem, *SIAM J. Appl. Math.*, **14**, p. 819–830, 1966.

[369] Portela, A. & Charafi, A., Trefftz boundary element-multi-region formulation, *Inter. J. Numer. Meth. Eng.*, **45**, pp. 821–840, 1999.

[370] Prager, W., Die Druckverteilung an Körpern in ebener Potentialströmung, *Physik Zeitschr.*, **29**, p. 865, 1928.

[371] Prenter, P.M. & Russell, R.D., Orthogonal collocation for elliptic partial differential equations, *SIAM J. Numer. Anal.*, **13**, pp. 923–939, 1976.

[372] Proskurwski, W. & Widlund, O., On the numerical solution of Helmholtz equation by the capacitance matrix method, *Math. Comp.*, **30**, pp. 433–468, 1976.

[373] Qin, Q.H., *The Trefftz Finite and Boundary Element Methods*, WIT Press, Southampton, Boston, 2000.

[374] Quarteroni, A. & Valli, A., *Numerical Approximation of Partial Differential Equations*, Springer-Verlag, Berlin, 1994.

[375] Quarteroni, A. & Zampieri, E., Finite element preconditioning for Legendre spectral collocation approximations to elliptic equations and systems, *SIAM J. Numer. Anal.*, **29**, pp. 917–936, 1992.

[376] Rankine, W.J.M., On plane water-lines in two dimensions, *Phil. Trans.*, 1864.

[377] Raviart, P.A. & Thomas, J.M., Primal hybrid finite element methods for 2nd order elliptic equations, *Math. Comp.*, **31(138)**, pp. 391–413, 1977.

[378] Rayleigh, J.W.S., *Proc. London Math. Soc.*, **4**, pp. 357–368, 1873.

[379] Rayleigh, J.W.S., *The Theory of Sound*, (originally published in 1877), 2nd ed., **1 & 2**, Dover, 1945.

[380] Rayleigh, R.J.S., *Life of John William Strutt, Third Baron Rayleigh*, University of Wisconsin Press, Madison, 1968.

[381] Reissner, E., On a variational theorem in elasticity, *J. Math. Phys.*, **29**, pp. 90–95, 1950.

[382] Reutskiy, S., A boundary method of Trefftz type with approximate trial functions, *Eng. Anal. Bound. Elem.*, **26(4)**, pp. 341–353, 2002.

[383] Rice, J.R., *Mathematical Computations and Mathematical Software*, McGraw-Hill, Inc., 1981.

[384] Ritz, W., Über eine neue methode zur Lösung gewissen variations–Problems der mathematischen physik, *J. Reine Angew. Math.*, **135**, pp. 1–61, 1908.

[385] Rizzo, F.J., *Some integral equation methods for plane problems of classical elastostatics,* Doctoral dissertation, University of Illinois, Urbana–Champaign, p. 34, 1964.

[386] Rizzo, F.J., An integral equation approach to boundary value problems of classical elastostatics, *Q. Appl. Math.*, **25**, pp. 83–95, 1967.

[387] Rizzo, F.J., The boundary element method, some early history – A personal view, *Boundary Element Methods in Structural Analysis,* ed. D.E. Beskos, ASCE, pp. 1–16, 1989.

[388] Rizzo, F.J. & Shippy, D.J., Formulation and solution procedure for the general non-homogeneous elastic inclusion problem, *Int. J. Solids Structures*, **4**, pp. 1161–1179, 1968.

[389] Rizzo, F.J. & Shippy, D.J., A method for stress determination in plane anisotropic bodies, *J. Composite Mater.*, **4**, pp. 36–61, 1970.

[390] Rizzo, F.J. & Shippy, D.J., A method of solution for certain problems of transient heat conduction, *AIAA J.*, **8**, pp. 2004–2009, 1970.

[391] Rizzo, F.J. & Shippy, D.J., An application of the correspondence principle of linear viscoelasticity theory, *SIAM J. Appl. Math.*, **21**, pp. 321–330, 1971.

[392] Roos, H.G., Stynes, M. & Tobiska, L., *Numerical Methods for Singularly Perturbed Differential Equations, Convection-Diffusion and Flow Problems*, Springer, Berlin, 1996.

[393] Rosser, J.B. & Papamichael, N., A power series solution of a harmonic mixed boundary value problem, MRC, Technical report, University of Wisconsin, 1975.

[394] Ruge, P., The complete Trefftz method, *ACTA Mechanica*, **78(3–4)**, pp. 234–242, 1989.

[395] Russell, R.D., Collocation for systems of boundary value problems, *Numer. Math.*, **23**, pp. 119–133, 1974.

[396] Russell, R.D., A comparison of collocation and finite difference methods for two-point boundary value problems, *Numer. Math.*, **14**, pp. 19–39, 1977.

[397] Russell, R.D. & Shampine, L.F., A collocation method for boundary value problems, *Numer. Math.*, **19**, pp. 1–28, 1972.

[398] Russell, R.D. & Sun, W., Spline collocation differentiation matrices, *SIAM J. Numer. Anal.*, **34**, pp. 2274–2287, 1997.

[399] Russo, R., A Green formulas for plane crack problems, *Boll. Un. Mat. Ital.* A(7), No. 1, pp. 59–67, 1987.

[400] Sanchez-Sesma, F.J., Herrera, I. & Aviles, J., A boundary method for elastic wave diffraction, Application to scattering of AH-waves by surface irregularities, *Bull. Seismol. Soc. Am.*, **72**, pp. 473–490, 1982.

[401] Saranen, J., The modified quadrature method for logarithmic-kernel integral equations on closed curves, *J. Inter. Equation Appl.*, **3**, pp. 575–600, 1991.

[402] Saranen, J. & Sloan, I., Quadrature methods for logarithmic-kernel integral equations on closed curves, *IMA J. Numer. Anal.*, **12**, pp. 167–187, 1992.

[403] Schaback, R., Error estimates and condition numbers for radial basis function interpolation, *Adv. Comput. Math.*, **3**, pp. 251–264, 1995.

[404] Schiff, B.D., Fishelov, D. & Whiteman, J.R., Determination of a stress intensity factor using local mesh refinement, *The Mathematics of Finite Elements and Applications III*, eds. J.R. Whiteman, Academic Press, London, pp. 55–64, 1979.

[405] Sharlin, H.I., *Kelvin, William Thomson, Baron*, Encyclopædia Britannica, 2003.

[406] Shaw, R.P., Diffraction of acoustic pulses by obstacles of arbitrary shape with a Robin boundary condition, *J. Acoust. Soc. Am.*, **41**, pp. 855–859, 1967.

[407] Shaw, R.P., Boundary integral equation methods applied to water waves, *Boundary-Integral Equation Method: Computational Applications in Applied Mechanics*, AMD **11**, eds. T.A. Cruse & F.J. Rizzo, New York, pp. 7–14, 1975.

[408] Shaw, R.P., Huang, S.C. & Zhao, C.X., The embedding integral and the Trefftz method for potential problems with partitioning, *Eng. Anal. Bound. Elem.*, **9(1)**, pp. 83–90, 1992.

[409] Shen, J., Efficient Spectral-Galerkin method I. Direct solvers of second- and fourth- order equations using Legendre polynomials, *SIAM J. Sci. Comput.*, **14**, pp. 1489–1505, 1994.

[410] Shen, J., Efficient Spectral-Galerkin method II. Direct solvers of second- and fourth-order equations using Chebyshev polynomials, *SIAM J. Sci. Comput.*, **15**, pp. 74–87, 1995.

[411] Shen, J., Efficient Spectral-Galerkin method III. Polar and cylindrical geometries, *SIAM J. Sci. Comput.*, **18**, pp. 1583–1604, 1997.

[412] Shippy, D.J., Application of the boundary-integral equation method to transient phenomena in solids, *Boundary-Integral Equation Method: Computational Applications in Applied Mechanics*, AMD **11**, eds. T.A. Cruse & F.J. Rizzo, New York, pp. 15–30, 1975.

[413] Sidi, A. & Israrli, M., Quadrature methods for periodic singular Fredholm integral equation, *J. Sci. Comp.*, **3**, pp. 201–231, 1988.

[414] Sladek, J., Sladek, V. & van Keer, R., Global and local Trefftz boundary integral formulation for sound vibration, *Adv. Eng. Softw.*, **33**, pp. 469–476, 2002.

[415] Sloan, I. H. & Spence, A., The Galerkin method for integral equations of first-kind with logarithmic kernel: Applications, *IMA J. Numer.*, **8**, pp. 123–140, 1988.

[416] Sneddon, G.E., Second-order spectral differentiation matrices, *SIAM J. Numer. Anal.*, **33**, pp. 2468–2487, 1996.

[417] Sobolev, S.L., *Application of Fundamental Analysis in Mathematical Physics*, AMS, Providence, Rhode Island, 1963.

[418] Somigliana, C., Sopra l'equilibrio di un corpo elastico isotropo, *Nuovo Cimento*, **3**, pp. 17–20, 1885.

[419] Somigliana, C., Sulla teoria delle distorsioni elastiche, *Note I e II. Atti Accad. Naz. Lincei. Classe Sci. Fis. Mat. e Nat.*, **23**, pp. 463–472, 1914.

[420] Somigliana, C., Sulla teoria delle distorsioni elastiche, *Note I e II. Atti Accad. Naz. Lincei. Classe Sci. Fis. Mat. e Nat.*, **24**, pp. 655–666, 1915.

[421] Stein, E., An appreciation of Erich Trefftz, *Computer Assisted Mechanics & Engineering Sciences* (*CAMES*), **4**, pp. 301–304, 1997.

[422] Stephan, E.P. & Wendland, W.L., An augmented Galerkin procedure for the boundary integral method applied to two-dimensional screen and crack problems, *Appl. Anal.*, **18**, pp. 183–219, 1984.

[423] Sternberg, W.J. & Smith, T.L., *The Theory of Potential and Spherical Harmonics*, University of Toronto Press, Toronto, 1944.

[424] Stojek, M., Least-squares Trefftz-type elements for the Helmholtz equation, *Inter. J. Numer. Meth. Eng.*, **41**, pp. 831–849, 1998.

[425] Stokes, G.G., *A Smith's Prize Paper,* Cambridge University Calendar, Cambridge, 1854.

[426] Strang, G. & Fix, G.J., *An Analysis of the Finite Element Method*, Prentice-Hall, New York, 1973.

[427] Strenger, F., *Numerical Methods Based on Sinc and Analytic Functions*, Springer-Verlag, Berlin, 1993.

[428] Sun, W., Block iterative algorithms for solving Hermite bicubic collocation equations, *SIAM J. Numer. Anal.*, **33**, pp. 589–601, 1996.

[429] Swartz, B. & Wendroff, B., The relation between the Galerkin and collocation methods using smooth splines, *SIAM J. Numer. Anal.*, **11**, pp. 994–996, 1974.

[430] Swedlow, J.L. & Cruse, T.A., Formulation of boundary integral equations for three-dimensional elasto-plastic flow, *Int. J. Solids Structures*, **7**, pp. 1673–1683, 1971.

[431] Symm, G.T., Integral equation methods in potential theory, II, *Proc. R. Soc., Lond., Ser A*, **275**, pp. 33–46, 1963.

[432] Symm, G.T., *Integral equation methods in elasticity and potential theory,* Doctoral dissertation, Imperial College, London, 1964.

[433] Tang, L.D., *The cracked-beam and related singularity problems*, Master thesis, Department of Applied Mathematics, National Sun Yat-sen University, Kaohsiang, 2001.

[434] Tazzioli, R., Green's function in some contributions of 19th century mathematicians, *Historia Mathematica*, **28**, pp. 232–252, 2001.

[435] Thatcher, R.W., The use of infinite grid refinement at singularities in the solution of Laplace's equation, *Numer. Math.*, **25**, pp. 163–178, 1976.

[436] Thatcher, R.W., On the finite element method for unbounded regions, *SIAM J. Numer. Anal.*, **15(3)**, pp. 466–477, 1978.

[437] Thompson, S.P., *The Life of William Thomson, Baron Kelvin of Largs*, **1 & 2**, Macmillan, Oxford, 1910.

[438] Tikhonov, A.N. & Samarskii, A.A., *Equations of Mathematical Physics*, The MacMillan Company, New York, 1973.

[439] Timoshenko, S.P., *History of Strength of Materials*, Dover, New York, 1983.

[440] Tomlin, G.R., *Numerical analysis of continuum problems in zoned anisotropic media,* Doctoral dissertation, University of Southampton, 1973.

[441] Trefftz, E., Ein Gegenstuck zum Ritz'schen Verfahren, *Proc. 2nd Int. Cong. Appl. Mech.*, Zurch, pp. 131–137, 1926.

[442] Trefftz, E., Konvergenz und fehlerabschatzung beim Ritz'schen verfahren, *Math. Ann.*, **100**, pp. 503–521, 1928.

[443] Tricomi, G.F., Matematici italiani del primo secolo dello stato unitario, *Memorie dell'Accademia delle Scienze di Torino, Classe di Scienze fisiche matematiche e naturali*, **IV,I**, 1962.

[444] Twaites, B. (ed), *Incompressible Aerodynamics*, Oxford University Press, Oxford, 1960.

[445] Vainberg, D.V. & Sinyavskii, A.L., Application of the potential method for numerical analysis of deformation in cylindrical shells, *Reports of the Ukrainian Academy of Sciences*, **7**, pp. 907–912, 1960.

[446] Vandrey, F., *On the calculation of the transverse potential flow past a body of revolution with the aid of the method of Mrs. Flügge-Lotz*, Astia AD40 089, 1951.

[447] Vandrey, F., *A method for calculating the pressure distribution of a body of revolution moving in a circular path through a prefect incompressible fluid*, Aeronautical Research Council R & M, No. 3139, 1960.

[448] Vekua, I.N., *New Method for Solving Elliptic Equations*, North-Holland, Amsterdam, New York, 1967.

[449] Vekua, N.P., *Systems of Singular Integral Equations and Some Boundary Value Problems*, Noordhoff, Amsterdam, 1967.

[450] Veriuzhskii, Y.V., *Numerical Potential Methods in Some Problems of Applied Mechanics*, Kiev State University Publishers, Kiev, 1978.

[451] Volkov, E.A., An exponentially convergent method for the solution of Laplace's equation on polygons, *Math. Sobonìk*, **109(151)**, pp. 323–354, 1979, *English Trans. in Math. USSR Sobonik*, **37**(**3**), 1980.

[452] Volkov, E.A., *Block Method for Solving the Laplace Equation and for Constructing Conformal Mappings*, CRC Press, Boca Raton, Ann Arbor, London, Tokyo, 1994.

[453] Volkov, E.A. & Kornoukhov, A.K., On solving the Motz's problem by a block meth, *Comp. Meths. Math. Phys.*, **43**, pp. 138–1391, 2003.

[454] Volterra, V., Sulla inversione degli integrali multipli, *Atti Accad. Naz. Lincei Rend. Cl. Sci. Fis. Mat. Natur.*, **5**, pp. 289–300, 1896.

[455] Volterra, V., Sur l'équilibre des carps élastiques multiplement connexes, *Anna. Sci. de l'École Norm. Super., Paris*, **24**, pp. 401–517, 1907.

[456] von Kármán, T., Berechnung der druckverteilung an luftschiffkörpern *Abhandl. Aerodynamischen Inst. Tech. Hoch. Aachen*, **6**, pp. 3–13, 1927.

[457] von Kármán, T., *Calculation of pressure distribution on airship hulls,* NACA TM 574, 1930.

[458] von Kármán, T. (with Edson, L.), *The Wind and Beyond*, Little Brown and Co., Boston, 1967.

[459] Wahlbin, L.L., Local behaviour in finite element method, Chap. VII., Superconvergence, *Finite Element Methods* (*Part I*), eds. P.G. Ciarlet & J.L. Lions, North-Holland, Amsterdam, pp. 501–522, 1991.

[460] Wahlbin, L.L., *Superconvergence in Galerkin Finite Element Methods*, Lecture Notes in Mathematics, Springer, Berlin, 1995.

[461] Washizu, K., *Variational Methods in Elasticity and Plasticity*, Pergamon, Oxford, 1968.

[462] Watson, J.O., *The analysis of stress in thick shells with holes, by integral representation with displacement*, Doctoral dissertation, University of Southampton, 1973.

[463] Wendland, H., Meshless Galerkin methods using radial basis functions, *Math. Comp.*, **68(228)**, pp. 1521–1531, 1999.

[464] Wendland, W.L., Boundary element methods and their asymptotic convergence, *Theoretical Acoustics and Numerical Techniques*, Springer, Vienna, pp. 135–216, 1983.

[465] Wheeler, M.F., A C^0- collocation-finite element method for two-point boundary value problems and one space dimensional parabolic problems, *SIAM J. Numer. Anal.*, **14**, pp. 71–90, 1977.

[466] Wheeler, M.F. & Whiteman, J.R., Superconvergent recovery of gradients on subdomains from piecewise linear finite-element approximation, *Numer. Meth. PDE*, **3**, pp. 65–82 & 357–374, 1987.

[467] Whiteman, J.R., Numerical treatment of a problem from linear fracture mechanics, *Numerical Methods in Fracture Mechanics*, eds. D.R.J. Owen & A.R. Luxmoore, University of Wales, Swansea, pp. 128–136, 1978.

[468] Whiteman, J.R. & Papamichael, N., Treatment of harmonic mixed boundary by conformal transformation method, *ZAMP*, **23**, pp. 655–664, 1972.

[469] Wigley, N.M., On a method to subtract off a singularity at a corner for the Dirichlet or Neumann problem, *Math. Comp.*, **23**, pp. 395–401, 1968.

[470] Wigley, N.M., An efficient method for subtracting off singularities at corners for Laplace's equation, *J. Comp. Phys.*, **78**, pp. 368–376, 1988.

[471] Williams, L.P., *Helmholtz, Hermann von*, Encyclopædia Britannica, 2003.

[472] Wilkinson, J.H., *The Algebraic Eigenvalue Problems*, Oxford University Press, Oxford, 1965.

[473] Wu, Z. & Schaback, R., Local error estimates for radial basis function interpolation of scattered data, *IMA J. Numer. Anal.*, **13**, pp. 13–27, 1993.

[474] Xu, J. & Zhou, A., Local and parallel finite element algorithms based on the two-grid discretization, *Math. Comp.*, **69**, pp. 881–909, 2000.

[475] Yan, Y., The collocation method for first-kind boundary integral equations on polygonal regions, *Math. Comp.*, **54**, pp. 139–154, 1990.

[476] Yu, D.H., *Natural Boundary Integral Method and its Application*, Science Press/Kluwer Academic Publishers, Beijing, Dordrecht, 2002.

[477] Yin, G., Sinc-collocation method with orthogonalization for singular Poisson-like problems, *Math. Comp.*, **62(205)**, pp. 21–40, 1994.

[478] Yoon, J., Local error estimates for radial basis function interpolation of scattered data, *J. Approx. Theory*, **112(1)**, pp. 1–15, 2001.

[479] Zheng, X.P. & Yao, Z.H., Some applications of the Trefftz method in linear elliptic boundary-value problem, *Adv. Eng. Softw.*, **24**, pp. 133–145, 1995.

[480] Zhou, T.X. & Feng, M.F., A least squares Petrov-Galerkin finite element method for stationary Navier-Stokes equations, *Math. Comp.*, **60**, pp. 331–543, 1993.

[481] Zieliński, A.P., Trefftz method: Elastic and elastoplastic problems, *Comput. Meth. Appl. Mech. Eng.*, **69(2)**, pp. 185–204, 1988.

[482] Zieliński, A.P., On trial functions applied in the generalized Trefftz method, *Adv. Eng. Softw.*, **24(1–3)**, pp. 147–155, 1995.

[483] Zieliński, A.P., Note on the First International Workshop: Trefftz method: recent development and perspectives, *Computer Assisted Mechanics and Engineering Sciences* (*CAMES*), **4**, pp. 299–300, 1997.

[484] Zieliński, A.P., Special Trefftz elements and improvements of their conditioning, *Commun. Numer. Meth. Eng.*, **13(10)**, pp. 765–775, 1997.

[485] Zieliński, A.P. & Herrera, I., Trefftz method-fitting boundary conditions, *Int. J. Numer. Meth. Eng.*, **24(5)**, pp. 871–891, 1987.

[486] Zieliński, A.P. & Zienkiewicz, O.C., Generalized finite element analysis with T-complete boundary solution functions, *Int. J. Numer. Meth. Eng.*, **21**, pp. 509–528, 1985.

[487] Zienkiewicz, O.C., Trefftz type approximation and the generalized finite element method, *Computer Assisted Mechanics and Engineering Sciences* (*CAMES*), **4**, pp. 305–316, 1997.

[488] Zienkiewicz, O.C., Kelley, D.W. & Bettess, P., The coupling of the finite element method and boundary solution procedures, *Int. J. Numer. Meth. Eng.*, **11**, pp. 355–375, 1977.

[489] Zienkiewicz, O.C. & Taylor, R.L., *Finite Element Method*, 5th ed., Butterworth-Heinemann, Oxford, 2000.

[490] *A survey of Muskhelishvili's scientific heritage*, http://kr.cs.ait.ac.th/~radok/mus/mus10.htm, 2003.

[491] *Niko Muskhelishvili, Curriculum Vitae*, http://www.rmi.acnet.ge/person/ muskhel/, 2003.

[492] *Science Citation Index Expanded, the Web of Science*, Institute for Scientific Information, http://www.isinet.com/isi/products/citation/scie/index. html, 2003.

Glossary of symbols

BAM: the boundary approximation method.

BEM: the boundary element method.

BIE: the boundary integral equation.

BIEM: the boundary integral equation method.

CM: the collocation method.

CTM: the collocation Trefftz method.

D–D, N–N, D–N: the Dirichlet and Neumann conditions are assigned to two edges of a sector.

DDM: the domain decomposition method.

FDM: the finite difference method.

FEM: the finite element method.

FEM–CM: the combinations of the finite element and the collocation methods.

FEM–RBFCM: the combination of the FEM and the RBFCM.

FVM: the finite volume method.

GRB: the Gaussian radial basis functions.

GTM: the generalized Trefftz method.

I.B.C.: the interior boundary condition.

IMQRB: the inverse multiquadratic radial basis functions.

LSM: the least squares method.

Models I, II: the models of the biharmonic equations with the crack singularities.

Model III: the model of Schiff, Fishelov, and Whiteman [404] in the rectangle $S = \{(x,y) | -1 < x < 1, 0 < y < 1\}$ *with the Dirichlet boundary condition.*

PDE: the partial differential equation.

RBFCM: the collocation method using the radial basis functions, or the radial basis collocation method.

RBF: *the radial basis functions.*

RGM: *the Ritz–Galerkin method.*

SAM: *the Schwarz alternating method.*

TM: *the Trefftz method.*

C: *a positive bounded constant.*

D_l: *the expansion coefficients.*

$\widetilde{D}_l$: *the approximate coefficient of* D_l.

P_c: *the penalty constant.*

S: *the solution domain.*

S^+: *the subdomain of S possibly involving solution singularities.*

S^-: *another subdomain of S with smooth solutions.*

$S_h = (\cup_{ij} \square_{ij}) \cup (\cup_{ij} \Delta_{ij})$: *the partition of S.*

u: *the true solution.*

u_I: *the interpolant of u.*

$u_h, \widetilde{u}_h$: *the approximate solutions.*

V_h: *a finite-dimensional subspace,* $V_h \subset H^1(S)$.

V_h^0: *a finite-dimensional subspace,* $V_h^0 \subset H_1^0(S)$.

ω: *the weight constant.*

Γ: *the exterior boundary of the solution domain.*

Γ_D: *the exterior boundary with the Dirichlet boundary condition.*

Γ_N: *the exterior boundary with the Neumann boundary condition.*

Γ_0: *the interface boundary, artificial or material.*

α, β: *coupling parameters.*

σ: *penalty power.*

λ: *Lagrange multiplier.*

$\lambda_{\max}$: *the maximal eigenvalue.*

$\lambda_{\min}$: *the minimal eigenvalue.*

$\widehat{\int}, \widehat{\iint}$ (*or* $\widetilde{\int}, \widetilde{\iint}$): *approximation of the integrals* $\int$, $\iint$.

$u = f_L + R_L$: *the solution expansion.*

$f_L = \sum_{i=1}^{L} a_i \phi_i$: *the leading terms.*

$R_L = \sum_{i=L+1}^{\infty} a_i \phi_i$: *the remainder.*

$\text{Cond}(\mathbf{A}) = \left\{ \frac{\lambda_{\max}(\mathbf{A})}{\lambda_{\min}(\mathbf{A})} \right\}$: *the traditional condition number of the symmetric and positive definite matrix* $\mathbf{A}$.

$I_\mu(r) = \sum_{k=0}^{\infty} \frac{1}{\Gamma(k+1)\Gamma(k+\mu+1)} \left(\frac{r}{2}\right)^{2k+\mu}$: *the Bessel functions for a purely imaginary argument.*

$K_n(\rho) = \frac{1}{2} \int_{-\infty}^{\infty} e^{-\rho \cosh \eta - n\eta} d\eta$: *the Hankel functions for a purely imaginary argument.*

$J_\mu(r) = \sum_{k=0}^{\infty} \frac{(-1)^k}{\Gamma(k+1)\Gamma(k+\mu+1)} \left(\frac{r}{2}\right)^{2k+\mu}$: *the Bessel functions of the first kind.*

$H_\mu^{(1)}(r) = \frac{i}{\sin(\mu\pi)}\{\exp(-\mu\pi i)J_\mu(r) - J_{-\mu}(r)\}$: *the Hankel functions of the first kind.*

$v = \frac{1}{2\pi}\ln\frac{1}{r}$, $r = \sqrt{(x-x_0)^2+(y-y_0)^2}$, $M_0 = (x_0, y_0) \in S$: *the fundamental solution of the Laplace equation.*

$K(r,\theta,\xi) = \frac{1-r^2}{2\pi}\frac{1}{1-2r\cos(\theta-\xi)+r^2}$: *Poisson's kernel.*

$T_i(x) = \cos(i\arccos(x))$, $-1 \le x \le 1$: *the Chebyshev polynomial.*

$\mathbf{Ax} = \mathbf{b}$: *the linear algebraic equation, where* $\mathbf{A} \in R^{n\times n}$ *is symmetric and positive definite,* $\mathbf{x} \in R^n$ *and* $\mathbf{b} \in R^n$ *are the unknown and known vectors, respectively.*

$\mathbf{Fx} = \mathbf{b}$: *the overdetermined system:* $\mathbf{F} \in R^{m\times n}$ $(m \ge n)$ *is the stiffness matrix,* $\mathbf{x} \in R^n$ *and* $\mathbf{b} \in R^m$ *are the unknown and known vectors, respectively.*

$\mathbf{F} = \mathbf{U\Sigma V^T}$: *the singular value decomposition for matrix* $\mathbf{F} \in R^{m\times n}$*, where* $\mathbf{\Sigma} \in R^{m\times n}$ *is a diagonal matrix with the singular values* $\sigma_i \ge 0$, $i = 1, 2, \ldots, n$, $\mathbf{U} \in R^{m\times m}$ *and* $\mathbf{V} \in R^{n\times n}$ *are two orthogonal matrices.*

$\text{Cond} = \frac{\sigma_1}{\sigma_n}$: *the traditional condition number in the 2-norm.*

$\text{Cond\_eff} = \frac{\|\mathbf{b}\|}{\sigma_n\|\mathbf{x}\|} = \frac{\|\mathbf{b}\|}{\sigma_n\sqrt{\sum_{i=1}^{n}\left(\frac{\beta_i}{\sigma_i}\right)^2}}$: *the effective condition number, where* $\beta_i = \mathbf{u}_i^T\mathbf{b}$, $\mathbf{u}_i$ *are given in* $\mathbf{U} = \{\mathbf{u}_1, \mathbf{u}_2, \ldots, \mathbf{u}_n\}$ *of the singular value decomposition for matrix* $\mathbf{A}$ *(or* $\mathbf{F}$*),* σ_n *is the minimal singular value,* $\mathbf{x}$ *and* $\mathbf{b}$ *are unknown and known vectors, respectively, in* $\mathbf{Ax} = \mathbf{b}$ *(or* $\mathbf{Fx} = \mathbf{b}$*).*

$\text{Cond\_}\tilde{\text{eff}} = \frac{\|\mathbf{b}\|}{\sigma_n\|\tilde{\mathbf{x}}\|}$: *the a posteriori effective condition number, where* $\tilde{\mathbf{x}}$ *is the approximate solution from* $\mathbf{Ax} = \mathbf{b}$ *(or* $\mathbf{Fx} = \mathbf{b}$*).*

$\text{Cond\_E} = \frac{\|\mathbf{b}\|}{\sqrt{\frac{\|\mathbf{b}\| - \beta_n^2}{\text{Cond}^2} + \beta_n^2}}$: *the simplified effective condition for* $\mathbf{Ax} = \mathbf{b}$*, where* $\beta_n = \mathbf{u}_n^T\mathbf{b}$.

$\text{Cond\_E} = \frac{\|\mathbf{b}\|}{\sqrt{\frac{\|\mathbf{Fx}\|^2 - \beta_n^2}{\text{Cond}^2} + \beta_n^2}}$: *the simplified effective condition number for* $\mathbf{Fx} = \mathbf{b}$*, where* $\beta_n = \mathbf{u}_n^T\mathbf{b}$.

$\text{Cond\_EE} = \frac{\|\mathbf{b}\|}{|\beta_n|}$: *the simplest effective condition number, where* $\beta_n = \mathbf{u}_n^T\mathbf{b}$.

$L + 1 = O(|\ln h|)$: *the coupling relation in combinations.*

span $\{\psi_i\}$: *to be spanned by the basis functions* ψ_i.

$u_\nu = \frac{\partial u}{\partial \nu}$: *the normal derivative of* u.

$\Delta = \frac{\partial^2}{\partial x^2} + \frac{\partial^2}{\partial y^2}$: *the Laplace operator.*

$\Delta\sqrt{\lambda_{\min}} = \sqrt{\lambda_{\min}} - \sqrt{\tilde{\lambda}_{\min}}$.

$\Delta D_i = D_i - \tilde{D}_i$.

$\varepsilon = u - u_n$.

$u^+ = u\,|_{S^+}$ *or* $u^+ = u\,|_{S_2}$.

$u^- = u\,|_{S^-}$ *or* $u^- = u\,|_{S_1}$.

$H^1(S) = \{v, v_x, v_y \in L^2(S)\}$.

$H_0^1(S) = \{v, v_x, v_y \in L^2(S),\ \ \text{and } v\,|_{\Gamma_D} = 0\}$.

$H^1_*(S) = \{v, v_x, v_y \in L^2(S),\ \ and\ \ v\,|\,\Gamma_D = g\}$.

$\max\limits_S |\epsilon| = \|\epsilon\|_{\infty,S}$.

$\delta = \min\limits_i\ |\frac{k^2-\lambda_i}{k^2}|$.

$$\|v\|_{l,\partial S} = \left[\sum_{|\alpha|\le l} \int_{\partial S} \left(\frac{\partial^\alpha v}{\partial s^\alpha}\right)^2 d\ell\right]^{\frac{1}{2}}.$$

$$|v|_{l,\partial S} = \left[\sum_{|\alpha|=l} \int_{\partial S} \left(\frac{\partial^\alpha v}{\partial s^\alpha}\right)^2 d\ell\right]^{\frac{1}{2}}.$$

$$\|v\|_{m,S} = \left\{\sum_{|\alpha|\le m} \int_S |D^\alpha v|^2\, dx\right\}^{\frac{1}{2}}.$$

$$|v|_{m,S} = \left\{\sum_{|\alpha|=m} \int_S |D^\alpha v|^2\, dx\right\}^{\frac{1}{2}}.$$

$\|v\|_1 = \{\|v\|^2_{1,S^+} + \|v\|^2_{1,S^-}\}^{\frac{1}{2}}$.

$|v|_1 = \{|v|^2_{1,S^+} + |v|^2_{1,S^-}\}^{\frac{1}{2}}$.

$|v|_H = \{|v|^2_{1,\Omega^+} + |v|^2_{1,\Omega^-}\}^{\frac{1}{2}}$.

$\|v\|_H = \{\|v\|^2_{1,\Omega^+} + \|v\|^2_{1,\Omega^-}\}^{\frac{1}{2}}$ *or* $\|v\|_H = \left(\|u\|^2_{1,S_1} + \|u\|^2_{1,S_2} + \frac{1}{h^\sigma}\|u^+ - u^-\|^2_{0,\Gamma_0}\right)^{\frac{1}{2}}$.

$\|v\|_{\mathrm{B}} = \{\int_{\Gamma_D} v^2\, d\ell + \int_{\Gamma_N} v_\nu^2\, d\ell\}^{\frac{1}{2}}$.

$\|v\|_{\widetilde{\mathrm{B}}} = \{\widetilde{\int}_{\Gamma_D} v^2\, d\ell + \widetilde{\int}_{\Gamma_N} v_\nu^2\, d\ell\}^{\frac{1}{2}}$.

$|\varepsilon|_2 = \{|\varepsilon|^2_{0,\overline{BC}} + \sigma^2|\varepsilon|^2_{0,\overline{AB}}\}^{\frac{1}{2}}$.

$|\varepsilon|_{\mathrm{II}} = \{|v^+ - v^-|^2_{0,\Gamma_0} + \sigma^2|u_\nu^+ - u_\nu^-|^2_{0,\Gamma_0}\}^{\frac{1}{2}}$.

$\overline{\|v^+ - v^-\|_{0,\Gamma_0}} = \left(\widehat{\int}_{\Gamma_0}(v^+ - v^-)^2 d\ell\right)^{\frac{1}{2}}$.

$\|v\|_{\frac{1}{2},\Gamma_0} = \left(\|v\|^2_{0,\Gamma_0} + \int_{\Gamma_0}\int_{\Gamma_0} \frac{(v(P)-v(Q))^2}{(P-Q)^2} d\ell(Q)\, d\ell(P)\right)^{\frac{1}{2}}$.

$\|u\|_{-\frac{1}{2},\Gamma_0} = \sup\limits_v \frac{\left|\int_{\Gamma_0} uv\, d\ell\right|}{\|v\|_{\frac{1}{2},\Gamma_0}}$.

Index